Handbook of Research on Resource Management and the Struggle for Water Sustainability in Africa

Innocent Simphiwe Nojiyeza
University of Zululand, South Africa

Oliver Mtapuri
University of KwaZulu-Natal, South Africa

Prosper Bazaanah
University of Johannesburg, South Africa

Edzisani Ellen Netshiozwi
Office of the Chief Justice, South Africa

A volume in the Practice, Progress, and Proficiency in Sustainability (PPPS) Book Series

IGI Global
PUBLISHER of TIMELY KNOWLEDGE

Published in the United States of America by
IGI Global
Engineering Science Reference (an imprint of IGI Global)
701 E. Chocolate Avenue
Hershey PA, USA 17033
Tel: 717-533-8845
Fax: 717-533-8661
E-mail: cust@igi-global.com
Web site: http://www.igi-global.com

Library of Congress Cataloging-in-Publication Data

Names: Nojiyeza, Innocent, DATE- editor. | Mtapuri, Oliver, editor. | Bazaanah, Prosper, DATE- editor. | Netshiozwi, Edzisani, DATE- editor.
Title: Handbook of research on resource management and the struggle for water sustainability in Africa / Innocent Nojiyeza, Oliver Mtapuri, Prosper Bazaanah, and Edzisani Netshiozwi, editor.
Description: Hershey : Engineering Science Reference, 2022. | Includes bibliographical references and index. | Summary: "Access to water and sanitation remains a critical challenge in various countries in Africa, especially in Sub Saharan Africa, so this book explores water resources management, water supply and sanitation and how climate change mitigation and adaptation as well as energy sources affect that region"-- Provided by publisher.
Identifiers: LCCN 2021026993 (print) | LCCN 2021026994 (ebook) | ISBN 9781799888093 (hardcover) | ISBN 9781799888116 (ebook)
Subjects: LCSH: Water-supply--Africa--Management. | Water resources development--Africa. | Sustainable development--Africa.
Classification: LCC TD315 .R47 2022 (print) | LCC TD315 (ebook) | DDC 333.9115096--dc23
LC record available at https://lccn.loc.gov/2021026993
LC ebook record available at https://lccn.loc.gov/2021026994

This book is published in the IGI Global book series Practice, Progress, and Proficiency in Sustainability (PPPS) (ISSN: 2330-3271; eISSN: 2330-328X)

British Cataloguing in Publication Data
A Cataloguing in Publication record for this book is available from the British Library.

All work contributed to this book is new, previously-unpublished material. The views expressed in this book are those of the authors, but not necessarily of the publisher.

For electronic access to this publication, please contact: eresources@igi-global.com.

Practice, Progress, and Proficiency in Sustainability (PPPS) Book Series

Ayman Batisha
International Sustainability Institute, Egypt

ISSN:2330-3271
EISSN:2330-328X

Mission

In a world where traditional business practices are reconsidered and economic activity is performed in a global context, new areas of economic developments are recognized as the key enablers of wealth and income production. This knowledge of information technologies provides infrastructures, systems, and services towards sustainable development.

The **Practices, Progress, and Proficiency in Sustainability (PPPS) Book Series** focuses on the local and global challenges, business opportunities, and societal needs surrounding international collaboration and sustainable development of technology. This series brings together academics, researchers, entrepreneurs, policy makers and government officers aiming to contribute to the progress and proficiency in sustainability.

Coverage

- Strategic Management of IT
- Outsourcing
- Eco-Innovation
- Environmental informatics
- Sustainable Development
- Green Technology
- Global Business
- ICT and knowledge for development
- Intellectual Capital
- Innovation Networks

IGI Global is currently accepting manuscripts for publication within this series. To submit a proposal for a volume in this series, please contact our Acquisition Editors at Acquisitions@igi-global.com or visit: http://www.igi-global.com/publish/.

The Practice, Progress, and Proficiency in Sustainability (PPPS) Book Series (ISSN 2330-3271) is published by IGI Global, 701 E. Chocolate Avenue, Hershey, PA 17033-1240, USA, www.igi-global.com. This series is composed of titles available for purchase individually; each title is edited to be contextually exclusive from any other title within the series. For pricing and ordering information please visit http://www.igi-global.com/book-series/practice-progress-proficiency-sustainability/73810. Postmaster: Send all address changes to above address.

Titles in this Series

For a list of additional titles in this series, please visit: www.igi-global.com/book-series

Achieving Sustainability Using Creativity, Innovation, and Education A Multidisciplinary Approach
Ziska Fields (University of Johannesburg, South Africa)
Information Science Reference • © 2022 • 260pp • H/C (ISBN: 9781799879633) • US $215.00

Handbook of Research on Sustainable Development Goals, Climate Change, and Digitalization
Rui Alexandre Castanho (WSB University, Poland)
Engineering Science Reference • © 2022 • 684pp • H/C (ISBN: 9781799884828) • US $315.00

Sociological Perspectives on Sustainable Development and Poverty Reduction in Rural Populations
Ladislaus M. Semali (Pennsylvania State University, USA)
Information Science Reference • © 2022 • 303pp • H/C (ISBN: 9781799846468) • US $195.00

Handbook of Research on Novel Practices and Current Successes in Achieving the Sustainable Development Goals
Cristina Raluca Gh. Popescu (University of Bucharest, Romania & The Bucharest University of Economic Studies, Romania & University of Craiova, Romania & The National Institute for Research and Development in Environmental Protection (I.N.C.D.P.M.), Romania & National Research and Development Institute for Gas Turbines COMOTI, Romania)
Engineering Science Reference • © 2021 • 461pp • H/C (ISBN: 9781799884262) • US $295.00

Advancements in Sustainable Architecture and Energy Efficiency
Roberto A. González-Lezcano (Universidad CEU San Pablo, Spain)
Engineering Science Reference • © 2021 • 337pp • H/C (ISBN: 9781799870234) • US $195.00

Management and Conservation of Mediterranean Environments
Rui Alexandre Castanho (WSB University, Poland) and José Martín Gallardo (Universidad de Extremadura, Spain)
Engineering Science Reference • © 2021 • 336pp • H/C (ISBN: 9781799873914) • US $195.00

Handbook of Research on Institution Development for Sustainable and Inclusive Economic Growth in Africa
Evans S. Osabuohien (Covenant University, Nigeria) Emmanuel A. Oduntan (Covenant University, Nigeria) Obindah Gershon (Covenant University, Nigeria) Olaronke Onanuga (Covenant University, Nigeria) and Oluyomi Ola-David (Covenant University, Nigeria)
Business Science Reference • © 2021 • 507pp • H/C (ISBN: 9781799848172) • US $235.00

IGI Global
PUBLISHER of TIMELY KNOWLEDGE

701 East Chocolate Avenue, Hershey, PA 17033, USA
Tel: 717-533-8845 x100 • Fax: 717-533-8661
E-Mail: cust@igi-global.com • www.igi-global.com

Editorial Advisory Board

List of Contributors

Bazaanah, Prosper / *University of Johannesburg, South Africa* 1, 47
Belaid, Mohamed / *University of Johannesburg, South Africa* 368
Charis, Gratitude / *Botswana International University of Science and Technology, Botswana* 277
Danha, Gwiranai / *Botswana International University of Science and Technology, Botswana* 277
Debwe, Thomas / *Midlands State University, Zimbabwe* 239
Kamuti, Tariro / *Center for Gender and Africa Studies, University of the Free State, South Africa* 354
Kusena, Winmore / *Midlands State University, Zimbabwe* 225, 239
Marley, Gordon / *University for Development Studies, Ghana* 47
Marson, Marta / *University of Turin, Italy & Turin Centre on Emerging Economies (OEET), Italy* 109
Mnguni, Asanda Buliswa / *University of Zululand, South Africa* 179
Muzenda, Edison / *Botswana International University of Science and Technology, Botswana & University of Johannesburg, South Africa* 368
Ndimande, Nokuphila / *University of Zululand, South Africa* 149
Netshiozwi, Edzisani Ellen / *Management College of Southern Africa, South Africa* 253
Nhubu, Trust / *University of Johannesburg, South Africa* 368
Nkhoma, Tafadzwa / *Manicalland State University of Applied Science, Zimbabwe* 277
Nojiyeza, Innocent Simphiwe / *University of Zululand, South Africa* 76, 96, 202, 297
Oppong, Patricia / *University for Development Studies, Ghana* 47
Tshinu, Germain Miteu / *University of Zululand, South Africa* 318, 337
Zulu, Nomfundo Nomcebo / *University of Zululand, South Africa* 127

Table of Contents

Foreword *by John Victor Mensah* xviii

Foreword *by Mogomme Alpheus Masoga* xix

Preface xxii

Chapter 1
Ecological Governance and the Sustainability of Rural Household Water Conservation Systems in the Savannah Region of Ghana 1
Prosper Bazaanah, University of Johannesburg, South Africa

Chapter 2
Examining the Role of NGOs in Community Water and Sanitation Improvement: A Case of the Tunayilli Community in the Northern Region, Ghana 47
Gordon Marley, University for Development Studies, Ghana
Prosper Bazaanah, University of Johannesburg, South Africa
Patricia Oppong, University for Development Studies, Ghana

Chapter 3
Mending Malawi's Water Institutions and IWRM Solutions 76
Innocent Simphiwe Nojiyeza, University of Zululand, South Africa

Chapter 4
Integrated Water Resource Management, Sustainability, and Pollution Abatement in Malawi: An Ecological Economics Perspective 96
Innocent Simphiwe Nojiyeza, University of Zululand, South Africa

Chapter 5
Water Demand Management and Household Level Metering in Low-Income Areas of Urban Africa 109
Marta Marson, University of Turin, Italy & Turin Centre on Emerging Economies (OEET), Italy

Chapter 6
Water Scarcity and Household Food Security: A Case of Ulundi Local Municipality in KwaZulu-Natal, South Africa 127

Nomfundo Nomcebo Zulu, University of Zululand, South Africa

Chapter 7
Equitable Access to Water at Alfred Duma Local Municipality 149
Nokuphila Ndimande, University of Zululand, South Africa

Chapter 8
Assessment of the Realization of the Fundamental Human Right to Water at the Umalalazi Local Municipality, KwaZulu-Natal 179
Asanda Buliswa Mnguni, University of Zululand, South Africa

Chapter 9
Durban Community's Experiences of Class-Based Urinary Diversion Toilets 202
Innocent Simphiwe Nojiyeza, University of Zululand, South Africa

Chapter 10
The Political Ecology of the Decentralized Water Management in Zimbabwe: Theory and Empirical Evidence From Sanyati Catchment Area 225
Winmore Kusena, Midlands State University, Zimbabwe

Chapter 11
An Analysis of Sector-Based Water Supply and Demand: Mapping the Possible Water Development and Management Ways for Sanyati Catchment, Zimbabwe 239
Winmore Kusena, Midlands State University, Zimbabwe
Thomas Debwe, Midlands State University, Zimbabwe

Chapter 12
Addressing Energy Poverty Through Ecological Governance of Solar Home Systems in South Africa 253
Edzisani Ellen Netshiozwi, Management College of Southern Africa, South Africa

Chapter 13
The Scope and Potential for Mini-Grid Power Systems Based on Biomass Waste for Remote Areas in Sub-Saharan Africa: Off grid 277
Gratitude Charis, Botswana International University of Science and Technology, Botswana
Tafadzwa Nkhoma, Manicalland State University of Applied Science, Zimbabwe
Gwiranai Danha, Botswana International University of Science and Technology, Botswana

Chapter 14
Climate Justice in Durban: State Contradictions and Grass-Roots Action 297
Innocent Simphiwe Nojiyeza, University of Zululand, South Africa

Chapter 15
Unpacking the Resource Curse and Realism Challenges on Economic Development in the

Democratic Republic of Congo (DRC): Case of Gecamines 318
Germain Miteu Tshinu, University of Zululand, South Africa

Chapter 16
Contribution of Mining Operations Towards Education, Healthcare, Food Security, Housing, Sports, and Recreation in Katanga Province of the DRC 337
Germain Miteu Tshinu, University of Zululand, South Africa

Chapter 17
Mining and Water Supply in Marginalised Communities: The Case of Anglo-Platinum Mogalakwena and Ga-Molekane Village, Limpopo Province, South Africa 354
Tariro Kamuti, Center for Gender and Africa Studies, University of the Free State, South Africa

Chapter 18
The Water-Waste Nexus: Sustainability Implications for Integrated Water Resources and Waste Management in Harare 368
Trust Nhubu, University of Johannesburg, South Africa
Edison Muzenda, Botswana International University of Science and Technology, Botswana & University of Johannesburg, South Africa
Mohamed Belaid, University of Johannesburg, South Africa

Compilation of References 388

About the Contributors 431

Index 435

Detailed Table of Contents

Foreword *by John Victor Mensah* xviii

Foreword *by Mogomme Alpheus Masoga* xix

Preface xxii

Chapter 1
Ecological Governance and the Sustainability of Rural Household Water Conservation Systems in the Savannah Region of Ghana 1
Prosper Bazaanah, University of Johannesburg, South Africa

This chapter examined the link between ecological governance and water conservation as sustainable pathways for enhancing rural livelihoods in the Savannah Region. Designs adopted were post-positivist and cross-sectional. Probability sampling techniques were used to sample 450 household and official respondents. Questionnaires were administered, while descriptive statistics and chi-square test were utilised to analyse the data. Findings showed significant relationship between conservation initiatives, finance, rehabilitation/maintenance, and gender inclusion and domestic water conservation. Therefore, with commitment to maintenance, funding, and gender inclusion in water decisions, there is the likely for water to be locally sustainable in rural communities of the region. Democratic, decentralised, and participatory approaches to ecological governance and empowerment of the local communities are recommended as essential preconditions for achieving ecologically self-governing communities and sustaining domestic water systems in the rural areas of the region.

Chapter 2
Examining the Role of NGOs in Community Water and Sanitation Improvement: A Case of the Tunayilli Community in the Northern Region, Ghana 47
Gordon Marley, University for Development Studies, Ghana
Prosper Bazaanah, University of Johannesburg, South Africa
Patricia Oppong, University for Development Studies, Ghana

This chapter examined the role of NGOs in water and sanitation improvement and the effects on the residence of Tunayilli in the Sagnarigu District. The design was descriptive. Questionnaires and interview guides were administered to household heads and key informants. Findings revealed that the water and sanitation condition in Tunayilli is generally poor. NGOs play diverse roles including the provision of water and sanitation facilities, community mobilization, and facility maintenance to ensure their sustained usability. However, inadequate funding, weak community mobilization, poor maintenance culture, and low education are challenges to water and sanitation improvement programmes of NGOs in the community. Measures to mitigate these includes increased central government funding, public education, community involvement, enforcement of sanitation by-laws, and regular maintenance of water and sanitation systems in the community. Stakeholders should intensify and sustain their educational campaigns. Meanwhile, by-laws on water and sanitation should be enforced and offenders prosecuted.

Chapter 3
Mending Malawi's Water Institutions and IWRM Solutions 76
Innocent Simphiwe Nojiyeza, University of Zululand, South Africa

The aim of this chapter is to explain the challenges of decentralisation and management of water as economic good principles of integrated water resources management (IWRM) in Malawi in the rural areas of Ntcheu, Balaka, and Mangochi. Semi-structured interviews, focus group discussions, community meetings, and observation of water points were conducted. It was established that households and other state institutions prefer to receive services from Mpira-Balaka Water Users Association, which subsidizes households, rather than Southern Regional Water Board, which operates along commercial lines. The chapter questions the focus in the implementation of IWRM, which remain the establishment and fixing of decentralised institutions whilst recognising water as an economic good.

Chapter 4
Integrated Water Resource Management, Sustainability, and Pollution Abatement in Malawi: An Ecological Economics Perspective 96
Innocent Simphiwe Nojiyeza, University of Zululand, South Africa

With the introduction of IWRM, local government institutions are expected to control water-related pollution, which is often associated with new water resource management, supply, and sanitation responsibilities devolved to them as part of decentralisation. The aim of this study was to look into eco-efficiency, environmental performance, physical scarcity and economic scarcity trends, sustainable variables, and externalities associated with the introduction of IWRM in Malawi. Focus group discussions and interviews were conducted with policymakers and households in Ntcheu, Mangochi, and Balaka. Findings confirmed that the challenges local levels of government face in managing water infrastructure and formulation of pollution control measures are a result of the fact that adoption of IWRM happened without any readiness to do so, especially in terms of water-related pollution control.

Chapter 5
Water Demand Management and Household Level Metering in Low-Income Areas of Urban Africa 109
Marta Marson, University of Turin, Italy & Turin Centre on Emerging Economies (OEET), Italy

Increasing the level of water metering is an objective of most initiatives for the operational restructuring of African water utilities promoted by donors and development agencies from the 1990s. Water metering penetration is a common benchmarking indicator to measure the performances of water utilities. In contrast with other benchmarks and targets set for the African water sector, which remain largely unmet, water metering at household and at water point levels are quite successful. The study discusses the arguments behind the widespread acceptance of the target of 100% metering, focusing on the suitability of household level metering for low-income settlements of urban Africa. An empirical analysis shows that metering is not an effective water demand management tool for domestic consumption, probably due to the fact that average consumption is already low, and it can hardly be reduced further. The case study shows that universal metering ambitions might discourage household level connections.

Chapter 6
Water Scarcity and Household Food Security: A Case of Ulundi Local Municipality in KwaZulu-Natal, South Africa 127
Nomfundo Nomcebo Zulu, University of Zululand, South Africa

The study employed the post-positivist epistemology to examine the impact of water scarcity to food security in three rural areas of Ulundi Local Municipality. A stratified random sampling technique was utilised to sample 400 respondents. Data was collected through self-designed questionnaires. Correlation analysis was used to determine the relationship between the variables. The findings showed a strong positive correlation between economic and social development, water access, and scarcity. There was also a positive relationship between water access and household food security. Water access also had a strong positive relationship with public health and development. The study concludes that water access influences food security, public health, human growth, and development. It is recommended that the municipality should improve water access in order to improve food security, public health, and development in the rural communities.

Chapter 7
Equitable Access to Water at Alfred Duma Local Municipality 149
Nokuphila Ndimande, University of Zululand, South Africa

Water plays a central role in the life of society. However, factors such as population growth, pollution, and poor allocation and distribution mechanisms place severe pressures on adequate and equitable water supply. The aim of this chapter was to look into equitable water access in the Alfred Duma Local Municipality as well as the ecological governance framework that supports water access in local areas. The chapter also looked at the position of local municipality in water access and the impact of ecological scale on water provision. Many people are still unable to exercise their constitutional right to water in Alfred Duma Local Municipality, where most women feel disempowered, marginalized, and excluded from the process of making water access decisions. This brings challenges to disadvantaged and arginalized groups socially, economically, and environmentally where vulnerable and marginalised groups have no opportunity to equitably benefit from water access.

Chapter 8
Assessment of the Realization of the Fundamental Human Right to Water at the Umalalazi Local Municipality, KwaZulu-Natal 179
Asanda Buliswa Mnguni, University of Zululand, South Africa

The aim of the study was to examine the municipal water resource governance and fundamental human rights to water in the UMlalazi Local Municipality of the KwaZulu-Natal Province. The study employed the post-positivist paradigm and closed ended questionnaires for collection of data. Stratified random sampling technique was used to sample 375 respondents, who were the households (head of households). Chi-square was used to determine relationships between the variables. The findings showed that most of the respondents strongly agreed that the 6kl of monthly water supply was not enough for meeting their domestic, consumptive, and hygiene needs. The municipality has a youthful population structure.

Females were mostly affected by household water scarcity compared to their male counterparts. For the households, a day would be terribly difficult without this life-enabling commodity.

Chapter 9
Durban Community's Experiences of Class-Based Urinary Diversion Toilets 202
Innocent Simphiwe Nojiyeza, University of Zululand, South Africa

The aim of this chapter is to look at the integrated water resources management (IWRM)-related institutional shift in governance of water and sanitation in Durban and the extent to which class is considered in providing rural inhabitants with facilities they regard as sub-standard. The key finding is that rural communities remain uncomfortable with nearly 90,000 UD toilets constructed by eThekwini Municipality, despite current efforts being made to relieve them of the most hated chores of emptying vaults when they fill up. In a study conducted in rural areas of Durban through interviews with 70 households and focus group discussions, communities continue to reject UDs on the grounds of lack of equity, proximity to areas with flush toilets, lack of communication, and lack of sustainable institutional mechanisms associated with this project.

Chapter 10
The Political Ecology of the Decentralized Water Management in Zimbabwe: Theory and Empirical Evidence From Sanyati Catchment Area 225
Winmore Kusena, Midlands State University, Zimbabwe

The chapter assesses the notion of local scale and decentralization that emanates from the IWRM principles. Evaluation of the benefits of decentralization was done through the political ecology lens that critically examines fairness and power struggles across spaces. Sanyati catchment was used to draw empirical evidence in light of the theoretical expectations of decentralization towards catchment water security. Qualitative approach was used to collect data. Interviews were the main sources of data collection. The findings showed that decentralization has failed to produce the desired results compared to what is assumed in the dominant narrative that highly esteems the decentralization management approach. The chapter showed that what brings results are not local scales and suggests that probably fair and transparent resource distribution and allocation at any scale may bring about water protection that does not trigger the tragedy of the commons.

Chapter 11
An Analysis of Sector-Based Water Supply and Demand: Mapping the Possible Water Development and Management Ways for Sanyati Catchment, Zimbabwe 239
Winmore Kusena, Midlands State University, Zimbabwe
Thomas Debwe, Midlands State University, Zimbabwe

This chapter sought to assess sector-based current water use levels in Sanyati catchment, assess a possibility of new ways of managing the catchment water using inter-basin transfer systems, and assess future water demands projection using WEAP model. The study was both quantitative and qualitative in nature. Interviews and observations were the main primary data collection instruments. Findings showed that the main sectors and institutions that utilised water in the catchment were mining, agriculture, manufacturing industry, and local authorities. Local authorities and farming sectors were the major users of water. WEAP model results showed a steady increase in demand for water as a result of both population growth and agricultural activities water demand. The study mapped possible inter-basin transfer routes and recommends water harvesting as part of development and management in Sanyati catchment to address the problem of water shortage.

Chapter 12
Addressing Energy Poverty Through Ecological Governance of Solar Home Systems in South Africa 253
Edzisani Ellen Netshiozwi, Management College of Southern Africa, South Africa

South Africa has long recognised the need to eliminate energy poverty, and significant progress has been made since 1996. With the recent global outcry about the use of sustainable and clean energy sources for human and ecological development, it has become essential for countries to upscale the use of non-grid electrification in order to effectively and sustainably eradicate energy poverty. South Africa implements different non-grid electrification programmes which focuses on ensuring universal access to clean energy by all, with one of these initiatives being the Solar Home Systems Programme. This study aims to contribute to academic discussion and knowledge about the role that solar home systems can play in bringing change on the current status of the energy poverty eradication initiatives which negatively affect the environment and the people. The study used a qualitative approach in assessing how the Solar Home Systems Programme contributes to energy poverty eradication and environmental preservation if managed in line with ecological governance principles. Semi-structured interviews were conducted with a total of 40 households in Limpopo and officials from government departments in the province as well as a service provider responsible for the roll-out and maintenance of solar home systems in Limpopo. Framed within the ecological economics theory which advocates for ecological and environmental solutions that take into consideration the future generation, the study established that the solar home systems can eradicate energy poverty and contribute to the protection of the environment if managed in line with ecological governance principles. The study further revealed that the Solar Home Systems Programme in South Africa failed due to lack of proper governance systems that provide solutions for the future as the programme was short-term focused and lacked proper ecological governance systems. In order for the solar home systems to contribute to energy poverty eradication and environmental protection, the study recommends a total overhaul of the programme which includes ensuring that the provision of SHSs is not viewed as a temporary measure that is only meant for poor rural households but a long-term and sustainable initiative.

Chapter 13
The Scope and Potential for Mini-Grid Power Systems Based on Biomass Waste for Remote Areas in Sub-Saharan Africa: Off grid 277
Gratitude Charis, Botswana International University of Science and Technology, Botswana
Tafadzwa Nkhoma, Manicalland State University of Applied Science, Zimbabwe
Gwiranai Danha, Botswana International University of Science and Technology, Botswana

Sub-Saharan Africa (SSA) has been experiencing an energy crisis. Socio-economic balances depend on access to clean, convenient, and dependable energy. This is critical for remote areas which are off the national grid, necessitating the installation of renewable energy sources such as bioenergy plants. These plants could valorize waste using combustion and gasification or biogas plants. The challenge is to produce a competitive levelized cost of electricity (LCOE). Nations like Germany and Sweden have successfully launched these. SSA can benchmark from these and valorize its biomass wastes. Key issues to consider would be cost-effective supply chains, sustainable harvest rates, after sales support, and good regulatory frameworks. This study was mostly a desktop review with a few field study observations. It was concluded that the stoker fired boiler and landfill gas 'biomass only' technologies would have the least capital costs, although gasification and anaerobic digestion are also competitive in terms of LCOE.

Chapter 14
Climate Justice in Durban: State Contradictions and Grass-Roots Action 297
Innocent Simphiwe Nojiyeza, University of Zululand, South Africa

This chapter situates itself in the climate justice discourse and unpacks the paradoxes in state and grassroots action. It argues that due to market-orientated and neo-liberal water and sanitation governance, the failure to take into account water and climate change governance as well as ecological and institutional economics resulted in climate change chaos. Due to this paralysis, communities failed to adapt to climate variability. The illogicalities and ambiguities in what the municipality says and does leaves local communities at the vagaries of nature and to deal with the resulting chaos albeit with some doing so in quite inspiring ways. The empirical data is collected through 120 interviews and four participatory action workshops conducted with civil society actors, government officials, academics, and communities of AmaQadi situated in North Central Durban, Umbumbulu in South Durban, Wood Glen and Zamani B in Durban Outer West, Ntuzuma G and Soweto in North Central Durban. The key finding is that the municipal Climate Change Protection Unit adopted a 'climate-resilient' strategy on the one hand, but on the other, city electricity officials built power stations in flood plains, the Economic Development Unit promoted a shopping mall in a wetland, and eThekwini Water is introducing unpopular UD toilets to deal with water scarcity, without looking at externalities in wetlands and the discharge of faecal waste from broken sewers that compromise biodiversity and ecosystem balances. This chapter is contributing knowledge of climate change resilience and adaptation by communities within the framework of new institutional and ecological economics.

Chapter 15
Unpacking the Resource Curse and Realism Challenges on Economic Development in the Democratic Republic of Congo (DRC): Case of Gecamines 318
Germain Miteu Tshinu, University of Zululand, South Africa

The third theme of the book seeks to understand the role of mineral resources in the African continent's socio-economic development. The case study of the DRC is used to understand the role played by the state-owned mining company Gecamines to socio-economic conditions of employees and the country's development at large. Resource curse and realism theories are utilised in this case study to unpack the role of mineral resources to Gecamines employees' access to education and healthcare in particular and to the entire country in general. The chapter employed a case study design with a qualitative approach research in its endeavour of exploring Gecamines' socio-economic contribution to its employees. Semi-structured interviews were conducted with Gecamines' managers, unskilled mine workers, and the Provincial Department of Mines' officers.

Chapter 16
Contribution of Mining Operations Towards Education, Healthcare, Food Security, Housing, Sports, and Recreation in Katanga Province of the DRC 337
Germain Miteu Tshinu, University of Zululand, South Africa

Some resource-rich countries face challenges when it comes to the socio-economic development of its employees and the community at large. This study seeks to examine Gecamines' contribution to the socio-economic conditions of its employees based in the Lubumbashi area and to some extent the province of Katanga under the rubric of resource curse theory. Through this study, lessons can be learnt from the findings on how state-owned mining companies could work on improving their socio-economic contributions in the interest of the employees and the entire population through generated revenues for the state. The study employed a case study design with a qualitative approach research in its endeavour of exploring Gecamines' socio-economic contribution to its employees in the city of Lubumbashi and the development of Katanga Province as a whole. Semi-structured interviews were conducted with Gecamines' managers, unskilled mine workers, and the Provincial Department of Mines' officers in the city of Lubumbashi. The key findings from these interactions are that the company's low level of production caused by poor management, political interference, corruption, and poor institutions had negative ramifications on the socio-economic conditions of the workers. A remedy to these challenges lies in strengthening the country's institutions to fight corruption and poor management of parastatals. Autonomy of parastatals from political influence and control is also important to enable the company to play a crucial role in the socio-economic development of employees.

Chapter 17
Mining and Water Supply in Marginalised Communities: The Case of Anglo-Platinum Mogalakwena and Ga-Molekane Village, Limpopo Province, South Africa 354
Tariro Kamuti, Center for Gender and Africa Studies, University of the Free State, South Africa

The extractive industry and water security nexus for vulnerable communities that surround a mine has been a subject of debate in the Global South. The study aimed to identify and explain the impact of mining on the ability of vulnerable and marginalized communities to secure adequate and safe water. The study focused on Anglo-Platinum Mogalakwena Platinum on Ga-Molekana Village in Mogalakwena Municipality, Limpopo Province of South Africa. Key informant interviews were administered with community leaders, local community-based organizations, local councilors, and public officials at the national level. Focus group discussions were also conducted for older women and youth to share current experiences. The Mogalakwena Municipality provides water to the Ga-Molekane Community and the Mogalakwena Platinum Mine also intervenes by providing support in this regard. Perceptions about the adequacy of water to meet their household needs were varied, with some being positive considering their situation to be better than others, while some said water had become inadequate.

Chapter 18
The Water-Waste Nexus: Sustainability Implications for Integrated Water Resources and Waste Management in Harare 368
Trust Nhubu, University of Johannesburg, South Africa
Edison Muzenda, Botswana International University of Science and Technology, Botswana & University of Johannesburg, South Africa
Mohamed Belaid, University of Johannesburg, South Africa

The management of water resources and waste is amongst the major challenges facing the majority of urban environments within developing nations due to the rapid population growth and urbanisation as well as improved lifestyles. The Greater Harare Metropolitan Province is not spared by these management challenges. This chapter proposes a number of scenarios that could be implemented in the short to medium terms to address these challenges within the GHMP from a water-waste nexus approach. The water-waste nexus approach is anticipated to significantly reduce the human health impacts and environmental impacts, specifically the water resources pollution which is responsible for the increase in cost of potable water production subsequently resulting in intermittent potable water supply in the GHMP. The study regards the recovery of energy from the waste generated in the GHMP as an integral component of the approach with the energy recovered used for water, wastewater, and waste treatment.

Compilation of References 388

About the Contributors 431

Index 435

Foreword

The effort of making water, sanitation and energy more accessible and sustainable remains a challenge to many developing countries, with some of the worst cases found in sub-Saharan Africa (SSA). This book shows that human survival is dependent on their environment but the sustainability of human life, biological and ecological systems have come under threat, requiring the need for action by actors. It uniquely discusses cross-cutting research themes on critical matters affecting African economies before, during and after decolonization and transitioning into democracy.

The book demonstrates that the medium to long-term measures to escape from future resources scarcity to environmentally friendly exploitation of nature for human consumption. It also emphasises on the importance of institutions in driving water, sanitation and energy agenda, policy processes and reconfigurations in SSA. The synergy of efforts of the state, the private sector and civil society groups in the management of water, sanitation and renewable energy is critical for driving sectoral, structural, and institutional reforms.

The conceptualizations in the book are drawn from diverse disciplines including development studies, environmental science, history and political economy. Case studies from across countries are used to highlight the various attempts and strategies to manage natural resources. The book provides diagnosis and prognosis of the strategies for managing environmental resources, which are key issues confronting different countries.

The book is organised into eighteen chapters. Chapters one to twelve succinctly discusses water resources management, water supply and sanitation. The cases are drawn from Ghana, Malawi, South Africa and Zimbabwe. Chapters twelve to sixteen unpack climate change mitigation, renewable energy and mineral resource cures with particular focus on South Africa and Zimbabwe.

Drawing cases from Democratic Republic of the Congo, South Africa and Zimbabwe, chapters sixteen to eighteen examine the effects of mining on the environment. Certainly, this is a useful book for policy makers, development practitioners, academics and professionals who seek to gain deeper understanding on the theory and practice of resource management in Africa.

John Victor Mensah
University of Cape Coast, Ghana

Foreword

The book *Resource Management and the Struggle for Water Sustainability in Africa* is compact and critical. It is a multi-layered and insightful and visionary offering about a basic element indispensable in human life, and a global crisis which has since set the world on a disastrous course. Water and climate change. Then another interesting issue, equally important, is engaged in this book as well, *mineral resources curse in Africa.* As it were, the book engages three critical subjects which are: water resources management, water supply and sanitation; climate change mitigation and adaptation as well as energy; and mineral resources curse in Africa, which is captured in a microcosmic representation through the case of Gécamines (GECAMINES), which is situated in Katanga Province of the Democratic Republic of Congo (DRC). Therefore, critically and skilfully engaging these important issues by locating them within African nation states, the authors bring forth an impactful analysis and discussion on resource (water) management, where they make a wonderful engagement with a social phenomenon (struggle for water sustainability), and environmental phenomenon, and a case that edges on the environmental and political; the authors adroitly locate these issues on the broad landscape of the African continent. This is a fiercely brilliant and robustly fruitful engagement with a social event that continues to trouble and burden both governments and the citizens. In the first part of the book (the case studies of water supply and sanitation), one of the more exciting aspects is the firm and solid contribution it makes to the field of 'Water Sustainability in Africa', while simultaneously linking it with other scholarships on the issues of resource management. Thus, the book makes a great contribution to a growing and crucial field.

The book excellently presents a pivotal view on how best can the management of resources such as water could be realised in society/Africa. The view is distilled in the following quotation: "constant participation and engagement of households", and added to it is the book's guide on inclusivity of people, especially those in rural areas. Pertaining the issue of climate change, all across the world we watched as the 'lungs of the world', the Amazon Rain Forest in Brazil caught fire on 15 August 2019 and typified the fatal problem we are facing on Earth. And for example, in South Africa, the Limpopo Province Ebenezer Dam which supplies Many parts of Mopani District (and supplies water for farmers for irrigation in the area), Turfloop and parts of Polokwane (Capricorn District) had its water level drop to 28.5% in 2019 and 15.6% in December 2020 (see Times Live, 2020). The dam dropped in water level because there was no rain, the absence of rain is due to climate change. These are some of the issues that the book represents. Chapter 6 of this book entitled 'Water Scarcity and Household Food Security: A Case of Ulundi Local Municipality, KwaZulu-Natal, South Africa' deals with the issue of water scarcity superbly. And Ulundi Local Municipality is a microcosm of many regions on the African continent burdened with this issue. Let me sample a morsel from what authors of chapter 6 say about water scarcity in correlation to food security, to illustrate the problem (ordinary) African people are faced with:

[...] the lack to access acceptable quantities of water has a negative impact on food security in rural areas of Ulundi Local Municipality (ULM) where agricultural activities are perceived to be responsible for poverty alleviation. Food security in rural areas is mostly secured through subsistence farming.

The issue of access to clean and sustainable water in the hinterlands of Africa is a one of the key arguments of this book, and it is inspiring to read scholars who are agitated by the problems that ordinary people face – water accessibility and scarcity. Submissions such as "The tricky issue in rural areas remains an issue of access which is often linked to the willingness and ability to pay for the installation and daily consumption", which can be construed as one of the common threads across the chapters in this book; and it gathers a collective cry that calls for solutions. The authors create a significant interweaving of issues such as "approaches such as Integrated Water Resources Management (IWRM), community water management, supply and demand led water resources development" and "impacts of climate change". It critically reviews the commitment of institutions such as the African Development Bank, the African Council for Water Ministers (AMCOW), the New Partnership for Africa's Development (NEPAD), the African Union (AU) and the African Union Commission (AUC) thereby creating a comprehensive index of the phenomena access to water, water scarcity, national and continental resource management techniques and programmes and sustainability. Furthermore, the book adds to the indices of best navigating and perceiving and understanding the issue of water, access and sustainability by reviewing even "UN agencies such as UN-Water in Africa" and "regional, economic and trade organisations such as" the Economic Community of West African States (ECOWAS), Southern African Development Community (SADC) and others to show that there are measures and initiatives put in place to manage resources and eradicate the social problem of access to and scarcity of water. Sustainability is a common theme that ties all the chapters of the book, coalescing them into a vibrant scientific production of a thoroughly researched and presented whole. A magnificent synthesis of the three components is fluidly and impeccably realised.

Notable is the book's take on the "adverse impacts of climate change". The book engages scientific climate change and resource management and sustainability narratives and views, and unpacks ideas such as the one captured in chapter 1, to give just a morsel for a sample, where the authors – Prosper Bazaanah and Innocent Simphiwe Nojiyeza – perfectly apply the "stakeholder salience theory" so as to fittingly "analyse the extent of power, legitimacy and urgency attached to stakeholders who are responsible for the formulation and implementation of environmental policies". This chapter, like all the others in the book, rigorously scrutinises the linkage between ecological governance and water conservation as sustainable conduits for enhancing rural livelihoods. It is its descriptive and inferential statistics which strike works out a broad and insightful view of the issue at hand, thereby rendering critically important the findings arrived at through the use of perfectly designed chi-square tests. The presentation of the findings regarding the significant symbiosis between "conservation initiatives, finance, rehabilitation/ maintenance and gender inclusion and domestic water conservation" is laudable as it is original and novel – the authors have revolutionised the issues. One can also turn the page to chapter 4, where the authors engage eco-efficiency, environmental performance, physical scarcity and economic scarcity trends by interweaving their visionary voices with various responses and readings on these issues, thereby allowing the reader to explore and feast on the multiplicity and clarity of perceiving and comprehending these issues.

The issue of water resources management, water supply and sanitation, and the issue of climate change mitigation and adaptation as well as energy (a critical issue all across the world), and the issues of mineral resources curse in Africa are, collectively, a perfect choice to produce a book on. These issues are close to the heartbeat of African developmental and transformation discourse and action. This book has adequately presented alternative perspectives pertaining to energy, and make the reader understand the need to reimagine energy production to favour the sustainability of resources. The negative and positive impacts of the mining industry on water and the climate are acutely engaged in Chapter 17. The author of this said chapter chiselled an argument around the growing demand for "socio-economic justice and thus questioning the impact that extractive industries have on surrounding communities". Excellent is the author's engagement with responses and readings from scholars such as Mehta, whose work on this is seminal. And the author makes a solid 'dig' at the failure of mines to help eradicate poverty and bring about developmental benefits to communities. The author engages the double-edged issue of poverty and inequality. Outstanding in the same chapter is the notion of stakeholder engagement, about which the author writes:

Issues of engagement among stakeholders such as the mine, communities surrounding the mine and local government are critical when it comes to the provision of adequate and safe water. This usually hinges on how amenable the mine and the local authority are in ensuring the demand for adequate delivery of water to the communities.

The authors clearly spent time researching and thinking through the chapter theoretically. This is well done. I admire the agency with which the project was realised; the rigour of knowledge production; the excellent application of theories.

I enjoyed reading this almost all-embracing, this concretely worked out intellectual cauldron of ideas. I recommend that this book be made seminal in academia, have student read and engage with it in lecture halls and their research essays. Once more congratulations to both editors and authors!

Mogomme Alpheus Masoga
University of Zululand, South Africa

Preface

INTRODUCTION

Access to water and sanitation remains a critical challenge in various countries in Africa, especially in Sub Saharan Africa. The crisis presents as a crisis of governance rather than the physical and economic scarcity. In most countries, water is realised as a human right while subsidies are provided for indigent households. The management of water resources from dams, treatment plants, purification and reticulation processes require constant participation and engagement of households. The tricky issue in rural areas remains an issue of access which is often linked to the willingness and ability to pay for the installation and daily consumption. Globally, many countries adopted approaches such as Integrated Water Resources Management (IWRM), community water management, supply and demand-led water resources development. Institutions such as the African Development Bank, the African Council for Water Ministers (AMCOW), the New Partnership for Africa's Development (NEPAD), the African Union (AU) and the African Union Commission (AUC) and UN agencies such as UN-Water in Africa and regional economic and trade organisations such as Economic Community of West African States (ECOWAS), Southern African Development Community (SADC), Common Market for Eastern and Southern Africa (COMESA) and others committed themselves to dedicate 1% of each of the AU member states' GDP to water supply and half a percent of the GDP of each of the AU member states to sanitation. There is no doubt that there are many initiatives in Africa which are championed by African countries.

Due to adverse impacts of climate change, African Ministers responsible for energy established a body similar to AMCOW, which is a platform for energy ministers to discuss strategies to mitigate the impacts of climate change and in particular the reliance on coal and other fossil fuels as sources of energy. In most African countries, the strategies adopted include a shift to renewable energy sources such as solar home systems, solar water heaters, bio-digesters, small hydroelectricity dams and other technologies. Most of the transboundary and shared water sources involve cooperation between various African countries in either water resources development or energy for the purposes of mitigating and improving the resilience and adaptive capacity of each of the member states. The recent hurricanes which affected South Africa, Zimbabwe, Mozambique and previous hurricanes which affected most of the SADC countries, indicate that African countries need to improve their resilience and adaptive capacities, while mitigating climate change through adopting renewable energy sources as part of the energy mix. Countries like Denmark and Germany are leaders in renewable energy sources, but there is potential for Africa to provide leadership in the production of clean and renewable sources. The Conference of the Parties (COP) 26 that took place in Glasgow in 2021 focused on funding the adaptive capacity of landlocked countries, small island states and the developing countries. Tanzania, in East Africa is embarking on building four hydroelectricity powered dams with an aim of mitigating the adverse impacts of climate change, improving clean energy generation as well as the storage of water resources. These initiatives in most of the Sub-Saharan countries are geared towards energy security, water security and strengthening the resilience of these countries, as reflected in this book.

There is no doubt that most of the African countries became Belgian, French, British, Portuguese, German and Spanish colonies because of the abundance of mineral resources found in most of these countries. As such, the scramble for Africa was, in essence, a scramble for mineral resources. Most of the civil wars which affected African countries after these countries attained independence from colonial powers was inspired by the looting of these resources. In African countries, mineral resources have become a curse instead of a blessing. This is because the economies of most African countries remained small despite having the abundance of mineral resources. In South Africa and Botswana, the communities are benefitting in terms of getting royalties, access to education, water, the construction of sporting facilities and health facilities such as clinics and hospitals. In the DRC, despite civil wars and Ebola which ravaged that country and affected the entire Central African Region and neighbouring East, North and Southern African countries, state-owned mining companies provided sporting, educational and health facilities to employees and communities surrounding these mines in provinces such as Katanga, Lubumbashi and Kinshasa. There is a possibility for mineral resources such as oil in Nigeria, Equatorial Guinea and other African states to become a blessing instead of a curse.

OBJECTIVES AND THE SCOPE OF THE BOOK

The book is divided into three components, namely (1) water resources management, water supply, and sanitation; (2) climate change mitigation and adaptation as well as energy; and (3) mineral resources curse and the case study of GECAMINES situated in Katanga Province of the DRC together with a case study of the mining and water nexus in the Limpopo Province of South Africa serve as examples of a link between mining and socio-economic development. The case studies concerning water supply and sanitation are drawn from Ghana, Malawi, South Africa, Zimbabwe and selected Sub Saharan countries. The theories considered in these case studies range from Political Ecology, Ecological Economics and Water Governance. These are practical examples of IWRM implementation in each of these countries. The second part of the book addresses contemporary issues of alternative energy as part of climate change mitigation and the other case study looks at how communities adapt to climate change. It reflects on the local responses by NGOs and communities to floods and other climatic conditions. In addition to these South African case studies, there are several Sub Saharan case studies of mini-grid power generation derived from biomass. The sources of alternative and clean energy are solar home systems, solar water heaters and biomass. The theories considered in these chapters include Ecological Economics and Environmental Economics. The final part of the book unpacks the contribution of GECAMINES to the socio-economic development of the DRC. The indicators looked at in these case studies include education, sports, food security and the provision of health care facilities to the employees of GECAMINES as well as the community of Katanga province. The theory that is considered in this case study is the resource curse and realism. Related to this case study of the DRC is the link between water supply and mining within the mining communities of the Limpopo Province of South Africa and the water and waste nexus in Harare City of Zimbabwe. The book paints a picture of Sub-Saharan Africa on issues of water resources management, water supply and sanitation, altenative sources of energy considered for the mitigation of climate change, water-waste nexus including the role of mines in the promotion of socio-economic development and mining-water nexus.

Water Supply and Sanitation

The case studies on water supply and sanitation in Africa featured in this book, address issues of governance, water quality, river basin management, IWRM, ecological sanitation, participation in water resources management such as catchment management authorities/ forums, transboundary and shared water resources. The first two chapters reflect on the governance of water resources in Northern Ghana and the role of non-state actors such as NGOs in providing water supply and sanitation services in Northern areas of Ghana. Water resources management in Ghana were in the hands of the state under the era of Kwame Nkrumah, the founding Pan Africanist who shifted the Gold Cost colony towards a Marxist Leninist economic development since 1947. When Nkrumah was ousted from office in 1966, several military and democratic governments shifted the country's economy towards Western capitalism, and currently towards neoliberalism. The water sector witnessed the separation of urban from rural water when Ghana Water Company (GWC) and Community Water and Sanitation Agency (CWSA) were established to provide water supply and sanitation services in urban and rural areas respectively. Most of these reforms were influenced by structural adjustment programmes (SAPs) of the World Bank and the adoption of IWRM. Other institutions established include CSIR, WRC, PURC, district assemblies and river basin authorities such as Volta River Authority and others. The shifts towards neoliberalism are characterised by the privatisation of water in urban areas and community management of water in rural areas. These chapters are framed according to the political ecology lenses and posit that the ecological governance and sustainability of water resources in rural and small town areas focus on conservation and water demand management including non-state operators such as Water Aid, Catholic Relief Services and others. The chapters paint a picture of decentralization of a state model of water supply and sanitation delivery and the role of NGOs in water supply and sanitation, community mobilization and strengthening the participation of communities in ecological governance. In both chapters, descriptive and inferential statistics were deployed to explain the various models of water governance adopted in Ghana.

The next three chapters focus on decentralization, cost recovery, recognition of water as an economic good, failing pollution abatement and the installation of water meters in various parts of Malawi championed by Lilongwe Water Board. The chapter argues that the installation of water meters did not prove to be effective in water demand management, but demonstrated the effectiveness of water utilities. In Malawi, since the dawn of democracy in 1994 which ended the one-party state led by the late Kamuzu Hastings Banda of the Malawi Congress Party, there were several attempts to mending Malawi's water institutions. The 1969 Water Resources Act of the Banda era, remained intact during the 10-year period of the United Democratic Front (UDF) rule and the floor crossing regimes of Bingu wa Mathuraka of the DPP.

It was only when PP of Joyce Banda came to power that Malawi managed to repeal the Water Resources Act of 1969 and replaced it with the Water Resources Act of 2013 which promoted IWRM solutions in the governance of water in Malawi. Reformed water institutions such as Mpira Water Authority, Southern Regional Water Board, Northern Regional Water Board, Lilongwe and Blantyre Water Boards were established and reformed during Kamuzu Banda, Bakili Muluzi, Bingu wa Mutharika, Joyce Banda and Peter Mutharika's administrations. The return of the Malawi Congress Party to office under Lazarus Chakwera is expected to deepen water reforms initiated by Kamuzu Banda. Malawi remains one of the few countries where a liberation movement, like the MCP was voted out and in office in different eras. The water supply and sanitation reforms, the sustainability of water resources, attempts to abate pollution and mitigation of climate change are some of the highlights of the water reforms unfolding in Malawi since independence in 1964 and several democratic dispensations including UDF, DPP, PP and MCP. The three case studies adopted qualitative as well as quantitative approaches.

In South Africa, most municipalities adopted ecological sanitation measures to deal with the physical scarcity of water. In Durban, the Metropolitan Council introduced the Urinary Diversion Toilets (UDT) in the rural areas. Communities that are beneficiaries of UDT regard such toilets as discriminatory, promote inequity and inequalities. In rural districts such as Zululand, King Cetshwayo and uThukela, communities are grappling with limited amounts of free basic water allocated to indigent households. As much as Free Basic Water (FBW) advances section 27 of the SA constitution in terms of the right to water, the 6kl provided to indigent households since 2001 is limited for domestic consumption as compared to eThekwini Metro Council, which increased the FBW allocation to 9kl and all households regardless of their social and economic class initially benefitted from such state subsidies, up until 2017 when the universal access to free basic water was reduced to 6kl and only then provided to indigent households. In Rustenburg, a Royal Bafokeng community receive unlimited amounts of water for free from the Royal Bafokeng Administration. This illustrates that mineral resources extracted by the Royal Bafokeng Nation could be used as a catalyst for water resources development and realisation of water as a human right.

These four case studies that look at water equity, political ecology, ecological economics and realisation of the human right to water in SA, are adding value in this book by illustrating water governance challenges facing South Africa. Like all other countries, South Africa adopted IWRM and other water governance models including privatisation of water during Nelson Mandela, and Thabo Mbeki administrations and strong state centred infrastructure development approaches under Kgalema Motlanthe, Jacob Zuma and Cyril Ramaphosa administrations. The river basin approaches of the Nelson Mandela era continued in all ANC administrations, with different emphasis such as eradication of bucket latrines, the introduction of ecological sanitation technologies and the introduction of free basic water under Mbeki and water infrastructure development under Motlanthe, Zuma and Ramaphosa Administrations. The four case studies were written using both qualitative and quantitative approaches.

Zimbabwe grappled with land reforms since gaining independence from Britain in 1980, whilst adopting several water reforms. The contribution of Zimbabwe in adopting innovative approaches to water governance is overshadowed by the dominant narrative on land reforms which resulted in the confrontation between this country and its former colonial master, the UK and its allies. This section of the book focus on the water reforms adopted since independence in 1980 by the ZANU-PF administrations of the late Robert Mugabe and the current administration. Zimbabwe also adopted IWRM and championed water reforms characterised by the establishment of river basin authorities and catchment councils and sub councils. Zimbabwe adopted decentralisation of water resources management and catchment management authorities were established. Examples of catchment management agencies in Zimbabwe are Sanyati Catchment and others. The two book chapters unpack the political ecology of decentralised water management in Zimbabwe and the mapping of future water demand in the Sanyati Catchment. The book chapters unpack the theory and empirical evidence from a catchment management agency e.g. Sanyati Catchment and others and use inter-basin transfer and the WEAP model to project future water demand scenarios in the 10 sub councils within the Sanyati catchment.

Mitigating and Adapting to Climate Change

There are three case studies presented in section B. As part of mitigation of climate change strategies, the South African government is investing in solar home systems and solar water heaters. In districts such as Vhembe and Polokwane in Limpopo and the City of Johannesburg in Gauteng province, solar home systems and solar water heaters are being rolled out in both urban and rural areas. In countries like the DRC, the Great Inga Dam is being developed to provide hydroelectricity power in the same manner as Kariba Dam did for countries in Southern Africa. Gas production is also being explored by many members of the Southern African Power Pool (SAPP). There is therefore a need to address energy poverty through ecological governance using solar home systems, solar water heaters and other renewable energy technologies in African countries. Whilst governments across the globe approach climate change mitigation and adaptation in the Conference of the Parties through renewable energy sources and improving adaptive capacities of states, civil society organisations are investing in promoting climate justice. The case study of Durban is looking at climate justice and the role of the state in promoting ecological and climate justice. The third case study is on off grid biomass electricity generation in Sub-Saharan countries.

Sub-Saharan Africa (SSA) has been experiencing an energy crisis. Socio-economic balances depend on access to clean, convenient and dependable energy. This is critical for remote areas which are off the national grid, necessitating the installation of renewable energy sources such as bioenergy plants. These plants could valorise waste using combustion and gasification or biogas plants to plants. The challenge is to produce a competitive levelized cost of electricity (LCOE). Nations like Germany and Sweden have successfully launched these LCOE. SSA can benchmark from these and valorise its biomass wastes. Key issues to consider would be cost-effective supply chains, sustainable harvest rates, after sales support and good regulatory frameworks. This study was mostly a desktop review with a few field study observations. It was concluded that the stoker fired boiler and landfill gas 'biomass only' technologies would have the least capital costs, although gasification and anaerobic digestion are also competitive in terms of LCOE. The examples presented in this chapter are drawn from Kenya, Tanzania, Malawi and Zimbabwe.

Mining, Water-Mining Nexus, and Water-Waste Nexus

The third theme of the book is looking into the role of mineral resources in Africa in strengthening economic growth and development. The DRC case study is looking into the effect of mineral resources on employee's access to education and health care. The Case of Gecamines in the Democratic Republic of the Congo is investigated in terms of the extent to which mineral resources are utilised to promote the wellbeing of employees and the community at large. The case study also unpacks the resources curse and realism challenges on economic development in the Democratic Republic of Congo. There is no doubt that countries like Botswana, South Africa, Equatorial Guinea, Nigeria, Ghana and others, are endowed with mineral resources and the experiences of the employees and communities are different from those of the DRC. The case study of the DRC sheds light on the contribution of the mining sector in promoting education, sporting facilities, health care and other amenities which are provided to employees and the mining communities in the mining towns of Katanga.

The South African case study traces the nexus between the mining sector in Limpopo and the provision of water supply and sanitation services within the mining towns. The extractive industry and water security nexus for vulnerable communities that surround a mine has been a subject of debate in the global South. The study aimed to identify and explain the impact of mining on the ability of vulnerable and marginalized communities to secure adequate and safe water. The study focused on Anglo-Platinum Mogalakwena Platinum on Ga-Molekana Village in Mogalakwena Municipality, Limpopo Province of South Africa. Key informant interviews were administered with community leaders, local community-based organizations, local councillors and public officials at the national level. Focus group discussions were also conducted with older women and youth to share current experiences. The Mogalakwena Municipality provides water to the Ga-Molekane community and the Mogalakwena platinum mine also intervenes by providing support in this regard. Perceptions about the adequacy of water to meet their household needs were varied, with some being positive considering their situation to be better than others, while some said water had become inadequate.

The final chapter looks at the water and waste nexus in Harare, Zimbabwe. The management of water resources and waste is amongst the major challenges facing a majority of urban environments within developing nations due to the rapid population growth and urbanisation as well as improved lifestyles. The Greater Harare Metropolitan Province (GHMP) is not spared by these management challenges. This Chapter proposes a number of scenarios that could be implemented in the short to medium term to address these challenges within the GHMP from a water-waste nexus approach. The water-waste nexus approach is anticipated to significantly reduce the human health and environmental impacts specifically the water resources pollution which is responsible for the increase in cost of potable water production subsequently resulting in intermittent potable water supply in the GHMP. The study regards the recovery of energy from the waste generated in the GHMP an integral component of the approach with the energy recovered used for water, wastewater and waste treatment.

TARGET AUDIENCE

The book is targeting policy makers, and scholars within Social Sciences in particular Development Studies, Sociology, International relations and Politics and natural sciences such as electrical engineering, environmental sciences, geography, hydraulics, chemistry, and physical sciences.

Chapter 1
Ecological Governance and the Sustainability of Rural Household Water Conservation Systems in the Savannah Region of Ghana

Prosper Bazaanah
https://orcid.org/0000-0002-9948-2147
University of Johannesburg, South Africa

ABSTRACT

This chapter examined the link between ecological governance and water conservation as sustainable pathways for enhancing rural livelihoods in the Savannah Region. Designs adopted were post-positivist and cross-sectional. Probability sampling techniques were used to sample 450 household and official respondents. Questionnaires were administered, while descriptive statistics and chi-square test were utilised to analyse the data. Findings showed significant relationship between conservation initiatives, finance, rehabilitation/maintenance, and gender inclusion and domestic water conservation. Therefore, with commitment to maintenance, funding, and gender inclusion in water decisions, there is the likely for water to be locally sustainable in rural communities of the region. Democratic, decentralised, and participatory approaches to ecological governance and empowerment of the local communities are recommended as essential preconditions for achieving ecologically self-governing communities and sustaining domestic water systems in the rural areas of the region.

INTRODUCTION

The governance and conservation of water resources is considered essential for effective water delivery and enhancement of rural livelihoods (Grimm, Grove, Pickett & Redman, 2000; Grimm & Redman, 2004). Hence, sustaining local development can be achieved by ensuring that water utilisation is linked

DOI: 10.4018/978-1-7998-8809-3.ch001

with ecological, human and social systems (Alberti et al., 2003). This is considered essential for planning, sustainable use, control over and protection of water resources (Grimm & Redman, 2004). Similarly, for Nelson and Shilling (2019), supplementing scientific data with local and traditional knowledge is imperative for broadening the information base needed for better governance and local decision making regarding the ecological management and sustainable use of water resources. The participation of rural dwellers in water decisions, access to potable water and utilisation is perceived to be a minority privilege (Jiménez et al., 2019). Water conservation is an ecological imperative in the face of rural water scarcity (Pickett, Cadenasso & Grove, 2004). The sustainability of water is perceived as the efficient utilisation of water, reduction in unnecessary water usage, losses and wastage of potable household water (Manu, 2015).

The governance of water and ecological resources relates to the understanding of "the complex relations between nature and society through a careful analysis of what one might call the forms of access and control over resources and their implications for environmental health and sustainable livelihoods" (Watts, 2000: 257). In line with this assertion, ecological governance embraces all the policies, strategies and activities to sustainably manage the natural resource of fresh water, to protect the environment, and to meet the current and future human demands of water (Goldratt & Cox, 2016). Yanmaz (2013) and Pennington (2008) perceives ecological governance as embracing the policies, legislations, capacities and skills required for the efficient planning and management of water resources available for consumption by human population. These water resources are composed of piped water, dug-out wells, surface water and underground water that can be treated/filtered for human use (Neil, 2018).

Embedded in ecological governance are traditional governance systems designed to conserve nature. Osunade (1994) contends that the traditional ecological governance systems, rites and rituals and customary practices do not only protect traditional water catchment areas but also serve as possible measures for minimising flooding and soil erosion. In general, water is considered to be a limited natural resource and a public good fundamental for life and health (United Nations, 2015). Therefore, the human right to drinking water is indispensable for leading a dignified socio-economic life and an improved standard of living. Undeniably, water and sanitation are essential prerequisites for the realization of other human rights (Todaro & Smith, 2014). The General Comment 15 of the United Nations (2003) argues that the rights to water entitle every human being to sufficient, safe, acceptable, physically accessible and affordable water for personal and domestic uses. As the population of the world increases with potable water becoming scarce, power (influence) and affluence of people affect water demand, allocation, consumption and conservation of water (Sanoff, 2000). Nonetheless, the United Nations (2003:1) have observed that "over one billion persons lack access to basic water supply, while several billions do not have access to adequate sanitation, which is the primary cause of water contamination and diseases linked to water". A continuing global contamination and unequal distribution of water is not only exacerbating global poverty but also, depleting the quality of ecological systems necessary for human survival. Therefore, both state and non-state parties have to adopt effective measures to ensure sustainable use and conservation of water and ecological resources. For Hyden and Neef (2004), key mechanisms for ensuring human-environment equilibrium and guaranteeing sustainable development is to ensure better ecological governance, effectively strong community participation and inclusion of local constituencies who are the ultimate beneficiaries of development. For Todaro and Smith (2014), water and sanitation are at the very core of sustainable development and critical to the survival of people and their ecosystems. As a result, civil society actors are advocating for better ecological governance (Kramer & Pahl-Wostl, 2014). Human rights and democratic theorists (Hoppe, 2009; Hoppe, 2010) extend the narrative further by arguing that opportunities to participate in environmental governance and water conservation should be expanded to include local constituents.

In addition, to achieve a green state, Eckersley (2019) and Béné and Arthur (2006) recommend the need for a participatory process to engage potential users/ beneficiaries in decision making and feedback through monitoring and evaluation processes, which engages rural dwellers in all aspects of development which affect their wellbeing. The idea of ecological governance is to create self-conscious communities inhabited by ecologically friendly, free, fair, participatory and more ecologically responsible citizenry (Neil, 2018). Nonetheless, creating spaces and structures for participation in decision making (Gershman, 2013) on eco-governance and water conservation has not been adequately researched and articulated particularly at the lowest /rural level of development (Manu, 2015). Therefore, Wanjiru (2014) contends that water conservation and governance must encompass the engagement and policy participatory processes, strategies and activities developed to manage potable water and sustainable exploitation of human ecological systems.

This is because, as ecological citizens, it is the responsibility of all societies and individuals to adopt attitudes and practices that conserve, protect and sustain water resources. To achieve this, democrats have offered the decentralisation (Oates, 1972; Hoppe, 2010) approach to water, where powerless and excluded households are empowered to participate (Arnstein, 1969) in identifying strategies for resolving the social, political, economic and ecological constraints (Goldratt & Cox, 2016) to their water conservation. This ensures that the utilisation of water resources proceeds along a sustainable trajectory (Munasinghe, 1992; Herman, 2006). Even though the rights to water are universally acknowledged, in most African countries, attempts to address poor sanitation services and water supply constraints have focused on hydrology and management at the expense of political considerations and ecological governance dimensions and have mostly taken a fragmented approach (FAO, 2017; Iribarnegaray & Seghezzo, 2012). In Ghana, the use of unsafe drinking water has been found to be the cause of diseases such as cholera and dysentery, resulting in increased death among rural-poor populations in the sub-region of Africa (WHO, 2014). This chapter is interested in exploring the nexus between the existing ecological governance systems, conservation of water and ecological resources in rural areas of the Savannah Region. This is necessary because a lack of power, unaccounted political ecological experiences and untold unbalance of power relations, participatory narratives and the governance of the rural water sector can have adverse effects on sustainable development and conservation of ecological endowments at the rural level.

BACKGROUND

Africa is regarded to be the second-driest continent in the world. Despite being critical to all aspects of human growth and development (Awoke, 2012), Africa suffers from severe water scarcity problems (Paulson, 2015; Misra & Malhotra, 2011). Moreover, the implementation policies and strategies towards the provision of affordable and universal access to drinking water and curbing sanitation constraints such as the disposal of both liquid and solid waste, appear to be challenging in most countries in Africa (UNESCO, 2017). Evidence reveals that in sub-Saharan African countries, in every ten persons, four lack access to improved drinking water and proper sanitation (WHO & UNICEF, 2017). The enforcement of policies, management of domestic waste and household water supply in rural sub-Saharan Africa (SSA) have generally been either ineffective or weak at best. This accounts for water scarcity, ill-health and resource conflicts in parts of the sub-region. Again, the rates of access to potable domestic water services in rural SSA are among the lowest globally, with approximately one in two rural dwellers, or 278 million people in total, lacking access to an improved water source (WHO & UNICEF, 2010). The lingering and acute water scarcity could possibly be controlled in most SSA countries with better water and sanitation delivery (Todaro & Smith, 2014) approaches which are founded on the principles of ecological governance (Naik, 2016).

In the Savannah Region of Ghana, access to water appears to be geographically, politically and socially constrained (Bazaanah, 2019). Comparatively, the region is both meteorologically and hydrologically drier environment than the southern and middle part of Ghana due to its proximity to the Sahara Desert, which is the largest and hottest desert in the world (Ghana Meteorological Agency, 2019). To reverse the prevailing inequalities of sanitation services and pervasive water scarcity at the rural level, the local government's Act 1993 (Act 462) [revised as Act 936 in 2016] established the Municipal, Metropolitan and District Assemblies as the lowest political, socio-economic and ecological governance structure in Ghana (GoG, 2016). As development agents, the MMDAs are funded by government and partner with other state and non-state agencies including the Community Water and Sanitation Agency (CWSA), Non-governmental Organisations (NGOs) and the private sector to mobilize local resources towards the delivery of improved potable water and sanitation services to local populations. To ensure inclusive and integrated management of water and sanitation, the decentralised and Community Management Model (CMM) is adopted by the MMDAs (CWSA, 2014). This entails the establishment of decentralised sub-structures including the Water and Sanitation Committees (WATSAN) and Water and Sanitation Management Teams (WSMTs) to manage water and sanitation delivery at the local level (CWSA, 2015; *Nyarko et al., 2006*). Nonetheless, rural poverty, power asymmetries and pervasive scarcity of water (Eguavoen, 2006) tend to perpetuate the narratives of water and ecological decision-making between power-holders and the powerless. This compels rural residents to extensively depend on surface water sources for personal and domestic uses being oblivious of the health risk of disease infections linked to poor sanitation and water contamination (Umesh & Nagarkatte, 2017).

MAIN FOCUS OF THE CHAPTER

This chapter articulates the praxis and effects of absence of space and participatory power and the effects of such vacuum on policies related to rural environments. Drawing empirical evidence from rural districts of Ghana, the chapter addresses the critical gap in the literature on whether power and participatory spaces in the context of rural ecological governance influences sustainability and domestic conservation of water. The chapter tests these ecological narratives by postulating and highlighting that participation creates spaces for local people and communities to become educated on their environment, be informed and empowered with ecological consciousness and this enables them to adopt ecologically friendly attitudes (less disruptive) and moral obligations towards the sustainability of their water and ecological systems.

Water and Ecological Governance Systems of Ghana

The ecology and governance of the territorial waters of Ghana are legally vested in the president of Ghana. The institutional foundation and theoretical impetus for eco-self-governance are established by Article 240 (1 & 2) of the 1992 Constitution and Act 462 (revised as Act 936 in 2016), which adopts a decentralised approach as a pathway to democratize and sustain water delivery at the micro level of development (GoG, 1992; GoG, 2016). The structure (see figure 1) of ecological and water resource governance is divided into two, namely urban/peri-urban water which is managed by the Ghana Water Company (GWC) and rural water sector, which is under the management of the Community Water and Sanitation Agency (CWSA) in collaboration with the District Assemblies, external donors (World Bank, 2008*)* and civil society organisations. In terms of regulation, GWC Ltd is regulated by the Public Utility Regulatory Commission (PURC), Water Resource Commission (WRC), Ministry of Water Resource, Works and Housing (MWRWH) (renamed as the Ministry of Sanitation and Water Resources-MSWR, in January 2017), Environmental Protection Agency (EPA), research institutions and international organization. Until the establishment of MSWR, both the urban and rural water sector were part of the MWRWH and the sanitation sector was under the Ministry for Local Government and Rural Development (MLGRD). The separation of MSWR from MWRWH was to formulate and implement water and ecological policies for sustainable management of national water resources, provision of safe, adequate and affordable water, environmental sanitation facilities, and sustainable management of both solid and liquid waste for the growth and wellbeing of all people in Ghana (GoG, 2019). Despite these institutional setups and mandates, water scarcity and ill-health arising from unsanitary environment (Todaro & Smith, 2014), are still prevalent in rural Ghana and settlements continue to live in squalid environmental conditions (Central Gonja District Assembly, 2017). This is due to multi-sectorial constraints including inadequate infrastructure, financial deficits and poor-culture of maintenance of facilities which continue to beset the rural water and sanitation sector of Ghana (WHO & UNICEF, 2017; GICG, 2018). Meanwhile, ecological governance and water delivery at the rural, small towns and community level (see figure 1) is provided by the CWSA, Ministry for local government, the Metropolitan, Municipal and District Assemblies (MMDAs) and Regional Coordinating Councils (RCCs) in collaboration with other civil society actors including EPA, Non-governmental organisations (NGOs), traditional authorities and individuals.

In figure 1, though the water sector is legally well constituted, major problems are still prevalent in rural Ghana because of lack of proper governance systems and measures to clean and conserve water (MWRWH, 2014). These problems are caused by both natural and anthropogenic factors including unplanned building, excessive deforestation, farming along water courses, inadequate drainage systems and poor waste management by the Assemblies. In addition, illicit bush burning destroys the vegetative cover meant to protect water bodies and ecological sources and therefore, further exacerbate the problem in the districts (CWSA, 2017). The non-compliance with buffer systems of ten meters along water bodies is resulting in contamination and drying up of surface water sources in the districts (GSS, 2018). Among the rural settlements of Ghana, human activities related to land use such as effluent released by growing small and medium scale industrial sectors, domestic and commercial sewage, acid mine drainage, faecal contamination linked to insufficient infrastructure and domesticated animals grazing too close to water source or sharing same water sources with the communities, are critical water quality and ecological management constrains (GICG, 2018).

Effective policies and plans for water conservation and ecological governance must be implemented in deprived areas if rural settlements are to continue to sustain their water and ecological systems (Wanjiru, 2014). Moreover, our knowledge on how spaces are or can be created to allow transparency, participation in ecological self-governance and how the local people's inputs are solicited in ecological and water governance to influence their development and conservation choices remains scanty. It is the prevalence of inadequate access to space and information for ecological engagement (Gaventa, 2006) and capacities of rural districts (Ahwoi, 2010) and CWSA which appear to be discouraging local people from contributing towards water conservation (CWSA, 2015). This present study is a step towards filling this gap, by adding space for finance, accountability, and participation in the existing debate on rural ecological governance and district level development programmes. In Figure 1, the narratives and theoretical framing on rural ecology and water conservation in this paper are founded on a top-down principal-agency model (Gaventa, 2006). The contribution of this study is to bring to the forefront the "principal-agency" relationship and articulates perspectives on participatory ecological governance and sustainability of household water in rural settings. This study perceives the "principals" as being hegemonic and power wielding (government, departments) with oversight responsibility for ecological conservation, yet failing to address development and disparities of sustainable water delivery in rural Ghana.

Figure 1. Institutional Set-up of Ghana's Water Sector
Source: Author's Construct (2019).

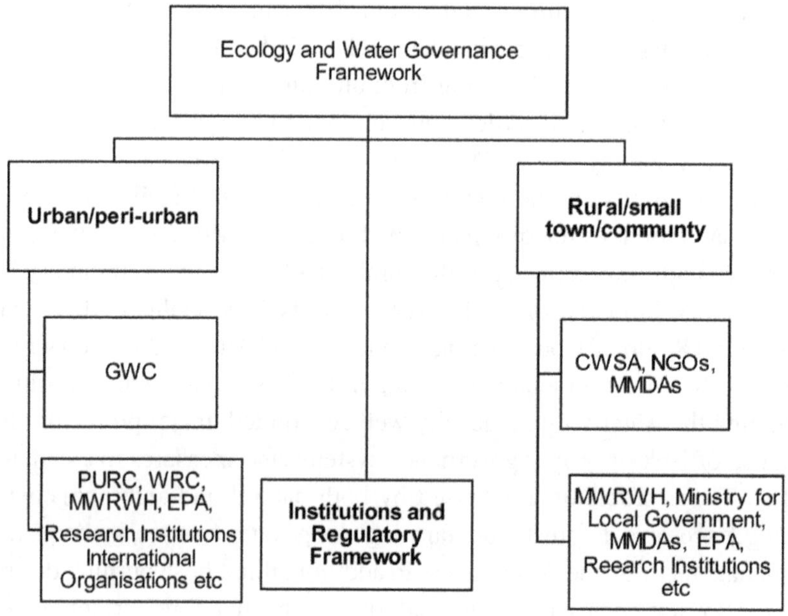

The "agent" in development discourse is highly subjective and political, hence makes this Chapter, in part, an unmasking critique of MMDAs and CWSA as autonomous and independent development agents at the community level. This study argues that although the agents constitute the legitimate development beneficiaries as they are empowered as promoters of local development (Zanu, 1996), without capacities to voice their concerns on constraints bedeviling the sustainability of water resources, they tend to lack adequate capacities to execute their mandate effectively at the local level (GoG, 1993). Moreover, to what point can one say community members are actively involved in rural ecology, water conversation and allocation. This study argued that as long as powerful actors make decisions on behalf of locals, there will be no democratisation in ecological governance and water conservation. These considerations are important in showing the power-relations among those involved in the design, implementation and administration of rural water and ecological policies (Pennington, 2008). Moreover, the empowerment of localities to participate fairly, equally and meaningfully would ensure equity, social justice (Walker,

2006) and pro-poor inputs are articulated and effectively considered in water and rural ecological decision making. The challenges with water sustainability in rural Ghana are therefore a governance problem, not that of resource scarcity (GWP, 2004; MWRWH, 20014; Brundtland Commission, 2014). For Lautze, De Silva, Giordano and Sanford (2011), the underlying constraints to water governance includes lack of openness, transparency, broad participation by actors, predictability, ethics and integrity. Therefore, the Dublin Statement stipulates that "decisions be taken at the lowest appropriate level, with full consultation and involvement of local users of water in the designing, planning, implementation, monitoring and evaluation of water projects" (ICWE, 1992:4). This idea has been criticized as being an oversimplification of the challenges in eco-governance. The question of who would represent large groups of stakeholders is a highly political issue which remains unaddressed. Moreover, the relationship of the people participating in any multi-stakeholder eco-governance process is problematic, especially when multi-parties are involved (Castro, 2007). Since water is an essential human right and a public good, this study argues that integrated approaches are required for dealing with the challenges of water conservation and governance at the micro level of development. Moreover, the participation of citizens is no longer indisputable, but a critical approach for rural water conservation.

The study maintains that in the face of rural water scarcity and the need for local inputs and support, the engagement of local citizens in direct democratic participation in eco-governance could become a feasible alternative to water conservation. A participatory system where citizens or community groups are offered an opportunity to articulate their collective societal interest, preference and needs in decision making processes could perhaps become the most feasible development option and ecological imperative (Christen & Robinson, 1980). From the rural and traditional governance context, Osunade (1994) found that the sacredness of water sources, traditions, religious practices and norms on the environment tend to shape attitudes, influence perceptions and regulate human behaviour towards the use of water and ecological resources. In the Savannah Region of Ghana, traditional authorities such as the local chiefs and traditional priests play critical complementary traditional governance roles in the establishment and enforcement of traditions and norms towards the utilisation of ecological resources. A synergy between these traditional and western-style democratic governance systems and structures are necessary in order to promote eco-friendly attitudes and conserve ecological endowments including water catchments of rural communities. Among the Marakwets, Lssozi (2012) found that it was taboo to wash in or nearer to a water collection points for fear of incurring the fury of the river spirits. In the Eastern Cape of South Africa, Palma (1996) and Penny (2001) found that perceptions of mermaids, rituals and norms surrounding water sources motivated community compliance and avoided pollution of water resources. Similarly, in Ghana, Nukunya, (1992) found that the Akans and other ethnic groups are noted to forbid discharge of industrial and human waste into water sources, lest the culprit incurs the wrath and punishment from traditional authorities and government agencies.

Statement of the Problem

The governance of water and ecological resources is to create a just and sustainable society, yet, offering spaces for decision making on eco-governance and water conservation has not been adequately researched and articulated particularly at the lowest level of development (Manu, 2015). Moreover, participation, rights to water decisions, access to potable water and utilisation is perceived to be a minority privilege (Scoones, 1999). There appears to be a gap in the literature explicitly addressing power and space for participation as vital aspects of rural ecology, sustainability and water conservation. The problem as further reported by GSS (2011) and Sanoff (2000) is that high-quality, safe and sufficient water is inadequate to support rural livelihoods. Moreover, lack of voice and active participation of localities appear to be increasingly becoming a threat to rural endowments and drinking water systems (GSS, 2011). With rural population increase and potable water becoming scarce, power (influence) and affluence of people affect water demand, allocation, consumption and conservation of water (Sanoff, 2000). This is a problem because rural ecological conservation designs and development activities devoid of participation may be inequitable to beneficiaries (Castro, 2007), negatively affecting the livelihoods, health (Munasinghe, 1992), productivity of rural dwellers (Herman, 2006), and result in policies that ignore the existence of the implicit needs and interest of the consumer (Neil, 2018), which in rural ecological context, are the beneficiary households (Hoppe, 2010). It appears, based on an extensive search for updated literature within this area that the gap still exists, particularly in the Savannah Region of Ghana despite decades of independence and implementation of decentralised approaches to governance (GoG, 1992, GSS, 2011; Wanjiru, 2014; Manu, 2015).

An inquiry into space and power for participation as measures for water conservation in deprived areas is necessary if rural settlements are to continue to sustain their water and ecological systems (MWRWH, 2014). The development of the rural areas in the study districts is facilitated through the establishment of traditional and District Assembly structures. The two structures are interrelated and interdependent yet complement each other to promote the development of the communities (Mohammed, 2015; GoG, 2010). The Assemblies have the same internal political structures (see Figure 1), with their legal basis originating from the Local Government's Act 1993 (Act 462) [revised as Act 936 in 2016] and the 1992 Constitution [Article 241 (2)] (GoG, 2016; Ahwoi, 2010; GoG, 1992). As the highest development agents, the districts collaborate with stakeholders to promote healthy environment, education, commerce, planning and public safety (GSS, 2014a; Ayee, 2008).

Though the Assembly structures are well constituted, the decentralised departments lack resources (Addo, 2010), technical and managerial capacities locally to effectively manage service delivery and this affects the quality and sustainable management of water facilities at the community level (WGDA, 2017; Central Gonja District, 2017; Central Gonja District, 2016; WGDA, 2016). The local chiefs are custodians of the land who formulate and enforce customary practices to safeguard natural resources, promote peace and development in the settlements. Unlike the southern sector of the country, chiefs in the Savannah Region of Ghana are traditionally enskinned and, hence, their authority, power and legitimacy are symbolized by their rights to sit on skins (Haruna, 2017). Even though traditions and local governance structures co-exist in rural spaces which ideally could be exploited to protect the environment, ensure water conservation and sustainability, the synergy between western-style local governance and traditional systems appear to be ineffective (Nukunya, 1992). This affects the formulation and enforcement of policies, local bye-laws and traditional rituals to shape attitudes and behaviour of localities towards the conservation of water and ecological endowments of localities in the districts (Pennington, 2008). Since post-independence, rural communities in the districts continue to experience deterioration and marginalization in access to socio-economic amenities and resource allocation (Awedoba, 2015). In this malady, the rural water and sanitation sector of Ghana has been the most affected. This study explored the relationship between the ecological governance systems, conservation of water and ecological resources in the rural areas of the study districts.

Objective, Hypotheses and Justification of the Study

The objective of the study was to examine the link between participation in ecological governance and water conservation among historically marginalised and peripheral rural households in the Bole, West Gonja and Central Gonja Districts of the Savannah Region. To achieve this objective, the study hypothesized (H_0) that:

$H_{0:}$ *there is no statistically significant relationship between the participation of rural households in water governance (conservation, funding, gender inclusion, labour support, maintenance), the distribution and conservation of drinking water in their communities.*

$H_{A:}$ *there is a statistically significant relationship between the participation of rural households in water governance (conservation, funding, gender inclusion, labour support, maintenance), the distribution and conservation of drinking water in their communities.*

The participation of beneficiaries has become de rigueur in rural ecology and sustainability discourses. Nevertheless, creating participatory spaces and empowering localities for participation in the governance and conservation of water have not been adequately embraced in the rural water sector of the selected districts. The study hypothesis was developed to test the unsettled perception that participation of rural dwellers in ecological governance of the districts could become the pathway for democratisation of water and ecological policy formulation, water conservation and sustainability of water resources within the districts. The governance and conservation of water at the rural level require a study in order to generate policy discourse and adaptation of essential rural-based measures for minimizing water losses and waste through preservation, care and securitization of potable water sources, ensuring efficiency and effectiveness of households' water use.

The hypothesis extends the frontiers of knowledge on rural political ecology and sustainability by investigating the influence of participation in rural ecological governance on the conservation of rural households' water systems. It further tests the power relationship between stakeholders in the rural water sector of the study districts. This offer reliable information for rural actors in the Savannah Region to develop appropriate strategies for achieving positive water and environmental conservation compliance among rural dwellers. Households require power to influence water decisions and basic information geared towards accountability on the part of government and its agencies in the districts through which the government is required to inform the citizens on water allocation in a more open manner and at citizens' level of understanding. It is expected that participation in ecological governance would have a positive and direct influence on sustainable utilization of water among individuals and rural households in the study districts. This study defined and measured participation in ecological governance and water conservation based on the Dublin statement which stipulates that "decisions be taken at the lowest appropriate level, with full consultation and involvement of local users of water in initiating/designing, planning, implementation, monitoring and evaluation of water projects" (ICWE, 1992:4).

Theoretical and Empirical Literature

The participation and power to have control over water has uniquely been found as a prerequisite for the governance and sustainable management of natural resources (Enserink et al., 2007). There have been diversified opinions on the manifestation of power within the processes and stages of participation in a self-governing ecologically driven community. This study employed the traditional political ecological governance principles of participatory democracy and decentralisation (Aotes, 1972; Hoppe, 2010) to argue that access to drinking water should be treated as a human right and the approach to water supply and ecological conservation must necessarily be inclusive, decentralised and democratic. This assertion contrast with the elitist (Nylen and Dodd, 2003) and technocratic approaches to ecological governance which argue that political elites, engineers and water specialists/ technocrats know better (Nelson & Shilling, 2019; Bertrand-Krajewski, et al., 2000) and therefore, decision making on water and ecological policies should rather be centralized and state-centered (Herman, 2006; Eckersley, 2004). Similarly, Lencha (2013) found that in rural settlements of Ethiopia, technical know-how is related to water service reliability, cost of production, water access and capacity to maintain water infrastructure. From the gender perspective, Karl (2000) and Sen and Smith (2015) postulates that education empowers and equips people with knowledge to participate in development process.

Since females play domestic roles in maintaining sanitary environment and collection of water, Christen and Robinson (1980) suggest that equality of gender representation and participation is imperative for rural water supply and ecological conservation. Nonetheless, in many parts of sub-Saharan Africa, patriarchal social arrangements continued socio-economic marginalization, powerlessness and unequal gender participation in water governance and environmental hygiene continue to impact sanitation and water sectors (Osunade, 1994; GSS, 2014a). Drawing from Eckersley (2019) and Yanmaz (2013), this study perceives political ecological governance as embracing the policies, legislations, capacities and skills required for the efficient planning and management of ecological systems and water resources available for consumption by human population.

The study argues that water is a basic resource necessary for human survival, hence, there is the need to ensure equity in water supply in rural settings (Walker, 2006). In addition, a political ecological framing of the study is meant to treat water and ecological resources as human needs requiring a decentralised and participatory approach to ensure effective distribution and conservation. For Neil (2018), political ecology addresses the synergy between human-environment and decision making in order to maintain ecological equilibrium and sustainable livelihoods. It lays emphasis on the continuous and balanced interaction between society-environment in order to promote sustainability. In addition, adopting a political ecological governance framework enables this study to set agenda for water discourses, shape ecological policy formulation and promote understanding of the linkages between the socio-economic and political interactions between anthropogenic actors and ecological systems in order to achieve an eco-friendly and sustainable environment.

The ecological endowments and water related resources of rural societies are composed of piped water, dug out wells, surface water and underground water that can be treated/filtered for human use. From ecological democrat's perspective, informing citizens of their water rights (Neil, 2018), ecological responsibilities and conservation choices can be the most significant first step toward genuine citizen participation in water supply and conservation of ecosystems (Berry & Mollard, 2010). However, in underdeveloped settings, too frequently the emphasis is placed on a one-way flow of information (usually from officials to citizens) with no conduit provided for feedback and no power for negotiation and consensus building. The unrestricted access to information on water improves the knowledge level of citizens, hence, their capacity to participate in decision making on conservation and water delivery projects (Karl, 2000; Sen & Smith, 2015). In the rural water sector, the most frequent methods used for consulting people are attitude surveys, neighbourhood meetings and public hearings (Dungumaro

& Madulu, 2003). When powerholders restrict the inputs of citizens' ideas solely to this level, community participation remains just a window-dressing ritual (Arnstein, 1969; Burke, 1968). Therefore, Arnstein (1969) suggests a theoretically prearranged model for citizen participation which indicates the most ideal form of participation is where citizens have power, spaces and control over decision making structures at the national and local level (Roberts, 2015). In addition, Poplin, Pereira & Rocha (2013) participatory-cub model draws from Fung (2006) who presents three dimensions of direct participations, namely, scope of participation, mode of communication and decision-making, and extent of authority. In Fung's (2006) participatory analogy, although the vast majority of public participation is open to all who wish to participate, those who choose to participate are often not the representatives of a large audience.

From the utilitarian and egalitarian perspective, the utilisation of ecological endowments (water) become sustainable (Munasinghe, 1992) when citizens have power, control and own-manage water systems with minimal external influence (Herman, 2006). Again, political ecologists (Eckersley, 2019; Béné and Arthur 2006) and ecological democrats argues that due to population increase, complexity of issues and knowledge gap on water conservation, creating spaces for end-user engagement should become the panacea for effective governance of rural environmental resources (Hoppe, 2009; MWRWH, 2014). Carrying these debates forward, Gaventa's (2006) view of the principal-agency model postulates that water sustainability depends on effective principal-agency relationships. The 'principals' are hegemonic and power wielding national actors with oversight responsibility for ecological conservation. From classical liberalism, the 'agents' constitute legitimate development beneficiaries and promoters of micro development (Gaventa, 2006). However, due to capacity gaps, they are unable to execute their mandate effectively at the local level (Zanu, 1996). Moreover, Sen and Smith (2015) argue that development should be human-centred and the invitation and engagement of citizens' opinions, through such activities including consultations, consensus building, referendum, voting and opinion polls (Marshall, 2008) can be legitimate and genuine steps toward full participation in ecological governance and micro water conservation activities (Coleman & Ferejoh, 1986). Through this, power is redistributed through negotiation and consensus building between powerless/ordinary citizens and powerholders (Burke, 1968). Besides, stakeholders mutually agree to share planning and decision-making responsibilities through such structures as joint decision-making, planning committees and mechanisms for resolving bottlenecks and disagreements. After the ground rules have been established through some form of a give-and-take interaction, such rules are not subjected to unilateral modifications without approval of both parties (Pateman, 1970). The engagement of rural people in ecological governance is a commitment to power sharing, where communities can take custody over decision-making processes, assume management roles, actively control, supervise, coordinate and monitor activities of managing their own water resources, ensuring efficiency and effectiveness of water supply and sustainability of water delivery facilities (Neil, 2018). There is minimal or no-excessive external interference by government, unbridled use of power is controlled, and human society attains eco-equilibrium (Barber, 2003). The literature search on participation has shown that in the context of rural ecological governance, space and power for participation are indispensable, yet they are either empirically missing, asymmetrical or ineffective and this tends to adversely affect water and ecological conservation.

Study Area and Methodology

The study draws on fieldwork conducted in the Savannah Region, which is one of the six new regions of Northern Ghana, created by Constitutional Instrument (C.I. 115) in 2019. Damongo is the capital of the region, which is also the traditional seat of the overlord of the Gonja-land. The Savannah Region is strategically located as it serves as the gateway which links the Southern sector of Ghana to the Northern Regions (i.e. Upper East, Upper West, Northern, Oti, and North East). In terms of the physical boundaries, the region is bordered on the north by the Upper West Region, on the west by the Ivory Coast, on the south by the Bono and Bono East Regions and the west by the North East and Northern Regions respectively. The region is drained by two major rivers in Ghana, namely the White and the Black Voltas, which are potential sites for tourist attraction, large-scale irrigation schemes along their valleys, freshwater fishing and fish farming. They also provide good waterways for water transport from Buipe and Yapei respectively to Akosombo via the Volta Lake in Yeji. There is an inland port and harbour facility at Buipe, along the Black Volta, and the region is a promising industrial hub, housing both sheanuts processing and cement factories. Agriculture, trade and commerce are the main economic activities in the region, yet the rural settlements are much drier and therefore, mostly challenged with access to water due to their geographic proximity to the Sahel and the Sahara Desert (WGDA, 2016; GSS, 2012b). The natural vegetation is guinea savannah and the soils in the region are fertile for the cultivation of cash crops, cereals, legumes and vegetables including sheanut, dawadawa, *"Parkia biglobosa"*, baobab, acacia, nim, maize, yam, millet, sorghum, groundnuts, rice etc. Although the Mole National Park, Larabanga historic Mosque and mystical stone, the Bui National Park, the Wechaiu Hippo Sanctuary and the Yakumbo forest reserve are some of the landmark national reserves and tourist potentials of the Savannah Region, the original vegetations of these parks and ecological reserves are being destroyed by uncontrolled human activities (WGDA, 2017; CGD, 2017; WGDA, 2016). The practice of bush fires, hunting, overgrazing, charcoal burning and fetching of firewood have reached alarming proportions which need to be checked to avoid water and environmental problems. Although the region is multi-religious and predominantly composed of Gonjas (i.e. aboriginal settlers) and other minority ethnic groups, there are inter-tribal marriages and peaceful co-existence which enhances unity in diversity (Haruna, 2017).

Figure 2. Demarcation of the Savannah Region and location of the study districts
Source: Ghana Meteorological Agency (2019)

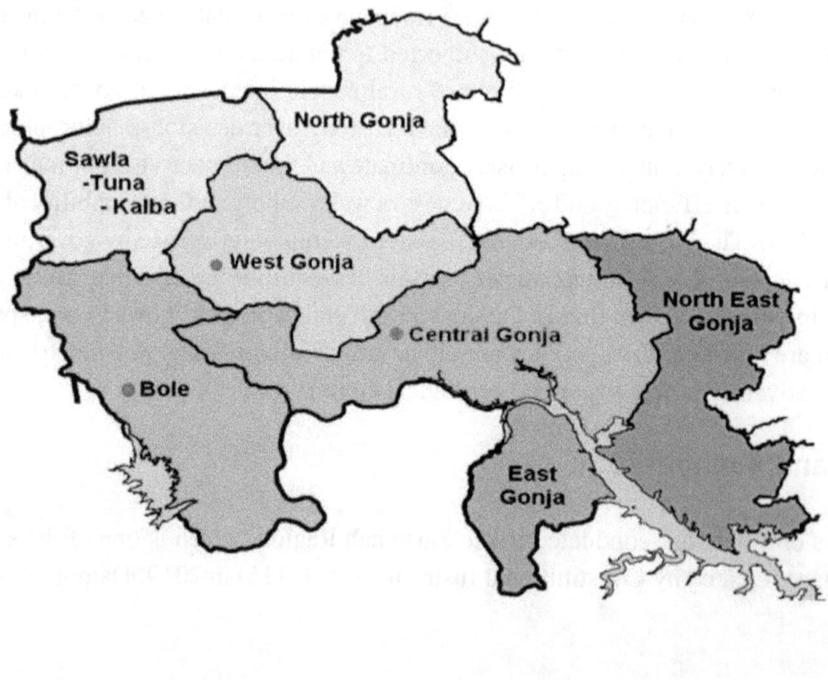

The multi-ethnicity and religious homogeneity of settlements tend to promote socio-cultural organization and religious tolerance necessary for inclusive growth and socio-economic development. In terms of local governance (see figure 2), the Bole, Central and West Gonja Districts form part of the seven districts of the Savannah Region, created by the Local Government Act (Act 936) (GoG, 2016). The creation of the districts was envisioned to improve accountability, transparency and responsive governance at the local level through the transfer of power and resources from central government to the districts (Ayee, 2008). It was further meant to empower the districts, enable them exercise political control at the local level, enhance local economic development, foster local participation, and improve service delivery and local capacity for management of water and ecological resources at the rural level (Ahwoi, 2010). Although three districts i.e. Bole, Gonja West and Gonja Central (see figure 2), have both surface water sources including dugouts, rivers and dams with pipe-born water sources, water security remains pervasive (GCD, 2017; Bole District, 2016). The practice of open defecation, environmental degradation, unplanned human settlements and the relatively dry nature of the vegetation constitute critical challenges which affect the conservation of ecological resources of the districts (GSS, 2014). In this study, the three districts (figure 2) were purposively selected as they are centrally located with common boundaries, demographic elements, ecological endowments and local governance structures, hence, made compelling and interesting cases for analysis of the nature of water and ecological governance in the Savannah area of Ghana. The map (figure 2) shows the seven districts which composed the Savannah Region and the location of the survey districts in the context of Ghana. The map is based on the geography and re-demarcation of Ghana after the 2018 referendum which resulted in a further division of the country into sixteen regions instead of the already known 10 regions which existed since the 90s'. The Savannah Region was one of the newly created regions, carved out of the Northern Region. In determining the target population, sampling and sample size, the study obtained the sample frame based on the total rural household's population threshold in each of the three study areas, namely, Bole District, Gonja Central District, and the Gonja West District (see figure 2).

Paradigm, Design and Sampling Procedures

The quantitative approach, founded on the post-positivist paradigm and cross-sectional design (Babbie, 2012) were appropriate for exploring the participants' emic views and to understand how they structure and give meaning to their experiences on participation in ecological governance and conservation of water. Although quantitative studies are often noted to be rigid and structured (Pallant, 2012), the objective for this study and nature of the hypothesis test required the study to adopt quantitative methods. The study was generally interested in ensuring precision in sample estimation, quantification of responses, deductive interpretations and generalization of the research findings. This required a carefully constructed, structured and controlled experiment devoid of bias and subjective judgements which are often associated with qualitative methods. The probability sampling approaches (Babbie, 2016), that is proportionate, systematic and simple random techniques (Bernard, 2011) were used to sample 450 adults (aged 18-50+), composed of 392 rural household heads and 58 officials in order to address the link between eco-governance and conservation of water at the household level.

The sampling frame for the rural households in each of these districts was a summation of all household populations who dwell in the rural areas of the districts. The sample size determination and sample selection were then based on the rural households, with the household heads as the units of analysis. The total rural household population in the three study areas is 19, 646, composed of 7,765 residents in Bole District, 8,905 residents in Gonja Central District and 2,976 residents in Gonja West District (GSS, 2014). In terms of the official respondents, 68 water related officials, including CWSA, research institutions, District Assemblies, local authorities etc were targeted as part of the stakeholders and the eligible study population. The sample selection was based on a statistical model established by Miller and Brewer (2003) {$n=N/ [1+N (\alpha)^2]$}, where, n= sample size, N = sampling frame (19,646), at 95% confidence interval with α = error margin, set at (0.05) and 1= constant value, hence $n = 19,646/[1+19,646 (0.05)^2]$, n = 392 households. Similarly, determination of the 58 sample size for the institutional officials was based on the same sampling model. The selection of elements at household and institutional levels were proportionally determined based on population strata of each target group. Moreover, a simple proportional formula {P x n/N}, where P = rural household population per each stratum or district, n = total sample size and N = total rural household population, was used to select the respondents per the population strata. Using the above formula, the proportionate sample for Bole District was determined as {7,765x 392/19, 646, = 155 respondents), Gonja Central {8,905 x 392/19, 646= 178 respondents), and Gonja West District {2,976 x 392/19, 646 = 59 respondents}, a summation of which was 392. Utilizing the same formula, the 58 institutional respondents were drawn as follows {CWSA= 17, research institutions (Council for Scientific & Industrial Research-CSIR) = 4, District Assembly officials = 9, local authorities (chiefs) including NGOs staff = 13, leaders of community representatives = 13 composed of (WATSAN committees =11, WSMTs = 4), and finally, government agencies (MWRWH) = 2}. In this study, the Water and Sanitation (WATSAN) committees and Water and Sanitation Malmanagement Teams (WSMTs) are recognised as part of the official respondents.

As leaders and community representatives, the WATSAN committees are part of the District Assemblies governance structure, constitutionally mandated by ACT 462 of 1993 (revised as ACT 936 in 2016), as development agencies at the local level (GoG, 2016). The WATSAN and WSMTs are part of the rural water governance structure in Ghana, responsible for managing water supply and sanitation facilities at the community level *(Nyarko et al., 2006)*. Therefore, in determining the sampling and sample size of the institutional respondents, the study included the leaders of community representatives which are made of the WSMTs and WATSAN committee members located at the three Districts (i.e. Bole, Gonja Central and Gonja West Districts). After establishing homogeneity and obtaining the total sample size for each of the population strata, the systematic technique was utilised to select the households while simple random sampling technique (Babbie, 2016) i.e. the lottery method was utilised for selecting the units of analysis from the various sub-groups/strata of the institutional staff.

These techniques allowed respondents in each sub-group to have equal and fair chances of being selected for the study. To start with the systematic selection of the individual households, i.e. sample units, a sample fraction *(K^{th})* was predetermined as a fraction of the rural household population by the total sample size, which was mathematically expressed as; $K^{th} = N/n$, where N= total rural population (19,646) and n= sample size (392), hence, $K^{th} = 19,646/ 392$, therefore, $K^{th} = 50^{th}$. A random household number between the interval of one and 50th was then chosen and the element on the household list (collected from each district) that corresponds to this number was included in the sample. This randomly selected number was then utilised as the starting point for choosing all the other elements until the total sample size of 392 households were selected for the study.

Instruments, Data Collection and Analysis

Drawing from Chilisa (2011), self-designed structured questionnaires were administered to both the officials and household respondents. The study instruments went through a process of quality assurance and validation which offered an opportunity for pretesting in a neighbouring Savelugu District to ensure that the instruments were reliable for the large scale data collection. The aggregation of responses is typical with quantitative studies where the instruments are structured to elicit for quantifiable data. Babbie (2016), Sarantakos (2011) and Pallant (2012) have observed that a standardized data randomly collected from structured questionnaires which asks same questions can be combined, easily quantified and analysed as a uniform sample. Drawing from this notion, a reasonably large and randomly selected samples were determined by the study. To mitigate the possibility for error, the study followed Pallant (2012) who argues that structured questionnaires which are designed to elicit uniformed data across a randomly predetermined sample is essential where the researcher is interested in obtaining data from a cross-section of a population or universe. Moreover, Sarantakos (2011) adds that to enable aggregation of data, the instruments must have uniform structure, the nature and type of questions should elicit the same information and the instruments administered at the same time period.

Therefore, in this study, the basic idea was to analyse the data as a single unit. The questionnaires were structured (close-ended), standardized, pilot tested, elicited the same information and the data was collected at the same period. These pre-planned and strictly controlled measures made it possible to aggregate the responses in line with the purpose, objective and hypothesis of the study. Descriptive statistics, correlation analysis and chi-square test (Pallant, 2003) were the techniques utilised to analyse the data and test the hypothesis. With 95% confidence interval and error margin (e)= 0.05, the decision rule for hypothesis test was; accept the null hypothesis (Ho) if the p-value is greater than (˃) the alpha level (α)= 0.05. The study was approved, and ethical clearance was obtained from the University of Johannesburg and local institutions with oversight responsibilities at localities where the fieldwork was conducted. As cautioned by Neuman (2010), ethical research standards prevailed at every stage of the study, while voluntary participation, security of data and respondent's confidentiality were maintained before, during and after the field activities.

Results of the Study

The age and educational attainment of a population are important in promoting ecological awareness, promoting participation and water conservation practices in rural settings. The success of enforcing local bye-laws, ecological policies and water conservation strategies at the district level require education of localities across, gender, income categories and all age groupings. Awareness of the local people is important to change perceptions and attitude of residents towards the use of ecological resources in the districts. From Table 1, this study found that in terms of age, awareness and participation in water and ecological conservation activities, respondents between 29-39yrs constituted the largest populations (58%) who were not participating at all in water conservation activities while respondents aged below 29years (21%) and those above 39yrs (21%), actively involved in decision making on water and ecological conservation.

Table 1. Age and awareness of water and ecological conservation

Age	Awareness of water and ecological conservation	
	Frequency	Percentage (%)
18 – 28	95	21
29 – 39	261	58
40 & above	94	21
Total	450	100

Source: Fieldwork (2019)

On average, respondents with senior high school (representing 22%) and tertiary education (representing 9%), showed the highest level of awareness and participation in water conservation activities (see Table 2), while those with basic education (primary/junior high school/middle school, representing 44%), non-formal and no-education (representing 24%) showed no knowledge on water conservation and were, therefore, most likely not to be able to participate effectively in water conservation related activities in the districts.

Table 2. Education and awareness of water and ecological conservation

Education level	Awareness and participation in water and ecological conservation		
	Yes (%)	No (%)	Total (%)
Non-formal/no-education	0	24.44	24.44
Primary/JHS/Middle School	0	44.44	44.44
Senior High	22	0	22
Tertiary	9	0	9
Total	31	69	100

Source: Fieldwork (2019), n = 450

The results indicate that future ecology and water related conservation interventions should rather target the less/least educated and households without any form of formal education as they constitute the critical group which requires information, awareness and modern knowledge on water and environmental conservation in the districts. In addition, the results highlight that water and ecological conservation education at all ages, gender and educational levels, needs to be improved in order to promote ecologically friendly attitudes among the rural residents of the districts (see Tables 1-3). The strategies for water education and ecological awareness creation must go beyond announcement via local radio stations to embrace multi-sectoral and community based approaches.

This approach should include advancing awareness and knowledge through integrated activities such as workshops and training aimed at water professionals and decision-makers. In addition, community education strategies including training, ecological durbars and public sensitization efforts are essential in order to promote districtwide water and ecological conservation as well as enhance skills in local co-management of water resources in the districts. From Table 2, the results further imply that water and education campaigns should be designed for media professionals, the WATSAN, WSMTs and Assembly members so that they can reach out or communicate water policies and ecological issues accurately and effectively to residents without formal or no-education in the districts. As illustrated by Table 3, when the data was correlated along with income and education variables, it became apparently clear that households with financial capacities (i.e. higher incomes) and high education levels (tertiary) tend to have more economic power, social recognition and capacity to influence decision making than non-formally educated rural households. The correlated results in Table 3 show a strong positive relationship between income (r = 0.759), education levels (r = 0.754) and gender (r = 0.769) and the capacity of the local people to influence ecological governance and decision making on water in the districts.

The nature of water and environmental education should focus on disseminating information on what water users can do in order to reduce waste of drinking water, water consumption and reduce the impact of activities of residents on their water supply systems. Even though females constitute 51% of national population and approximately 52% of the population in the Savannah Region (GSS, 2014), Table 4 convey to the forefront the persistent phenomenon of female marginalisation, powerlessness and unequal participation in decision-making. From Table 4, the results could be due to the prevailing patriarchal system, the culture of the people, the traditional and belief systems among residents in the districts which uphold the "male" as the decision maker, bread winner and the family/ household head. Therefore, women who participated at water conservation meetings could have been either nominated by their households or elected by their communities to serve as representatives in decision making activities. Meanwhile, geographically, this study discovered variations in terms of the location of the districts and the resident's participation in water conservation activities. Comparatively, Gonja Central residents fairly showed the highest interest (42%) in terms of awareness and participation in water conservation decision making while Bole (36%) and Gonja West (22%) settlements showed the least awareness and interest in participating in water conservation activities in the communities. As illustrated by Table 4, the results reveal that in terms of responding to challenges in the governance and conservation of ecological resources including water, there are no uniform or one-size-fits-all solutions in the districts. However, this study argues that active participation is essential to guarantee understanding and effective enforcement of local ecological decisions. Hence, the ecological decision making process will require the need for adequate information, consultation, dialogue and partnership with the rural residents of the districts. The results show that decentralisation of power to the districts do not necessarily translate into effective ecological governance and sustainability of water resources at the local level.

Table 3. Socio-economic status and influence on ecological governance decisions in the districts (Spearman rank ordered)

Socio-economic status	Influence on ecological governance		
	Correlation Coefficient (r)	p-value	n
Income	.759*	.001	450
Education	.754*	.022	450
Gender	.769*	.042	450

**Correlation is significant at the 0.01 level (2-tailed).* Source: Fieldwork (2019), n= 450

Table 4. Gender, awareness and participation in ecological governance decisions

Gender	Participation in water and ecological governance		
	Frequency		**Percent**
Male	315		70
Female	135		30
Total	450		100
District Location	Frequency		Percent
Bole District	163		36
Gonja Central District	187		42
Gonja West District	100		22
Total	450		100

Source: Fieldwork (2019), n = 450

The findings suggest that a democratic process of creating or inviting voluntary participation, gender inclusion through the formation of village water and sanitation cooperatives or election of leaderships would perhaps ensure effective popular representation, participation and ownership of ecological projects at the districts level. Again, the sustainability of rural water sources in the districts means that the environment must be embedded in all levels of governance and decision making. As illustrated by Figure 3, the findings showed that a little above half of the respondents (55%) never participated in decision making activities on household water conservation, 19% rarely participated and about 16% occasionally participated in water conservation decision making. Not surprisingly, however, only 4% of the respondents very often participated while 6% always participated in meetings and decision-making surrounding ecological governance decision making and distribution of drinking water in the districts.

In addition, Figure 4 revealed the reasons which accounted for the failure to participate in ecological governance decision-making. The study found that about half (50%) of the respondents were unaware of the decision-making schedule and structures, 19% did not have trust in the local authorities,13% lacked knowledge in the subject matter, 10% did not have adequate information while 8% claimed that they were not involved by the local decision-makers. The results are indications that the households were either unaware of the existing participatory structure or they lacked adequate information, knowledge or capacities required to influence discourses and decision making on water allocation.

The results demonstrate that participation of local people perhaps depends on the politics, power and nature of decision-making on water and the environment. If indigenous people believe that their inputs would be critically considered and have adequate information and knowledge on subjects of decision-making, they would be most likely inclined to participate in decision-making and offer support to policies on water. However, popular participation in rural ecological decisions is likely to fail at the micro level if local people have less faith and trust in decision-making systems. They could be demoralised, become apathetic, disinterested and therefore less inclined to participate in water and rural ecological conservation decisions. The conservation of ecological resources should become a common concern for all humanity, the decentralised institutions and structures of the districts. Since water is necessary for human survival, there is the need for the districts to encourage joint participatory efforts on measures for indigenous technologies to ensure a clean environment and conserve water resources. In Figure 5, the study presents the decision-making stages, power relationships and participatory governance structure of the districts.

Figure 3. Participation of households in decision making on water conservation
Source: Fieldwork (2019), n = 450

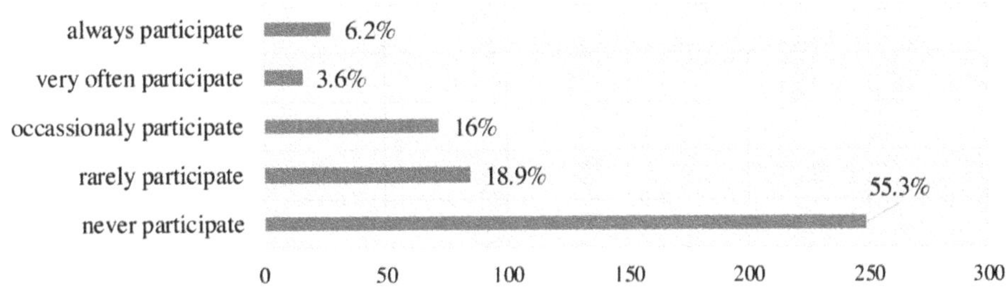

Figure 4. Factors responsible for non-participation in ecological decision-making
Source: Fieldwork (2019). n = 450

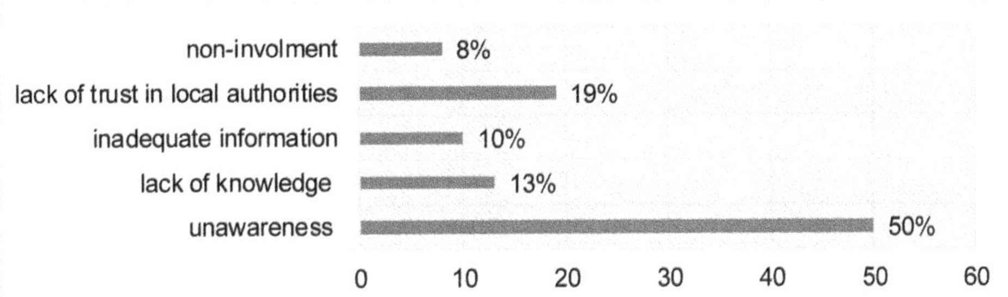

The results (see Figure 5) are based on the views of the respondents regarding their extent of involvement in the decision making processes regarding water supply and sanitation service delivery within the districts. The results show that ecological governance in the districts was established on a five pillared participatory structure namely, information, communication feedback, consultation, planning and citizen control. As illustrated by figure 4, about 80% of the respondents confirmed that the target was to inform "more" residents in the communities through a variety of publicity/information channels including the use of posters, newsletters, radio announcements and press releases. In addition, about 10% of the respondents indicated that the "community feedback" mechanism was meant to achieve a "demand-responsive approach" for water use and conservation. Moreover, the study found that at the state of "community feedback", the beneficiary communities of water facilities in the districts were required to hold regular meetings with their Assembly members to discuss the content of application forms in respect of the request for new water and sanitation amenities, funding modalities, make decisions on water and sanitation needs and complete the water storage application forms with support from their Assembly members. The purpose for water conservation or storage applications was to solicit subsidies from either the MMDAs or the CWSA in respect of the provision of community water reservoirs/ tanks, drenching rivers and lakes, dam impoundments, construction of long-term security fences in order to protect surface and groundwater source zones including dams/rivers and dugout-wells in communities which are hydrologically challenged. Nonetheless, in figure 4, the findings show that about (10%) of the respondents indicated that there is lessor participation during community feedback meetings, either because they were never aware or they possibly perceived such meetings to be mere "window-dressing"

as decisions were possibly marred with political party colourations, hence, disregarding/excluding the interest and needs of powerless citizens and poor-households in the communities. In effect, this could possibly result in discriminatory siting of water projects to favour communities with greatest party supporters for incumbent governments, lack of community ownership, inability to sustain and provide adequate drinking water to households in the communities.

Figure 5. Water and ecological governance structure in the districts.
Source: Author's Construct (2019), n = 450.

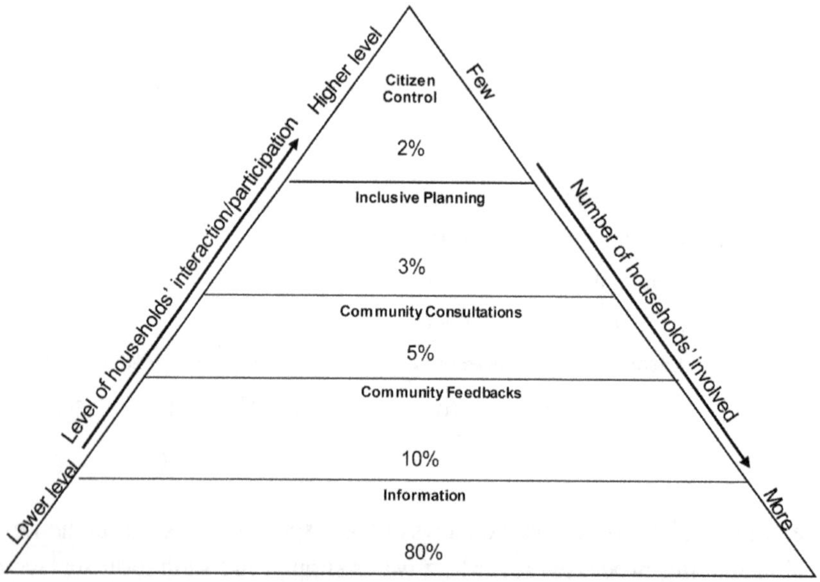

Attempts to express their dissatisfaction/grievances with the unfair distribution and denial of powerless settlements to decision making and water projects can result in violent confrontations between local citizens, the District Assemblies and the CWSA. Moreover, there is the need to close funding gaps in rural water and sanitation delivery by revisiting the sustainability implications for removing the (5%) Compulsory Community Contribution to Capital Cost (CCCC), particularly in deprived communities. In Figure 5, the "consultation" and planning stages of conservation activities entailed the pre-selection, district investment approval, community mobilisation, participatory planning, project appraisal, tendering and award of contracts. Albeit the aims for these stages were to achieve transparency and equitable distribution of investments in water and sanitation facilities and promote district-wide economic development, this study found that the actual level of participation of households (see Figure 5) at these stages were disappointingly low as only a few households showed interest and actively participated in decision-making on water and ecological conservation in the districts. The study further discovered that the highest level of household's participation in water and sanitation service delivery was at the stage of "citizen control (2%)", where local citizens are empowered to take control over the management and distribution of water and developing local strategies to resolve controversies on environmental sanitation at the rural community level. The findings of this study revealed that although 5% of the respondents indicated that they were consulted, only 3% were included at the planning stages of water and ecological conservation projects. At the topmost level (i.e. citizen control), the results showed that only 2% of the residents had decision-making power and greater management control and were thus always participating in water activities in the districts. The results implied that probably concrete support for water and ecological sustainability activities often truncated after handing over facilities and this has adverse implications for monitoring, evaluation, rehabilitation and long-term conservation of facilities post-construction.

In order to understand the power dynamics among water and ecological actors in the districts, Table 5 presents an analysis of the stakeholder participation and power relations in the governance of ecological resources. This is because the study maintains that asymmetrical power relation preserves human dignity and creates legitimate spaces for participation, accountability and equality of access to development opportunities. The study draws from Mitchell et al. (1997) stakeholder salience theory in order to analyze the extent of power, legitimacy and urgency attached to stakeholders who are responsible for the formulation and implementation of environmental policies in the districts. From Table 5, the stakeholders are categorized into five typologies, namely, definitive, dominant, dependent, dormant and discretionary stakeholders. As depicted by Table 5, the findings showed that the institutions such as the CWSA, the District Assemblies, traditional leaders, NGOs, etc. were considered as definitive actors, who had high power to control/ influence (33%) local decisions, with high legitimacy (34%) and high urgency (33%) attached to water and ecological issues which required swift action at the district levels.

Though key actors, quite surprisingly, the beneficiary communities and end-users of water resources were not involved at the highest and strategic levels of water and ecological governance. In addition, the respondents indicated that the dominant strategic actors included central government, ministries and research institutions who are responsible for setting water quality standards, regulating and certifying water for consumption. The dominant actors though had high power (40%) and high legitimacy over setting water regulations and standards, there was low sense of urgency (20%) attached to the needs of water consumers. The end-users of water were generally dependent on the District Assemblies and NGOs for water supply and sanitation maintenance, with low power (5%) over water supply and ecological governance decisions. Though water consumers were considered legitimate (30%) beneficiaries and their demands for potable water were considered urgent (65%) and prioritised, they had low power to influence their local authorities nor control over the management and distribution of water. This has implications on water-related disease outbreaks and ecological conservation. Furthermore, the findings show that external development partners, including the media, investors and donors, though were acknowledged to be physically absent (external) from the communities, the districts and the region, they had high power (60%), low legitimacy (20%) and low urgency (20%) attached to decision making on water supply in the rural settlements. Moreover, the discretionary views of researchers, logistic providers, contractors and engineers were acknowledged and considered legitimate. However, these stakeholders had low power (20%) to compel or effect modifications/ changes to definitive (cast-in-stone) rural water policies and decisions.

They were highly recognized as legitimate (60%) institutions, yet their decisions were not urgently (20%) considered and prioritized in the rural water sector. In Figure 6, the study discovered that about a little above half of ecological engagement processes were undertaken through popular representation (52%), involving the selection or election of community representatives to represent the interest of constituents across the rural settlements in the study districts. About 11% were engaged through water conservation workshops and fora, 7% were engaged through hotline and telephone conversations and 9.11% of the respondents indicated that they were engaged via contractual engagements. Other minority engagement processes deployed by the districts included opinion polls, referendum and surveys (5%), experts' panel and focus group discussions (5%), while only 2% were engaged through the media, including web-communications, local radios (referred to as "court of public opinion" by the local people), posters and community-centre notices and publications. From the results in Figure 6, this study argues that the participation of households is central to the sustainable water supply and equitable access to rural ecological endowments in the districts. The decentralisation strategy can only remain meaningful to the rural people where the ecological engagement processes are multi-dimensional and locally tailored to the socio-economic and political context of the districts. The definitive and most dominant water and ecological actors should ensure power balance in the decision making and planning activities of the districts. The rights of the people to water, sanitary environment and human dignity can better be preserved where deliberate legitimate spaces are created to enable less powerful actors to participate equally and have access to development opportunities in the districts.

Table 5. Stakeholder power relations in ecological governance decision making

Stakeholder type	Attributes and Rating			Stakeholder classification (internal & external)	Stakeholder salience
	Power (%)	Legitimacy (%)	Urgency (%)		
Definitive	High (33%)	High (34%)	High (33%)	DAs, CWSA, traditional leaders, NGOs, investors/ donors/financial institutions	Prioritised as essential
Dominant	High (40%)	High (40%)	Low (20%)	Standard setting, regulatory & and certifying authorities (EPA, research institutions), central government, Ministries (LGRD & MSWR).	Prioritised as essential
Dependent	Low (5%)	High (30%)	High (65%)	Individual consumers, households, small water distributor providers/tanker service providers, retailers of borehole water	Prioritised as essential
Dormant	High (60%)	Low (20%	Low (20%)	External development partners, Media (radio stations), Investors and donors	Acknowledged locally
Discretionary	Low (20%)	High (60%)	Low (20%)	Researchers, logistic providers, Contractors, Engineers and technicians	Acknowledged locally

Source: Fieldwork (2019), n = 450

Figure 6. Procedures for stakeholder engagement in ecological governance and decision making.
Source: Fieldwork (2019), n = 450.

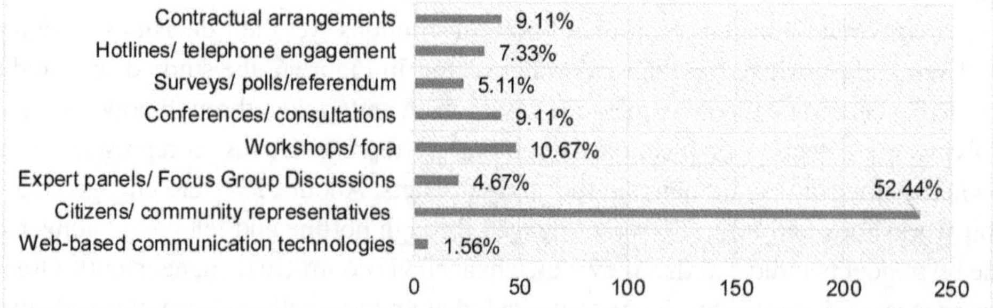

Again, the study inquired from the respondents on the nature of the rural ecology and strategies utilized by the residents for domestic water conservation. From Table 6, it appears that there are no remarkable discrepancies among the three districts (Gonja Central-32%, Bole-34%, Gonja West- 34%) in domestic conservation of water. The commonest strategies for water conservation were rainwater harvesting and storage (37%), ponds and dug-outs (33%), underground/overhead water polytanks (6%), water rationing and control of water pollution (8%), collection, recycle and reuse of wastewater (5%) and adoption of measures to prevent water leakages (11%). About 11% of the households harvested rainwater from Gonja Central District area while 13% each from Bole District and Gonja West District settlements harvested rainwater. The results in Table 6 further attest that the construction of pond and dug-outs structures in Gonja Central (11%), Gonja West (11%) and Bole district (10.5%) as constituting surface depressions, natural or created, which were utilised by households for collecting and storing rainwater, runoff and seepage from the surrounding areas. Due to high overhead cost, the use of underground/overhead polytanks were unpopular and was restricted to about 2% each of the households across the localities. Besides, there were no appreciable differences in wastewater collection and utilisation among the three communities (Gonja Central District-1.8%, Bole District-2.2% and Gonja West District-1%). Despite the government policy on controlling water leakages (MWRWH, 2014), the findings showed that control of water leakages was poorly practiced among the communities (Gonja Central District-2.3%, Bole district- 4% and Gonja West District-4%). The findings could perhaps be attributed to the low knowledge and capacities of the people on wastewater reuse, lack of systems for water accounting, reporting water leakages and lack of maintenance of outdated water systems. The results suggest that the construction of ponds and dug-outs structures in the districts act as productive and agronomic catalyst which lessens the intensity of drought, helps in recharging groundwater, provides crucial irrigation for multi-crop production. The study found that ponds and dug-outs function as sources of multiple uses for the villages and communities like washing, fishing, bathing and reservoir of water for livestock and maintenance of a good natural environment.

Table 6. Methods for domestic water conservation in the districts

Households water conservation strategies	Area Location							
	Gonja Central District		Bole District		Gonja West District		Total	
	n	%	n	%	n	%	n	%
Rain harvesting	50	11	60	13	58	13	168	37
Ponds & dug-outs	49	11	48	10.5	50	11	147	33
Underground/overhead polytanks	10	2.25	9	2	8	2	27	6
Ration water use and pollution control	15	3.4	10	2.2	12	3	37	8
Wastewater collection, recycle and reuse	8	1.8	10	2.2	5	1	23	5
Check water leakages and proper plumbing	10	2.3	18	4	20	4	48	11
Total	142	32	155	34	153	34	450	100

Source: Fieldwork (2019), n = 450

As a result of these benefits, this study concludes that rainwater harvesting, ponds and dug-outs could become essential growth poles and should be built in the rural community, based on land and topographical features of each area. In addition, the use of ponds, dug-outs structures, rainwater harvesting, and pollution control does not only imply opportunities for multi-cropping agriculture and fish production in the communities but also these conservation measures are priority resource-base for many other activities such as the collection of fodder, the making of bricks for construction, baskets etc., with women offering labour and assistance in these processes. Therefore, pollution control, dug-outs and pond structures on community land should become a very important part of the socio-economic system in the districts, affecting significant income opportunities, checking flood activities and conservation of water for use in lean periods.

Furthermore, in order to examine the link between participation in ecological governance and water conservation among the rural households in the districts, this study hypothesized that there is no significant relationship between the sample respondents and the magnitude of their participatory levels which could influence decision making on ecological governance and conservation of drinking water within the districts. The study design and sample distribution required that the chi-square test (χ^2), a form of non-parametric test, be carried out. A non-parametric statistical analysis is a distribution-free approach which would not require normality distribution. Consistent with Pallant (2003: 255), "non-parametric tests are ideal for use when the data is measured on nominal (categorical) and ordinal (ranked ordered) scales". Therefore, since the variables in this study were measured using the Likert rank ordered scales, the chi-square test was applied in analyzing the variables. The decision rule applied in testing the hypothesis was: {accept Ho, if "p-value" $>$ $\alpha = 0.05$ and do not accept Ho, if "p-value" = or $<$ $\alpha = 0.05$}.

In Table 7, the findings from the chi-square (χ^2) values showed that water conservation initiation, water project design/planning, financial contribution towards water facilities, labour support, supervision and evaluation of conservation activities, repair/maintenance/rehabilitation of water facilities, setting water tariffs, and {gender representation were statistically significant in explaining variations in participation in ecological governance and domestic water conservation at the district level. From Table 7, the Phi and Cramer's V systematically measured the strength of the association between participation in eco-governance and the conservation of drinking water. Based on Cohen's (1998: 23) proposition that "(r) = 0.30 to 0.49 or (r) = -0.30 to -0.49 = Medium correlation", the results as illustrated by Table 7, showed a positive and average association between participation in ecological governance and water conservation at the rural level.

Moreover, since the computed 'p-values = 0.00' were statistically less than the significance level, that is, alpha value of *($\alpha < 0.05$)* with 16df (degree of freedom), we fail to accept the null hypothesis (Ho), rather we accept the alternative hypothesis (H_A). That is, households prefer to participate in ecological governance to ensure effectiveness and efficiency in the distribution, consumption and conservation of water in their communities. The results affirmed that the inclusion of water consumers at the household level is essential at all levels of ecological projects cycle, beginning with the initiation, designing, planning, implementation, water reticulation and monitoring and evaluation (M&E) of water facilities. The participation process is thus indispensable for engineering local support in terms of households' financial contributions, labour support, ensuring water security and gendered approach, with active women representation in rural water conservation. Ensuing the hypothesis test outcome, the study conducted an investigation into the indigenous strategies for water sustainability in the three districts in order to establish whether the indigenous knowledge system of the communities plays a role in sustaining water resources in the districts. As illustrated by Table 7, the findings revealed that the majority of the households (33.3%) regards some of the water sources (rivers, lakes, dams etc.) to be sacred, hence, traditional taboos were governing such water bodies. In addition, 22% confirmed that water facilities were sustained through the construction of fence walls around point sources including dams and boreholes, 20% indicated that the Assemblies formulated bye-laws in order to govern the operation of water facilities and 11% indicated that the communities initiated sustainability measures such as the establishment of community watchdogs around water sources in the localities.

Table 7. Effects of participation in ecological governance on conservation of households drinking water in the districts

Participatory stages	Chi-Square (test results)			Symmetric Measures			
				Nominal by nominal			
	Pearson Chi-Square Value (χ^2)	df	Asymp.sig (p-value)	Phi	Cramer's V	Asymp.sig (p-value)	Number of valid cases
Design &planning: Water conservation initiatives	86. 231[a]	16	.001	0.441	.235	.000	450
Water conservation projects design/ planning	77.587[a]	16	.001	0.415	.208	.000	450
Implementation: financial contribution towards water conservation facilities	84.060[a]	16	.001	0.432	.216	.000	450
Labour support towards water conservation	56.311[a]	16	.001	0.354	.177	.000	450
Monitoring and evaluation of water conservation projects in the communities	71.866[a]	16	.001	0.400	.200	.000	450
Post-construction sustainability: repair/maintenance/rehabilitation of water facilities in the communities	70.117[a]	16	.001	0.395	.197	.000	450
Setting water tariffs within the communities	79.537[a]	16	.001	0.477	.239	.000	450
Gender inclusion	55.061[a]	16	.001	0.309	.155	.000	450

a. 0 cells (.0%) have expected frequencies less than 5. The minimum expected cell frequency is 10.0. Source: Fieldwork (2019), n = 450

Table 8. Traditional strategies for ecological governance and water sustainability

Type of respondent	Indigenous Strategies									
	Community watchdogs around water sources		Traditional taboos and sacredness of surface water sources		Construction of fence walls around water dams & boreholes		Bye-laws of the Assemblies		Total	
	n	%	n	%	n	%	n	%	n	%
Households	50	11	150	33.3	100	22.19	92	20.4	392	87
Official respondents	13	3	5	1.1	10	1.3	30	6.7	58	13
Total	63	14	155	34.44	110	24.44	122	27.11	450	100

Source: Fieldwork (2019), n = 450

In terms of the official respondents, about 7% indicated that water sources were sustained through bye-laws of the Assemblies and the CWSA, 3% confirmed that there were community watchdogs (such as the WATSAN, WSTM) committees which ensured security around water facilities, while 1% each confirmed that there were taboos on water and that the Assemblies constructed defensive walls and fences around water sources situated in the communities. From the perspective of the households, the majority (90%) indicated that the purity and sustainability of water entail no bathing, washing, no dumping of refuse into water sources and being able to maintain a continuous supply of drinking water in the communities. However, the official/ institutional respondents perceived hygienic and sustainable water from the perspective of the pH level (7) of water which is readily available for human consumption.

Discussion of Findings

Respondents Background

The demographic analysis of the districts reveals that the youth are essential actors in the development of water resources and maintenance of hygienic environments. If young people are educated or made aware of their responsibilities towards ecological systems, there is the possibility for the districts to explore their talents and energies towards the implementation of social and eco-friendly reforms in local efforts towards the conservation of water resources. From Table 1, the findings show that more than half of the respondents, who constituted the most youthful population (i.e., 29-39years), were not participating in water and ecological conservation activities in the districts. Meanwhile, highly educated respondents were aware and most likely to participate in conservation activities than residents with low or non-formal education (see Table 1 & 2). Consistent with GSS (2014), the incomes earned appear to determine local economic capacity and power of the rural residents to influence decision making on water and sanitation delivery in the districts. Although females play domestic roles in maintaining sanitary environment and collection of water (Christen and Robinson, 1980), the findings showed that they remained largely marginalised and powerless in water and ecological decision making (see Table 4). Efforts should be made by the Assemblies and CWSA to overcome the existing patriarchy in water and ecological decision making. The inclusion of women and minority groups in water education and participatory efforts is essential in order to democratize planning and ecological policy making in the districts. The findings agree with Karl (2000) and Sen and Smith (2015) who postulates that education and information empower and equip people with knowledge to participate in the development process. Similarly, the findings affirm Christen and Robinson's (1980) argument that equality of gender representation and participation is imperative for rural ecological conservation. The findings of this study further suggest that the most essential areas of concern for policy intervention and strategy formulation include public health education, awareness creation, sensitisation and law enforcement (Umesh & Nagarkatte, 2017), to promote citizens' interest, sustain water conservation activities and achieve participatory local level development.

Again, Oates (1972) and Hoppe (2010) postulates that decentralisation is the panacea to effective and sustainable local development, however, the evidence from this study rather suggest that despite decentralizing rural water and sanitation service delivery (GoG, 2016), citizen participation in decision making (Ahwoi, 2010) on environmental governance in the districts is constrained by socio-economic factors including education, income and gender. Therefore, although the decentralisation and local governance strategy was envisioned to address rural-urban service delivery inequalities in Ghana (GoG,1992), these noble goals are not translating into sustainable outcomes in the rural water and sanitation sector due to prevailing participatory inequalities confronting the districts. Similarly, Manu (2015) found that low education and non-participation in ecological decisions is positively correlated with water and sanitation delivery. Again, Goldratt and Cox (2016) attribute the widespread delays and lack of ownership of water conservation projects to lack of engagement, broader consultation and lack of awareness/ knowledge to engender active citizen participation.

Hence, water education, and public awareness on ecological conservation should therefore seek to promote efficient utilisation of water, reduction in water usage, losses and waste of potable water (Manu, 2015). The findings of this study further highlight the need for water actors and stakeholders in the districts to plan and prioritise water and ecological conservation intervention with foreknowledge on the existing socio-economic and demographic variations in their jurisdictions (Manu, 2015). The patriarchal social arrangements continued socio-economic marginalisation, powerlessness and unequal gender participation in water governance and environmental hygiene should be redressed (GSS, 2014) in order to sustain the rural water and sanitation sector. This is because the complexity of societies and the need to ensure human-ecological balance requires that rural ecological governance and water conservation practices should be pursued cognizant of the culture, norms, traditions and indigenous knowledge systems prevalent in rural society settings (Karl, 2000). From the stakeholder theory perspective, the study reinforces the argument that the extent of power, legitimacy and urgency of dormant rural actors (Mitchell et al.,1997; Suchman, 1995) such as households and individual residents, is essential for fostering a democratic culture and inclusive ecological decision making in the districts. This is because, ecologically friendly attitudes and effective political ecological climate (Yanmaz, 2013), can only be realized where the Assemblies restructure the WATSAN and WSMTs to refocuses on co-management approaches which are inclusive, participatory and bottom-up oriented (Addo, 2010).

Ecological Governance and Distribution of Drinking Water

In Figure 3, more than half of the respondents were not participating in decision-making on ecological governance and drinking water supply in the districts. The findings indicate that among the rural dwellers, even where water infrastructure is in place in the districts, with low or non-participation of the rural households, it could become either a common practice or inadvertent occurrence of denying low-income households (particularly in hinterlands), their legitimate rights of access (Suchman, 1995) to free basic water supply and or disconnect them from water services in the districts. With reference to the findings, the study argues that the interest, capacity and level of participation of households depend on access to reliable information and empowerment among the communities to be able to take control over the decision-making processes on water allocation. The low/non-participation of residents means that the political ethos and participatory approaches of the districts require modification to effectively respond to water conservation challenges at the micro level. Drawing from Berry and Mollard (2010), low/non-participation could be due to limited access to information, community empowerment, unawareness of existing participatory structure and capacities to influence discourses and decision making on water supply.

From the findings, this study argues that since water is a public good (United Nations, 2015), decentralising power and administrative controls to the districts will not be enough for guaranteeing the rights of all residents to sanitary environments, potable drinking water and capacity to participate equally in decisions locally. Again, figure 3 highlights that the ecological imperative for collaborative management approaches and capacities of the decentralised environmental governance structures in the districts including the environmental protection department, District Works Department, the WATSAN and WSMTs have to be taken into account in order to comprehend the individual, communities and districtwide constraints to participatory ecological governance and water conservation. Since water is recognised by the United Nations (2003) as a basic human right, the governance of the districts' ecological resources including water must be guided by the principles of sustainability. In this effort, this study suggests the need for inclusion of policy makers and authorities, producers and water users, the scientific community and civil society (Kramer & Pahl-Wostl, 2014) to co-manage water resources, hold service providers accountable and enforce binding rules related to household's participation in water conservation decisions of the districts.

The low participatory efforts of the rural residents are not peculiar to the study districts. In sub-Saharan Africa, in every ten persons, four lack access to improved drinking water and proper sanitation (WHO & UNICEF, 2017). This is because participatory efforts necessary for co-management and inclusive governance of potable water and curtailing sanitation challenges, seem to be evading most developing countries (UNESCO, 2017), hence, exposes rural poor environments to avoidable public health risks. This study suggests that to promote environmental governance and water conservation, politicians, contractors and officials of the District Assemblies should adopt precautionary measures through a careful analysis of the irreversible environmental cost of development projects they implement in the districts. The findings disagree with the purely elitist (Nylen & Dodd, 2003) and centralised governance approaches to water. Nonetheless, the study agrees with the democrats and green ecologist perspectives on ecological governance (Neil, 2018) which lay emphasis on human rights to water, democratization of decision making and active participation in water conservation (Berry & Mollard, 2010). Moreover, the study reinforces Karl (2000) and Sen and Smith (2015) perception that unrestricted access to information and education of the people on water improves citizens' knowledge and capacity to participate in decision making on water conservation. Therefore, although the decentralisation approach in Ghana provides for macro (top-down) and micro (bottom-up) level development (Ahwoi, 2010), this study maintains that the decision making process should involve the participation of a myriad of local actors. This is because, if the socio-economic interest and water needs of micro-level (less powerful) actors are not adequately articulated and sustainably addressed through participatory and integrated approaches, it becomes impossible to enact effective policies which are relevant to the conditions of rural communities and households.

Participatory-Cube and Implications for Sustainable Ecological Governance

In the Savannah Region, governance of ecological resources, particularly water sources such as dams and rivers are affected due to increased human activities including poor farming and fishing methods, encroachment by real estate developers and indiscriminate lumbering of trees for fuel wood and charcoal (WGDA, 2017). Moreover, continued practice of washing in point sources such as rivers, dams and lakes tend to pollute water sources (GSS, 2018), making them impure for drinking and contributing to disease infections (micro-organisms, chemical substances and radiological hazards) which constitute a threat to the health of rural populations (CGD, 2017). Besides, forest and trees which grow along these water bodies have been critical in avoiding increased evaporation as they serve as protective cover for surface water and the capacity for such water sources to provide continued water supply to the communities, particularly during harmattan periods (WGDA, 2016). To sustain continues utilization of ecological resources, the Dublin Statement admonishes ecological actors to make decisions "at the lowest appropriate level, with full consultation and involvement of local users of water in the designing, planning, implementation, monitoring and evaluation of water projects" (ICWE, 1992:4).

As a result, this study argues that strategies for protecting the environment around water sources which appear to be threatened by increased human activities should be critically reconsidered through planning, legislation and appropriate ecological conservation approaches. Similar to Gershman's (2013) assertion on the need for local participatory structure, this study found that ecological governance activities of the districts were established on a five pillared participatory structure namely, information, communication feedback, consultation, planning and citizen control (see Figure 5). At the "lowest level" of citizen participation was the preliminary "promotional stage" for rural ecology and water conservation projects. The aim for promotion phase of water conservation was to sensitize, disseminate information, educate and create stakeholder awareness at the district and rural levels. Stakeholders were informed on financial arrangements, benefits of water and ecological conservation, expected roles and responsibilities of beneficiaries. At the promotional phase, though environmental and water conservation initiatives were promoted, there was no participation from the local citizens (households) for whom such projects were to be implemented. The findings showed that though more citizens were "informed" about the need for water and ecological conservation, the promotional phase represented the lowest and minimal level of citizen participation (Herman, 2006) in ecological governance and development activities of the districts. In line with Arnstein (1969) and Roberts' (2015) observation, this study discovered the promotional phase of ecological projects to be the simplest form of participation between the Assemblies, investors, water service providers and households to keep water consumers "informed" about water security and ecological preservation but without any opportunity for comments nor the involvement of households.

Although informing citizens on participatory spaces in the districts is essential, this level of participation represented the simplest form of communication between district assemblies, investors, water service providers and community members/households. The aim was to keep water consumers informed about water projects, development plans and decision making, but without any opportunity for comments nor involvement of the households. At this phase of ecological projects initiation and implementation, the study found that the households were not involved, and no specific roles were played by the beneficiaries of water projects. Therefore, without active citizen participation at this phase, it would be difficult for the communities to offer local inputs and initiate actions to support and own-manage water and ecological conservation projects at the local level.

Hence, as observed by Herman (2006), such conservation projects could be designed to fail from the very beginning of project implementation. From Figure 5, the second stage of the participatory process involved the solicitation of "community feedbacks" on water projects (CWSA, 2015). The aim for soliciting community feedback was to ensure "some level" of transparency, accountability and equity in the distribution of water investment and ecological endowments across the rural settlements of the districts. At this stage, the communities which required new water facilities and sanitation amenities were required to obtain forms from the CWSA, complete the application forms and have them submitted to their District Assemblies and the CWSA through a community representative or Assembly member for support.

The communication of feedback was meant to elicit opinions from the rural communities. The district planner, investors, water service providers, government departments solicit information or invite comments from the public to supplement knowledge or better understand issues. Despite most of the residents being informed on their rights to participation and creation of participatory spaces as mandated by Act 936 (GoG, 2016), the findings in Figure 5 show that only a few of the rural residents provided feedback and thus made inputs to ecological decision-making. In terms of funding and participation, the arrangement for rural water and sanitation amenities (Todaro & Smith, 2014) and service delivery in the districts required external support from donor agencies (90%), MMDAs (i.e. government-5%), with the remaining 5% paid by the local communities as contribution to capital cost of installing facilities (World Bank, 2008*)* and setting up of water boards (usually non-professional community members) to be in charge of managing the water and sanitation facilities (*Nyarko et al., 2006).* While the National Community Water and Sanitation Plan (NCWSP) recognizes the (5%) beneficiary contribution as community participation, demonstration of goodwill, ability to pay for installation, own and maintain such facilities (CWSA, 2015), evidence from this study suggested a different narrative. In rural communities where public water and sanitation utilities existed, households considered such facilities to be owned either by the government (i.e. MMDAs, WATSAN, WSMTs and CWSA) or donor partners, because either the Assemblies, government, CWSA or external partners constructed them and paid for their operation and maintenance (World Bank, 2008).

Similar to Eguavoen (2006) who argues that poverty is a limitation to participation in decision making, this study found that while the beneficiary contribution could be functional in better-off regions, it could constrain participation of rural settlements in decision making particularly in the Savannah Region where poverty levels are found to be widespread (CGD, 2016; WGD, 2017). This study further highlights that the District Assemblies and CWSA would equally require empowerment and integrated governance approaches to enable them to identify innovative and comprehensive pathways for planning and sector investment (i.e. micro-credit schemes, affordable technologies etc). The findings in Figure 5 agree with Walker (2006) who proposes the need to ensure greater equity and redistributive justice in the rural water sector. As admonished by Castro (2007) and Pennington (2008), this study found that equity in participation can be achieved through advocacy, responding to request for public submissions on projects, policies and by-laws, as well as through deeper coordination between social movements, including the District Assemblies, Non-governmental Organisations (NGOs), Community-based Organisations (CBOs), CWSA and powerless and impoverished settlements.

The study found that actual participation of communities in water and ecological governance activities commenced at the "consultation" and "planning" stages of water projects implementation (see Figure 5). The consultation stage is a critical engagement between planners, water service providers, public and water consumers to dialogue on project progress, identify issues/concerns, aggregate community needs and interest. Meanwhile, the stage for "inclusive planning" was an extended involvement of households with mutual responsibility for planning and resolving complex and vexed matters on water & sanitation delivery through arbitration, negotiations, mediation advisory group (Marshall, 2008). Nevertheless, Figure 5 reveals that only 5% of the respondents were consulted and 2% were included in the planning of water and ecological activities in the districts. Without effective engagement and broader consultations (ICWE, 1992), the selection processes for investments in household's water conservation and prioritisation of communities in dire need of water could be compromised by the most powerful local actors (Goldratt & Cox, 2016). The result resonates with Oates (1972), Eckersley (2019), Béné and Arthur (2006) who argue for greater decentralisation and engagement of beneficiaries in all aspects of development. This study admonishes a courteous, complementary and decentralised strategy for rural ecological decision making. This is because decentralisation into dysfunctional and power laden local structures could adversely polarize and marginalize poorly endowed communities. Moreover, the communities are likely to lack professionals and competent water facility management personnel and gender balanced WSMTs and this could adversely affect quality service delivery at the rural level. In Figure 4, the findings show that the stage of "citizens control" represented the highest level of household's participation. Although at this stage, citizens were expected to be empowered, trained, and their capacities developed to own-manage and control over water supply and ecological resources, the findings showed only 2% of the residents in the districts participated at this stage of ecological decision-making. The findings contrast with the decentralization and democratization principles (Zanu, 1996; GoG, 1992; GoG, 2016) which underpin Ghana's water and sanitation sector. The findings in Figure 5 further suggest the existence of considerable polarization and lingering inequalities of water governance which could have adverse bearing on the stability and sustainability of rural ecological endowments. Similarly, the findings highlight the enormity of capacity and local empowerment constraints involved in promoting a stable participatory space as preconditions necessary for development of the rural settlements in the districts (MWRWH, 2014; Brundtland Commission, 2014).

Besides, decentralised decision making should be context specific and have regard for rural peculiarities and ecological endowments of the districts. This is because decentralised rural water and ecological systems could only be made effective when existing principal-agency relationships are strengthened (Gaventa, 2006) with local people empowered and involved at the highest level of strategy formulation and ecological policy implementation. Moreover, this study deduced the need to prioritize, mainstream and or strengthen the concepts of "community ownership" and "co-management" through training and empowerment of the local citizens, inclusion and representation of all stakeholders (Mitchell et al., 1997) in the decision-making, planning and governance processes of ecological resources and water service delivery at the district level.

In line with this assertion, Sen and Smith (2015) advocate a human-centred approach to development while Roberts (2015) postulates that this is attainable when development beneficiaries are empowered to participate in ecological policy making. In Figure 5, the findings extend the dimensions of participation and the analysis of space in which any decentralised and democratic mechanisms of public decision making can be located, particularly in rural settings. Similarly, Fung's (2006) participatory-cub analogy presents three dimensions of direct participation, namely, scope of participation, mode of communication and decision-making, and extent of authority. In extending Poplin et al. (2013) participatory-cub model, Figure 5 shows that although the vast majority of public participation was opened to all who wish to participate, those who choose to participate were often not the representatives of a large audience especially at the level of citizen control.

Stakeholder Participatory Analysis and Power Relationship in Ecological Governance

The participation of stakeholders in maintaining ecological sustainability (Mitchell et al., 1997) in rural settlements remains a central part of achieving equitable access to water and conservation of rural ecosystems. This is because stable and balanced power relations between development stakeholders are often linked to improved service quality and ensuring the dignity of the poor and peripheral segments of society who are historically denied legitimate spaces and decent living conditions (Coleman & Ferejoh, 1986). The participation of stakeholders creates an enabling space for holding municipalities to their legitimate obligations of delivering services to the satisfaction of end-users of water at the rural level. However, from Table 5, this study found discrepancies of power, legitimacy, urgency and capacity to influence water distribution (Suchman, 1995). The study identified and categorized the stakeholders into two, namely internal and external stakeholders. The categorizations were based on the geographic location, power, legitimacy and responsibilities assigned to the various stakeholders in the delivery of water and ecological conservation projects within the districts. The findings revealed that the power relations between the stakeholders were either definitive, dominant, dependent, dormant or discretionary. At the highest driving seat (strategic) and prioritised level of water and ecological governance saw the participation of key stakeholders who had vested interest in water and ecological conservation, including, District Assemblies, traditional leaders, investors, NGOs and financial institutions who were either located within or outside the districts.

As depicted by Table 5, the rural consumers of water were not involved in determining water safety and quality standards. The end-users of water were generally dependent on their districts and the CWSA for sustaining water supply and they had limited power over water supply and capacity to influence ecological governance decisions. Although water consumers were considered to be the legitimate beneficiaries, with their water needs considered urgent and prioritised, they had little or no capacity to influence their local authorities. This has implications on water-related disease outbreaks and jeopardizing ecological conservation efforts of the districts. Furthermore, the findings show that external development partners, including the media, investors and donors, though were acknowledged to be physically absent (external) from the communities, the districts and the region, they had high power, low legitimacy and low urgency attached to decision making on water supply in the rural settlements.

Moreover, the discretionary views of researchers, logistic providers, contractors and engineers were acknowledged and considered legitimate. However, these stakeholders had low power to compel or effect modifications/changes to definitive (cast-in-stone) rural water policies and decisions. They were highly recognised as legitimate institutions, yet their decisions were not urgently considered and prioritized in the rural water sector. In effect, the WHO (2014), GSS (2018) and CWSA (2015) standards of rural water quality could be compromised if the views and standards established by water quality control and regulatory institutions are ignored by dominant and powerful actors in the rural water and sanitation sector. In Table 5, the findings agree with Mitchell et al. (1997) claim that dominant stakeholders have powerful legal support to impose their will and discretions. Similarly, it resonates Suchman's (1995) contention that discretionary and dependent stakeholders have urgent claims but lacks power and are perceived as merely passive recipients of corporate philanthropy. In effect, a predominating political elite (Nelson & Shilling, 2019) and power-lessness of the local people (Arstein, 1969) in ecological governance defeats the purposes of decentralisation (Hoppe, 2010) which is a pillar for local self-governance and co-management of rural water and sanitation in Ghana. The findings of this study suggest that due to capacity gaps, the local agents in the districts are unable to execute their mandate effectively at the rural level and this stifles respect for civil rights and liberties of the people. This discovery disagrees with the notion of classical liberalism which advocates for civil liberties and see 'agents or stakeholders' as constituting legitimate development beneficiaries and promoters of micro development (Gaventa, 2006). The study further analysed the procedures or existing protocols utilised for engaging stakeholders in the formulation and implementation of water and ecological policy decisions in the districts. This is because the rational for Act 936 of 2016, is to promote development through local self-governance and participation (GoG, 2016). From Figure 6, the findings showed that the water and ecological engagement processes of the districts included popular representation, workshops and fora, hotline and telephone conversations, contractual engagements, opinion polls, referendum and surveys, experts' panel, focus group discussions, local radio stations, posters and community-centre notices and publications.

The results in Figure 6 agree with political ecologists (Eckersley, 2019; Béné and Arthur, 2006) and ecological democrats who argue that as a result of population increase, complexity of issues and knowledge gap on water conservation, creating spaces for end-user engagement (Hoppe, 2009; MWRWH,

2014) should become the panacea for effective governance of rural ecological endowments. The findings in Figure 6 imply that among the three study districts, water is recognized by all stakeholders as essential resources for securing rural livelihoods. Therefore, priority in the formulation of policies and allocation of water and sanitation maintenance must be given to the local people who constitute the ultimate beneficiaries of local development (Sen & Smith, 2015; Todaro and Smith, 2014). Moreover, attention should be given to ensuring that disadvantaged and marginalized rural populations, including women are adequately represented in policy formulation and that they have equitable access to water and management services.

The state parties and local structures responsible for water governance including the District Assemblies, CWSA, WATSAN committees and WSMTs, should ensure that there is adequate representation of the indigenous people in decision making and equity of access to water for agriculture and for securing livelihoods. Again, there is the need for inclusive and non-discriminatory engagements to prevent threats to water and ecological resources in the districts. This will strengthen partnerships, prevent starvation and eminent disease outbreak and enable rural populations to assert their human rights of access to sufficient potable and affordable water (Nelson & Shilling, 2019; UN, 2003; Osunade,1994).

Even though this study agrees with Oates (1972) and Hoppe (2010) contention of the decentralisation strategy which seeks to bring governance closer to the people, the findings expand this assertion further by suggesting that for decentralisation to become a useful development tool, the powerful agencies at the districts level should afford localities the opportunities to participate in ecological conservation and co-management of water supply under a democratic environment. This is because, without empowerment, equal access to information and enabling capacities, citizen representatives would either be reduced to mere symbolic representation or placated (Arstein, 1969) at the strategic level of ecological governance. Without power, tangible voice and lacking understanding, these representatives could either be coerced or morally persuaded into accepting decisions which may not accurately represent the common good and priorities of rural settlements. The findings agree with the advocates for citizen participation who contend that inviting citizens' opinions, through such activities including consultations, consensus building, referendum, voting and opinion polls can be legitimate steps toward their full participation in community development activities (Marshall, 2008). Conversely, the findings appear to have disagreed with Eckersley's (2004) state-centric, Hoppe's (2010) centralized and Nylen and Dodds' (2003) elitists decision making approaches to water and ecological policy formulation as these approaches are no longer feasible particularly in efforts towards public participation, democratisation and sustainability of rural water and ecological systems.

Rural Ecology and Domestic Water Conservation Strategies

The narratives on political ecology address the synergy between human-environment and decision making in order to maintain ecological equilibrium and sustainable water access (Neil, 2018). This study considered that an important dimension for maintaining stable rural environments is the need for effective strategies to conserve water in the districts. In the face of increasing water scarcity, domestic conservation of water is no longer an option but a considered priority particularly in rural settings of the districts which are confronted with challenges of access to potable water. This study found that multiple approaches and strategies were adopted by the households in order to preserve environmental resources, particularly water. From Table 6, the findings showed that there were no remarkable discrepancies among the rural settlements of the districts in terms of domestic conservation of water. The commonest strategies for water conservation were rainwater harvesting and storage, ponds and dug-outs, water polytanks, water rationing and control of water pollution, collection, recycling and reuse of wastewater and adoption of measures to prevent water leakages. The study found that the activities of rainwater harvesting in the districts included the capturing and storage of rainwater, efficient utilization, conservation, controlling rainwater runoff, evaporation and seepages. The residents of the districts utilized two broad strategies of harvesting rainwater, namely surface water runoff harvesting which is normally utilized for farm irrigation and backyard gardening and rooftop rainwater harvesting.

The residents used rainwater for multi-purposes including human consumption, bathing, washing, breeding of domestic livestock, laundry, construction of buildings and flushing toilets. The prolonged dry season coupled with increased need for water in the districts requires that rainwater is stored either in dugouts, vessels, barrels, or into large cisterns to supply households water demands. The findings meant that rainwater has the capacity to improve not only the quality of groundwater through dilution but also increased water availability for domestic and productive purposes. However, due to poverty, the use of polytanks was generally unpopular among the rural households in the districts. Similar to this study, other scholars have argued that supplementing scientific data with local and traditional knowledge on the environment are imperative for better governance (Nelson & Shilling, 2019; Osunade, 1994), local decision making on ecological management (Hyden & Neef, 2004) and sustainable use of water resources (Manu, 2015). The decision making on the location of ponds and dug-outs and their physical conditions and maintenance are however a matter of much concern to the people, particularly women, in carrying out their domestic and economic activities. The maintenance of natural resources through participation and in the continuous process of use and conservation means not merely the assurance of livelihoods to the people from multi-ethnic, traditional and environmental settings, but of equal consideration is the preservation of ecological balance which is critical for human livelihoods, growth and development. Drawing from Pickett et al (2004) advocacy on water conservation, this current study suggests that rural water conservation is no longer an option, but an ecological imperative, particularly in the face of climate variabilities (Neil, 2018; Arnstein, 1969) which predispose deprived settlements to water scarcity. Moreover, it has further been argued that conservation ensures that the utilization of ecological endowments, i.e. water resources, proceeds along a sustainable path (Munasinghe, 1992), which is imperative for the preservation of rural ecosystems for future purposes (Herman, 2006).

Hypothesis Test and Implications

In order to achieve the general goal and purpose for this study which was to examine the link between participation in ecological governance and water conservation among the peripheral households of the districts, the study hypothesized (H_0) that "there is no statistically significant relationship between the participation of rural households in water governance, the distribution and conservation of drinking water in their communities". As illustrated by Table 7, the findings from the chi-square (χ^2) values showed that water conservation initiation, water project design/planning, financial contribution towards water facilities, labour support, supervision and evaluation of conservation activities, repair/maintenance/rehabilitation of water facilities, setting water tariffs, and gender inclusion in decision making were statistically significant in explaining variations of participation in ecological governance and domestic water conservation at the district level. The 'p-values = 0.00' were found to be statistically less than the significance level, that is, alpha value of $(\alpha < 0.05)$ with 16df (degree of freedom), and therefore, we fail to accept the null hypothesis (H_0), which means that the alternative hypothesis (H_A) remained valid and was accepted by the study. The study generally concluded that a democratic political ecological system is essential for engendering inclusion of rural water consumers in the co-management of ecological endowments of the districts. Therefore, the participation of household should be considered essential at all levels of ecological governance activities in the districts.

The results were corroborated upon interaction with officials from the District Assemblies, traditional authorities, the WATSAN and the WSMTs, who confirmed that access to water was considered critical and thus, the participation of water consumers at the household level would be essential for ensuring reliability and sustainability of water supply to local communities. Moreover, the results corroborate the Dublin Statement which stipulates that "decisions be taken at the lowest appropriate level, with full consultation and involvement of local users of water in the designing, planning, implementation, monitoring and evaluation of water projects" (ICWE, 1992:4). In theory, the results confirmed the theoretical framing of the study which argued that where decentralisation (Oates, 1972; Hoppe, 2009) is effectively pursued with the principal-agency relationships clearly specified, powerless and excluded households tend to be empowered to participate (Arnstein, 1969) in democratic process (Goldratt & Cox, 2016) of identifying sustainable measures for water conservation. The findings agree with Munasinghe (1992) and Herman (2006) who argue that such legitimate participatory processes are critical for sustainable use of rural endowed water resources. The results of this study further implied that creating spaces for participation in ecological governance at the local level would afford powerless households the opportunity for interest aggregation, representation and collective bargaining power to enable powerless households influence decision making, development planning and resource allocation at the community level. This means that access to water and sustainable development tends to be compromised in a community where participatory governance is ineffective or significantly lacking.

Moreover, to achieve this would inevitably require effective and democratic ecological governance structure anchored on local stakeholder inclusivity and participatory decision making on community water production, distribution and conservation activities. This is because, through participation and feedback communication, local people come to understand the socio-economic and ecological conditions under which they live and how their activities impact environmental resources in the districts. In addition, local citizens possess relevant local knowledge critical for the solution of the myriad of development problems in their communities. Consistent with Hyden and Neef (2004), the findings of this study affirmed that free, fair and equal participation would allow the agencies and service providers to acquire a better understanding of local needs than would centralize systems (Oates, 1972). This could contribute to reduction in over-dependence on outside experts and enable the formulation of locally tailored-water solutions since local people would have been adequately consulted and empowered to conserve their own water systems sustainably and independently. Drawing from Hoppe (2009) and Hoppe (2010), the findings of this study highlight that space for participation in ecological governance should be expanded to enable popular engagement in all aspects of development. In like manner, Poplin et al (2013) suggest the "space" and "participation processes" (Béné & Arthur, 2006) should engage beneficiaries in decision making and feedback through monitoring and evaluation processes.

Traditional Governance Systems and Effects on Rural Water Sustainability

Humanity's entire life support system, well-being and healthy living depend on sustainable water access and hygienic environments. As depicted by Table 8, the findings of this study indicate that water facilities in the districts are sustained through a variety of measures including the construction of fence walls around water sources, implementation of bye-laws, establishment of community watchdogs on water, the traditions, myths and rituals systems governing water systems in the districts. In this instance, the sacredness of the water sources in the districts was associated with either the whole or part of the water sources and the attitudes of the local people towards the use of water from such sources. In terms of scientific explanations, Osunade (1994) contends that these practices do not only protect traditional water catchment areas but also serve as possible measures for minimizing flooding and soil erosion in agric-based rural settings. Hence, violating the customary practices, norms and bye-laws attracted sanctions from either the traditional leaders, the WATSAN and WSTMs committees, the District Assemblies or the CWSA. In addition, it could equally be possible that the local water actors enforced bye-laws on conservation of vegetation along lakes and riverbanks. Moreover, from the traditional perspective, it could be inferred that perhaps, the households and communities perceived these surface water sources as sacred areas. Similar to the findings of this study, the traditional demography of sub-Saharan countries reveals that during prolonged drought seasons or delays in the onset of rains, it is a common practice for elders and farmers to visit shrines to pray for rains (Haruna, 2017). It was probably a common belief among the households who fetched directly from these surface water sources that they were sacred and so they saw the need to keep drinking water sources pure from contamination. From the findings, the study deduced that effective enforcement of the existing taboos could deter farmers from cultivating fields along rivers and lake courses or dam catchment areas. The communities should be made aware that eroded soils from crop fields could possibly increase siltation, resulting in the depletion of water sources and punitive measures should be enforced to serve as deterrent. In addition, strict adherence to the traditional norms could possibly discourage pollution of surface water and unwarranted destruction of wetland areas which could potentially become sources of water during drought as the rural communities can easily sink traditional dugout-wells in such wetland areas. Similarly, in the Eastern Cape, Palma (1996) and Penny (2001) discovered that rivers, lakes and dams were considered abodes for spirits such as mermaids; hence, these norms motivated community compliance, discouraged pollution and secured water resources.

From the bye-laws of the CWSA and the Assemblies, it is prohibited to defecate in open areas within the catchment area of water sources in the districts. Although challenging, the study revealed that the WATSAN committees ensure that defecating must be done in a toilet or a hole dug in the ground which must be buried after use while the social norms also forbid the communities from unhealthy fishing practices and throwing of objects into watersheds. Likewise, among the Marakwets, Lssozi (2012) found that it was taboo to wash in or nearer to water collection points. In the Southern part of Ghana, Nukunya, (1992) also found that the Akans and other ethnic groups are similarly noted to forbid discharge of industrial and human waste into water sources, lest the culprit incurs the wrath and punishment from traditional authorities and government agencies. Hence, this study concludes that even though customary laws on water founded on indigenous knowledge systems may lack scientific explanations, application of such systems could still be relevant for sustainability of water delivery.

To promote effectiveness in water sustainability, evidence from this study suggests that water actors can learn from context-specific indigenous/local knowledge and institutional mechanisms prevalent in the rural areas of the districts. Clearly, genuine synergy between scientific and local knowledge systems towards water and environment sustainability is important for effective governance and sustainability of communities' water systems. Hence, policy makers (MMDAs, CWSA, government) etc., should seek to integrate rural systems into water and ecological policies, take advantage to minimize ecological degradation and sustain water systems. The findings disagree with the purely technical notions (MWRWH,

2014), and elitist (Herman, 2006) sustainability approaches being adopted by water actors. This study maintains that perhaps, the enforcement of indigenous customs and bye- laws on water in rural settings is likely to mandate users to keep point sources such as lakes, rivers and other sources sustainable and pure for continued human use. In the specific case of water and ecological management in the districts, this study suggests that purely technical approaches seem insufficient to adequately respond to the demands of a constantly growing population and mounting water consumption pressure (Nelson & Shilling, 2019; Bertrand-Krajewski, et al., 2000). Furthermore, the findings disagree with Lencha (2013) who discovered that in rural settlements of Ethiopia technology (technical know-how) is positively correlated with water service reliability, cost of production, and quality of water, infrastructure and distribution trends.

SOLUTIONS AND RECOMMENDATIONS

The findings of this study convey to the forefront the following proposals in order to ignite agenda setting and policy discourses among actors in rural ecology; Foremost, there is the need for integrated efforts and multi-stakeholder participatory approaches towards water and ecological conservation. The study proposes integrative approaches which address capacity building and local empowerment, systemic and institutional constraints, gender considerations, attitudes and behaviour towards participatory decision processes. This approach should create legitimate *spaces* for effective community engagement in ecological governance and conservation of water at the micro level. Through popular participation, powerless rural households could come to understand and appreciate the social, political, economic and ecological conditions under which they live and measures for preservation of their ecosystems. This should be part of rural ecological policy making because local people possess relevant local knowledge and their inputs should matter in local efforts towards the solution of the myriads of development and ecological problems in their communities. In addition, water conservation and ecological development plans should be home-grown, people-centered and community-led with support from external development actors. Water conservation plans should reconsider the rurality of the local people and broader consultation of the beneficiary households. The study admonishes that the more powerful water actors (government, MMDAs, CWSA, NGOs, traditional authorities) who are critical decision makers, to create enabling preconditions for open discourse; interest articulation, aggregation of local needs and making collective decisions. The local people must be at the forefront of planning and focal reference for all ecological activities. The engagement of local citizens at all stages in water dialogue and planning is specifically encouraged and should be priorities as essential for ownership and sustainability of rural water.

The study further suggests a bottom-up approach to ecological development initiatives which provide technical and legal support, environmental spaces, training and gender empowerment needs. To achieve this, indigenous knowledge systems and inputs of the people at the micro or lowest level should matter in local efforts towards the solution of the myriads of development and ecological problems in the rural communities. Moreover, a gender-based approach, greater knowledge through equal access to information and commitment of community beneficiaries in decision making on water could create enabling preconditions for open discourse, interest articulation, aggregation of local needs and making informed decision for development planning. The participation of local people in allocation of ecological endowments and conservation of resources should be heralded as panacea for environmental safety, public health, sustainability of water and hygiene development at the local level. Water conservation activities of the state, local governments and civil society organisations should afford rural dwellers opportunities to be effectively engaged in ecological governance to ensure transparency, collective accountability and better water service provision at the community level.

FURTHER RESEARCH DIRECTION

The study proposes that an in-depth exploration of how local people become committed to public participation in water allocation and how to avoid agnostic or skeptical principal-agency relationships in the rural water and sanitation sector would be very helpful. Hence, further research might compare, for example, conflicts among national, local governments, civil society actors and local communities in engendering commitments towards participation in ecological governance. Moreover, other researchers could explore traditional leadership roles, and how collective leadership and traditional institutional structures might support effective public participation in water and ecological conservation in rural settings. In addition, further research to develop conceptual approaches and carry out a full-scale socio-economic, political and ecological cost-benefit analysis of public participation in ecological governance would be beneficial. Even though methodologically challenging, it would be very beneficial to use a mixed method or conduct longitudinal or longer-term study with the aim of quantifying the impact of public participation on key indicators such as participant recruitment and retention in clinical trials relative to the implementation of long-term water projects in rural settlements.

CONCLUSION

The study concludes that spaces for participation reinforce the principal-agency model, re-echoes participatory democrats' perspectives and strengthens the top-down decentralised approach to eco-governance as pathways for rural water conservation. Space and power for participation provides opportunities for the poor to be part in water and ecological initiatives designed for their benefit, with the motive of enhancing sustainability. However, without enabling politico-ecological preconditions in place, decentralisation and participation of rural people would not produce sustainable ecological outcomes. Active, free, and meaningful participation of rural settlements is a human right and a precondition for sustainability of rural water and ecosystems. The decision making structures in the study communities are dominated by a patriarchal social arrangement system. The marginalization, powerlessness and unequal participation of females in eco-governance decisions adversely affect ecological plans and sustainability of rural water conservation systems. As such water initiative should be home-grown and include the existing socio-economic and demographic variations of the communities. Moreover, the lack of appropriate community consultation and feedback meant inability of the localities to achieve a "demand-responsive" and inclusive approach to ecological planning.

These are likely to affect the community's sense of ownership, indigenization and local control over eco/water initiatives. Therefore, sustainability and conservation initiatives of the communities are most likely to fail either before, during or post-implementation. The stakeholders power relations were either definitive, dominant, dependent, dormant or discretionary. Even though key ecological actors, beneficiary communities and end-users of water were minimally involved at the highest and strategic levels of decision making. Though symmetrical power relation preserves human dignity and creates legitimate spaces for participation and equal access to development opportunities, the stakeholder's saliences, inclusion and capacity to participate in policy making were based on the power, legitimacy and urgency of the stakeholders in decision making. There were noticeable variations among the three districts in terms of domestication and conservation of water. The commonest strategies for water conservation in the localities were rainwater harvesting, ponds and dug-outs, underground/overhead water polytanks, water rationing and control of water pollution, collection and reuse of wastewater and adoption of measures to prevent water leakages. The domestic conservation of water is no longer an option but a considered priority in rural settings which are water dependent, yet economically, socially and hydrologically challenged.

From the hypothesis test, the study failed to accept the null hypothesis but rather accepted the alternative hypothesis. Therefore, with maintenance, labour support, funding and gender inclusion in water and ecological decisions, there is the possibility for water to be locally sustainable. The study reinforces the principal-agency model, re-echoes and strengthens the top-down decentralised approach to ecological governance by offering *spaces* for participation as pathway for rural water conservation. This implied the need for consensus building, consultation and engagement of local citizens at all stages in water dialogue and planning, providing them with technical, legal, environmental spaces, training/empowerment needs and ultimately, preserving their rights to reserve or withhold consent to decisions at the local level. Where participatory processes fail to redress marginalization, entrenched power structures and injustices, they tend to jeopardize, reinforce and legitimizes rural inequalities. However, where participatory eco-governance initiatives are inclusive, voluntary, transparent and motivated by meaningful improvements and empowerment of local people, ecological sustainability become significantly evidenced among rural populations. The participation of rural people in ecological governance essentially entails decision making and engagement processes. Ecological decision making should therefore not only involve the most powerful government officials, civil society organizations, water managers and the wealthy class, but equally important are interested but powerless parties such as households.

ACKNOWLEDGMENT

The author expresses his sincere gratitude to the local government administrators in the Bole, Gonja Central and Gonja West Districts, local water supply authorities, data enumerators and the study participants whose overwhelming support and collective contributions enabled the successful completion of this research. My sincere thanks also go to Dr. Raphael Ane Atanga of the University of Johannesburg, Ms Julia Masego Tabane, University of Johannesburg, Mr. Paaga Dominic, University of Development Studies (Wa Campus) and Dr Fidelia Dake, University of Ghana for proofreading the manuscript and for their constructive criticisms. Finally, the financial assistance of the National Institute for the Humanities and Social Sciences (NIHSS) and the Council for the Development of Social Science Research in Africa (CODESRIA) towards this research is hereby acknowledged. However, opinions expressed herein, and conclusions arrived thereof are exclusively those of the author and are not necessarily to be attributed to the NIHSS and CODESRIA.

REFERENCES

Addo, L. Y. (2010). Institutional analysis of urban water supply in Ghana: The case of Accra Metropolitan assembly. Academic Press.

Ahwoi, K. (2010). *Local government and decentralisation in Ghana*. Unimax Macmilan.

Alberti, M., Marzluff, J. M., Shulenberger, E., Bradley, G., Ryan, C., & Zumbrunnen, C. (2003). Integrating humans into ecology: Opportunities and challenges for studying urban ecosystems. *Bioscience, 53*(12), 1169–1179. doi:10.1641/0006-3568(2003)053[1169:IHIEOA]2.0.CO;2

Arnstein, S. R. (1969). A ladder of citizen participation. *Journal of the American Planning Association, 35*(4), 216–224.

Awedoba, A. K. (2015). *The peoples of Northern Ghana*. National Commission on Culture.

Awoke, Z. A. (2012). *Assessment of challenges of sustainable rural water supply: Quarit Woreda, Amhara region: A Project Paper Presented to the Faculty of the Graduate School of Cornell University*. Cornell University.

Ayee, J. R. A. (2008). The balance sheet of decentralisation in Ghana. In F. Saito (Ed.), *Foundations for Local Governance*. Physica-Verlag HD. doi:10.1007/978-3-7908-2006-5_11

Babbie, E. (2012). *The practice of social research* (13th ed.). Wadsworth Cengage Learning.

Babbie, E. (2016). *The practice of social research* (14th ed.). Wadsworth Cengage Learning.

Barber, B. (2003). Strong democracy: Participatory politics for a new age (20th anniversary ed.). University of California Press.

Bazaanah, P. (2019). Sustainability of Rural Communities Drinking Water Systems and Local Development Projects in the Bole, West and Central Gonja Districts of the Savannah Region, Ghana. *Review of Social Sciences, 4*(1), 16–36.

Béné, C., & Arthur, E. N. (2006). *From participation to governance: A critical review of the concept of governance, co-management and participation, and their implementation in Small-Scale Inland Fisheries*. World Fish Center.

Bernard, H. R. (2011). *Research methods in anthropology: Qualitative and quantitative approaches*. Altamira Press.

Berry, K. A., & Mollard, E. (2010). *Social participation in water governance and management*. Earthscan.

Bertrand-Krajewski, J. L., Barraud, S., & Chocat, B. (2000). Need for improved methodologies and measurements for sustainable management of urban water systems. *Environmental Impact Assessment Review, 20*(3), 323–331. doi:10.1016/S0195-9255(00)00044-5

Bole District Assembly-BDA. (2016). Composite budge of the Bole District Assembly for the 2016 fiscal year. Bole District Assembly.

Brundtland Commission. (2014). *Defining sustainable development: The World Commission on Environment and Development*. Earthscan/Routledge.

Burke, E. M. (1968). Citizen participation strategies. *American Institute of Planners, 15*(5), 290–291.

Castro, J. C. (2007). Water governance in the twentieth -century. *Ambiente & Sociedade, 10*(2), 1809–4422. doi:10.1590/S1414-753X2007000200007

Central Gonja District Assembly. (2016). *Composite budge of the Central Gonja District Assembly for the 2016 fiscal year.* Central Gonja District Assembly.

Central Gonja District Assembly. (2017). *Draft medium-term development plan (2014-2017) based on national medium-term development framework.* Central Gonja District Assembly.

Central Gonja District Assembly. (2017). Annual progress report. Central Gonja District Assembly.

Chilisa, B. (2011). *Indigenous research methodologies.* Sage.

Christenson, J., & Robinson, W. (1980). Community development in America. Iowa State University Press.

Coleman, J., & Ferejohn, J. (1986). Democracy and social choice. *Ethics, 97*(1), 6–25. doi:10.1086/292814

Community Water and Sanitation Agency-CWSA. (2014). *2014 annual report.* Ministry of Water Resources, Works and Housing.

Community Water and Sanitation Agency-CWSA. (2015). *Fact sheet: Rural and small towns water service.* Ministry of Water Resources, Works and Housing.

Community Water and Sanitation Agency-CWSA. (2017). *Environmental and social management framework (ESMF) for the sustainable rural water and sanitation project- Additional financing addendum report final.* Ministry of Water Resources, Works and Housing.

Dungumaro, E., & Madulu, N. (2003). Public participation in integrated water Resources Management: The case of Tanzania. *Journal of Physics and Chemistry of the Earth,* 1009–1014.

Eckersley, R. (2004). *The green state: Rethinking democracy and sovereignty.* MIT Press. doi:10.7551/mitpress/3364.001.0001

Eckersley, R. (2019). Ecological democracy and the rise and decline of liberal democracy: Looking back, looking forward. *Environmental Politics, 29*(2), 214–234. doi:10.1080/09644016.2019.1594536

Eguavoen, I. (2006). Drinking water policy: Water rights and allocation practice in rural Northern Ghana. Centre for Development Research (ZEF).

Enserink, B., Patel, M., Kranz, N., & Maestu, J. (2007). Cultural factors as co-determinants of participation in river basin management. *Economia e Sociologia, 12*(2), 24. doi:10.5751/ES-02096-120224

Food and Agriculture Organisation- FAO. (2017). *Coping with water scarcity: Challenge of the 21st Century.* https://www.fao.org/3/a-aq444e.pdf

Fung, A. (2006). Varieties of participation in complex governance. *Public Administration Review,* (66), 66–75.

Gaventa, J. (2006). Finding the space for change: A power analysis. *Institute of Development Studies Bulletin, 37*(6), 23–33. doi:10.1111/j.1759-5436.2006.tb00320.x

German Industry and Commerce in Ghana (GICG). (2018). *Access to clean drinking water and water management in Ghana.* Ghana Statistical Service.

Gershman, S. D. (2013). *An evaluation of public participation techniques using Arnstein's ladder: The Portland plan*. University of Florida.

Ghana Meteorological Agency. (2019). *Regional weather, Savannah Region.* Accessed on 05/10/2019 from; https://www.meteo.gov.gh/gmet/regional-weather-savannah-region/

Ghana Statistical Service. (2012b). *2010 Population and Housing Census: Summary of final results.* Ghana Statistical Service.

Ghana Statistical Service (GSS). (2011). *Ghana-multiple indicator cluster survey-2011*. Government of Ghana.

Ghana Statistical Service (GSS). (2014). *2010 population and housing census: District analytical report, West Gonja District*. Ghana Statistical Service.

Ghana Statistical Service (GSS). (2014a). *2010 population and housing census: District Analytical report of the West Gonja District*. Ghana Statistical Service.

Ghana Statistical Service (GSS). (2018). *Ghana multiple indicator cluster survey; Snapshots on key findings*. Ghana Statistical Service.

Global Water Partnership (GWP). (2004). *Integrated water resources management (IWRM) and water efficiency plans by 2005. Why, What and How? TAC Background Papers No. 10*. Global Water Partnership.

Goldratt, E. M., & Cox, J. (2016). *The goal*. Routledge. doi:10.4324/9781315270456

Government of Ghana. (1992). 1992 Constitution of the republic of Ghana. Tema: Ghana Publishing Corporation.

Government of Ghana. (1993). Local government Act, 1993 (Act 462). Tema: Ghana Publishing Corporation.

Government of Ghana. (2010). *Community water and sanitation agency sector guidelines-general (rural communities & small towns)*. CWSA.

Government of Ghana. (2016). *Local government Act, 2016 (Act 936)*. Ghana Publishing Corporation.

Government of Ghana. (2019). *Ministry of Sanitation and Water Resources: Medium-term expenditure framework (2017-2019)*. Ministry of Finance.

Grimm, N. B., Grove, J. M., Pickett, S. T. A., & Redman, C. L. (2000). Integrated approaches to long-term studies of urban ecological systems. *Bioscience*, *50*(7), 571–584. doi:10.1641/0006-3568(2000)050[0571:IATLTO]2.0.CO;2

Grimm, N. B., & Redman, C. L. (2004). Approaches to the study of urban ecosystems: The case of Central Arizona – Phoenix. *Urban Ecosystems*, *7*(7), 199–213. doi:10.1023/B:UECO.0000044036.59953.a1

Haruna, M. (2017). *Chieftaincy controversies over the Bole and Mandari skins since 1908*. https://mynewsgh.com/chieftaincy-controversies-over-the-bole-and-mandari-skins-since-1908/

Herman, D. (2006). *Sustainable development: Definition, principles and policies*. Springer.

Hoppe, K. P. (2010). Decentralisation and local governance: Theory and the practice in Botswana. *Development Southern Africa, 17*(4), 519–534. doi:10.1080/03768350050173912

Hoppe, K. R. (2009). Capacity development for good governance in developing Countries: Some lessons from the field. *International Journal of Public Administration, 32*(8), 728–740.

Hyden, G., & Neef, J. (2004). Comparing governance across Countries and over time: Conceptual challenges. In D. Olowy & S. Sako (Eds.), *Better governance and public policy: Capacity building and democratic renewal in Africa*. Bloomfield Kumarian Press Inc.

International Conference on Water and Environment. (1992). *The Dublin statement and report of the conference*. Author.

Iribarnegaray, M. A., & Seghezzo, L. (2012). Governance, sustainability and decision making in water and sanitation management systems. *Sustainability, 4*(11), 2922–2945. doi:10.3390u4112922

Jiménez, A., LeDeunff, H., Ricard Giné, R., Sjödin, J., Cron, R., Murad, S., Takane, M., & Bartram, J. (2019). The enabling environment for participation in water and sanitation: A conceptual framework. *Water (Basel), 11*(308), 1–21. doi:10.3390/w11020308

Karl, M. (2000). Monitoring and evaluating stakeholder participation in agriculture and rural development projects: A literature review. Food and Agriculture Organization (FAO).

Kramer, A., & Pahl-Wostl, C. (2014). The global policy network behind integrated water resources management: Is it an effective norm diffusor? *Ecology and Society, 19*(4), 11.

Lautze, J., De Silva, S., Giordano, M., & Sanford, L. (2011). Putting the cart before the horse: Water governance and IWRM. *Natural Resources Forum, 35*(1), 1–8. doi:10.1111/j.1477-8947.2010.01339.x

Lencha, A. (2013). *Rural water supply management and sustainability in Ethiopia with special emphasis on water supply schemes in Adama Area*. Swedish University of Agricultural Sciences.

Lssozi, A. (2012). Values and participation: The role of culture in nature preservation and environmental education among the Baganda. *The Journal of Sustainability Education, 7*(3), 88–99.

Manu, P. K. (2015). *Challenges of water supply in urban Ghana: An assessment of the coping strategies in Adentan Municipality*. University of Ghana.

Marshall, G. (2008). Nesting, subsidiarity, and community-based environmental governance beyond the local level. *International Journal of the Commons, 2*(1), 75-98.

Miller, R., & Brewer, J. (2003). *The A-Z of social research*. Sage. doi:10.4135/9780857020024

Ministry of Sanitation and Water Resources. (2019). *Enforce sanitation by-laws on sanitation*. MSWR.

Ministry of Water Resources, Works, and Housing (MWRWH). (2014). Water sector strategic development plan (2012-2025). Ministry of Water Resource, Works, and Housing.

Misra, K., & Malhotra, G. (2011). Water management: The obscurity of demand and supply in Delhi, India. *International Journal of Management of Environmental Quality, 23*(1), 23–35. doi:10.1108/14777831211191575

Mitchell, R., Agle, B., & Sonnenfeld, J. (1999). Who Matters to CEOs? An Investigation of Stakeholder Attributes and Salience, Corporate Performance, and CEO Values. *The Academy of Management Review, 22*(5).

Mitchell, R., Agle, B., & Wood, D. (1997). Toward a Theory of Stakeholder Identification and Salience: Defining the Principle of Who and What Really Counts. *Academy of Management Review, 22*(4), 853–886. doi:10.2307/259247

Mohammed, A. K. (2015). The politics of municipal fragmentation in Ghana. *Commonwealth Journal of Local Governance, 16*(17), 123–145. doi:10.5130/cjlg.v0i0.4493

Munasinghe, M. (1992). *Environmental economics and sustainable development.* Paper presented at the UN Earth Summit, Rio de Janeiro.

Naik, P. K. (2016). Water crisis in Africa: Myths or reality? *International Journal of Water Resources Development, 33*(2), 326–339. doi:10.1080/07900627.2016.1188266

Neil, C. (2018). *The politics of the environment: Ideas, activism, policy* (3rd ed.). Cambridge University Press.

Nelson, N. K., & Shilling, D. (2019). *Traditional ecological knowledge: Learning from indigenous practices for environmental sustainability.* Cambridge University Press.

Neuman, W. L. (2010). Social research methods: Qualitative and quantitative approaches (7th ed.). Pearson.

Nukunya, G. K. (1992). *Tradition and change in Ghana: An introduction to sociology.* Ghana Universities Press.

Nyarko, K. B., Oduro-Kwarteng, S., & Adama, I. (2006). Cost recovery of community-managed piped water systems in the Ashanti region, Ghana. *Water and Environment Journal: the Journal / the Chartered Institution of Water and Environmental Management, 21*(2), 92–99. doi:10.1111/j.1747-6593.2006.00051.x

Nylen, W., & Dodd, L. (2003). *Participatory democracy versus elitist democracy: Lessons from Brazil.* Palgrave Macmillan. doi:10.1057/9781403980304

Oates, W. E. (1972). *Decentralisation theorem.* Harcourt Brace Jovanovich.

Osunade, M. A. (1994). Community environmental knowledge and land resource surveys in Swaziland. *Singapore Journal of Tropical Geography, 15*(2), 157–170. doi:10.1111/j.1467-9493.1994.tb00059.x

Pallant, J. (2003). *SPSS survival manual: A step by step guide to data analysis using SPSS for Windows (Version 10 and 11).* Open University Press.

Palmer, R. (1996). *Residue of tradition or adaptation to an arid environment—cosmology, gender and the struggle for water in two rural communities in the Eastern Cape* (Unpublished paper). Anthropology Department, Rhodes University, Grahamstown, Eastern Cape, South Africa.

Pateman, C. (1970). *Participation and democratic theory.* Cambridge University Press. doi:10.1017/CBO9780511720444

Paulson, L. D. (2015). What is water scarcity? *RWL Water, 5*(2), 1–7. PMID:25220873

Penny, S. B. (2001). *Ecological implications of water spirit beliefs in Southern Africa: The need to protect knowledge, nature, and resource rights*. Science and stewardship to protect and sustain wilderness values: Seventh World Wilderness Congress symposium, Port Elizabeth, South Africa. https://www.fs.fed.us/rm/pubs/rmrs_p027/rmrs_p027_148_154.pdf

Pickett, S. T. A., Cadenasso, M. L., & Grove, J. M. (2004). Resilient cities: Meaning, models, and metaphor for integrating the ecological, socio-economic, and planning realms. *Landscape and Urban Planning*, *69*(4), 369–384. doi:10.1016/j.landurbplan.2003.10.035

Poplin, A., Pereira, G. C., & Rocha, M. C. F. (2013). The participatory cube: A framework for analysis of online participation platforms. In S. Geertman, F. Toppen, & J. Stillwell (Eds.), *Planning support systems for sustainable urban development. Lecture Notes in Geoinformation and Cartography* (Vol. 195). Springer. doi:10.1007/978-3-642-37533-0_23

Roberts, N. C. (2015). The age of direct citizen participation. London: Routledge, Taylor and Francis Group.

Sanoff, H. (2000). *Community participation methods in design and planning*. John Wiley & Sons.

Sarantakos, S. (2013). *Social research* (6th ed.). Macmillan: Red Glob Press. doi:10.1007/978-1-137-29247-6

Scoones, I. (1999). New ecology and the social sciences: What prospects for a fruitful engagement? *Annual Review of Anthropology*, *28*(1), 479–507. doi:10.1146/annurev.anthro.28.1.479

Sen, A., & Smith, C. S. (2015). *Economic development* (20th ed.). Pearson.

Suchman, M. C. (1995). Managing legitimacy: Strategic and institutional approaches. *Academy of Management Review*, *20*(3), 571–610. doi:10.2307/258788

Todaro, P. T., & Smith, C. M. (2014). *Economic development* (12th ed.). Pearson.

Umesh, P., & Nagarkatte, N. O. (2017). *Theory of constraints: Creative problem solving*. Taylor and Francis Group.

United Nation Children's Emergency Fund. (2017). *Progress on drinking water, sanitation & hygiene*. https://data.unicef.org/wp-content/.../safely-manageddrinking-water-JMP-2017-1.pdf

United Nations. (2003). *General comment No.15: The right to water (Article 11 and 12 of the covenant)*. Adopted at the twenty-ninth session of the Committee on Economic, Social and Cultural Rights. United Nations.

United Nations. (2015). *Transforming our world: The 2030 agenda for sustainable development*. United Nations.

Wanjiru, M. M. (2014). *Determinants of sustainability of community water projects in Kieni East District, Nyeri County: Kenyan*. University of Nairobi.

Watts, M. J. (2000). Political ecology. In E. Sheppard & T. Barnes (Eds.), *A companion to economic Geography*. Blackwell.

West Gonja District Assembly. (2016). Composite budge of the West Gonja District Assembly for the 2016 fiscal year. Damango: West Gonja District Assembly.

West Gonja District Assembly. (2017). Composite budge of the West Gonja District Assembly for the 2017 fiscal year. Damango: West Gonja District Assembly.

World Health Organization. (2014). *Preventing diarrhoea through better water, sanitation and hygiene: Exposures and impacts in low- and middle-income countries*. WHO Press.

World Health Organization (WHO), United Nations Children's Fund (UNICEF), & Joint Monitoring Programme (JMP). (2010). Progress on drinking water, sanitation and hygiene: World Health Organization.

World Health Organization (WHO), United Nations Children's Fund (UNICEF), & Joint Monitoring Programme (JMP). (2017). Progress on drinking water, sanitation and hygiene: 2017 Update and SDG Baselines. Geneva: WHO and UNICEF.

Yanmaz, A. M. (2013). Applied water resources engineering (3rd ed.). Odtu: Metu Press.

Zanu, S. Y. Y. M. (1996). *The institutionalization of local self-government in Ghana's present-day constitutional processes: A paper presented at conference on traditional and contemporary forms of local participation and self-government in Africa*. Nairobi: Konrad Adenauer Foundation.

KEY TERMS AND DEFINITIONS

Ecological Governance: It embraces all the policies, strategies, and activities to sustainably manage the natural resource of fresh water, to protect the environment, and to meet the current and future human demands of water (Goldratt & Cox, 2016).

Legitimacy: The legitimacy of the stakeholders is generally the perception or assumption that the actions of an entity are desirable, proper, or appropriate within some socially constructed system of norms, values, beliefs, and definitions (Suchman, 1995, p. 574).

Power: Power is defined by this study as the extent to which a party has or can gain access to coercive (physical means), utilitarian (material means) or normative (prestige, esteem and social) means to impose their will (Mitchel et al., 1997, p. 864).

Rural Community and Households: Rural community refers to settlements with populations less than 5000, whereas a household is a group of people who live together and share the same or common housekeeping arrangements (GSS, 2018).

Stakeholder: Stakeholder is used with reference to state and non-state actors. In this study, stakeholder classification, power dynamics and ecological governance participatory experiences are based on Mitchell et al. (1997) stakeholder theory and salience model.

Urgency: Is considered by this study as the degree to which stakeholder claims call for immediate attention by authorities and decision-making structures (Mitchell et al., 1997, p. 867).

Water Conservation: Encompasses the engagement and policy participatory processes, strategies and activities developed to manage potable water and sustainable exploitation of human ecological systems (Wanjiru, 2014).

Chapter 2

Examining the Role of NGOs in Community Water and Sanitation Improvement:
A Case of the Tunayilli Community in the Northern Region, Ghana

Gordon Marley
University for Development Studies, Ghana

Prosper Bazaanah
https://orcid.org/0000-0002-9948-2147
University of Johannesburg, South Africa

Patricia Oppong
University for Development Studies, Ghana

ABSTRACT

This chapter examined the role of NGOs in water and sanitation improvement and the effects on the residence of Tunayilli in the Sagnarigu District. The design was descriptive. Questionnaires and interview guides were administered to household heads and key informants. Findings revealed that the water and sanitation condition in Tunayilli is generally poor. NGOs play diverse roles including the provision of water and sanitation facilities, community mobilization, and facility maintenance to ensure their sustained usability. However, inadequate funding, weak community mobilization, poor maintenance culture, and low education are challenges to water and sanitation improvement programmes of NGOs in the community. Measures to mitigate these includes increased central government funding, public education, community involvement, enforcement of sanitation by-laws, and regular maintenance of water and sanitation systems in the community. Stakeholders should intensify and sustain their educational campaigns. Meanwhile, by-laws on water and sanitation should be enforced and offenders prosecuted.

DOI: 10.4018/978-1-7998-8809-3.ch002

INTRODUCTION

One of the most debilitating shortfalls of rural poor communities, is inadequate access to water and improved sanitation. Good sanitation and potable water are very important to human existence, and the absence of these is detrimental to human survival. Societies without adequate water and sanitation systems will not have a favourable chance of prospering. The relevance of achieving adequate water and sanitation is highlighted in several studies globally. The advantages of attaining global access to water and sanitation exceed the costs by a factor of 5.5 to 1 (United Nations; 2015; UNDP and UN Water, 2014: 2). A 2014 study produced by Sanitation and Water for All (SWA) (United Nations, 2014: 2) estimates that there is a $5 return for every $1 dollar invested in sanitation, and universal access to water and sanitation services would result in a possible $170 billion in annual economic gains. The study further revealed that inadequate investment in water and sanitation services (WSS) costs developing countries in Asia and Africa around 4-6 percent in Gross Domestic Product (GDP) annually. Therefore, there is a strong urgency to achieve goal six of the Sustainable Development Goals (SDGs) which focuses of universal access to potable water and sanitation services.

Global statistics suggests that approximately 2.5 billion people are without access to water and proper sanitation (WHO/UNICEF, 2017). Countries in Southern Asia and sub-Saharan Africa are reported to be having the worst form of water and sanitation access due to the growing population, especially in the urban and rural areas (WHO/UNICEF, 2017). The United Nations Children's Fund (UNICEF, 2012) shows that an estimate of about 1.1 billion people practice open defecation, and the rest are using shared or unimproved water and sanitation facilities. About 27 percent and 13 percent of people in Africa and Asia respectively, face the problem of poor sanitation and unimproved access to water. With approximately 215 million people practicing open defecation, sub-Saharan Africa shoulders the greatest water and sanitation challenges (Galan, Kim, & Graham, 2013).

It is partly because of this trend that the UN describes this as likely to see the world missing out on the sanitation target by 600 million people (United Nations, 2015). Confronted by these challenges, at the Sharm el-Sheikh conference, in 2008, African Heads of States and governments committed themselves to contribute 1% of their GDP equivalent budget to water supply (African Union Assembly, 2008). In addition, the African heads of states and their water and sanitation ministers signed the eThekwini Declaration in 2008 to dedicate half a percent of their GDP to sanitation infrastructure and health and hygiene promotion (African Union, 2008). Despite these milestones, water and sanitation access appears to be worsening in rural communities. It was partly because of this trend that the UN describes this as likely to see the world missing out on the MDG sanitation target by 600 million people (United Nations, 2015). The challenges of providing unimproved water and poor sanitation have created an increasing need for improved sanitation and water supply systems in many areas in the world, particularly in West Africa. Countries in West Africa are struggling to meet the demands for improved water and sanitation. This region has witnessed relative stagnation in sanitation coverage since 1990, when total access to basic sanitation was 32 percent (WHO & UNICEF, 2006a). Improvements in water and sanitation coverage have been targeted by the UN Sustainable Development Goals (SDGs) because of their strong links to issues of environmental and public health, economy, and human dignity. It has been estimated that about 1.6 billion people must be able to access improved water and sanitation services in order to meet the SDGs target of halving the percentage of people without access to improved sanitation (United Nations, 2015).

Many countries in the world, and particularly in Africa, may not be able to meet these goals due to compounded problems of population growth, urbanization, and historically inefficient service provision. The inability of water and sanitation efforts to keep pace with population growth has resulted in an increase of over 110 million people in West Africa without access to water and improved sanitation (UNICEF, 2014). The provision of adequate water and management of human waste has wide-ranging benefits to a community, whether located in a rural village or in an urban slum. Mara, Lane, Scott, and Trouba (2010), have for example argued that, adequate sanitation, good hygiene and safe water, are fundamental to health and socio-economic development. Mara et al. (2010) further argues that having access to improved water and sanitation results in lower health system costs, fewer days lost at work/ school through illness and care for the sick, reduced queue time at shared water and sanitation facilities and eliminates open defecation. It is also estimated that poor sanitation, hygiene, and water accounts for about 50 percent of the consequences of childhood and maternal underweight and death, because it strengthens the synergy between diarrhea diseases and under nutrition (Bastien, Hetherington, Hatfield, Kutz and Manyama, 2017). This means that the exposure to one of these conditions increases vulnerability to the other. At the global, West African and particularly Ghana to be specific, the challenges of providing water and sanitation have resulted in the emergence of international and local non-governmental organizations (NGOs). These organizations bring to the fore, the need for capacity building, funding and advocacy for improved governance of water and sanitation supply. NGOs have been established to drive amongst other projects, rural water supply and sanitation initiative at national and local levels (Nathan et al., 2002).

In Ghana, the Community Water and Sanitation Agency (CWSA) and the Metropolitan, Municipal and District Assemblies are responsible for rural water supply and sanitation delivery. To deliver on this mandate, the CWSA and Assemblies partner with NGOs to drive many projects in urban and rural areas. In northern Ghana, NGOs such as WaterAid, USAID, World Vision and Catholic Relief Services etc, promotes rural water and sanitation delivery. In the rural water and sanitation sector, NGOs provide water and sanitation capacity programmes at different levels. These initiatives are undertaken through development partnerships, networks as well as community-based training programmes. NGOs have conducted Water, Sanitation and Hygiene (WASH) research, training for local participants, and volunteer advocacy programmes to sensitize and educate communities on water and sanitation challenges in local communities (Dangah, 2012). These initiatives have not only improved Ghana's Human Development Index (HDI) (ie 0.611 for 2019), it also placed the country in the medium human development category, positioning it at 138 out of 189 countries (UNDP, 2020). This has improved from a value of 0.465 in 1990, reflecting an increase of 31.4 percent for 2019. While the WHO and UNICEF Joint Monitoring Programme reports gradual improvements over the last 20 years in access to improved water and sanitation in Ghana, huge challenges remain with the provision of rural water and sanitation. Available statistics on sanitation in Ghana show that only 13 percent of the Ghanaian population have access to improved sanitation (GNA, 2018). Despite increased interventions by NGOs, studies have found that more than half of the population of Ghana (59%), the highest in the world, depend on shared sanitation facilities including public toilets and about 19% of Ghanaians practise open defecation while 8% depend on various forms of unimproved sanitation facilities options such as bucket latrines (Akphakli, Manyeh, Akpakli, Kukula and Gyapong, 2018). In addition, coverage of sanitation is 8 percent whilst urban coverage is 15 percent, with 20 percent of Ghanaians practicing open defecation (GNA, 2018; Ghana Statistical Service, 2014). In this chapter, we draw from the place-based political ecological lens (Leff, 2017) to examine the role of Non-governmental Organizations (NGOs) in water and sanitation improvement and the effects on the residents of Tunayilli in the Sagnarigu District. The chapter argues that an understanding of the roles of domestic sanitation and water users and other stakeholders can draw new meanings for sustainability, drive water and sanitation sector reforms and set agenda for rural water and sanitation policy formulation and implementation in rural communities. We highlight that place-based water and sanitation constraints require home-grown approaches and solutions. NGOs are better placed to lead in rural water and sanitation reconfigurations. The chapter is organized into different sections. Proceeding this introductory section, the next section and subsections cover the background, main focus of the chapter, reflections on the theoretical and empirical literature, description of the study area, results and discussions, conclusions and recommendations.

BACKGROUND

Clean water and good sanitation are necessities of life, whether rich or poor. National governments alone cannot provide these necessities. In Ghana, several commitments and efforts have been made by the government to fix the water and sanitation problem by implementing policies and programmes such as the setting up of the Community Water and Sanitation Agency (CWSA) to help provide potable water and better sanitation to the rural areas aimed at rooting out rural poverty and accelerating rural development. This has been made possible through the District Assemblies which represent the central government. However, in Ghana, the government recognizes that the public sector alone cannot provide adequate water and sanitation for all. Thus, NGOs fill the gaps left by governments by providing health care, education; water and sanitation facilities, economic opportunities and human rights advocacy to many communities in the developing world (Dangah, 2012). In carrying out this mandate, NGOs have become critical in ensuring human security (Nathan et al., 2002). It is in the recognition of the limited capacity of government to address these issues that a number of NGOs have come in to play a complimentary role in the provision of a wide range of social amenities including water and sanitation in the rural areas. It is in the recognition of the limited capacity of government to address these issues that a number of NGOs have come in to play a complimentary role in the provision of a wide range of social amenities including water and sanitation in the rural areas. Among the NGOs providing interventions in this part of the Northern Region of Ghana are the Catholic Relief Services, WATERAID, Suntaa Nuntaa and the Adventist Development and Relief Agency (ADRA). In 2017, about 23.5 million people in Ghana lacked access to basic sanitation services, and 5.2 million people practiced open defecation (WHO/UNICEF, 2019). As at 2015, only one rural household out of ten were using improved household toilets while three in every ten of them practiced open defecation and not a single district in Ghana has achieved an open defecation-free status (UNICEF, 2018). However, the joint cooperation between government and the NGOs has made some positive impact in the fight against poor water and sanitation service delivery in certain parts of the country. But in other parts of the country, the target of improving sanitation appears elusive. The Northern Region of Ghana continues to remain in this category. Water and sanitation assessments have consistently shown that the expected rate of improvement is yet to be achieved. Tunayilli is a rural community in the Sagnarigu District which was ranked 199 out of 216 districts (CDD/UNICEF, 2015). Meanwhile, the Sagnarigu District in the Northern Region is one of the districts that continue to present persistently poor drinking water access and sanitation results. A 2015/16 assessment of the 26 districts in the Northern Region, in a joint collaboration between the Centre for Democratic Development (CDD) and UNICEF scored the Sagnarigu District zero percent in potable water and sanitation delivery (CDD & UNICEF, 2015).

MAIN FOCUS OF THE CHAPTER

Problem Statement

Ghana has enacted policies to help in proper sanitation and water management. These include the Local Government Act (1994), Act 462, the Environmental Protection Agency Act (1994), Act 490, the Pesticides Control and Management Act (1996) Act 528, and the Environmental Assessment Regulations 1999 (LI 1652). Despite the existence of these laws, the problem of poor sanitation conditions and lack of potable water continue to persist. The problem is worse in rural communities such as Turnayilli. Governmental interventions have had minimal effect on drinking water supply and proper waste management to ensure potable water delivery and high sanitation practices. Non-Governmental Organizations (NGOs) have come in to support government efforts in combating poor water management and sanitation conditions in a number of communities. However, the required success has so far been difficult to achieve.

The water and sanitation situation in the Sagnarigu District of Northern Region, Ghana is even more challenging. According to the 2015/2016 District Water and Sanitation League Table, twelve out of the 26 districts in the Northern Region scored zero percent. As indicated earlier, the Sagnarigu District was one of the twelve districts that scored zero percent and ranked 199 out of 216 districts. The district attracts more attention because of its closeness to Tamale, the capital city of the region. Tunayilli, a community in the Sagnarigu District, typifies the problems faced by peri- urban communities in the Northern Region of Ghana in the area of water and sanitation access. According to the 2010 Population and Housing Census, most communities in the district including Tunayilli, use open spaces to dump waste and practice open defecation. In this case, the question people may want to find answers to may include, what kind of interventions are the NGOs providing? How are the interventions' programmes affecting the people? What are the challenges being faced by the NGOs and the way forward? Against this backdrop, this chapter investigated the role of NGOs in water and sanitation improvement in Tunayilli community in the Sagnarigu District of the Northern Region of Ghana.

Research Objectives

The main objective of the study was to assess the role of NGOs in community water and sanitation service delivery in the Tunayilli community of the Sagnarigu District in the Northern Region of Ghana. The specific objectives were to:

1. Identify the water and sanitation improvement programmes implemented by NGOs in Tunayilli community of the Sagnarigu District.
2. Examine the effects of the water and sanitation improvement programmes on users in the Tunayilli community of the Sagnarigu District.
3. Identify the challenges with water and sanitation improvement programmes in the Tunayilli community of the Sagnarigu District.
4. Determine appropriate measures for addressing the challenges of water access and sanitation improvement programmes implemented by NGOs in the study area.

EMPIRICAL LITERATURE

Water and sanitation are complex issues that have direct link to health. This section reviews the concept of water and sanitation, bringing into perspective the role of NGOs in sanitation improvements and how this has impacted on social and economic development of Ghana. This section provides conceptual clarity, drawing from narratives on water and sanitation, nexus between water, sanitation, hygiene and development. It provides evidence on global state of water and sanitation and its impact, water and sanitation conditions in Ghana and that of the Sagnarigu District. The section concludes with a narrative of the role of NGOs in water and sanitation improvement in the context of Ghana.

Unpacking the Concepts of Water and Sanitation

Globally, water and sanitation are recognized as ecological systems that promotes improved water access, proper disposal of wastes, use of toilet and avoiding open space defecation (ADB, 2011; Bartram & Cairncross, 2010). Ideally, a sustainable water and sanitary system should provide potable water and environment that protects and promotes human health, breaks the disease cycle and is socio-economically viable and appropriate (Okot-Okumu & Oosterveer, 2010). Overall, 2.5 billion people lack access to improved water and sanitation facilities (Unger, 2013), exposing them to fecal contaminants and a multitude of diseases. Estimation is made that 4.2 percent or more of the annual global mortality is preventable if all people had access to safe drinking water, reliable sanitation and decent hygiene practices (Prüss et al., 2002; Tumwebaze et al., 2013). This estimation may prove hard to achieve because across the globe, one in every five people habitually defecates in the open or use some form of shared sanitation facilities such as a basic hygienic latrine or a flush toilet (Bartram and Cairncross, 2010). Reports indicate that such shared toilets are often filthy and disgusting to use, making them unpleasant to clean, thus a provocation to relapse to open defecation (OD) (Barnard et al., 2013; Chambers and Myers, 2016). Recent debates concerning the post-2015 development agenda have drawn attention to neglected diseases and have focused on the importance of consensus regarding Water, Sanitation, and Hygiene (WASH) targets and indicators (Bastien et al., 2017). In sub-Saharan Africa, for example, access to basic water and sanitation was identified as one of the most off-track Millennium Development Goals (AfDB, 2011; Fewtrell et al., 2005; WHO and UNICEF, 2014). The anticipated health benefits of potable water and sanitation, including reduction in the direct cost of hospitalization, burial costs, increased household income and safe food for consumption are increasingly hard to meet (Esrey, et al., 1991; Mugambe et al., 2013). Numerous water and sanitation campaigns being implemented in rural areas do not often show the desired impact on communities, or plan for their large-scale replication across the regions of need (Okot-Okumu and Oosterveer, 2010).

Although the health-threatening effects of inadequate water and sanitation can be effectively solved by proper budgetary allocation, fewer resources are continually committed to water, sanitation and hygiene (Mugambe et al., 2013; WaterAid, 2011). Poor sanitation costs Ghana 420 million Cedis each year, equivalent to US$290 million (WSP, 2012). This sum is the equivalent of US$12 per person in Ghana per year or 1.6% of the national GDP. This cost is manifested through premature deaths, increased health costs, and productivity losses (Kulabako, et al., 2010) The UN Sustainable Development Goal six (SDG 6) is one of 17 Sustainable Development Goals established in 2015. It calls for clean water and sanitation for all people. The official wording is "Ensure availability and sustainable management of water and sanitation for all". The goal has eight targets to be achieved by 2030. SDG 6 is linked with other Sustainable Development Goals (SDGs). For example, improving sanitation helps make cities more sustainable (Goal 11), leads to more jobs and economic growth (Goal 8) and improves health (Goal 3) (United Nations 2015).

Sanitation, Water and Development Nexus: A Global and Local Focus

Sanitation is associated with hygiene (which describes a set of practices performed to preserve health) and development. According to Van Minh and Hung (2011), poor water and sanitation access is responsible for one of the heaviest existing disease burdens worldwide. Diseases associated with poor water and sanitation include diarrheal diseases, acute respiratory infections, under nutrition and other tropical diseases such as helminth and schistosomiasis infections (Stanwell-Smith, et al., 2012). Globally, about 1.7 million people die from diarrheal diseases annually, with 90 percent being children under five years, mostly in developing countries (Fewtrell et al, 2005). According to Roche et al. (2017), each premature death puts economic strain on the family and country of residence, calculated by the discounted potential income lost due to each life unlived to its full potential.

Aside from diarrhea-related mortality, millions of people are unable to maximize their economic productivity due to sickness and disease caused by water-borne diseases and pathogens which are found in human waste. Billions of hours of labor are lost every year, with a significant impact on the global economy (WHO, 2009). Furthermore, treatment for water and sanitation- related diseases, whether provided by the state or paid for by individuals, takes money out of the economy that could otherwise have been spent differently (WHO, 2009). Poor water and sanitation access impacts negatively on tourism development in the country. In Ghana, the Ministry of Tourism continuously bemoans the insanitary conditions at various tourist sites, especially that of the country's beaches. At the last tourism awards night in December 2017, the Honorable Minister in charge of the sector urged beach communities to improve on sanitation and hygiene practices for enhanced tourism development in Ghana (GNA, 2017). To appreciate the relationship between sanitation, water and development, it is important to first understand the economic cost of poor water, sanitation and hygiene. World Bank (2008) found that poor water and sanitation in Cambodia leads to economic losses of US$448 million per year which translates into per capita loss of approximately US$32. These economic losses are equivalent to 7.2 percent of Cambodia's GDP in 2005. This amount is roughly equivalent to the contribution of the fishery sector to the GDP, or twice the forestry's contribution. While these economic costs are not all tangible, the immediate financial losses amount to about US$160 million per year, which is roughly 2.5 percent of the GDP.

In addition, the cost of premature death is estimated to be the biggest source of health costs which amounts to US$169 million annually contributing to more than 90 percent of the health impacts. This huge cost is due to the high number of deaths among children from diarrhea, and from diseases related to malnutrition which results from diarrhea. Besides premature death, there are two other components of health costs making up total health cost. The cost of healthcare amounts to over US$13 million per year, including the costs of consultation, treatment, medicine, and transportation. The productivity cost, which refers to the productive time loss of patients both adults and children as well as their careers, amount to US$5 million per year (World Bank, 2008). Poor water and sanitation access also negatively impacts on tourism development in the country. Using the study of Cambodia, tourism impacts of poor sanitation are computed on the assumption that visitors to Cambodia are sensitive to sanitation conditions. The study estimated that poor sanitation and hygiene lead to the tourism loss of about US$74 million representing seven percent of the tourism income in 2006.

The potential growth of tourism may not be fully achievable if sanitation and hygiene conditions in the country are not properly managed. This estimation, nevertheless, does not account for the sanitation and hygiene-related sickness, and welfare loss due to the sickness. In other words, this loss is actually a result of under-exploitation of tourism capacity (below optimal capacity) and that the envisaged tourism growth is not achievable. The economic loss caused by the loss of time due to unimproved water and sanitation is roughly US$38 million representing nine percent of the total economic costs. This cost is incurred by those who use unimproved water, practice open defecation (journey time) and shared toilet users (waiting time) (World Bank, 2008). In terms of the impact of poor sanitation on environmental surroundings and related human welfare, the impact was assessed by focusing on the problem of household solid waste. Based on the empirical literature, it is clear that solid waste in urban and rural areas of Cambodia is not yet properly handled. Although in some urban areas where the collection service is available, the scattering of waste in public places such as markets and along the streets is still common. This practice damages the sight of the areas, and also produces bad odours to the people, affecting the welfare of the local people as well as attractiveness for tourists. Moreover, the designated dump site, particularly in Phnom Penh, is already operating at above capacity, thus impacting air pollution and water resources (World Bank, 2008). The World Bank (2008) further evaluated the benefits associated with improved sanitation and hygiene practices. The study linked better hygiene practices and improvements in toilet systems to a reduction in health costs, while improved physical access and treatment/disposal can reduce the other cost components. The study showed that improved hygiene practices such as hand washing can reduce health costs by approximately US$84 million. Improved physical access to sanitary toilets can reduce economic costs associated with time use by about US$38 million, and contribute US$0.3 million to input markets (latrine builders and materials). Improved toilet systems can reduce health costs by US$60 million and contribute to input market of US$1.2 million. Improvement in the treatment or disposal of waste has a large impact on water resources which can reduce costs totaling US$223 million where US$149 million is from mitigated water impacts and US$74 million is from tourism gain. The reviewed literature shows that studies have been conducted in some areas of Ghana, including some communities in the northern part of the country. There is however, no evidence of any study on sanitation in the Tunayilli community specifically. Hence the choice of Tunayilli Community for this study.

Demystifying Water and Sanitation Supply in Ghana

It has been reported by National Development Planning Commission (NDPC) of the Government of Ghana and the United Nations Development Programme (UNDP) that given with the current trend of water and sanitation development, the proportion of the population with access to improved sanitation missed the 2015 MDGs target of 52 percent (NDPC/UNDP, 2010). The report by the UNDP suggested that there must be approximately five times increase in coverage to be able to achieve the set target (UNDP, 2010). Ghana has a low coverage of improved sanitation, and this is partly caused by the fact that the Joint Monitoring Programme (JMP) does not classify improved facilities as if they are shared or not. The JMP does not classify shared toilet facilities as improved because they may not be hygienic, convenient and private enough for users (UNICEF, 2008). Many shared toilet facilities in Ghana provide all the necessary parameters that characterize an improved toilet facility (UNICEF, 2008). Again, there is a wide difference in access to sanitation between and within regions in the country, and between rural and urban areas. More than half the population of Ghana (59%), the highest in the world, depend on shared sanitation facilities including public toilets (Gyapong, 2013]) and about 19% of Ghanaians practice open defecation while 8% depend on various forms of unimproved sanitation facilities (Obeng, 2015). For example, 3 percent of the population in the Northern Region use improved (not shared) sanitation facilities as against 25 percent in the Greater Accra Region in 2008 (Mensah and Antwi, 2013). Development and major programmes in water and sanitation have been concentrated in urban centres and in the southern areas, while the poorest coverage is in the northern regions and rural communities (NDPC/UNDP, 2010). According to the JMP (WHO/UNICEF, 2012), Ghana has the highest proportion of the population using shared water and sanitation facilities in the world with a record of 58 percent. Compared to Bolivia, which has the second highest record in shared sanitation category, the difference is 22 percentage points, which illustrates the exceptionally high usage of shared facilities in Ghana. Globally, the sharing of the water and sanitation facility is an urban phenomenon and also in Ghana it is more common in the urban areas than in rural areas. For instance, 73 percent of the urban populations share sanitation facilities whereas in rural areas the corresponding proportion is 43 percent. Instead of sharing the toilet facility, the problem in rural areas is open defecation, as 33 percent of the rural population practice open defecation (WHO/UNICEF, 2012). According to WSP (2010), the predominant use of shared sanitation facilities in urban areas is principally due to residence patterns of several households living in compound houses. A more worrying development is the heavy reliance by many urban dwellers on public toilets.

Water and Sanitation Conditions in the Sagnarigu District

The 2010 population and housing census of Ghana reported that primary sources of water include surface water such as rivers, dams, and dug-out wells (GSS, 2011). On average, 80% of community members rely on polluted surface water sources for consumptive and domestic purposes. In terms of sanitation, the main methods of waste disposal in the Sagnarigu District is public dump (open space) which constitutes 29.3 percent. For the urban areas, public dump in open spaces is 30.6 percent, burnt by household is 22.5 percent and public dump in a container is 22.4 percent. For the rural areas, the three main methods are public dump in open spaces (27.0 percent), indiscriminate dumping (24.9 percent) and burnt by household (19.5 percent). Some of these practices are not healthy and have implications for public health, particularly the dumping of refuse in open spaces and indiscriminate dumping methods. The GSS (2011) also notes that the proportion of household waste that is collected is lower in the urban areas (9.3 percent) compared to the rural localities (10.2 percent). Again, in the urban localities, indiscriminate disposal of waste is 8.3 percent. The proportion of households that buries waste is also small at 5.9 percent in the urban and 3.9 percent in the rural areas. The 2010 population and housing census of Ghana further reported that there is limited use of pipe-borne water and modern methods of liquid waste disposal in the district in either the urban or rural areas, although it is relatively better in the urban areas (GSS, 2014). The modern methods of liquid waste disposal include the use of the sewerage system, drainage system into a pit/soak away or gutter. Also, 20.4 percent of urban households and 17.2 percent of rural households practices these modern methods. The main methods of liquid waste disposal in the district is by throwing onto the street which is 53.1 percent (50.5 percent for urban and 57.5 percent for rural) and throwing onto the compound (21.5 percent for urban and 20.3 percent for rural). These practices do have serious implications for the public health of the population. Tunayilli, the community under study typifies the statistics above. The term 'sanitation' as used in this chapter refers to 'a toilet facility,' as used in the indicator for measuring MDG progress for sanitation delivery. It is referred to as 'adequate' when one uses an 'improved toilet facility' (at least a ventilated improved pit latrine). Basically, sanitation means being free from germs.

Water and Sanitation Improvement: Role of NGOs

NGOs play various roles in technical, financial, educational and advocacy on the water and sanitation situation in Ghana, especially in the rural areas. Government's support to District Assemblies in the area of water and sanitation is normally not adequate to meet all the basic needs of the people in that area. NGOs have therefore stepped in to supplement the efforts of the central government. The role played by NGOs is evident in the numerous water and sanitation projects implemented around the country. For instance, Coalition of NGOs in Water and Sanitation (CONIWAS) in 2003 was able to advocate for an increase in government funds into water and sanitation in rural areas (Lane, 2005). The support given by NGOs has led to the construction and rehabilitation of safe water points and construction of improved toilet facilities for a number of communities. They have also been able to link it up with hygiene awareness creation. This has helped to maximize the full benefits of the water and sanitation facilities (Chambers & Myers, 2016). The maximization of these benefits eventually leads to an improved access to Water, Sanitation and Hygiene (WASH) services, changes in sanitation and hygiene practices and also improves institutional arrangements and capacity for support and sustainability. These have had positive impacts on the livelihoods of the people, improved female enrolment in school and also promote gender equity. According to Water Aid (2009), the impact of improved sanitation would lead to an improvement in general livelihoods in rural people. The District Assemblies and the Community Water and Sanitation Agency, however, need to give the NGOs full support and also monitor their activities effectively so that there would not be duplication of efforts in any particular area.

THEORETICAL FRAMEWORK

The continued constraints suffered by rural communities whose livelihoods depend on the availability of quality water and access to sanitation amenities (Braimah & Fielmua, 2011), decades after Ghana's transition into democratic dispensation (Ahwoi, 2010), has generated concerns and urgency to provide both theoretical analysis and practical strategies for redress (Watts, Peet & Robbins, 2010). Water and sanitation remain central part of the global economy due to its increasing significance in multiple scales (Igoe, 2013). Thus, an investigation into the role of NGOs in community water and sanitation improvement becomes imperative and an interesting perspective. Taking the place-based political ecological (PE) approach, this study digs a bit deeper into the water and sanitation programmes of NGOs, identifying the effects of such programmes and unpacking the challenges and measures for sustainability of such programmes at the local level (Robbin, 2012). Political ecologists tend to conduct place-based research where they engage with residents in communities to understanding the local meanings attributed or ascribed to environmental conservation and sustainability practices. Frequently wielding either ethnographic, descriptive, cross-sectional or explorative approaches, political ecologists create social bond with local community members that allow them privileged access to the more nuanced understandings of nature and the environment.

Critical to political ecological analyses that foreground the significance of 'place-based perspectives', are understanding of the social, economic and political environmental changes occurring within specific communities and the role of multiple actors/stakeholders in enabling sustainable exploitation of nature (Peet et al., 2010). Thus, a place-based investigation helps understand the unique socio-demographic elements (gender, income, education, occupation, ethnic characteristics etc) in which sanitation and water allocation discourses are often situated. Sociological, economic and demographic categories such as these play into access and control over water and natural resources and are critical to effective environmental planning and policy initiatives (Mostafanezhad et al., 2016; Watts, 2009). Political ecologists are noted to have often focused on "the ways in which environmental concerns are traced through gender roles, knowledge and practices" (Watts and Peet, 2004:13), through issues such as the exclusion and 'silencing' of women's knowledge. They highlight that there are complex and ongoing struggles over the discursive and material conditions of the livelihoods of individuals that are mediated by social categories within communities (Turner, 2014). In this chapter, these insights are critically relevant to water and sanitation improvement, where considerations of gender, ethnicity, income, etc. are at the epi-centre of water and sanitation distribution, yet at times either forgotten or deliberately overlooked. Also, significant here for embracing the place-based perspective is to understand how the voices of community members may still be (un)heard, though non-community participants may have equally valid needs and genuine interest in water affairs within their localities. In applying this theory, this study expands the debate on place-based PE analogy, by arguing that NGOs can contribute towards achieving an eco-friendly and sustainable water and sanitation improvement (Peet, Robbins and Watts, 2010).

Through the PE's place-based approach, this study shifted the narratives on water and sanitation in Ghana from demand-led to supply-driven, revealing that seeing the water and sanitation sector in Ghana only through the lenses of profit-motivation (Duffy, 2011), can cut off potentially productive lines of enquiry centred around subjects such as discourses on water and sanitation distribution, sustainability of ecological resources, welfarism and social justice principles at local scales (Watts, 2015). The positioning of this study in line with the place-based analogy stimulates NGOs sustainability imaginations to lead the way towards the construction of a sustainable future and different rural communities in Ghana. More than being a call for rebirth and rearranging the puzzles of the environment, this study summoned society to rethink about the environment from the perspective of water actors in which NGOs are inextricably inscribed. In this way, NGOs are able to facilitate the dialogues that can steer the construction of rural water sustainability and hygienic communities.

STUDY AREA AND METHODOLOGY

This section discusses the research methods used or applied in conducting this study. It includes research design, profile of the study area, sample and sampling procedure, sources of data, data collection instruments, ethical concerns, pretesting and data analyses.

Research Design

A research design is a framework for conducting research. It is the basic plan for conducting the data collection and analysis. This study used the descriptive research design. A descriptive research method is used when the researcher wants to describe specific behaviour as it occurs in the environment (Malhotra, 2007). This technique was suitable because it had the ability to fit into a mixed (both quantitative and qualitative) research methodologies and could demonstrate relationships and describe the state of the phenomenon investigated by the study. A mixed method research design is a procedure for collecting, analysing, and "combining" both quantitative and qualitative research and methods in a single study to understand a research problem (Creswell, 2013). In this study, the mixed research design was appropriate because it enabled the study to draw from a mix of strategies from within the qualitative and quantitative approaches in order to provide answers to the questions investigated by this study.

Profile of the Study Area

The Sagnarigu District was carved out from the Tamale Metropolis by a Legislative Instrument (LI) 2066, on the 24th of June, 2012 (Ghana Statistical Service, 2014). Sagnarigu is the capital of the Sagnarigu District. One of the reasons for the creation of the district was to redirect developmental projects to the communities North and West of the Tamale Metropolis which were relatively underdeveloped compared to the urban areas in the metropolis. The District is mainly rural, with the majority of the people being smallholder farmers. This population is totally dependent on subsistent agriculture.

The District has a total population of about 160,000. It has 76 communities comprising 20 urban, six (6) peri-urban, and 53 rural communities. The District Chief Executive is the political head. The district is divided into three Area Councils namely Choggu-Sagnarigu, Kalpohini and Kanvili (GSS, 2014). From Figure 1, Tunayilli, the study location, is a settlement located approximately 10 kilometres away from Sagnarigu township. The community has a population of roughly 2,103 people in 2010 (GSS, 2012). The majority of the people are Dagombas. Immigrants from other parts of the country are also found in the community.

Figure 1. Map of the Tunayilli Community
Source: Sanarigu District Assembly (2014)

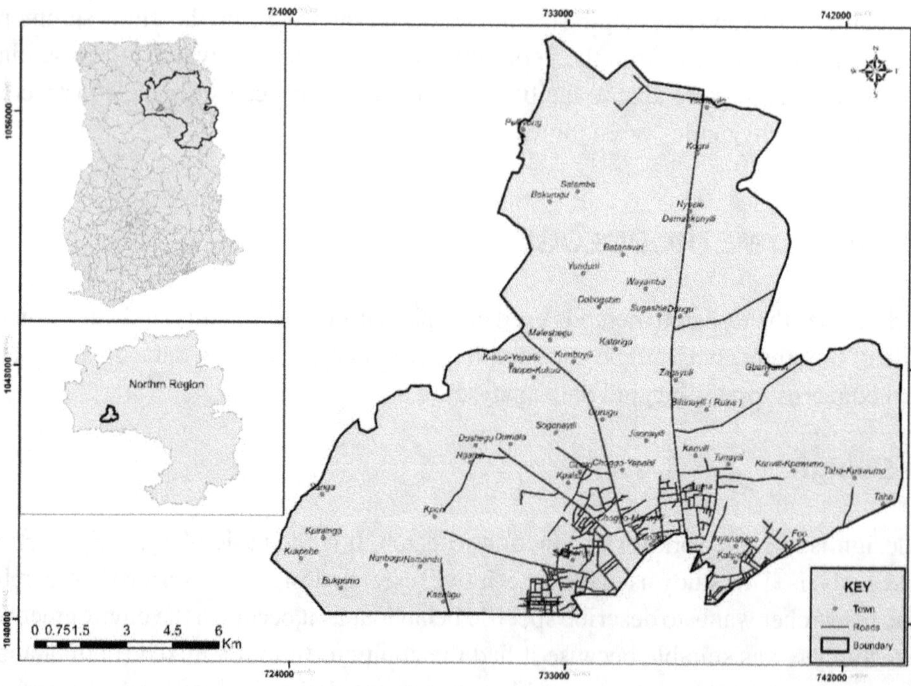

The marriage system is predominantly polygamous, while the dominant religion is Islam, but some still hold tight to traditional religion as they see any other religion to be alien, especially the older generation. Education has not yet been given much attention in this community. Tunayilli is home to a number of NGOs in the Sagnarigu District. Water and sanitation is a challenge that has plagued the community for quite a long time due to the fact that the area still has no rapid concentration of urban civilization to pull in standard public water and waste management practices and standard road network systems (GSS, 2014).

Sampling Procedure

The selection of the sample begun with the determination of the study population. Bryman et al (2003), describes a study population as the whole group that the research focuses on. This study targeted household heads, NGOs interested in promoting water and sanitation in the area and the District Assembly. The total number of houses in the town was 224 (GSS, 2012), out of which 142 were selected. Representatives from four selected NGOs (WATERAID, Catholic Relief Services-CRS, Suntaa Nuntaa and Adventist Relief Agency-ADRA) promoting sanitation in Tunayilli, as well as the Head of the Environmental Department of the Sagnarigu District Assembly representing the Assembly were selected for the study. Sampling is very critical for Social Science research. Researchers usually cannot make direct observations of every individual in the population they are studying, but instead they collect data from a subset of individuals, a sample, and use those observations to make inference to the whole group (Biju, 2011). The sampling was done at different stages of this study. The sample size was determined, survey respondents were sampled, as well as respondents from Institutions like NGOs working to promote sanitation in the area and the District Assembly. Data was collected from 142 sampled household respondents. The sample size was selected based on the technique developed by Krejcie and Morgan based on the formula:

$S=X^2 NP (1-P)/d^2 (N-1) +X^2 P (1-P)$ (Krejcie & Morgan, 1970)
S=required sample size
X= the table value of chi-square for 1 degree of freedom at the desired confidence level (3.841)
N= the population size.
P= the population proportion (assumed to be 0.50 since this would provide the maximum sample size).
D= the degree of accuracy expressed as a proportion (0.50).

The households were represented by their heads in this study. These household heads were selected using the systematic random sampling technique. In this case, households in the study community were listed to construct a sampling frame that was adopted for the systematic sampling technique. Random sampling provides each household an equal opportunity to be chosen among the lot. Systematic random sampling was used because it is appropriate for household surveys. The sample fraction or interval (k) was calculated after the sample size had been determined. The first respondent was determined through a simple random sampling technique that was conducted for the members within the first group of fractions. After this, the kth (sample interval) term was applied to select the rest of the respondents. Purposive sampling was used for the Head of the Environmental Department of the Sagnarigu District Assembly and a representative each from the four NGOs working to improve water and sanitation in the study area (Tunayilli). They were selected based on their in-depth knowledge of the subject matter.

Instrumentation

The study relied on both primary and secondary data sources. Primary data were obtained through the use of interview schedules and interview guides which were developed and used as data collection instruments. Primary data collected assisted in the objective presentation of the issues that were studied. Secondary data was also obtained from published literature and reports from Ghana Statistical Service, the Assembly and the NGOs. An interview schedule was developed and used to gather data from household respondents (household heads). Field assistants (5) were recruited and trained adequately to use the interview schedule to gather data from household heads. The questions were framed in English and the field assistants had to translate them into the local language (mainly Dagbani) to collect data through interviews. This was considered best because, it is not all the respondents who could read and write in English and at the same time helped in getting all the questions on the instruments responded to by the right people. It also ensured a very high retrieval rate of hundred percent.

Key informant interviews were conducted to gather data from the four selected NGOs and the Environmental Department of the Sagnarigu District Assembly. These interview guides were used to collect data from the key informants. The researchers then arranged for a face-face interview with the scheduled institutions. This was considered best because these groups of respondents have an appreciable understanding of the phenomenon based on their working experience. Qualitative data were collected, where the interviewer had the opportunity for probing for clarity, examples and also observed their gestures for additional details. The data were collected to complement that of the household respondents. The main challenges that these institutions faced in their intervention process were brought to the fore and some suggestions towards solving them were presented. The study addressed ethical concerns that pertained to the study.

Ethical Considerations

The consent of respondents was sought before being interviewed. Respondents were informed of the objectives of the research. This enabled them to understand what the study is about before consenting to partake in the study. The data gathered did not include their names. Importantly, they were also assured of confidentiality of responses obtained from them. Pre-testing is a very important assignment in survey research. It is an absolutely necessary process to ensure that all kinds of errors that are associated with survey instruments are reduced (Collins, 2003). It helped to improve the quality of data collected significantly. The pre-testing was done on a small sample of respondents. After the pre-test, the Field Assistants were asked series of questions regarding the survey as well as the process of data collection during the debriefing session. The debriefing sessions helped to detect potential problems with the instruments including ambiguity of words, misinterpretation of questions, inability to answer a question, sensitive questions, and many other problems associated with the instruments as well as the process of administering the survey. The pre-test provided the researchers with the opportunity to address these challenges in order to improve the reliability of data collected.

Data Analysis

Data collected from the field were analyzed quantitatively to achieve the set objectives of the study. As noted by Creswell (2013), when analyzing survey instruments, it is particularly important to refer to the original aims and questions of the study in order to guide the analysis. The field data collected were cleaned, edited, coded, keyed in into SPSS (v.17.0) and Microsoft Excel to generate results. Data were analyzed descriptively and both databases were created. Every response item on the instruments were carefully entered as a numbered code under the question's header in the SPSS and Microsoft Excel, as well as referring to the aims of the study. The findings from the survey were presented in the form of charts and tables. The graphs presented contain frequencies and percentages which are important in illustrating the findings from SPSS. The results were also presented descriptively into different sections. The results of qualitative data collected using key informant interviews were presented in narratives to describe the existing phenomenon that was studied. In some cases, it either strengthened the evidence or provided further explanations to the quantitative data.

RESULTS

The results of the study are presented under five major themes namely, the demographic characteristics of respondents: water and sanitation improvement programmes implemented by NGOs in the district; its effects on beneficiaries; its associated challenges; and possible measures to address the challenges. The demographic variables examined included the sex, age, level of education and number of years spent as a resident in the Tunayilli community. Of the 142 household heads that were sampled a response rate of 84.5 percent (120 respondents) was attained. Representatives of four NGOs, namely: The Catholic Relief Services (CRS); WATERAID; Suntaa Nuntaa; and the Adventist Relief Agency (ADRA) that were providing water and sanitation services in the District and the District Assembly were also interviewed for the study. The results from Table 1 show a marginal sex dominance (50.8%) in favour of female representation of the household respondents. The female domination in the survey could probably be the case that more adult females were found in the houses during the time of the survey.

Table 1. Sex distribution of respondents

Sex	Frequency	Percent
Male	59	49.2
Female	61	50.8
Total	120	100.0

Source: Fieldwork (2018)

The ages of respondents were considered important in this study. This is because age is an impact determinant of experience. Therefore, results from an experienced population have a lot of credibility. The age distribution of the survey respondents is presented in Table 2.

The ages of the respondents were categorized into six groups. The results from Table 2 show that the survey respondents who were between 18 and 20 years were in the minority (2.5%) group. More than a two-thirds (67.5%) of the survey respondents were found to be above forty years of age. The level of education can impact positively on responses. Hence, the educational level of respondents was assessed, and the results presented in Table 3.

Table 2. Age distribution of respondents

Age (Ages)	Frequency	Percent
Less than 20	3	2.5
21-30	16	13.3
31-40	20	16.7
41-50	28	23.3
51-60	32	26.7
Above 60	21	17.5
Total	120	100.0

Source: Fieldwork (2018).

Table 3. Educational level of respondents

Level of Education	Frequency	Percent
No formal education	11	9.2
Basic	28	23.3
Secondary	48	40.0
Tertiary	33	27.5
Total	120	100.0

Source: Fieldwork (2018).

Results from Table 3 show that less than a tenth of the respondents had no formal education, a little above one-fifth had only basic education, forty percent had secondary education and nearly three-tenth had tertiary education. This implies that the majority (90.8%) of the respondents have had some level of formal education. Occupational information in social surveys can be a key resource for understanding contemporary social life, and the occupation-based indicators are central to sociological investigations. The types of occupation of the respondents was surveyed and results are presented in Table 4. The results revealed that the majority (50.8%) of the respondents were engaged in trading, while a fifth (20.8%) of them were engaged in farming activities. However, a tenth of them reported to be unemployed.

Table 4. Type of occupation of respondents

Occupation	Frequency	Percent
Farmer	25	20.8
Trader	61	50.8
Formal sector employee	22	18.4
Unemployed	12	10.0
Total	120	100.0

Source: Fieldwork (2018).

The length of time a respondent has stayed in the community was considered vital because it can help in determining the reliability of responses. Results on the length of time a respondent has stayed in the community is presented in Figure 2. The results revealed that a few (7.0%) of the respondents have resided in the area for less than 4 years, while the majority (63.0%) have been there for more than ten years. About 21 percent of the survey respondents reported that they have lived in the community for a period ranging from 7 to 10 years. The results suggest that 84 percent of the respondents had spent a minimum of 7 years as residents of the Tunayilli community showing that they have deep knowledge and enough experience of water and sanitation conditions prevailing in the community.

Figure 2. Length of time respondents have stayed in the community
Source: Fieldwork (2018).

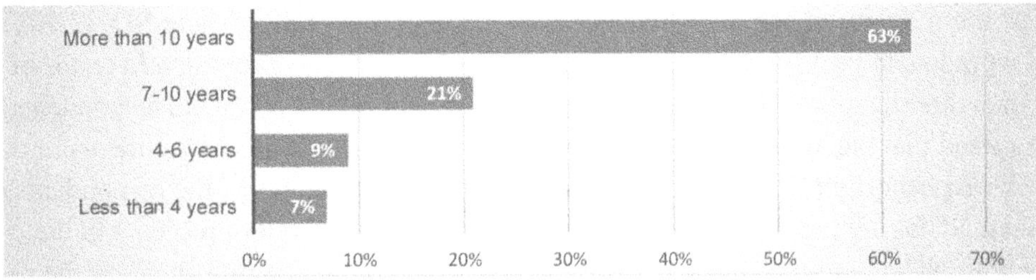

Water and Sanitation Conditions in the Community

The study sought to establish the prevailing water and sanitation condition in the community. A set of variables were used to establish the condition by assessing the knowledge of residents on the water and sanitation situation in the community. The knowledge assessment was done using the 5- point Likert scale to obtain and analyze the percentages of each of the variable responses of (Strongly Agree) SA, (Agree) A, (Not Sure) NS, (Disagree) D and (Strongly Disagree) SD and the results are presented in Table 5.

Table 5. Water and sanitation situation in Tunayilli community

Statement	SA (%)	A (%)	NS (%)	D (%)	SD (%)	Total (%)
There are proper water management systems in place	5.0	22.5	20.0	27.5	25.0	100.0
There are proper waste management systems in place	-	-	-	75.0	25.0	100.0
There is a gradual improvement in potable water provision in the community	27.5	37.5	15.0	17.5	2.5	100.0
There is a gradual improvement in sanitation in the community	-	73.3	10.0	11.7	5.0	100.0
There is generally a low level of awareness among residents on potable water provision	-	8.3	11.7	17.5	62.5	100.0
There is generally a low level of awareness among residents on proper sanitation practices	20.0	63.3	8.3	7.5	0.9	100.0
NGOs are the only organizations visibly addressing water problems in the community	-	5.0	12.5	22.5	60.0	100.0
NGOs are the only organizations visibly addressing sanitation problems in the community	33.3	25.8	12.5	25	3.3	100.0
Water challenges require greater effort from stakeholders	18.5	31.5	11.7	25.8	12.5	100.0
Sanitation challenges require greater effort from stakeholders	56.7	43.3	-	-	-	100.0

Source: Fieldwork (2018).

In Table 5, the findings showed that a little above half (52.5%, composed of SD-25% and D-27.5%) of the respondents confirmed that they have a problem with the existing water management systems in place. Nevertheless, the majority of the respondents (65%), constituting D-37.5% and SD-27.5%, confirmed that there is a gradual improvement in potable water provision in the community. In terms of awareness level, more than three quarters (80%), constituting D-17.5% and SD-62.5% of the respondents disagreed/strongly disagreed that the awareness level of potable water provision among residents in the community is low. A vast majority (82.5%, composed of SD-60% and D-22.5%) of the respondents confirmed to the fact that, NGOs are the only organisations visibly addressing water problems in the community. Interactions with NGOs and the District Assembly confirmed these positions held by the respondents, and further indicated that potable water provision in the community is mainly a government affair and the NGOs only complemented by providing boreholes, capacity training and hygiene sensitisation. The results indicate that there is poor waste management system in the community. However, the majority (83.3%, constituting SA-73.3% and A-10%) of the respondents confirmed that the sanitation situation in the community is gradually improving. The study further established that the level of awareness of proper sanitation practices among residents was low for the majority (83.3%, SA-20% and A-63.3%) of the respondents. In terms of the interventions to address the challenges, most of the respondents (59.1%, constituting SA-33.3% and A-25.8%) reported that NGOs were the only organisations visibly addressing sanitation problems in the community. The position of the household respondents on the sanitation situation in the community was confirmed by the representative of the Sagnarigu District Assembly who indicated that the sanitation situation in the community was poor and would require sustained efforts from all stakeholders to bring the situation to an acceptable level. Similarly, the NGOs working to improve the sanitation situation in the community unanimously described the prevailing sanitation situation in the community as a significant improvement compared to the situation over the past five years. All the respondents reported that sanitation challenges in the community require greater efforts from stakeholders to address it. The Assembly and NGOs contacted revealed that the NGOs have supported the community heavily in the areas of sanitation and hygiene improvement through capacity building, sensitisation of hygiene practices and equipment supplies. The findings confirmed reports by the Ghana Statistical Service (GSS, 2010) that the sanitation situation in the community is unacceptably high and requires immediate action.

Water and Sanitation Services Provided by NGOs

The role of NGOs in the improvement of water and sanitation conditions at the community level cannot be overemphasized. The District Assembly reported that water and sanitation services have been provided by NGOs such as CRS, WATERAID, Suntaa Nuntaa and the (ADRA) to enhance access to safe water and improved sanitation in the Tunayilli community. WATERAID provided the people with improved skills and rehabilitation of identified water supply schemes. The District Assembly revealed that these NGOs have provided the schools in the community with containers for hand washing in school; carried out education on hand washing especially after visiting the toilet; educated the people on the benefits of cleaning their surroundings regularly. The NGOs reported that they have supported many households within the community to construct improved toilet facilities such as the WCs, Enviro Loo, Bio-digester (bio fill), Bio-gas etc. This was confirmed by the District Assembly. These interventions have contributed to a reduction in the practice of open defecation in the community. The NGOs further reported that, they have helped the community with equipment like wheel burrows, cutlasses and shovels to help organise clean up campaigns. They also assisted community members to construct dumping sites to ensure improvement in sanitation conditions. All these support given by NGOs including hygiene awareness has helped to maximize the full benefits of the water and sanitation improvement practices in the community. All the household respondents who were contacted for the study indicated that the provision of good water and sanitation facilities have increased their productivity, has improved their business activities and above all led to an improvement in their living standards. To them, this was as a result of good health they have enjoyed from accessing good water and sanitation. All the survey respondents revealed that the NGOs have constructed water and sanitation facilities for them, which has resulted in an improved access to water and sanitation points. Therefore, less time is spent in water collection and this has eventually led to an improvement in their living standards. It has also resulted to an improvement in the water and sanitation coverage in the area. These were confirmed by the NGOs concerned and the District Assembly, that indeed their interventions have contributed to the improvements in the water and sanitation coverage but also indicated that the coverage will be further improved if more funds are secured.

Significance of NGOs Involvement in Water and Sanitation Improvements

Overall, there has been some level of improvement in the availability of water and sanitation facilities as well as sanitation and hygiene practices in the community. Hence, the study assessed the significance of NGOs involvement in the water and sanitation interventions in the community, using a Likert scale analysis. The findings are presented in Table 6.

Table 6. Significance of NGO's involvement in water and sanitation improvements

Statement	SA (%)	A (%)	NS (%)	D (%)	SD (%)	Total (%)
NGO-led water and sanitation interventions in the community are significant	25	57.5	9.2	6.7	1.6	100.0
There is good collaboration between residents and the NGOs	41.7	34.1	16.7	4.1	3.4	100.0
There is good collaboration between government agencies and the NGOs and their activities	28.4	54.1	9.1	7.5	0.9	100.0
NGOs have committed adequate resources to sanitation improvements	45.8	33.4	4.1	13.3	3.4	100.0

Source: Fieldwork (2018).

Respondents' knowledge of the significance of NGOs' involvement in the water and sanitation improvement intervention in the community was assessed to establish the influence of the NGOs activities in the community. From Table 6, a little above half (57.5%) of the respondents agreed that NGO-led water and sanitation interventions in the community were significant. The results show that a little above half of the respondents (54%) agreed while 28% strongly agreed that there exists a good collaboration between them and the NGOs. Similarly, 82.5 percent (ie A -54.1% and SA-28.4%) of the respondents indicated that there is a good collaboration between the NGOs and government agencies. About 79.2% (Ie.,

A-45.8% and SA-33.4%) of the survey respondents confirmed that the NGOs have committed adequate resources to water and sanitation improvement activities in the community. Therefore, residents were largely aware of NGOs involvement in water and sanitation improvement activities in the community. These views were also corroborated by the NGOs and the District Assembly.

Challenges Faced by the NGOs

The respondents were assessed on their knowledge on the challenges faced by the NGOs while carrying out their water and sanitation improvement activities in the community and the results are captured in Table 7. The majority (97.5%, composed of A- 43.3% and SA-54.2%) of the respondents agreed to the fact that inadequate funding was a major challenge faced by the NGOs.

Table 7. Challenges faced by NGOs in water and sanitation improvement activities

Statement	SA (%)	A (%)	NS (%)	D (%)	Total (%)
Inadequate funds	54.2	43.3	2.5	-	100.0
Logistical challenge	57.5	35.8	5.8	0.9	100.0
Weak community mobilization	40.0	42.5	17.5	-	100.0
Low level of education and awareness	48.3	42.5	9.2	-	100.0

Source: Fieldwork (2018).

Other major challenges that confronted the activities of the NGOs that respondents agreed and strongly agreed to include logistical challenges (93.3%, composed of A-57.2% and SA-35.8%); weak community mobilisation (82.5%, composed of A-42.5% and SA-40%); and low education level and awareness (90.8%, composed of A-42.5% and SA-48.3%). These challenges were confirmed by the District Assembly and the NGOs. They both identified inadequate funding as a serious hindrance in addressing the water and sanitation challenges confronting the community. It was observed that most of the challenges would require the intervention of the Central Government due to the magnitude of the cost involved.

Measures to Address Water and Sanitation Improvement Activities

The above challenges identified tend to impact negatively on the water and sanitation improvement practices. Therefore, the study sought to identify measures to address these challenges and the results are presented in Table 8. In Table 8, the study found that among the measures to address water and sanitation challenges in the community includes increasing central government funding allocation (A-58.3% and SA-41.7%), increase community education and sensitisation (SA-65.8%% and A-32.2%), increased community involvement (SA-48.3% and A-42.5%), enforcement of bye-laws (SA-59.2% and A-40.8%) and community support towards maintenance of water and sanitation project of NGOs (SA-50.8% and A-43.3%).

Table 8. Suggested measures to address water and sanitation challenges

Statement	SA (%)	A (%)	NS (%)	Total (%)
Increase central government funding for water and sanitation improvement activities	58.3	41.7	-	100.0
Increase community education and sensitization on proper water and sanitation management	65.8	34.2	-	100.0
Increase community involvement in water and sanitation improvement activities	48.3	42.5	9.2	100.0
Enforcement of water and sanitation bye-laws	59.2	40.8	-	100.0
Community assisted regular maintenance of water and sanitation systems	50.8	43.3	5.9	100.0

Key: SA-strongly agree, A-agree, NS-not sure, D-disagree, SD- strongly disagree
Source: Fieldwork (2018).

The results revealed that all the survey respondents agreed that there is a need to increase central government funding for water and sanitation improvement activities; there is also the need to increase community education and sensitization on proper water and sanitation management; and to enforce water and sanitation bye-laws. These views were in line with the views of the NGOs and Assemblies who strongly advocated for increase funding from central government, increase community participation through capacity building training and awareness and strict enforcement of the bye-laws.

DISCUSSION OF THE FINDINGS

The discussion covers water and sanitation improvement programmes implemented by NGOs in the district, its effects on beneficiaries, its associated challenges, and possible measures to address these challenges. The demographic characteristics of the respondents are critical since they give a fair idea of the calibre of people who responded to the instruments. The survey engaged more females than males which could be as a result of the patriarchal system which persists in the northern part of Ghana. The findings on gender of the respondents revealed more females were involved in WASH than males. Most of the participants were engaged in petty trading and farming. Thus, understanding the social, economic, and environmental context of the people would be relevant for basic water and improved sanitation access. The literacy rate remained low with most of the people engaged in petty trading and farming. This means that understanding the social, economic, and environmental context of the people would be relevant for basic water and improved sanitation access. The findings have affirmed that a place-based investigation helps understand the unique socio-demographic elements (gender, education, occupation etc.) in which sanitation and water allocation discourses are often situated. The findings confirmed the PE's place-based narrative which highlights that understanding prevailing socio-demographic conditions are necessary to change attitudes toward the environment and finally, improve water and sanitation conditions in rural settings (Peet, 2010; Watts, 2015).

Water and Sanitation Improvement Programmes by NGOs

There exists a strong link between poverty and access to safe drinking water and improved sanitation situation (GSS, 2014). However, central governments have not been able to meet the needs of the citizens due to financial constraints and other competing demands. This has inspired NGOs to assist communities in acquiring such facilities. Water supply and sanitation improvement programmes by NGOs in the community have increased access to safe piped water services and improve sanitation. The activities of these NGOs have to enhance the living conditions and social inclusion of beneficiaries by increasing the quality of services, reducing the prevalence of malnutrition and expanding social inclusion and equitable access. This finding is consistent with Dangah (2012) that NGOs' activities in water and sanitation have gone a long way to improve the lives of the people. Largely, all the NGOs in this sector seek to improve water supply and sanitation infrastructure. In the case of water, WATERAID supported in the expansion of piped water services. They also provided new and the rehabilitation/expansion of safe water supply facilities in close coordination with the District Assembly. The CRS, Suntaa Nuntaa and ADRA mainly provided sanitation improvement facilities. A key element of their intervention is to conduct hygiene, education and awareness programmes in the beneficiary areas to educate community members including school children on good hygiene practices. Institutional and capacity strengthening also provided. They provided support to strengthen other institutions that are required to deliver safe piped water access and sanitation to the people.

Effects of NGOs Involvement in Water and Sanitation Improvement

The effects of NGOs involvement in improving water and sanitation situation in the community was well appreciated by the people. Early findings of the NGOs involvement in water and sanitation has been emphasized on several body of available literature which agrees with this study. For instance, NGOs have motivated and increased their efforts in addressing water and sanitation issues at the community level (Lane, 2005*)*. Similarly, the NGOs have pioneered the support on the construction and rehabilitation of safe water points and construction of improved toilet facilities; and have subsequently been able to link it up with hygiene awareness creation (Chambers & Myers, 2016) to ultimately maximize the full benefits of the water and sanitation facilities in the community. Interaction with officials of NGOs and the Assembly revealed that there has been effective collaboration among the NGOs delivering on water and sanitation facilities in the District.

Challenges of Water and Sanitation Programmes Implemented by NGOs

The findings revealed the following challenges; Poor hygienic practices, Low level of solid waste management, Low access to good and acceptable toilets, Inadequate budgetary allocation for investment in sanitation infrastructure, Inadequate arrangements for cleaning and maintenance, Inadequate logistics, Encroachment on sanitary sites, Lack of appropriate final disposal sites, Inadequate and unreliable data for effective planning and implementation of sanitation programmes and projects, Lack of a central workshop for repair and maintenance of sanitation vehicles/equipment/tools, Lack of clear policies for managing public toilets, Lack of district specific sanitation bye-laws, Low level of home toilet ownership, No user fee charges for waste dumping, Poor drainage system leading to inter-house erosion, flooding and breeding of mosquitoes, Political interference in the enforcement of sanitation laws and management of public toilets, Inappropriate strategies/policies of donors in planning and implementation of sanitation facilities, Low rate of waste separation and re-use and Cultural beliefs/customs/superstition which encourage open defecation. This finding is in line with the finding of the 2015/16 assessment of the 26 Districts in the Northern Region, in a joint collaboration between the Centre for Democratic Development (CDD) and UNICEF scored the Sagnarigu District zero percent in potable water and sanitation delivery (CDD & UNICEF, 2015). The findings are further supported by a research by NDPC/UNDP (2010) which posit that; development and major programmes in sanitation have been concentrated in urban centres and in the southern areas, while the poorest coverage is in the northern regions and rural communities.

Measures to Address Water and Sanitation Improvement Challenges

The responses from the survey indicates that all respondents (agreed that, to address the challenges with sanitation improvement, the following measures are important: There should be increased central government funding to improve sanitation, increase in community education and sensitization on proper sanitation management and enforcement of sanitation bye-laws to deter offenders. Over 90 percent of respondents agreed that there should be an increase in community involvement in sanitation improvement and that the community should assist in regular maintenance of sanitation systems. The study has shown that as much as there are some improvements in sanitation through the interventions of NGOs in Tunayilli, there is still much to be done. The MDG report by NDPC/UNDP (2010) illustrates that with the current trend of sanitation development, the proportion of the population with access to improved sanitation missed the 2015 target of 52 percent.

RECOMMENDATIONS

The residents of Tunayilli are largely aware of the involvement of NGOs in the water and sanitation improvement programmes carried out in the community. The study recommended among other things that:

- There is the need for greater effort from all stakeholders, especially government agencies to address the water and sanitation problems in the community.
- The international bodies interested in issues regarding water and sanitation should provide more financial support to enable the NGOs to do more.
- The District Assembly should assist in strengthening the weak community mobilization situation existing in the area to enable all community members to come together to address the situation.
- All stakeholders must support the activities of the NGOs to intensify and sustain educational campaigns to drum home the need to keep proper sanitary conditions in the District.
- The Assembly and all the other government agencies in the zonal areas in the district should dedicate more resources to addressing sanitation challenges in the community and elsewhere.
- Finally, the District Assembly should endeavour to enforce all water and sanitation bye-laws and offenders duly prosecuted where necessary.

AREAS FOR FURTHER RESEARCH

There is the need for further research on funding water and sanitation service delivery in the community. In addition, research should focus on public participation in water and sanitation delivery.

CONCLUSION

The water and sanitation condition in Tunayilli is generally poor but with the intervention of the NGOs, there has been a marked improvement. There is now an improvement in the waste management systems of the community, and an increase in the level of awareness among residents on proper sanitation practices. The NGOs have collectively invested significantly in water and sanitation facilities in the Tunayilli community and have managed to keep most of the facilities in fairly good condition, thereby ensuring their sustained usability. To sustain these gains, the sensitization on proper sanitation and personal hygiene carried out by the NGOs in the community should be intensified to increase knowledge on water and sanitation management practices. Drawing from the political ecology theory, the measures used by NGOs to address the water and sanitation systems of the community should be home-grown and related to place-based conditions.

ACKNOWLEDGMENT

We acknowledge our colleagues and friends for their unmeasurable support and continued guidance throughout this research process. To our families, we salute you for keeping the fight while we engage in this research.

REFERENCES

African Union. (2008). *The eThekwini Declaration and AfricaSan Action Plan*. Accessed from; https://www.wsp.org/sites/wsp/files/publications/eThekwiniAfricaSan.pdf

African Union Assembly. (2008). *Sharm El-Sheikh Water and Sanitation Commitments*. Accessed from: http://ielrc.org/Content/e0825.pdf

Ahwoi, K. (2010). *Local government and decentralisation in Ghana*. Unimax Macmilan.

Akphakli, D. E., Manyeh, A. K., Akpakli, J., Kukula, V., & Gyapong, M. (2018). Determinants of access to improved sanitation facilities in rural districts of southern Ghana: Evidence from Dodowa health and demographic surveillance site. *BMC Research Notes*, *2018*(11), 473. doi:10.118613104-018-3572-6 PMID:30005694

Asian Development Bank. (2011). *Sanitation and Water Management in Asia and the Pacific: Asia-Pacific*. Asian Development Bank.

Barnard, S., Routray, P., Majorin, F., Peletz, R., Boisson, S., Sinha, A., & Clasen, T. (2013). Impact of Indian Total Sanitation Campaign on latrine coverage and use: A cross-sectional study in Orissa three years following programme implementation. *PLoS One*, *8*(8), e71438. doi:10.1371/journal.pone.0071438 PMID:23990955

Bartram, J., & Cairncross, S. (2010). Hygiene, sanitation, and water: Forgotten foundations of health. *PLoS Medicine*, *7*(11), e1000367. doi:10.1371/journal.pmed.1000367 PMID:21085694

Bastien, S., Hetherington, E., Eggers, M., Wamoyi, J., Hatfield, J., Manyama, M., & Kutz, S. (2017). Participatory science and innovation for improved sanitation and hygiene: Process and outcome evaluation of project SHINE, a school-based intervention in Rural Tanzania. *BMC Public Health*, *17*(1), 172. doi:10.118612889-017-4100-7 PMID:28173789

Biju. (2011). Sampling in Social Science Research: Principles, Methods, And Practices used in Social and Health Science Researches. *Nigerian Postgraduate Medical Journal, 22*(4), 195-250.

Braimah, I., & Fielmua, N. (2011). Community ownership and management of water and sanitation facilities: Issues and prospects in the Nadowli District of the Upper West Region of Ghana. *Journal of Sustainable Development in Africa*, *12*(2), 1520–1550.

Bryman, A. (2003). *Quantity and quality in social research*. Routledge. doi:10.4324/9780203410028

Centre for Democratic Development & UNICEF. (2015). District League Table on Sanitation. CDD and UNICEF Annual Report on Sanitation: Accra

Chambers, R., & Myers, J. (2016). *Sustainable sanitation for all: experiences, challenges and innovations*. Practical Action.

Collins, D. (2003). Pretesting survey instruments: An overview of cognitive methods. *Quality of Life Research: An International Journal of Quality of Life Aspects of Treatment, Care and Rehabilitation*, *12*(3), 229–238. doi:10.1023/A:1023254226592 PMID:12769135

Creswell, J. W. (2013). *Research design: Qualitative, quantitative, and mixed methods approaches* (3rd ed.). Sage Publications.

Dangah, M. M., (2012). *The role of NGOs in the provision of water and sanitation in Northern Ghana: the case of Pronet in the Nadowli District*. Academic Press.

Duffy, J. E. (2003). Biodiversity loss, trophic skew and ecosystem functioning. *Ecology Letters*, *6*(8), 680–687. doi:10.1046/j.1461-0248.2003.00494.x

Esrey, S. A., Potash, J. B., Roberts, L., & Shiff, C. (1991). Effects of improved water supply and sanitation on ascariasis, diarrhea, dracunculiasis, hookworm infection, schistosomiasis, and trachoma. *Bulletin of the World Health Organization*, *69*(5), 600–609. PMID:1835675

Fewtrell, L., Kaufmann, R. B., Kay, D., Enanoria, W., Haller, L., & Colford, J. M. Jr. (2005). Water, sanitation, and hygiene interventions to reduce diarrhea in less developed countries: A systematic review and meta-analysis. *The Lancet. Infectious Diseases*, *5*(1), 42–52. doi:10.1016/S1473-3099(04)01253-8 PMID:15620560

Galan, D. I., Kim, S. S., & Graham, J. P. (2013). Exploring changes in open defecation prevalence in sub-Saharan Africa based on national level indices. *BMC Public Health, 13*(1), 500–527. doi:10.1186/1471-2458-13-527 PMID:23721324

Ghana News Agency. (2017). *Challenges with eradicating illegal mining in Ghana: A perspective from the grassroots.* www.ghanadistricts.com

Ghana News Agency. (2018). *Nadowli District rural water coverage higher than MDGs target.* http://www.ghanabusinessnews.com

Ghana Statistical Service. (2012). *2010 Housing and population census report.* Ghana Statistical Service.

Ghana Statistical Service. (2014). *2010 Housing and population census report. District analytical report.* Ghana Statistical Service.

Government of Ghana. (1994). *Structure of Local Government, 1994.* Government of Ghana.

Gyapong, M., Sarpong, D., Awini, E., Manyeh, K. A., Tei, D. O. G., & Agyepong, I. A. (2013). Health and demographic surveillance system profile: The Dodowa HDSS. *International Journal of Epidemiology, 2013*(42), 1686–1696. doi:10.1093/ije/dyt197 PMID:24415607

Igoe, J. (2013). Consume, connect, conserve: Consumer spectacle and the technical mediation of neoliberal conservation's aesthetic of redemption and repair. *Cultural Geographies, 6*(1), 16–28.

Kulabako, R. N., Nalubega, M., Wozei, E., & Thunvik, R. (2010). Environmental health practices, constraints and possible interventions in peri-urban settlements in developing countries–a review of Kampala, Uganda. *International Journal of Environmental Health Research, 20*(4), 231–257. doi:10.1080/09603120903545745 PMID:20645199

Lane, J. (2005). *Non-Governmental Organizations: New Players in Water Policy.* http://www.wateraid.org/

Lane, J. (2012). Barriers and opportunities for sanitation and water for all, as envisaged by the New Delhi statement. *IDS Bulletin, 43*(2), 13–20. doi:10.1111/j.1759-5436.2012.00302.x

Leedy, P. D., & Ormrod, J. E. (2005). *Practical Research.* Pearson Custom.

Leff, E. (2017). Power-knowledge relations in the field of political ecology. *Ambiente & Sociedade, 20*(3), 1809–4422. doi:10.1590/1809-4422asocex0004v2032017

Malhotra, N. K., & Birks, D. F. (2007). *Marketing research: An applied approach.* Pearson Education. doi:10.1108/S1548-6435(2007)3

Mara, D., Lane, J., Scott, B., & Trouba, D. (2010). Sanitation and health. *PLoS Medicine, 7*(11), e1000363. doi:10.1371/journal.pmed.1000363 PMID:21125018

Mensah, J. V., & Antwi, K. B. (2013). *Bridging water and sanitation infrastructure gap in Ghana. Pennsylvania Clarion.* University of Pennsylvania.

Mugambe, R. K., Tumwesigye, N. M., & Larkan, F. (2013). Barriers to accessing water, sanitation and hygiene among people living with HIV/AIDS in Gomba and Mpigi districts in Uganda: A qualitative study. *Journal of Public Health, 21*(1), 29–37. doi:10.100710389-012-0515-x PMID:23703127

Nathan, S., Rotem, A., & Ritchie, J. (2002). Closing the gap: Building the capacity of non-government organizations as advocates for health equity. *Health Promotion International*, *17*(3), 69–78. doi:10.1093/heapro/17.1.69 PMID:11847140

National Development Planning Commission and United Nations Development Programme. (2010). *Ghana millennium development goals report*. Government of Ghana.

Obeng, P. A., Keraita, B., Oduro-Kwarteng, S., Abaidoo, R. C., Awuah, E., & Flemming, K. (2015). *Usage and barriers to use of latrines in a ghanaian peri-urban community*. Springer International Publishing. doi:10.100740710-015-0060-z

Okot-Okumu, J., & Oosterveer, P. (2010). Providing sanitation for the urban poor in Uganda. In *Social perspectives on the sanitation challenge*. Springer. doi:10.1007/978-90-481-3721-3_4

Peet, R., Robbins, P., & Watts, M. (2010), Global political ecology. London: Routledge.

Prüss, A., Kay, D., Fewtrell, L., & Bartram, J. (2002). Estimating the burden of disease from water, sanitation, and hygiene at a global level. *Environmental Health Perspectives*, *110*(5), 537–542. PMID:12003760

Robbins, P. (2012). *Political ecology: A critical introduction* (2nd ed.). John Wiley and Sons.

Roche, R., Bain, R., & Cumming, O. (2017). A long way to go–Estimates of combined water, sanitation and hygiene coverage for 25 sub-Saharan African countries. *PLoS One*, *12*(2), e0171783. doi:10.1371/journal.pone.0171783 PMID:28182796

Schreier, M. (2012). *Qualitative content analysis in practice*. Sage Publications.

Stanwell-Smith, R., Bloomfield, S. F., & Rook, G. A. W. (2012). *The hygiene hypothesis and its implications for home hygiene, lifestyle and public health*. International Scientific Forum on Home Hygiene. http://www. ifh-homehygiene. org/reviews-hygiene hypothesis

Tumwebaze, I. K., Orach, C. G., Niwagaba, C., Luthi, C., & Mosler, H. J. (2013). Sanitation facilities in Kampala slums, Uganda: Users' satisfaction and determinant factors. *International Journal of Environmental Health Research*, *23*(3), 191–204. doi:10.1080/09603123.2012.713095 PMID:22873693

Turner, M. D. (2014). Political ecology I. An alliance with resilience? *Progress in Human Geography*, *38*(4), 616–623. doi:10.1177/0309132513502770

UCLA Center for Communication Policy. (2005). *The UCLA Internet Report on sanitation Survey in Digital Future*. UCLA Center for Communication Policy.

Unger, C. C. (2013). Treating diarrheal disease in children under five: The global picture. Archives of disease in childhood. *Archives of Disease in Childhood*, *99*(3), 273–278. doi:10.1136/archdischild-2013-304765 PMID:24197873

United Nations. (2007). *Global monitoring of water and sanitation: History, Methods and future challenges*. United Nations.

United Nations. (2014). *United Nations water: Global analysis and assessment of Sanitation and Drinking Water (GLAAS) 2014 Report*. UNDP.

United Nations Development Programme and United Nations-Water. (2014). Implementing water sanitation and hygiene (WASH) to improve water quality and supply, sanitation and hygiene. *Cochrane Database of Systematic Reviews*, *8*(8), CD009382.

United Nations International Children's Emergency Fund. (2008b). *Use of improved sanitation facilities in Ghana*. UNICEF.

United Nations International Children's Emergency Fund. (2012). *Eliminating open defecation*. UNICEF.

United Nations International Children's Emergency Fund. (2014). *Water Sanitation and hygiene (WASH)*. UNICEF.

United Nations International Children's Emergency Fund and World Health Organization. (2010). *Progress on Drinking Water and Sanitation*. UNICEF.

United Nations-Water. (2015). *Water, sanitation and hygiene*. United Nations.

Van Minh, H., & Hung, N. V. (2011). Economic aspects of sanitation in developing countries. *Environmental Health Insights, 5*, EHI-S8199.

WaterAid. (2009). Menstrual hygiene in South Asia: A neglected issue for WASH (water, sanitation and hygiene) programmes. *Gender and Development*, *18*(1), 99–113.

WaterAid. (2011). A condensed version of the WaterAid research report Looking back: Participatory impact assessment of older projects. *African Journal of Environmental Science and Technology*, *5*(13), 1170–1176.

Watts, M. (2015). Now and then: the origins of political ecology and the rebirth of adaptation as a form of thought. In T. Perrault, J. McCarthya, & G. Bridge (Eds.), *Handbook of political ecology*. Routledge.

Watts, M., & Peet, R. (2004). Liberating policy ecology. In Liberation ecologies: Environment, development, social movements. New York: Routledge.

Watts, M., Peet, R., & Robbins, P. (2010). Global political ecology. London: Routledge.

William, N. (2010). *UNICEF – "At a glance: Ghana – Improved sanitation catches on in rural Ghana one latrine at a time*. http://www.unicef.org/infobycountry/ghana 55767.html

World Bank. (2008). *Project Paper on a Proposed Additional Financing Credit to the Republic of Ghana for a Small Towns Water Supply and Sanitation Project in Support of the Second Phase of the CWSP. Report No: 39782 – GH*. World Bank.

World Health Organization. (2009). *Towards universal access: Scaling up priority HIV/AIDS interventions in the health sector*. WHO.

World Health Organization and United Nations International Children's Emergency Fund. (2006a). *Water and sanitation coverage: Drinking water, sanitation and hygiene database*. WHO.

World Health Organization and United Nations International Children's Emergency Fund. (2010). *Progress on sanitation and drinking water: Update*. WHO.

World Health Organization and United Nations International Children's Emergency Fund. (2010). *Progress on sanitation and drinking water: 2010 Update*. WHO.

World Health Organization and United Nations International Children's Emergency Fund. (2012). *Trends in maternal mortality: 1990 to 2010*. WHO.

KEY TERMS AND DEFINITIONS

Improved and Sustainable Water and Sanitation Facility: Ability of a water and sanitation facility to provide standard water and basic sanitation service over a period of two years with minimal breakdown.

NGO Programmes: Interventions and initiatives of non-governmental organizations which are meant to address water and sanitation constraints confronting people living in rural and urban communities.

Non-Governmental Organizations: They are non-profit making organizations involved in the provision of services in the water and sanitation sectors of Ghana.

Northern Ghana: A geographic or land area which covers the northern region of Ghana.

Rural Community: A settlement with a population less than 5000.

Water and Sanitation Services: Activities of NGOs which promote water supply and improved sanitation delivery in rural and urban areas.

Chapter 3
Mending Malawi's Water Institutions and IWRM Solutions

Innocent Simphiwe Nojiyeza
https://orcid.org/0000-0002-9948-2147
University of Zululand, South Africa

ABSTRACT

The aim of this chapter is to explain the challenges of decentralisation and management of water as economic good principles of integrated water resources management (IWRM) in Malawi in the rural areas of Ntcheu, Balaka, and Mangochi. Semi-structured interviews, focus group discussions, community meetings, and observation of water points were conducted. It was established that households and other state institutions prefer to receive services from Mpira-Balaka Water Users Association, which subsidizes households, rather than Southern Regional Water Board, which operates along commercial lines. The chapter questions the focus in the implementation of IWRM, which remain the establishment and fixing of decentralised institutions whilst recognising water as an economic good.

INTRODUCTION

In the 1990s, there was an increase in the emphasis on governance of water and sanitation services as a participatory process, with various stakeholders negotiating water allocations for various uses. The Dublin Conference of 1992, the organisation of annual World Water Forums (WWF) by the World Water Council, the establishment of the African Council for Water Ministers (AMCOW) and many river basin authorities established by Regional Economic Communities (RECs) in Africa such as the Southern African Development Community (SADC) played an important role in strengthening the implementation of Integrated Water Resources Management (IWRM) in Africa (Allouche, 2016, van Koppen et al, 2016, Mehta, et al, 2016). The water and sanitation experts who participated in the crafting of IWRM, agreed on 4 principles commonly known as the Dublin Principles (Solanes and Gonzalez-Villarreal, 1999).

1. Freshwater is a finite and vulnerable resource, essential to sustain life, development and the environment;

DOI: 10.4018/978-1-7998-8809-3.ch003

2. Water development and management should be based on a participatory approach, involving users, planners and policy-makers at all levels;
3. Women play a central part in the provision, management and safeguarding of water; and
4. Water has an economic value in all its competing uses and should be recognised as an economic good (Global Water Partnership (1992) cited in Manzungu and Derman, 2016, Derman and Prabhakaran, 2016).

The implementation of the second and the fourth IWRM principles meant that the provision of water supply and sanitation services were devolved to lower levels of government. State water utilities were increasingly expected to commercialise their services and also to privatise as part of the Structural Adjustment Programmes (SAPs) implemented by the World Bank and the International Monetary Fund (IMF) in the 1990s as a condition to get loans from these institutions (Manzungu and Derman, 2016). In the SADC region, many countries began to implement IWRM as part of the conditions imposed by donors (van Koppen et al, 2016, Manzungu and Derman, 2016, Mehta, et al, 2016).

Malawi is one of the first countries assisted by international donors to implement IWRM by 2005 as agreed upon during the World Summit on Sustainable Development (WSSD) which took place in Johannesburg in 2002 (Mehta, et al, 2016, Jonch-Clausen, 2004, Manzungu and Derman, 2016). The focus was on establishing water and sanitation institutions, whilst ensuring that water is managed at the lowest level possible through decentralisation. This started in 1998 with the promulgation of the Decentralisation Act of 1998 and subsequently the election of the first group of local government councillors in 2014 since the country gained independence from Britain in the 1960s and recognition of water as an economic good in policy documents (van Koppen, et al, 2016). Mpira Water User Association (MWA) and Southern Region Water Board (SRWB) are some of the institutions established to devolve the management of water and sanitation services from the Ministry of Irrigation and Water Development (MIWD) to lower levels of government.

Although the Water Resources Act of 2013 was promulgated in April and legally established the National Water Resources Authority proposed in the National Water Policy of 2005, Water Resources Act of 1969 remain the most important piece of legislation when it comes to water resources management. Water and sanitation policies and water laws are being introduced as part of IWRM-related reforms (Mehta, et al, 2016). Like in Uganda and Zimbabwe international donors supported the implementation of IWRM (Nicol and Odinga, 2016, Derman and Manzungu, 2016). In Malawi, the implementation of IWRM is harmed by institutions that took over the responsibilities which were initially performed by MIWD, i.e. MWA and SRWB that contradict each other when it comes to the interpretation of decentralisation and recognising water as an economic good (van Koppen et al, 2016, Mehta, et al, 2016). Malawi also established Catchment Management Committees and Village Action Committees (Derman and Manzungu, 2016) in addition to tap committees which existed since 1987 when Mpira Balaka Water Scheme sponsored by the AfDB was handed over to the Government of Malawi.

Like South Africa which modelled all its laws according to IWRM and is in the process of establishing Catchment Management Agencies (CMAs) in all 9 provinces (Denby et al, 2016), Malawi is fixing institutions whilst continuing to implement IWRM. Malawi, just like other countries in the SADC is adapting institutions, promulgating laws, setting water prices, rationing water uses and providing equitable access to services without any blueprint (Movik, Mehta and Manzungu, 2016). There are contradictions, 'power struggles' and debates among water and sanitation institutions regarding the Malawian model of IWRM (Movik et al, 2016). The learning from the ground regarding the implementation of IWRM in Malawi (Allouche, 2016), set the tone of the popularity of IWRM at the local level in terms of fiscal decentralisation (Alba, Boulding and Ducrot, 2016). As much as the local authorities of Balaka, Ntcheu and Mangochi Districts were given authority to manage water supply and sanitation services, they consistently returned to the Ministry of Water for financial and institutional support. This was especially when SRWB disconnected households within these areas due to rationing and non-payment, municipalities asked for help from the MIWD.

The focus of the study was on the two principles of IWRM, decentralisation (Alba and Bolding, 2016, Bolding and Alba, 2016) and managing water as an economic good and the gender issues (Derman and Prabhakaran, 2016) were raised time and again by study participants (Solanes and Gonzalez-Villarreal, 1999). The next section will look into the methods adopted in collecting data in this article.

Methodology

In unpacking contradictions and continuous fixing of institutions purported to implement IWRM in Malawi, semi-structured interviews were conducted with 100 households in Balaka (40), Ntcheu (30) and Mangochi (30), policy makers and institutions (5 government officials and 4 NGO representatives) responsible for the implementation of IWRM. The policy making and NGO participants were chosen using stratified random sampling based on the membership or employment within an NGO or government agency (Sarantakos, 2011). Households were selected using a simple random sampling strategy whereby every household had an equal chance of participating in this study based on their availability (Sarantakos, 2011). Six stakeholders participated in 6 focus group discussions, namely MIWD, NGOs, SRWB, WRB, MWA, Balaka District Hospital, members of the water point committees, tap committees, village action committees or village development committees, WUAs and CMAs.

The themes dealt with in focus group discussions included catchment protection and related protection control, decentralisation, privatisation and commercialisation of water and sanitation, community management of water and sanitation projects, capacity building in decentralised institutions, catchment management forums and agencies, operation and maintenance of schemes, tariff system and cost recovery, environmental sanitation, public latrines, health promotion campaigns, capacity and willingness to pay, sustainable development and community participation. The interview questions were also structured according to these themes and other themes which emerged in the field (May, 2002 and Grey, 2009).

The epistemology and paradigms which shaped the choice of the methodology are critical theory, post-positivism and social constructivism. In this study, knowledge is socially constructed by participants (Denzin and Lincolin, 2008) and there are subjective meaning to the realities they face regarding the lack of access to water and sanitation services (Holstein and Gubrian, 2008). Mpira Balaka Water Scheme was chosen among 13 projects funded by the AfDB as part of the Rural Water Supply and Sanitation Initiatives. This scheme is one of these AfDB funded before they adopted IWRM policy as a conditionality for funding. The data was coded, conceptually organised using interrelated themes. The fixed model of qualitative analysis whereby written records of project documents, laws and policies of each country is analysed was adopted in this study (Sarantakos, 2005). The data was labelled according to conceptual patterns and open coding supported by the NVIVO version 9 package (Corbin and Strauss, 2008). Each participant filled an informed consent form prior to participation in the study. Community participants remained anonymous and pseudonyms were used when reporting findings. Government and NGO stakeholders who preferred to be cited were given opportunities to do so. All the ethically sound protocols were complied with in this study. The next section will look into the water and sanitation governance, collective action, principles of water and sanitation governance, institutional mechanisms, property rights, public and private provision, decentralisation and centralisation and the role of market mechanism in Malawi and the extent to which institutions were remodelled for the implementation of IWRM. The next section is the presentation and discussion of findings based on the above mentioned themes.

Water and Sanitation Governance

The water sector leader in Malawi is the MIWD and its key role is to facilitate the development of water resources in the country. In discharging its water resources functions, the Minister established the National Water Resources Authority in accordance with The Water Resources Act of 2013. The National Water Resources Authority is responsible for granting water rights for abstractions and the discharge of effluents as well as for monitoring adherence to water rights. At Balaka, Ntcheu, and Mangochi, water and abstraction rights are granted through the local office of the National Water Resources Authority. MWA which is a catchment management authority is expected to recommend applications for approval to the main committees of the NWRA. The MWA is governed by a board of trustees appointed by the Minister of Irrigation and Water Development and managed by the CEO and government employees seconded from the Ministry. The current CEO is Aaron Mapsere who also acts as the district water development officer responsible for the Balaka district. Each ministry with water-related functions will employ district water officers who will report directly to district commissioners and no longer to permanent secretaries of their respective ministries. The board of trustees represents the interests of the water user associations (WUAs) and traditional authorities such as Makwangala, Champiti, Ganya, and Phambala in the Ntcheu district (Interview with government officials on 20 August 2012).

The MWA is constituted of catchment committees, water point committees, and area committees. There are several water point committees in each of the water and sanitation schemes managed by MWA. The WUA is responsible for managing water and sanitation schemes, collecting tariffs from users, mobilising communities within the catchment area, and contributing towards the operation and maintenance of the scheme. WUAs work with local authorities in the operation and maintenance of schemes. They are assisted by the MIWD if the operation and maintenance work is major, but there is no capacity to conduct such work within catchment areas. The WUAs are also responsible for monitoring water quality in various water resources, but the day-to-day functions are performed by state employees on its behalf. Once a full staff compliment is appointed in the MWA they will be responsible for the collection of revenue as well as monitoring water quality. The SRWB is responsible for water reticulation in the urban areas of Ntcheu, Balaka, and Mangochi, whereas the MWA is responsible for water resource development and reticulation in rural areas in the above districts (Interview with state officials on 20 August 2012).

Decision-making on water developments rests with the MIWD, the MWA and the SRWB that operate at meso and formal levels, whilst informal and micro institutions such as traditional authorities and WUAs are also consulted. This institutional arrangement is based on the Water Resources Act of 1969, the Water Works Act of 1995 and Water Resources Act of 2013 at a formal level of water governance whereas the informal rules are embedded in the cultural and social milieu of the community (Livingston, 2005:22, Klein, 1999:463, Nabli and Nugent, 1989:1334).

Communities served by the MWA consistently had access to water while those served by the SRWB did not have access to water for more than three months after the water board instituted water rationing. Households living in the Kandengwe Village outside Balaka reported that there were instances when water only starts flowing at 1am, and the times in which water is scheduled to flow after a mechanical cut-off is not known. Sometimes they wake up at 12 mid-night to collect water, only to discover that there is nothing. According to an official in charge of the SRWB's Balaka water treatment plant, households in Balaka access water between 9 am and 12pm, and 11-2pm whilst Liwonde households access water between 2pm-3pm and 3-9pm, and disputed the claim that water only starts to flow during the night and that there are times when households do not get access to water at all for a whole month (Interview with SRWB official on 20 August 2012, Household Interviews).

The reasons for water rationing, according to the SRWB, is that the MWA does not allow them to buy enough bulk water to reticulate to the various users served by the water board. Water rationing is meant to ensure that all households get access to water. An official of the MWA indicated that water rationing and low water pressure resulted in the Balaka District Hospital opting to build their own pump station, to connect directly from the water authority, and to stop their SRWB connection, and that these problems are associated with the relationship between the water board and the hospital; the authority is not involved. He confirmed that the water authority took over from the water board to ensure that critical hospital wards such as theatres and intensive care units have sustainable access to water. It is clear however that rationing and low water pressure is not a technical problem but an institutional and market-related problem. Even the disconnection of households unable to pay for water is exclusively found in the areas managed by the SRWB; the households using water served through the MWA never experience water disconnections (Castro, 2007:98, Interview with government officials in 2012).

An SRWB official confirmed that the SRWB disconnects households to encourage a culture of payment, that they do not have an indigent policy, and that no subsidies are provided to poor households. An MWA official argued that the MWA intervenes if water is wasted at a communal water point which only costs K50 per household per month, whereas households served by SRWB pay between K2000-4000 per month and that is why disconnections are common. The different tariff systems and different views on limiting access to water through rationing and disconnections by both the SRWB and the MWA is a reflection of institutional failure manifested in markets, bureaucracies, redistribution, and remediableness (Williamson, 2000:183).

The SRWB views water as a commercial commodity whereas the MWA views water as a social and a public good. These two state institutions are caught up in hydro-politics and the fact that rationing and disconnections are viewed as techno-scientific solutions which erode institutional thought and socio-political paradigms of water and sanitation governance (Rodgers, 2002). The MWA engages water users through a process of substantive democratic participation in the form of traditional authorities, water user associations, tap committees and other local actors, whereas the SRWB do not engage stakeholders at all (Interview with state officials on 20 August 2012).

Collective Action

State institutions, NGOs, and communities are bridging the gap between the techno-scientific, socio-political, economic, and cultural aspects of water and sanitation governance. They are pursuing collective action that may be difficult for individuals to achieve acting on their own (Rodgers, 2002). Declaring areas open defecation free, providing households with chlorine to prevent waterborne diseases, and moving away from the service of a commercialised water board, are some of the examples of collective action taken by stakeholders in Malawi (Rogers, 2002:4, Bromley, 1978:46, Nabli and Nugent, 1989:1338).

There is also evidence of collective action in rural areas where communities use WUA structures to engage the MWA on water rights. The Balaka District Hospital refused to pay the SRWB because of the low pressure water supply in the main pipeline from Mpira to Liwonde which resulted in critical wards such as theatre and intensive care unit lacking access to water. They resorted to digging a borehole which is expensive because it involves removing salt from the water, and negotiated with the MWA for a transfer from the SRWB connection to the water authority's scheme. All primary health care centres in Kalembo, Mbera, Nandombo, Namenolo, Ulongwe, Kwitembo, Chienda Usilwo, Utale 1, Utale 2, Kankao, Phimbi, Pharula, Katshenga, and the Outpatients Department in Balaka were transferred from the SRWB to the MWA to ensure sustainable access to water of good quality and high pressure (Interview with a Department of Health official in 2012).

The collective action embarked upon by the Balaka District Hospital, together with the primary health care centres which refer patients to this hospital, was influenced by water rationing in most areas served through the SRWB and high incidences of cholera reported in the district due to lack of access to water and related hygiene promotion. The district hospital proposed abstracting water directly from the Shire River and reticulating it to all primary health care centres. The reason the Mpira Dam built on the Mpira River was chosen is because areas such as Katshenga which is located 70km away from the Mpira Dam has access to water while Balaka and Ntcheu are subjected to rationing by the SRWB. The hospital has the capacity to abstract water and operate pump stations that can increase water pressure in all primary health care centres. Water drawn from boreholes is not treated as there are no treatment facilities in health care centres. Only the Mbera clinic has access to adequate water in the correct high pressure because it is connected directly to the Mpira Dam through the MWA. This act of collective action by the Balaka District Hospital illustrate Rogers's (2002:4) assertion that interest groups with a stake in water use and allocations can meet their individual objectives in an existing structure of rights and laws and seek to change the rules which govern water and sanitation services in a manner that promotes their group interests.

When the then Deputy Minister of Health, Mrs Halima Dauda, visited the Balaka District Hospital on 10 August 2012, the meeting with hospital officials was dominated by water resource management issues. The state of the hospital facilities were so poor, that a pregnant woman gave birth in a toilet due to a lack of beds in the Ntcheu District Hospital (Malawi Broadcasting Corporation, News bulletins, 10 August 2012, an interview with a state official on 13 August 2012). In 2012, the Balaka and Ntcheu Hospitals were in the process of working with Water Missions to ensure that water abstracted through boreholes is treated before it is consumed by patients. The social capital achieved through changing a water allocation from a water board to a water authority is a reduction in the instances of cholera which have dominated Balaka in the past 3 years (an interview held with a state official on 13 August 2012, Livingston, 2005:24, Williamson, 2000:147).

The environmental health department of the Balaka District Hospital is working with NGOs such as World Vision and Concerned Universal to deal with open defecation in the district. The programme entails dealing with waterborne diseases such as cholera and dysentery, water-based diseases such as malaria, and watershed diseases such as scabies and other skin infections. The NGOs and the district hospital distribute tins of chlorine to households to treat water from unprotected sources in seasons when water rationing results in a lack of access. The chlorine powder is mixed with 20 litres of water; 1 teaspoon per 20-litre drum is kept for two weeks where after households are given additional bags of chlorine (an interview with a state official on 13 August 2012).

The Department of Health is also working with NGOs such as Goal Malawi, World Vision and Concerned Universal in promoting community-led total sanitation (CLTS) in 12 schools within the Balaka district. The district coordinating committee meets every two weeks to monitor open defecation in schools and the broader community (Interview with a state official on 13 August 2012). This initiative which is reducing the prevalence of waterborne diseases associated with open defecation and promoting school attendance by girl children, is premised on public or collective action goods that bring about joint benefits to society and also give voice, joint use, and public and common property rights to all regardless of wealth, power and social status (Nabli and Nugent, 1989, 1339).

Principles of Water and Sanitation Governance

The principles of water and sanitation governance will be elaborated on in the section below. Participants in the study raised concerns about the lack of openness, communication, and transparency on the part of the SRWB in terms of rationing and disconnections. In the village of Ntonya, households connected to the SRWB did not have access to water for three months, whilst their neighbours have continuous access to public water points provided for by the MWA. Even at Kandengwe Village, households connected to the SRWB did not have access to water for four months, whilst within the same village, those who got water through the authority continued to have access without interruption. Throughout these months of drought in the SRWB pipes, no meetings were convened to explain the problem to households. No notices were served to the communities advising them when the rationing and disconnection would end (Focus Group Discussion notes taken on 11 August 2012).

In community No. 8 in the Kandengwe Village, a participant in a focus group discussion stated:

"Water bills compiled by SRWB took more than three months to be delivered to us and there are no user platforms we can use to raise our concerns with the water board."

Lack of communication, transparency, and openness was also cited in focus group discussion in the village of Ntonya. A participant claim:

"SRWB did not consult us on tariff increases; in 2003 we were disconnected after alleged non-payment when households did not receive water bills for four months. After we were disconnected we negotiated the payment of outstanding tariffs and we were re-connected."

They also felt that there is no transparency in the water board since they are billed even if no water was coming out of the taps. Details of water rationing were also not communicated to the Ntonya community (Focus Group Discussion notes taken on 13 August 2012). Community participation was very strong during the construction phase of the Mpira-Balaka water scheme. A government official outlined the following problems regarding participation.

"During the construction phase, members of the community provided building material such as bricks and cement that was used to construct various water points, including boreholes. Government disseminated information regarding the project, but when the scheme was managed by SRWB as part of the implementation of IWRM, participation became tokenistic and eventually there was no participation at all."

It is only the households who are part of the MWA schemes which are engaged through WUAs, the community water point committees and borehole committees. IWRM institutional arrangements do not promote participation, but inhibited the popular participation of 1987 when the scheme was at a conceptual and construction phase. In community water points, constructed and managed by World Vision, communities are involved throughout the project cycle in planning, implementation, monitoring, and evaluation of the project. An NGO official recalls: "World Vision entered into agreement with respective communities to raise K15000 maintenance and operation fund prior to the construction of a borehole that is estimated to cost about K1-2 Million."

This was meant to ensure sustainability, but the results show that most schemes are not sustainable; the Mpira Dam is not enough to cater for all water supply needs which is why the SRWB and Balaka Hospital are requesting permission to divert additional water from the Shire River in the hope to end water rationing (Interview held with state officials on 13 August 2012). These arguments support the views of Hoekstra (2006:18), Heller (2007:3), UNESCO (2006), Spash (1999:432), and Pearce (2002:61), who argue that water and sanitation governance must serve future as well as present users of water and sanitation services.

In the eyes of the participants in this study the roles in executive and legislative processes are not clear and result in institutions failing to explain and take responsibility for what they do. There are no boundaries to the functions of the SRWB and MWA regarding the provision of water supply and sanitation services. In many villages there are no hydrological boundaries or scope of practice between these two institutions. For instance, the Mponda Primary School in Balaka is served by both institutions. During the era of water rationing the school used a communal standpipe established by the MWA. The situation is similar in the other schools in the Mponda Zone, namely Kapalamula Primary, Sasola Primary, Zambimba Secondary, St Tlasys, Ostia, Matola, Khwisa, Kabango, Toleza, Ncheza, St Louis Secondary, the Bakhita and Praisy private schools, Secot Primary, Kink Range Private Secondary School, and Sir Green Johns Primary. The government is not responsible for paying school water bills regardless of whether they are public or private. They all need to use school fees to pay for their water bills, and are not subsidised, even though primary education and access to water are regarded as human rights in Malawi. Most schools in the Mponda Zone pay about K1500 water bills per month to the SRWB and about K50 for the use of communal standpipes in times of water rationing or when there are disconnections for non-payment (Interview with the staff of Mponda Primary School on 11 August 2012).

There is also a lack of policy coherence, effectiveness and efficiency in water and sanitation governance. In theory, many government institutions are responsible for various elements of water and sanitation policy, but in practice the MIWD, the water boards, and Water Resources Board remain the main actors. School sanitation is implemented by the Ministry of Health, together with NGOs. The policies on water and sanitation are not in line with achieving the millennium development goals on education (Interview with state officials on 12 August 2012). IWRM advocates for the management of water at the lowest possible level, but none of the three districts in which the study was conducted has any capacity to manage water resources or even reticulate water to its citizens; it is a responsibility which both the MWA and the SRWB compete over. There is no evidence of cooperative governance at the local level as envisaged in the Water Policy of 2005 and Sanitation Policy of 2008. The pricing structure of the SRWB is based on full marginal cost, and disconnections are used as a tool to enforce water user efficiency in the neo-classical sense. Charging schools an average of K1500 per month and disconnecting them if they are unable to cover their water bills is a failure to balance economic efficiency with political efficiency (Klein, 1999:470, Weber, 2001:54).

Results show that allocations of water and technology used in sanitation is not equitable. The fact that hospitals are provided with low pressure when they require high pressure in theatres and intensive care units undermines the right to health. All schools, regardless of whether they are in rural areas, use unimproved pit latrines. The Mponda Primary school with 2 761 learners has 19 pit latrines (eight for girls and eight for boys, two for female teachers and one for male teachers). This inadequate number of latrines result in long toilet queues during intervals, yet the school is situated in Balaka town which is served with single septic tanks flush toilets. The school is in the same vicinity as the Catholic Church which uses single septic tanks flush toilets (Interview with Mponda Primary School Staff on 11 August 2012).

The water rationing affects the school since pupils spend long hours waiting for water in the communal standpipe in very long queues, instead of learning (Household Interviews, 13 August 2012). There are no water tankers to assist schools during the water rationing season. Even households in the city centre and in small towns use pit latrines. There is no level of service or standards that define adequate sanitation. Amidst this scarcity of water, rich people who live in estates enjoy water which they abstract and divert from rivers. In most estates, there is no water-rationing as neither the water boards nor the water authority serves water in these areas. In short, whilst there is a scarcity for poor households in rural and peri-urban areas, there is abundance in estates and affluent areas. This constitutes class discrimination and lack of equity (UNESCO, 2003, Hoekstra, 2006:18, Ghandhi and Namboordini, 2009:15, Livingston, 2005:26, Klein, 1999:474, Heller, 2007:3, UNESCO, 2006, Spash, 1999:423).

Results show that there are no agreed upon property rights that advocated limitation of access, use, and ownership of water and sanitation. There is no discrimination based on race, colour, gender, and disability when it comes to access, but there are class connotations. Access to water and type of toilet facility depends on whether one can afford to purchase it. Failure to pay for water connected by the SRWB results in disconnections and some water committees restrict access to users that are unable to pay. Discrimination may not be based on race or colour, but access to capital and affordability of water restricts poor households who reside in rural areas and lack money to pay water bills. Full cost recovery violates the fundamental right of the poor to access water and sanitation which is unethical since most of the poor would remain excluded if the SRWB was the only provider of water and sanitation services in Ntcheu, Balaka, and Mangochi (Household Interviews on 13-14 August 2012) This view is supported by Rogers, (2002:4), Fiani (2004:1003) Heller (2007:3), UNESCO (2006) and Spash (1999:432).

Institutional Mechanisms

The Water Resources Act of 2013, the Water Works Act of 1995, the Water Policy of 2005, and the Sanitation Policy of 2008 in Malawi outlines institutional mechanisms, organisations, institutions, formal rules, informal constraints, informal norms, and enforcement characteristics that govern water (Interview with MWA official in 2012). The formal rules that govern the water sector are found in legislations such as the Water Resources Act of 2013 and IWRM policies that advocate a regulatory role for the MIWD whilst the MWA, the SRWB and other commercialised state water utilities provide services (Mkandawire, 2011). Results show that the implementation of IWRM is facing difficulties because the MIWD is forced to intervene when local authorities in Ntcheu, Mangochi, and Balaka are unable to provide water and sanitation services. The MWA is supposed to be a local regulator to the SRWB and district assemblies and focus on the sale of bulk water to water boards and assemblies. Market failures of the commercialised water boards, however, result in the MWA supplying water directly to hospitals, households, schools, and even towns (Interview with MWA official in 2012).

These findings support the views of Pearce (2002:63) that IWRM philosophy which is based on neo-classical, revisionist and market-friendly theories, is flawed when it comes to institutional mechanisms through which water and sanitation projects are governed. The current institutional model is enforcing the allocation of water through market mechanisms which means that those who can afford to pay about K4000 per month will have a connection to the SRWB while those that are unable to pay will either join the queue in community standpipes managed by borehole committees on behalf of the MWA, or get water from unprotected rivers and shallow wells. According to Williamson (1995:189) the revisionists view market failures as pervasive and a justification for governments to lead markets in critical ways and this is the role played by the SRWB in the commercialisation of water access.

The SRWB uses price mechanisms and not institutions such as water user associations, borehole committees, subsidies, or indigent policies to increase sustainable access by the poor to water and sanitation services. If households in Ntonya Village can make a choice between a subsidised K50 communal standpipe and a K4000 SRWB connection, it is clear they will choose the former (Focus Group Discussion at Ntonya Village on 12 August 2012).

The rejection of market mechanisms and the pricing approach adopted by the SRWB have resulted in public hospitals such as the Balaka, Ntcheu and Mangochi District Hospitals choosing to get access to water through the MWA that puts more emphasis on getting the institutions and institutional mechanisms right. The view is supported by North (1993:5), Richter, (2005:1), Ghandhi et al (2007:5), and Klein (1999:468), namely that governance structures are a polarisation between market and hierarchy and determining the right price for water hardly solves the problem since the major problem lies in invoking the price and cost recovery.

Property Rights

In Malawi, abstraction rights are allocated to both government institutions and the private sector. These rights are allocated together with either private or public property rights. Farmers in large estates abstract water from rivers, lakes, and aquifers and store them in irrigation dams. They pay water rental fees yearly as stipulated in their abstraction licenses to the National Water Resources Authority. Non-governmental organisations such as World Vision, Goal Malawi, Water Aid, Concerned International, and the Presbyterian Church also apply for abstraction rights in the many water points which they manage in their respective jurisdictions (World Vision Interview, 2012). Once they get an abstraction license, it becomes a private property right. This private property right mechanism is in line with the views of Coase and North (1992), North (1993), and Quiggin (2000:17), that private property rights are exclusive (in the sense that property rights which are ownership title and use rights in the hands of one individual exclude that these rights can be claimed by another individual.

The Malawi government has jurisdiction over water points, irrigation, and multi-purpose dams constructed by the private sector and NGOs as stipulated in the Water Resources Act of 2013. This was not the case with the Water Resources Act of 1969. Within the private property rights discourse, there are water boards who are supposed to function as state institutions and their property rights are supposed to be public, but their commercial nature brings them closer to private property rights. In the dispute between the Balaka District Hospital and the SRWB, the MIWD was unable to intervene to correct the market distortions that resulted in low water pressure in various hospital wards due to their culture of non-payment. The only option for the MIWD was to remove Balaka District Hospital from the commercialised private property water allocation of the SRWB to a public property allocation of the MWA.

Public authorities such as the MWA, the Ministry of Agriculture, the Ministry of Health, and Ministries Responsible for Natural Resources such as Mines, Forestry, Fisheries, Lands, Environment, Parks, and

Wildlife, are also responsible for the abstraction of water. They are also subjected to stringent application conditions for licenses. Once the MIWD approves licenses, the extraction of water which is a public property right can commence. It is still difficult to know the exact amount of water that is abstracted in various rivers, lakes, and aquifers; the government is supposed to know how much is abstracted per day by at least the various state institutions. In 2012, Blantyre was experiencing water shortages despite being located next to lakes and the Blantyre Water Board claimed that the shortages were due to a lack of power in various pump stations. In the public property dispensation, water boards that operate as commercial entities have their own by-laws in accordance with Water Works Act of 1995 (Interview with MIWD official on 10 August 2012).

In the public property rights discourse, there are WUAs that are influential in water allocation and the granting of abstraction rights. They participate in approving abstraction licenses in river or catchment systems and ensure that there is due consideration for down-stream users in allocating such rights. In Zomba, for instance, WUAs play an active role in their local IWRM platform in the rehabilitation of the Mulunguzi Dam. It is through the active participation of WUAs that the government currently provides free mosquito nets to all pregnant women in areas such as Mangochi and Zomba. WUAs working in the structures of MWA are ensuring that public property rights are protected (Interview with MIWD official on 10 August 2012).

The public property right in the water sector supports the views of van den Bergh (2007:530), Quiggin (2000:19), Pearce (2002:59), and Kuks (2004:52) that common property occurs where rights reside jointly in the same group of actors who own undivided shares of the property. In this case, WUAs, the MIWD, and other state institutions are allocated abstraction rights within the parameters of public property rights. The role played by WUAs confirms the views of Gandhi et al (2007:6), Quiggin (2000:19), and Bromley (1978:53), that crafting institutions to manage water and sanitation governance is challenging and requires skill in understanding how the rules, property rights associated with physical, economic, and cultural environments, and appropriate institutions can promote maximum benefit for the beneficiaries. Beneficiaries of the Mpira Balaka Water Scheme work through WUAs to influence water allocations thus ensuring that it is agreed upon at the lowest level possible. Issues of water quality that are normally raised by water professionals are openly discussed in WUAs and actions are undertaken to enhance the good quality of water resources and the rehabilitation of rivers, lakes; and aquifers; and the process is driven by beneficiaries (Interview with MWA official on 14 August 2012 and interview with MIWD official on 10 August 2012).

Public and Private Provision

Results show that within the Balaka, Ntcheu, and Mangochi districts, there is both private and public provision of water and sanitation services. The two state providers of water and sanitation services in the above mentioned three districts are the MWA that sells bulk water to the SRWB which in turn reticulates it in urban and peri-urban areas. The MWA is responsible for selling bulk water to water boards and also reticulates water in rural areas within the three districts. The Environmental Health department operating in Balaka, Ntcheu, and the Mangochi district hospitals is spearheading health and hygiene promotion in schools, public centres, and communities. The various initiatives of the Department of Health which includes declaring certain areas "open defecation-free", distribution of chlorine to households, and prevention of water-borne diseases, are supported by a number of water and sanitation NGOs that operate in the three districts (Interview with a Department of Health official on 13 August 2012).

According to Dumisani Mfipha of the Malawi Red Cross, building pit latrines in schools produce difficulties associated with transaction costs since schools are expected to provide their own bricks and cement. In this arrangement, the provision of school sanitation is a private matter. Schools that are unable to provide bricks and cement have to wait for a longer time before NGOs are willing to assist them. Even though NGOs provide training in health and hygiene promotion, the continuation of this project depends on the availability of bricks and cement and other transaction-specific investments. Colex Champenduka, also from Malawi Red Cross, highlighted the lack of demand for sanitation facilities. If the prices of bricks and cement go up, a lower number of facilities will be demanded. Bernard Chiwalo, a beneficiary of toilet facilities constructed through Red Cross financial support confirmed that households and school communities do not have money to purchase bricks and cement because of the depreciation of the Malawi kwacha and a general lack of foreign currency (Malawi Broadcasting Corporation 10 August 2012 News Bulletin).

Households in estates and affluent areas build septic tanks and abstract water from rivers, aquifers, and lakes. In this private arrangement of water and sanitation delivery, there is a lack of balance between public and private provision. Commercial agriculture and other industrial users do not pay per cubic metre of water abstracted as they should since it is difficult to charge for privately abstracted water. They only pay for licenses when they are issued and when they are renewed, despite the fact that they are supposed to pay water license rental for the water they use as stipulated in Water Resources Act of 2013. The MIWD does not have the capacity to install water meters that can be used to monitor the amount of water diverted for private and commercial use by commercial agriculture and industries. If the capacity to collect revenue from private and commercial users of water can be developed, poor households could be subsidised with bricks and cement.

This institutional dilemma highlights the role of government in shaping the institutional environments of business and commons, as well as its administrative capacity to collect revenue from businesses that abstract water for commercial use (Richter, 2003:10). This is where Williamson's transaction cost economies differ from North's theories regarding the object of research: private goods in the case of transaction cost economy (TCE) and public goods in the case of North's theory. The poor pay for the construction of pit latrines and boreholes whereas the abstraction of water for commercial purposes remains a free service in Malawi (Interview with MIWD official on 10 August 2012). The need for a basic social structure, or rules of the game on how the poor households can be cross-subsidised by the private sector that abstract water which is a public good and maximise profits without incurring any transactional costs is relevant in Malawi (Richter, 2003:16).

Decentralisation and Centralisation

Results show that since the promulgation of the Decentralisation and Local Government Act of 1998, district assemblies have increasingly been expected to drive water and sanitation provision. District Assemblies are expected to assist in the preservation of the environment and natural resources. Planning of water supply and sanitation starts at the village action plan (VAP) level, and the plan crafted with the assistance of extension officers is submitted to the area development committee (ADC) which in turn forwards it to the district coordination team (DCT). The district coordination team is the sub-committee of the district executive committee. The former comprises district heads from the Ministries of Water Development, Health and Population, Women, Youth and Community Services, Natural Resources, and

NGOs. The most senior official in the water sector at the district level is the district water officer Aaron Mapsere, responsible for the Balaka, Ntcheu, and Mangochi district assemblies at the time of writing.

He is responsible for community-based planning and training, supervision of works, and the monitoring of water supply projects. The district water officer works together with the district development committee which comprises an MP, traditional authorities, representatives of business, and WUAs. This is the executive committee of the district assembly which allocates resources from the district development fund based on projects suggested and prioritised in the Village Action Plan (Ferguson and Mulwafu, 2004:1, Weller, 2005:12, Mkandawire, et al, 2011:10). In the three districts combined there are 400 sanitation schemes. The MWA constructs toilet slabs and the communities are responsible for digging pits, and mixing of the precast; the construction of the top structure is provided by the government. The WUAs are responsible for health and hygiene promotion and verifying the demand for sanitation facilities within various traditional authorities. They are also responsible for the election of the catchment, water point, and area committees. They participate in the board of trustees of the MWA and are elected by various communities to represent their interests in the general assembly of the authority (Interview with MWA official held on 14 August 2012).

Political patronage dominates the distribution of local development funds. On 11 August 2012, the then Deputy Minister of Finance, Ralph Jooma, used the event in which houses for principals were handed over at Mangochi to promote the then ruling People's Party:

"The event is making the People's Party of Joyce Banda popular, even members that defected from DPP constituency committees are welcome to join PP. PP is the future of Malawi and we are using this opportunity to strengthen our structures. The houses we are distributing today demonstrate our commitment as PP government to ensure that our school children learn in the atmosphere that is conducive to learning and we are increasing access to water and sanitation in our schools." (Malawi Broadcasting Corporation News Bulletin, 11 August 2012).

The houses that Ralph Jooma was referring to was a refurbished residence of the principal of the Mangochi Primary School and the event coincided with the launch of constituency committees at Mangochi. Participants were wearing the orange colours of the People's Party which defected from the DPP upon the death of the former president Bingu wa Mutharika. It is strange that the Deputy Minister of Finance was delegated to open one house and one borehole in the presence of over 5,000 members of the People's Party. The state function was, in fact, a People's Party rally rather than a government function. Ralph Jooma simply used the function and the resources of the local development fund for party political purposes. In a similar function Sam Kalanda, the Mangochi District Commissioner, announced that the training of natural resource committees responsible for Lake Malawi is sponsored by the AfDB. He was also wearing a People's Party uniform when he made this announcement. Members of Parliament dominate the allocation of local development funds and use political patronage to advance party political interests (Malawi Broadcasting Corporation News Bulletin, 11 August 2012).

In another function which was supposed to be hosted by the local development committee, the then Deputy Minister of Environment and Climate Change, Ebrahim Matola, announced that the UK government was assisting the Meteorological Department of Malawi with a supply of digital weather recording instruments. The Ministry of Agriculture also used a similar event to announce farm diversification programmes undertaken by the agricultural district commissioners of Ntcheu, Mangochi, and Balaka to adapt to climate change. The programme promoted climate resilient crops that do not consume much water cultivated through crop rotation. Dr Stanley Khaila from the Bunda Agricultural College was working with the Ministry of Irrigation and Agriculture in promoting a farm diversification programme. Normally, such announcements are made in press conferences convened by the government and not in party rallies (MBC, News Bulletin, 11 August 2012).

Results further show that village health and water committees lack the capacity to monitor water quality on a monthly basis, and that the decentralisation of administrative functions is not accompanied by fiscal decentralisation. District assemblies have limited control over boreholes and other schemes. Boreholes are constructed by NGOs such as World Vision, Concerned Universal, and Goal Malawi. The district assemblies in which such schemes are allocated have no control since it is the responsibility of the Water Resources Board to grant abstraction rights to NGOs; district assemblies are not involved in this arrangement. Other boreholes are constructed, operated, and maintained by the SRWB and the MWA (Interview with MWA official on 14 August 2012, Interview with SRWB official on 13th of August 2012).

The findings are contrary to the views of Indranil (2009), Batterbury and Fernando (2006:1851), Selee and Tulchin (2004:297), and Wunsch (2001), that decentralisation takes care of the needs and preferences of communities as powers and responsibilities are devolved to lower levels. There is no evidence that decentralisation increases accountability of the government in the delivery of services. Instead, communities that were involved in borehole committees from as early as 1987 when the Mpira Balaka water scheme was conceived, indicated that there was strong public participation when the MIWD was still involved, compared to the current period during which the SRWB simply cuts off water for non-payment without notification. Due to the failure of the water board to send water bills on time, communities are charged for water even when cut-offs have lasted for more than three months (Focus Group Discussion, 13th of August 2012).

There is also no evidence in the field that decentralisation promotes efficiency, enhancement of public services, and more open and accountable forms of government as propounded by Batterbury and Fernando (2006:1851), Selee and Tulchin, 2004:297 and The Sustainable Livelihoods in Southern Africa (SLSA) (2008:3). Failure to circulate water bills on time by the SRWB, disconnection of water without verification of non-payment, failure to consult households about water rationing, and the imposition of low water pressure on district hospitals, are some of the reasons households, government officials, and the private sector believe that decentralisation in Malawi has brought about inefficiency, lack of openness or enhancement of public services, and a lack of accountability.

The fact that Aaron Mapsere is the district water development officer responsible for Balaka, the project engineer for Mpira Balaka Water Scheme, and also the CEO of the Mpira Water Scheme clearly shows the weaknesses of decentralisation. How can one employee be given so many responsibilities? How can one employee manage three vast water authorities at a district level and still be in charge of the secretariat of the MWA? Results show that there is a lack of sufficient technical human resources to perform budgeting functions. There is also evidence of confusing roles among the locally elected representatives and their national counterparts, as well as the recapturing of authority and resources allocated to local government as postulated by Wunsch (2001:280) and SLSA (2008:8). In Malawi, the bylaws of water boards are ignored by the water resources authority and the MIWD. When granting licenses for the construction of boreholes water boards are consulted. They are even invited in meetings where approvals of water licenses take place. The allocation of water that is supposed to be agreed upon at a local level in a multi-stakeholder process is always considered. The challenge is to ensure that this is done consistently.

In 2012, the Balaka District Hospital applied for the abstraction of water through a pump station and the usage of the MWA's Water Treatment Plant. The negotiations were at an advanced stage but the SRWB claim that it was never consulted. The MWA which acts as the local office of the Water Resources Board was ready to take over the functions that are normally performed by the water boards. There were households in rural and urban areas that were part of both schemes. During water rationing, they continued to access water through the MWA's communal taps. The operation of both schemes in one area is not just a duplication of efforts but is rendering decentralisation useless in Malawi.

The Role of Market Mechanisms

Regarding consumption, Veblen's theory of the leisure class is confirmed in the findings whereby the Mpira Dam is situated in rural areas of Ntcheu but benefits the urban households of Ntcheu, Mangochi and Balaka. Households that enjoy flush toilets and connection to sewers and water are mainly located in urban areas and farm estates. The households who were responsible for the construction of the dam and production of water do not benefit, but only get access to water through communal standpipes (Hamilton, 1989:1100-1).

Chapman's (1973:1415) observation that consumption or quasi-consumption is localised in metropolitan areas and effluents are concentrated to the point of swamping the assimilative capacity of local environments holds true in Balaka Town. Dilapidated sewers are visible in the railway station and effluents are discharged into rivers and water points. In 2012, a perception was created by the SRWB about the scarcity of water in the water pipes from the Mpira Dam to Liwonde and used to justify rationing to various households. The fact that households directly connected to the MWA are not subjected to rationing clearly supports the views of Miller (2003:58), who rejects the neo-classical theory of production. In contrast, many institutional economists hold the view that an increase in scarcity values only serves private interests and is not in the interests of society; public interest is to increase production and not scarcity.

There is evidence that the water sector in Malawi is deregulated. The government has no record of water points, boreholes, and pump stations are controlled by the NGOs and private sector. Irrigation dams controlled by commercial agriculture and NGOs are not known to government (Interview with MIWD official on 10 August 2012). Even during the era of UDF from 1994-2008, various attempts were made to regulate water resource management and to harmonise existing water abstraction rights. To date, there are an estimated number of 700 dams in Malawi which is further endowed with water resources in the form of lakes, rivers, dams and aquifers. Water rationing imposed using scarcity is thus a reflection of distorted market size, diseconomies of scale, and an increase of market share and sales which characterises the use of the theory of production in the neo-classical sense (Miller, 2003:59, Klein, 1999:463).

Within the parameters of the theory of distribution, results show that when allocating resources such as access to water and sanitation, various water institutions in Malawi use orthodox economics to justify existing income distributions on both efficiency and equity grounds. To illustrate this theory of distribution, the Mponda Primary School is well placed since it is in the same area as the Catholic Church in Balaka. Mponda is connected to the SRWB, as well as having access to a communal water point. When it comes to sanitation, however the Catholic Church has a septic tank and the school uses pit latrines. The Malawi government encourages free education in primary schools but does not subsidise access to water and sanitation in schools. The SRWB ignored equitable distribution when allocating flush toilets to the Catholic Church and not the school in the same area. The school board used an income efficiency criterion, which stipulates that marginal revenue produced is equal to the marginal cost of employment, to justify denying the school access to flush toilets (Interview with Mponda Primary school staff on 12 August 2012, for theory refer to Hamilton, 1987:250, Miller, 2003:60).

Access to water and sanitation facilities in Malawi is determined by social and class ranking. In this regard social rankings and not marginal productivity are the basis of income differentials and therefore give access to various levels of water and sanitation standards. Linked to this view are high poverty levels in rural areas, low levels of education, poor shelter, and low consumption, which are used to justify pit latrines even in areas where sewer lines are available. Boundaries between the market centres of Balaka, Mangochi, and Ntcheu are artificial and invisible, but social ranking and standing determines the type of sanitation and water facilities provided. Those who fall into the category of rural poor are subjected to communal stand pipes or boreholes and pit latrines, and those who are classified as high income earners are provided with house connections to water and flush toilets through water board sewers or septic tanks (Focus group discussion at Ntonya Village on 14 August 2014, see Hamilton, 1971:73 for theoretical arguments).

CONCLUSION

The MIWD play an important role as the water sector leader in Malawi. Through the Water Resources Board, the MIWD grant water licenses in a similar manner as in most SADC countries (Manzungu and Derman, 2016, van Koppen et al, 2016, Mehta et al, 2016, Alba, Bolding and Ducrot, 2016). Unlike other SADC counterparts, Malawi is still relying on very old pieces of legislation such as the Water Works Act of 1995. The Water Resources Act of 2013, the Water Policy of 2005 and the Sanitation Policy of 2008 are some of the latest water and sanitation reforms taking place in Malawi. The IWRM-modelled water and sanitation policies were recently formulated, but their promulgation remained a challenge under the MCP,UDF, DPP, PP and DPP administration since the country gained independence from Britain in the 1960s (Mkandawira et al, 2011). There are new institutions established as a result of the implementation of IWRM such as MWA, SRWB, Water Resources Board and water supply and sanitation is devolved to Local and District Assemblies (Mkandawira, et al, 2011). The new water institutions created in Malawi defy the objectives of IWRM when it comes to cooperation and integration. The pricing structure and the notions of water as a social and economic good remained contested among the various institutions created to promote IWRM. In most countries water and sanitation institutions speak in one voice on issues of water rationing and disconnection of households who are not paying their bills. In Malawi, MWA openly disagreed with SRWB on the reasons for water rationing and prices which may discourage wasteful behaviour.

It became easy for public hospitals to ignore the commercialisation motives of the SRWB because they had an option when their water pressure is lowered by SRWB. When households, schools and other water users were experiencing water rationing and cut-offs, they could rely on MWA in terms of continuing to get access to water services. Consultation and participatory democracy in Malawi remain very basic. Communities which enjoyed participation during the MCP era are no longer participating in water governance after three administrations which claimed to be democratic and participatory (Mkandawira, et al, 2011). Even if communities were given water bills on a monthly basis and informed before SRWB institute water rationing, that could mean a lot to them to get information which is their right. Whilst the communities are being denied access to water and sanitation services due to new confusing institutional arrangements, it is appalling to know that other sectors continue to abstract and divert water resources from rivers and stored in dams, sometimes without licenses. The water withdrawals by the agricultural sector remains above sixty percent due to private property rights dispensation promoted through IWRM (Mkandawira, et al, 2011). As much as water and sanitation remain a human right and mostly provided by public institutions, lack of subsidies to poor households and commercialisation of water boards as a result of the promulgation of the Water Works Act of 1995 remain a challenge (Richter, 2003, North, 1990).

The political patronage demonstrated by various governments after the fall of the MCP dictatorship of Kamuzu Banda also rendered decentralisation difficult to implement. Local authorities were forever bailed out by MIWD. The implementation of the IWRM principle Number 2 calling for management of water resources at a local level faced bottlenecks because the new institutions remained not ready to perform this function. Unfunded mandates, poor performance and poor local leadership remain common challenges experiences since the implementation of IWRM in Malawi. Indranil (2009), Batterbury and Fernado (2006) and other supporters of decentralisation were all proved wrong in Malawi. Like in all developing countries rural areas continue to produce water and urban areas continue to consume water services (Miller, 2003, Klein 1999). Under the leadership of MIWD, Malawi continue to fix water institutions who openly contradict each other and therefore the implementation of IWRM is still expected to strengthen decentralisation and provision of subsidies to poor households in rural and peri-urban areas.

ACKNOWLEDGMENT

The author would like to extend gratitude to the National Research Foundation (NRF) of South Africa for funding this study, as well as comments received from Dr Shauna Mottiar (Univerty of KwaZulu Natal, South Africa), Prof Patrick Bond (University of Western Cape, South Africa), Dr Brigton Chunga (Mzuzu University, Malawi) and all the study participants from different communities and institutions in Malawi for their insights and information.

REFERENCES

Alba, R., & Bolding, A. (2016). IWRM Avant la Lettre? *Four Key Episodes in the Policy Articulation of IWRM in Downstream Mozambique Water Alternatives*, *9*(3), 549–568.

Alba, R., Bolding, A., & Ducrot, R. (2016). The Politics of Water Payments and Stakeholder Participation in the Limpopo River Basin, Mozambique. *Water Alternatives*, *9*(3), 569–587.

Allouche, J. (2016). *The Birth and Spread of IWRM-A Case Study of Global Policy Diffusion and Translation Water Alternatives*. Academic Press.

Batterbury, S. P. J., & Fernando, J. L. (2006). Rescaling Governance and the Impacts of Political and Environmental Decentralisation: An Introduction. *World Development*, *34*(11), 1851–1863. doi:10.1016/j.worlddev.2005.11.019

Bolding, A., & Alba, R. (2016). *Viewpoint-IWRM and I: A Reflective Travelogue of the Flows and Practices Research Team Water Alternatives*. Academic Press.

Bromley, D. W. (1978, March). Property Rules, Liability Rules, and Environmental Economics. *Journal of Economic Issues*, *12*(1), 43–60. doi:10.1080/00213624.1978.11503504

Castro, J. E. (2007, July-December). Water Governance in the Twentieth-First Century. *Ambiente & Sociedale Campinas*, *X*(2), 97–118. doi:10.1590/S1414-753X2007000200007

Chapman, D. (1973, November). Review: Environmental Economics. *Ecological Society of America*, *54*(6), 1414–1415.

Corbin, J., & Strauss, A. (2008). *Basics of Qualitative Research*. Sage.

Denby, K., Movik, S., Mehta, L., & van Koppen, B. (2016). The 'Trickle Down' of IWRM: A Case Study of Local-Level Realities in the Inkomati Water Management Area, South Africa. *Water Alternatives*, *9*(3), 473–492.

Denzin, N. K., & Linoln, Y. S. (Eds.). (2008). *Collecting and Interpreting Qualitative Materials*. Sage.

Derman, B., & Manzungu, E. (2016). The Complex Politics of Water and Power in Zimbabwe: IWRM in the Catchment Councils of Manyame, Mazowe and Sanyati (1993-2001). *Water Alternatives*, *9*(3), 513–530.

Derman, B., & Prabhakaran, P. (2016). *Reflections of the Formulation and Implementation of IWRM in Southern Africa from a Gender Perspective Water Alternatives*. Academic Press.

Ferguson, A. E., & Mulwafu, W. O. (2004) *Decentralization, Participation and Access to Water Resources in Malawi*. University of Wincosin-Madison.

Fiani, R. (2004, December). An Evaluation of the Role of the State and Property Rights in Douglas North's Analysis. *Journal of Economic Issues*, *38*(4), 1003–1020. doi:10.1080/00213624.2004.11506754

Gandhi, V., Crase, L., & Herath, G. (2007). *Determinants of Institutional Success for Water in India: Results from a Study across Three States*. Paper presented at the 51st Annual Conference of the Australian Agricultural and Resource Economics Society (AARES), Queenstown, New Zealand.

Gandhi, V. P., & Namboodiri, N. V. (2009). *Groundwater Irrigation in India: Gains, Costs and Risks*. Indian Institute of Management.

Gray, D. E. (2009). *Doing Research in the Real World* (2nd ed.). Sage.

Hamilton, D. (1989). Thorstein Veblen as The First Professor of Marketing Science. *Journal of Economic Issues*, *23*(4), 1097–1103. doi:10.1080/00213624.1989.11504976

Hoekstra, A. Y. (2006). *The Global Dimension of Water Governance: Nine Reasons for Global Arrangements in order to cope with local water problems*. United Nations Education, Science and Cultural Organization-IHE Institute for Water Education.

Holstein, J. A., & Gubrium, J. F. (2008). *Handbook of Constructionist Research*. The Guilford Press.

Indranil, D.E. (2009, Jan. 3). Can Decentralisation Improve Rural Water Supply Services. *Economic and Political Weekly*.

Jonch-Clausen, T. (2005). *Integrated Water Resources Management (IWRM) and Water Efficiency Plans by 2005 Why, What and How?* Global Water Partnership.

Klein, P. G. (1999). *New Institutional Economics*. University of Georgia.

Livingston, M. L. (2005). Evaluating Changes in Water Institutions: Methodological Issues at the Micro and Meso levels. *Water Policy*, *7*(1), 21–34. doi:10.2166/wp.2005.0002

Manzungu, E., & Derman, B. (2016). Surges and Ebbs. *National Politics and International Influence in the Formulation and Implementation of IWRM in Zimbabwe Water Alternatives*, *9*(3), 493–512.

May, T. (Ed.). (2002). Qualitative Research in Action. London: Sage Publications.

Mehta, L., Movik, S., Bolding, A., Derman, B., & Manzungu, E. (2016). *Introduction to the Special Issue- Flows and Practices: The Politics of Integrated Water Resources Management (IWRM) in Southern Africa Water Alternatives*. Academic Press.

Miller, E. S. (2003, March). Evolution and Stasis: The Institutional Economics of David Hamilton. *Journal of Economic Issues*, *37*(1), 1045–1059. doi:10.1080/00213624.2003.11506554

Mkandawire, T., Mulwafu, W., Chipofya, V., Bota, S., & Kainja, S. (2011). *The Road to a National Integrated Water Resources Management / Water Efficiency (IWRM/WE) Plan: Challenges and Lessons from Malawi*. University of Malawi.

Movik, S., Mehta, L., van Koppen, & Denby, K. (2016). *Emergence, Interpretations and Translations of IWRM in South Africa Water Alternatives*. Academic Press.

Movik, S., Mehta, L., & Manzungu, E. (2016). The Flow of IWRM in SADC: The Role of Regional Dynamics. *Advocacy Networks and External Actors Water Alternatives*, *9*(3), 434–455.

Nabli, M. K., & Nugent, J. B. (1989). The New Institutional Economics and its Applicability to Development. *World Development*, *17*(9), 1333–1347. doi:10.1016/0305-750X(89)90076-4

Nicol, A., & Odinga, W. (2016). *IWRM in Uganda- Progress after Decades of Implementation Water Alternatives*. Academic Press.

North, D. C. (1990). *The New Institutional Economics and Development*. Washington University.

Pearce, D. (2002). An Intellectual History of Environmental Economics. *Annual Review of Energy and the Environment*, *27*(1), 57–81. doi:10.1146/annurev.energy.27.122001.083429

Quiggin, J. (2000). *Environmental Economics and the Murray-Darling River System*. Australian Research Council.

Richter, R. (2005). The New Institutional Economics – Its Start, Its Meaning, Its Prospects. *The European Business Organisation Law Review*, *6*(2), 161-200.

Rogers, P. (2002). *Water Governance in Latin America and the Caribbean*. Inter-American Development Bank.

Sarantakos, S. (2005). *Social Research* (3rd ed.). Palgrave MacMillan. doi:10.1007/978-0-230-20901-5

Selee, A.D., & Tulchin. (2004). Decentralization and Democratic Governance: Lessons and Challenges. In *Decentralization, Democratic Governance, and Civil Society Perspective- Africa, Asia, and Latin America*. Woodrow Wilson Center Press.

Solanes, M., & Gonzalez-Villarreal, F. (1999). *The Dublin Principles for Water as Reflected in a Comparative Assessment of Institutional and Legal Arrangement for Integrated Water Resources Management Sweden*. Global Water Partnership/ Swedish International Development Cooperation Agency.

Spash, C. L. (1999). The Development of Environmental Thinking in Economics. *Environmental Values, 8*(4), 413–435. doi:10.3197/096327199129341897

The Sustainable Livelihoods In Southern Africa. (2008). *The Politics of Decentralisation in Southern Africa.* Future Agriculture Consortium Policy Process Briefing 2008: The Sustainable Livelihood in Southern Africa.

Van den Bergh, J. C. J. M. (2007). Evolutionary thinking in environmental economics. *Journal of Evolutionary Economics, 17*(5), 521–549. doi:10.100700191-006-0054-0

Van Koppen, B., Tarino, A. K. P. R., van Eeden, A., Manzungu, E., & Sumuni, P. M. (2016). *Winners and Losers of IWRM in Tanzania Water Alternatives*. Academic Press.

Weber, M. L. (2001). Markets for Water Rights under Environmental Constraints. *Journal of Environmental Economics and Management, 42*(1), 53–64. doi:10.1006/jeem.2000.1149

Welle, K. (2005). *Learning for Advocacy and good practice- Water Aid Water Point Mapping- Report of Findings based on country visits to Malawi and Tanzania*. Overseas Development Institute and Water Aid.

Williamson, O. E. (1995). *The Institutions and Governance of Economic Development and Reform*. The International Bank for Reconstruction and Development.

Williamson, O. E. (2000). The New Institutional Economics: Taking Stock, Looking Ahead. *Journal of Economic Literature, 38*(September), 595–613. doi:10.1257/jel.38.3.595

Wunsch, J. S. (2001). Decentralisation, Local Governance and "Recentralization" in Africa, Public Administration and Development. *Public Admin. Dev, 21*, 277–288.

Chapter 4
Integrated Water Resource Management, Sustainability, and Pollution Abatement in Malawi:
An Ecological Economics Perspective

Innocent Simphiwe Nojiyeza
https://orcid.org/0000-0002-9948-2147
University of Zululand, South Africa

ABSTRACT

With the introduction of IWRM, local government institutions are expected to control water-related pollution, which is often associated with new water resource management, supply, and sanitation responsibilities devolved to them as part of decentralisation. The aim of this study was to look into eco-efficiency, environmental performance, physical scarcity and economic scarcity trends, sustainable variables, and externalities associated with the introduction of IWRM in Malawi. Focus group discussions and interviews were conducted with policymakers and households in Ntcheu, Mangochi, and Balaka. Findings confirmed that the challenges local levels of government face in managing water infrastructure and formulation of pollution control measures are a result of the fact that adoption of IWRM happened without any readiness to do so, especially in terms of water-related pollution control.

INTRODUCTION

Studies on IWRM in the third world and Malawi in particular such as Chiluwe and Nkhata (2014), and Gleick (1998) focus on water governance and institutional mechanisms as part of the ecosystem in order to ensure sustainability of the resource. The missing link in critics of IWRM is recognising that water is an intricate part of the ecosystem and therefore sources of water such as rivers and aquifers should be treated as a natural capital (Gleick, 1998, Wackernagel and Rees, 1997). Wackernagel and Rees (1997)

DOI: 10.4018/978-1-7998-8809-3.ch004

define natural capital as 'a stock of natural assets that is capable of producing a sustainable flow'. Water is a catalyst for development, a contributor to national and natural income and a replenishable current (asset) stock which is renewable (Wackernagel and Rees, 1997). It is thus imperative to prevent water pollution, ensure that rivers continue to flow and also to prevent groundwater depletion (Gleick, 2000, Postel, 2000). Local authorities and agencies responsible for water distribution are therefore expected to ensure the sustainability of this ecological service through sound management of water schemes, availability of treatment and purification facilities in order to prevent water related and water borne diseases which are symptomatic of developing countries such as Malawi (Gleick, 1998, Postel, 2000, Gleick, 2000).

The Dublin Principle of "managing water at a lowest level possible" need to factor in the management of water as part of the land. The other consideration is to treat water as part of the vital ecosystem. This will assist in strengthening pollution abatement measures and thus deal with externalities whilst ensuring that physical and economic scarcity is not perpetuated (Gleick, 1998, Chiluwe and Nkhata, 2014, Postel, 2000). Ecological scarcity is deepened by water withdrawals for agricultural and industrial use. The construction of large dams which are more than 15 metres high, negatively affect 'ecological functions of rivers and freshwater ecosystems' (Postel, 2000). Poorly prepared and unfunded local authorities face challenges in dealing with 'ecological degradation' of water resources which affect sustainability and results in ecological suicide. Agricultural use of water often results in fertiliser spillages to groundwater sources, aquifers and rivers whilst domestic water use may result in faecal matter being discharged into rivers when waste water treatment facilities break (Gleick, 1998). There are no substitutes for equitable, renewable, ecologically sound and sustainable water resources (Wackernagel and Rees, 1997).

Water is biophysical in nature and pollution reduces it to just another factor of production as postulated by classical economists (Christensen, 1989), and Malthus (1815-1836), Ricardo (1817) and other neoclassical economists tend to include water as a subject of diminishing return in its various stages of the water cycle (Stern, 1997). Christensen (1989) challenges the classical theory of production by including 'the low-entrophy energy and materials extracted from environmental systems and eventually returned as waste' in his definition of the biophysical nature of water. This approach allows one to deal with diseconomies and double externalities which have an impact on sustainable development (Rennings, 2000). The value of water within sustainable development debates (Farber, Costanza and Wilson, 2002), ecosystem functions, goods and services (de Groot, Wilson and Boumans, 2002) and weaknesses of ecological footprint (Fiala, 2008, van den Bergh and Verbruggen, 1999) are some of the arguments which add light to poor pollution abatement and controls in the water and sanitation sector in Malawi.

The high levels of pollution highlighted in this study are traced to the IWRM notion of valuing water and therefore manage it as an economic and not public good (Hornborg, 1998) which result in marketization of pollution control measures and collective action undertaken by some actors as a response to ecological degradation (Paavola and Adger, 2005). I will be using ecological efficiency, sustainability, externality, scarcity and environmental performance in shedding light on unsustainable management of water resources, poor pollution control measures and the role of economic scarcity of water resource management in Malawi. In this chapter, I use various themes drawn from Ecological Economics to highlight the weaknesses of IWRM implementation especially decentralisation and marketization to highlight unsustainable and inefficient pollution abatement strategies adopted by local authorities.

Methodology

I conducted focus group discussions (2), household interviews (30) and engaged with policy makers (8) in Ntcheu, Mangochi and Balaka areas located within the Southern Region of Malawi. The community participants were chosen using a systematic random sampling. They participated based on their availability during the days in which interviews were conducted. Policy makers who participated in the study were chosen using a purposive sampling strategy. The criteria used was their knowledge of IWRM and the work they do in the water and sanitation sector in Malawi. The Civil Society participants in each of the two focus group discussions were chosen using snowballing method. All CSO participants were referrals by WES NGO (Water, Environment and Sanitation NGO) network. The language used in most interviews were Chichewa for most of the households, English in focus group discussions and interviews conducted with various state officials. As a long serving environmental justice practitioner, it was easy to identify state officials as well as CSO participants for my study. I also used observation as a data collection tool. I kept a field diary where I wrote all notes from interviews and focus group discussions. I used semi structured interviews with state officials as well as community participants. I analysed my data using thematic analysis and my findings and discussion is based on these themes. All participants got involved in the study voluntarily, they were assured of their privacy and could withdraw at any time without any repercussions. All participants signed an informed consent forms prior to either participation in interviews or focus group discussions. The experiences of state officials, members of the community and Civil Society Organisations (CSOs) are presented in the next section of this chapter.

Findings and Discussions

The findings section of this chapter is divided into four themes, namely eco-efficiency and sustainability, environmental performance, physical and economic scarcity and externalities. The emerging cross-cutting themes are further sub-divided into four themes (1) the water and pollution situation in Malawi in the form of borehole operation and maintenance, water-related and water-borne diseases, pollution abatement, treatment and purification protocols (2) the role of the national and local government in each of the above mentioned issues (3) the role of the communities who are at the receiving end of clean, purified and treated water supply and diseases burdens (4) environmental issues. The theories are often interwoven and thus not necessarily presented in this order.

Eco Efficiency and Sustainability

The criteria used to judge whether a water scheme is sustainable, is to look at whether it is supply-led or demand-led (Gleick, 1998). In 1980s, the Government of Malawi constructed Mpira Dam as part of Mpira-Balaka water scheme. Communities of Ntcheu, Mangochi and Balaka were expected to benefit from getting access to water. The water scheme was constructed on a supply measures and in the late 1990s when the United Democratic Front (UDF) took over the governance of Malawi from the Malawi Congress Party (MCP), they introduced demand-led measures as part of the Structural Adjustment Programme (SAP). The challenge which faced the UDF government was whether to continue providing water as part of a supply-led approach and fully sponsored by the government. The other challenge that they faced was regarding the sustainability of the scheme (State Official Interview). The Mpira Dam was constructed for a population that is far smaller than the current populations of Balaka, Ntcheu, and Mangochi.

The government of Malawi is grappling with sustainable development, as advocated by scholars such as Schaltegger and Synnesvedt (2002:343). ‘Sustainable development is development that meets the needs of the present without compromising the ability of future generations to meet their own needs’ (WCED, 1997 cited in Gleick, 1998, see also Wackernagel and Rees, 1997). The biggest sustainability challenge in Malawi is the lack of operation and maintenance budgets in many schemes. “The fact that the number of people who have access to water has declined by 32% since 2001 due to the breakdown of various systems, clearly indicates that sustainability remains a big challenge. In government boreholes there are no tariffs levied to households, but whenever there is a breakdown, communities contribute towards the maintenance of the borehole through their committees (An interview held with a state official).” The borehole at Ntonya Village was constructed in 1988 and is still functioning to design capacity (Community participant associated with the scheme since its inception). Since there is no alternative source of drinking water, households collect funds if there is a breakdown in the facility. Since 1988 households have been saving K10 per household per month to meet operation and maintenance needs. However, there has been instances where government conducted repairs without community member’s contributions (NGO Focus Group 1).

When households are looking for spare parts, they are readily available in the Balaka market centre. After purchasing parts, the Mpira Authority is invited to assess the problem and recommend technical solutions. Engineers at the Mpira Water Authority provide advice regarding the purchasing of spare parts or conducting an overhaul of the water infrastructure if it is beyond repair (State Official Interview). Unlike the Ntonya Village households, Community No. 8 and Community No. 26 do not have a bank account in which operation and maintenance funds can be kept (community participant interview). The borehole committee for No. 8 Borehole advised government to make seed money available for communities to sustain water and sanitation projects. Members of the committee have been volunteering since 1987 and suggested that an incentive in the form of stipends can contribute towards sustaining committee members who manage the public taps (NGO Focus Group Discussion 2). In other schemes, households benefit from their association with international NGOs such as World Vision.

World Vision officers based in Balaka acknowledge that even though spare parts are available, they are expensive since they are imported from South Africa, and sometimes difficult to find. Most NGOs that operate water and sanitation projects help with the cleaning and maintenance of pipes, but do not provide plumbing which requires advanced technical skills. In order to ensure that a project is sustainable, World Vision requires that each community raises an operation and maintenance fee of K15 000. Fundraising prior to the construction of a borehole contributes to community ownership of the project. The K15 000 is kept in an account and used for operation and maintenance while World Vision contributes between K1 and K2 million towards the construction of infrastructures. Projects such as the Mpira Water Scheme contributes to the reduction of waterborne diseases, and projects established to complement the schemes have reduced cholera prevalence over the past three years (State and NGO Official Interviews, See Scanlon et al, 2016 on the role of non-state actors in water service provision).

In Balaka, there are incidences of waste being dumped in water resources, and incidences of environmental degradation are very common. The Balaka District Assembly does not collect waste from households. Similar situations were reported at Ntcheu and Mangochi. Such unsustainable practices arose as the result of inappropriate choices of marginal discount rates. Therefore, sustainability became oriented to restoring water flows, ecological reserve and water quality in the case of dams and rivers. The

Mudi River in Blantyre is heavily polluted with faecal bacteria counts as high as 20 000 faecal coliforms per 100 millimetres (Interview held with a state official on 08 August 2012 and another interview held with a state official on 13 August 2012).

Through observation it was discovered that accumulated sludge from pit latrines and septic tanks has often not been removed, leading to overflow problems in sewerage systems that suffer from operational problems associated with vandalism. Water-related vector-borne diseases such as malaria, typhoid, cholera, and bilharzia have infected most people around the irrigation schemes in Malawi. It is difficult to spend an evening in Ntcheu, Mangochi, and Balaka without using a mosquito net (Community Participant Interview). There are no civil society structures to engage government on issues of environmental degradation in Malawi. Within local authorities, there are no departments advocating Local Agenda 21, monitoring water quality, and promoting river health, and even the Ministry of Irrigation and Water Development has not developed or agreed upon criteria for local authorities to account for environmental degradation (State Official Interview).

Field visits to the treatment and purification plant at Mpira in 2008 and 2012, revealed that there was a lack of chlorine in the water purification process (State Official Interview). The state official responsible for water resource management at the Ministry of Irrigation and Water Development, downplayed the lack of chlorine as a normal problem in developing countries, and blamed local authorities for failure to perform duties that were devolved to them as a result of the Decentralisation Act of 1998. Exactly 10 years after the promulgation of the Decentralisation Act of 1998 and two years after the adoption of IWRM (2010), not a single local authority had established water and sanitation departments or environmental or catchment management departments. In the absence of such institutions, ecosystems are irreversibly lost to future generations. The findings also confirm the viewpoint advocated by Soderbaum (1990:484) and Pindyck (2007:49) that environmental problems are extended in time often across administrative boundaries and sectors. Environmental problems caused by a municipality that mismanages a waste water treatment plant could affect other municipalities, agriculture, the economy, industrial users, and domestic users and may not be confined to one sector; sustainable development is therefore critical.

The fact that the MIWD was hoping that local authorities would raise enough money to purchase the chlorine required for water purification and that district water and environmental officers who report to principal secretaries within their national ministries would drive environmental protection and prevent water degradation, illustrate the faults in the system. There is no agreement on equitable sharing of revenue between local authorities and national government for functions that are devolved. The Southern Regional Water Board is expected to treat water before it is distributed to households, but there are no cooperative governance arrangements agreed upon between water boards and the National Water Resources Board on how to deal with environmental degradation. Water boards are expected to monitor water quality within the water supply systems and promote catchment management and pollution control (Ng'ong'ola, 1999:11).

The Ministry of Irrigation and Water Development is also responsible for the regulation of water quality standards that need to be adhered to by water boards. The process of how this is done in terms of benchmarks and reporting to the legislature and holding water boards accountable, is not clear. The ministries in charge of natural resources are responsible for conservation, protection of catchment areas, protection and rehabilitation of river catchments, eradication of aquatic weeds, and coordinating all cross-cutting environmental activities required for water resources management and water services. The implementation of IWRM is still at an early stage in Malawi because promotion of health and hygiene education in water and sanitation services, and monitoring and providing guidance concerning the quality of drinking water, is the responsibility of the Ministry of Health and not Irrigation and Water Development. The Ministry of Health is also responsible for providing appropriate interventions to prevent the prevalence of water-related diseases and water demand management for public health institutions which is also a function of the Ministry of Environment (Chipofya, et al, 2010:265, Mkandawire et al, 2011:11). The roles of various national ministries are not clear and it is even worse when it comes to environmental degradation and over-consumption of water by one stakeholder at the expense of others.

Environmental Performance

Studies show that the quality of water in Malawi is poor. The majority of households who participated in the study draw water from boreholes and consume untreated water. Only households who are part of projects managed by NGOs and the Ministry of Health (in particular environmental health offices in the Balaka, Mangochi, and Ntcheu District Hospitals) who benefit from doses of chlorine. Schalteggar and Synnestvedt (2002) postulate a link between economic performance and environmental performance and results show that economic growth and development is low in the areas in which the study was conducted and therefore environmental performance is also poor. In market centres such as Balaka, Ntcheu, and Mangochi one can expect general cleanliness and maintenance of public toilets and water points by the local assemblies. But in the public transport ranks there is no maintenance staff, no planted trees, and unkempt verges.

There are no parks and gardens departments in all three local assemblies responsible for the market centres. The public toilets that households use at a cost of K5 per person are not cleaned at all. People resort to visiting bread and breakfasts and motels to access clean public toilets in market centres. The argument put forward by Schaltegger and Synnesvedt (2002:340), Sagoff (2000:1426), Munasinghe (1993:127) and Pindyck (2007:46), that willingness to pay for environmentally friendly services depends on environmental and health regulations and stakeholder pressure to maintain acceptable quality standards, developed and enforced at country level, is equally true in Malawi. People who use public latrines in market centres are doing so because there are no other alternatives available. In 2012, there was water rationing in most of the areas serviced by the Southern Regional Water Board and a lack of water for over 4 weeks with effect from 10 July 2012. Water only flowed during the night and hospitality industries were unable to provide water to flush toilets; buckets were used to bath and flush toilets. Hand washing facilities are still available in most motels and public toilets and the use of buckets for hand washing and flushing toilets did not affect the occupancy rate in both these facilities.

There is no evidence of stakeholder pressure to maintain acceptable quality standards regarding river health, and spillage of sewage into water bodies such as groundwater, aquifers, and boreholes. There are no environmental action groups in Balaka, Ntcheu, and Mangochi putting pressure on government to adhere to acceptable water quality standards. The few NGOs that participated in the study are more involved in financially supporting water point construction, latrine construction, and health and hygiene promotion. They are equally not putting any pressure on either the Mpira Water Authority or the Southern Regional Water Board to deal with effluent discharges into river systems. The viewpoint articulated by Synnestvedt (2002:342) and Sederbaum (1990:484) that there are economic disincentives for corporate environmental protection which results in companies putting more focus on short-term profit maximisation, is clearly visible in Balaka, Ntcheu, and Mangochi. Both the Southern Regional Water Board and the Mpira Water Authority are not active in environmental protection; their main focus is selling water to households. The Mpira Water Authority is mainly preoccupied with selling bulk water to the Southern Regional Water Board and reticulates directly to rural households within its jurisdiction. Even the operation and maintenance of the water infrastructure is not receiving attention as it is not contributing to profit maximisation.

No environmental cost accounting, investment appraisal, other environmental management systems, or ecological control systems are followed by the Southern Regional Water Board. There is no evidence in the jurisdiction of the Southern Regional Water Board of catchment protection, stream cleaning, and removal of alien plants and invader weeds as part of river health restoration. Even the Water Resources Board is confined to water licensing, and the catchment protection duties of the Mpira Water Authority are not performed due to lack of funds and the assumption that the Ministries Responsible for Natural Resources such as Mines, Forestry, Fisheries, Lands, Environment, Parks, and Wildlife will include the state of water resources in the state of natural resources report and act on the findings regarding environmental degradation (Chipofya, et al, 2010:264).

There is a lack of what Schaltegger and Synnesvedt (2002:345) and Soderbaum (1990:484) regard as optimum environmental performance which is achieved when a water and sanitation utility establishes ecological control measures. This should include ensuring that production processes are optimised to reduce costs through environmental tools which support the identification of eco-efficiency potentials and the successful implementation of eco-efficiency enhancing measures. There are no ecologically sound sanitation facilities that use less than six litres of water per flush or ecologically sound sanitation proposals in the Sanitation Policy of 2007. The latrines constructed in all 400 projects implemented by the Mpira Water Authority are mere pit latrines with ventilation and not constructed according to ecological sanitation principles.

Physical and Economic Scarcity

There are various views in Malawi regarding the scarcity of water. There is general consensus that Malawi is considered to be endowed with relatively rich natural water resources in the form of rivers, lakes, and aquifers (Njalam'mano, 2007:4, FAO, 2007, Ng'ongo'la, 1999:9, Ferguson and Mulwafu, 2004:3, Chipofya, Kainja and Bota, 2010:261, Mkandawire, et al, 2011:1). Scarcity is therefore not linked to the production possibility curve traced by Massarutto (2007:2) in the neo-classical economic sense, nor to the total volume of water needed divided by country's total renewable water resources as outlined by Hoeksta (2006:20), Spash (1999:415) and Munasinghe (1993:126), but to economic scarcity.

There is no evidence of physical scarcity in Malawi. The country's total renewable water resources are estimated at 30 cubic kilometres per year. Ng'ongo'la (1999:9), Ferguson and Mulwafu, (2004:3) estimate renewable freshwater resources of about 3,000 cubic metres per capita per year, but distribution is irregular and varies according to class and economic standing. Physical scarcity in Malawi is linked to water availability constrained by hydrological factors, opportunity costs of water and sanitation, the costliness of water services, the construction of reticulation water pipes, the availability of waste water treatment plants, the availability and management of purification plants, sufficient water to flush toilets, and the availability of water storage facilities with sufficient quantities of bulk water that can be reticulated for domestic consumption.

Regarding the costs of water, it is clear that households served by the Southern Regional Water Board pay more for water than those served by the Mpira Water Authority. The water pipes used to reticulate water to households are old and require rehabilitation since the scheme came into operation in 1987. An under-resourced purification plant at Mpira Dam results in un-purified water being sent to households in some instances. Employees who manage the purification plants are not adequately qualified to perform their duties; for instance, the Mpira Water Authority relies on Engineers Without Borders (EWB) for additional capacity to perform project management functions. Households in the areas in which the study is conducted have very few flush toilets which made it difficult to establish whether there is enough water to flush toilets or not. According to the National Statistical Office (2009:64) the number of households who have access to water in Ntcheu, Balaka, and Mangochi are estimated at 78%, 86%, and 91%, respectively. The time taken to get to the nearest supply of drinking water is estimated at between 0-14 minutes for about 81% of Ntcheu households, 81% of Balaka households, and 92% of Mangochi households respectively.

The estimates of the NSO (2009:64) were confirmed in the field where similar results were established, but the NSO underestimated water rationing which results in longer distances and more than 60 minutes taken by households looking for water. In the Shire River system there are 62 dams which include the Mpira Dam. In the other 16 catchment areas in Malawi there are 687 dams. Generally, there are adequate water storage facilities in Malawi; the challenge is capacity to increase water supply, purify and treat water due to lack of infrastructures, and to find the human resources to do so.

Saleth and Dinar (2004) and Pearce (2002:60) highlight economic scarcity which is also visible in Malawi. This refers to the set of institutional rules, property rights and shared cultures in any given historical context, appropriate regulatory systems, and institutional mechanisms. The institutional mechanisms in Malawi are the Ministry for Irrigation and Water Development responsible for water supply and sanitation in rural areas, and water boards responsible for market and urban centres. With the introduction of IWRM in 2010, the functions of the MIWD are now devolved to the Mpira Water Authority. The system works well in rural areas where households have access to water even if they are disconnected by the Southern Regional Water Board, but peri-urban and urban households have no access to water if they are disconnected. Rural communities who are only served through the Southern Regional Water Board lack access during water rationing and disconnections. It is not the lack of water as a physical resource that results in scarcity, but institutional mechanisms and rules that result in lack of water in households in the Ntonya village. Community No. 8, for example, is not experiencing scarcity.

Externalities

Poor quality of water resources is caused by agricultural activities such as the application of agrochemicals, and poor farming practices and land husbandry. The other cause is the disposal of industrial and human waste, particularly in urban areas. In 2012, there was a sewage blockage next to Community No. 8 and the disposal of human waste was visible in the street. The observation by Mkandawire et al (2011:2) regarding pollution from agricultural activities and human waste is confirmed by findings in the Ntcheu Market Centre where public toilets are sometimes blocked for more than two weeks due to water rationing. Human waste was visible in different parts of the city and no authority accepts the responsibility for removing the waste. This is in line with views shared by many scholars (Stavins, 2004:1, Weber, 2001:53, Van den Bergh, 2007:529, Diets and van der Straaten, 1992:28) that government should take action to deal with externalities and come up with laws and penalties for visible diseconomies caused by the private sector.

There is evidence in Malawi that the government is not considering imposing taxes on emissions equal to the cost of the related damages at an efficient level of control as proposed by Pindyck (2007:53), Cropper and Oates (1992:680), Stavins (2004:1) and Pearce (2002:58). There is no evidence of companies paying pollution taxes. In the absence of taxes and levies, the government is subsidising the water treatment and purification system. Even private providers of treatment facilities are not subjected to additional taxes and levies associated with externalities and in particular effluents discharged in the river sources. Even the agricultural sector cited as the main source of externalities is immune to pollution taxation. Households who bear the brunt of pollution end up being the only ones paying for externalities. In the Coasian framework, both polluters and victims are supposed to pay for externalities, in the sense that it may make broader economic sense to allow an efficient producer to impose these costs on others. In Malawi, polluters are high income agents and the victims are low income agents as postulated by Coase (Weber, 2001:53, Cropper and Oates, 1992:680, Pearce 2002:61); in any case, there are no mechanisms to hold polluters accountable and make them pay for the externalities.

CONCLUSION

The implementation of IWRM in Malawi especially pollution control measures, faces many challenges covered in this article. There is no debate about water resources availability in the form of lakes, rivers, and aquifers. I conclude that the scarcity some households experience is not related to lack of water as a physical resource, but economic one experienced due to institutional challenges. Water resource development infrastructures that include dams, policies in water resource development, irrigation, and sanitation are steps in the right direction. The lack of operation and maintenance of water and sanitation infrastructures remains a setback in IWRM implementation. The roles of various stakeholders outlined in national water and sanitation policies are clear, however overlaps create confusion. Financial management and sustainable development, enforcement of the Water Resources Act of 1969 and 2013, and the implementation of the Decentralisation Act of 1998 prior to the establishment of local authorities, are setbacks in the implementation of IWRM. The challenges of waterborne diseases, poor coverage in rural areas and of schools, weak sector leadership, poor data, hydrological and climate change, are matters which require further attention. The unique context of IWRM implementation in Malawi is also a matter which policy makers need to factor in when scaling up decentralization of water resources management, especially pollution abatement.

The role of the MIWD, SRWB, Mpira Water Authority, WUAs, and tap and borehole committees were clearly outlined in various policies and laws as part of an IWRM institutional framework with special focus to their roles in pollution control and sustainable development. The challenge remains lack of communication and lack of integration by these various institutions. The different pricing mechanisms used by the SRWB and the Mpira Water authority, and steps taken by health centres were also examples of this lack of integration. Lack of participation of households in tariff reforms were cited as examples of poor IWRM implementation which tend to undermine sustainable development. There are areas such as water and sanitation governance, institutional frameworks and mechanisms that need improvement. Property rights skewed in favour of private property and market mechanisms, have negative impacts on water supply and sanitation delivery. Even though the number of households that have access to water and sanitation is estimated to be above 70% in Ntcheu, Balaka, and Mangochi, water rationing and competing institutional mechanisms undermine sustainable access to water for everybody, always. High levels of pollution and lack of equity also poses problems.

Local government structures are new in Malawi (they were elected in 2014) and therefore it will be important to conduct a follow-up study when such institutions are well established and new institutional mechanisms on pollution abatement reach a matured stage.

ACKNOWLEDGMENT

The author would like to extend gratitude to the National Research Foundation (NRF) of South Africa for funding this study, as well as comments received from Dr Shauna Mottiar (Univerty of KwaZulu Natal, South Africa), Prof Patrick Bond (University of Western Cape, South Africa), Dr Brigton Chunga (Mzuzu University, Malawi) and all the study participants from different communities and institutions in Malawi for their insights and information.

REFERENCES

Ayres, R. U. (1998). Eco-thermodynamics: Economics and the second law. *Ecological Economics*, *26*(2), 189–209. doi:10.1016/S0921-8009(97)00101-8

Biswas, A. K. (2008). Integrated Water Resources Management: Is it Working? *International Journal of Water Resources Development*, *24*(1), 5–22. doi:10.1080/07900620701871718

Chipofya, V., Kainja, S., & Bota, S. (2010). Policy harmonisation and collaboration amongst institutions- A strategy towards sustainable development, management and utilisation of water resources: Case of Malawi. *Desalination*, *252*, 261–266.

Chirwa, E., & Junge, N. (2007). *Poverty and Social Impact Analysis: Private Sector Participation in the Distribution and Management of Water Services, A study of Low income Areas in Blantyre and Lilongwe*. Ministry of Economic Planning and Development and Ministry of Irrigation and Water Development.

Christensen, P. P. (1989). Historical Roots for Ecological Economics- Biophysical Versus Allocative Approaches. *Ecological Economics*, *1*(1), 17–36. doi:10.1016/0921-8009(89)90022-0

Chuluwe, Q., & Nkhata, B. M. (2014). Analysis of Water Governance in Malawi: Towards a Favourable Enabling Environment? *Journal of Water, Sanitation, and Hygiene for Development: a Journal of the International Water Association*, *1*(2), 313–323. doi:10.2166/washdev.2014.087

Cropper, M. L., & Oates, W. E. (2002). Environmental Economics. A Survey. *Journal of Economic Literature*, *30*(2), 675–740.

De Groot, R. S., Wilson, M. A., & Boumans, R. M. J. (2002). A Typology for the Classification, Description and Valuation of Ecosystem Functions, Goods and Services. *Ecological Economics*, *41*(3), 393–408. doi:10.1016/S0921-8009(02)00089-7

Dietz, F. J., & van der Straaten, J. (1992, March). Rethinking Environmental Economics: Missing Links between Economic Theory and Environmental Policy. *Journal of Economic Issues*, *26*(1), 27–51. doi: 10.1080/00213624.1992.11505261

Farber, S. C., Costanza, R., & Wilson, M. A. (2002). Economic and Ecological Concepts for Valuing Ecosystem Services. *Ecological Economics*, *41*(3), 375–392. doi:10.1016/S0921-8009(02)00088-5

Ferguson, A. E., & Mulwafu, W. O. (2004) *Decentralization, Participation and Access to Water Resources in Malawi.* University of Wincosin-Madison.

Fiala, N. (2008). Measuring Sustainability: Why the Ecological Footprint is Bad Economics and Bad Environmental Science. *Ecological Economics*, *67*(4), 1–7. doi:10.1016/j.ecolecon.2008.07.023

Gleick, P. H. (1998). Water in Crisis: Paths to Sustainable Water Use. *Ecological Applications*, *8*(3), 571–579. doi:10.1890/1051-0761(1998)008[0571:WICPTS]2.0.CO;2

Gleick, P. H. (2000). The Changing Water Paradigm- A Look at Twenty-first Century Water Resources Development. *Water International*, *25*(1), 127–138. doi:10.1080/02508060008686804

Hoekstra, A. Y. (2006). *The Global Dimension of Water Governance: Nine Reasons for Global Arrangements in order to cope with local water problems*. United Nations Education, Science and Cultural Organization-IHE Institute for Water Education.

Hornborg, A. (1998). Towards an Ecological Theory of Unequal Exchange: Articulating World System Theory and Ecological Economics. *Ecological Economics*, *25*(1), 127–136. doi:10.1016/S0921-8009(97)00100-6

Martinez-Alier, J., Munda, G., & O'Neill, J. (1998). Weak Comparability of Values as a Foundation for Ecological Economics. *Ecological Economics*, *26*(3), 277–286. doi:10.1016/S0921-8009(97)00120-1

Massarutto, A. (2007). *Urban water conflicts: an ecological – economics approach*. United Nations Educational and Cultural Organisation.

Mkandawire, T., Mulwafu, W., Chipofya, V., Bota, S., & Kainja, S. (2011). *The Road to a National Integrated Water Resources Management / Water Efficiency (IWRM/WE) Plan: Challenges and Lessons from Malawi*. University of Malawi.

Munasinghe, M. (1993). Environmental Economics and Biodiversity Management in Developing Countries. *Ambio,* Vol. 22, No 2/3, *Biodiversity: Ecology, Economics. Policy*, (May), 126–135.

National Statistical Office of Malawi. (2008). *2008 Population and Housing Census. Zomba*. National Statistical Office.

National Statistical Office of Malawi. (2009). *The Welfare Monitoring Survey. Zomba*. National Statistical Office.

National Statistical Office of Malawi. (2010). *Malawi Demographic and Health Survey 2010: Preliminary Report. Zomba*. National Statistics Office.

Ng'ong'ola, D. H. (1999). *Policies Influencing Patterns of use of Water Resources in Malawi*. Ministry of Irrigation and Water Development.

Njalam'mano, J. B. J. (2007). *Current Status and Management of Hand Pump Equipped Water Facilities in Blantyre Rural District, Malawi: a case study of Kapeni and Lundu Traditional Authorities*. Linkoping University.

Paavola, J., & Adger, W. N. (2005). Institutional Ecological Economics. *Ecological Economics*, *53*(3), 353–368. doi:10.1016/j.ecolecon.2004.09.017

Pearce, D. (2002). An Intellectual History of Environmental Economics. *Annual Review of Energy and the Environment*, *27*(1), 57–81. doi:10.1146/annurev.energy.27.122001.083429

Pindyck, R. S. (2007). *Uncertainty in Environmental Economics*. Oxford University Press.

Postel, S. L. (2000). Entering an Era of Water Scarcity: The Challenges Ahead. *Ecological Applications*, *10*(4), 941–948. doi:10.1890/1051-0761(2000)010[0941:EAEOWS]2.0.CO;2

Rennings, K. (2000). Redefining Innovation- Eco-Innovation Research and Contribution from Ecological Economics. *Ecological Economics*, *32*(2), 319–332. doi:10.1016/S0921-8009(99)00112-3

Sagoff, M. (2000). Environmental Economics and the Conflation of Value and Benefit. *Environmental Science & Technology*, *34*(8), 1426–1437. doi:10.1021/es990674d

Saleth, R. M. (2004). *Understanding Water Institutions: Structure, Environment, and Change Process*. International Water Management Institute.

Saleth, R. M., & Dinar, A. (2003). *Water Institutions and Sector Performance: A Subjective Theory of Institutional Change*. Paper presented at the 25th Congress of the International Association of Agricultural Economics, Durban, South Africa.

Scanlon, T., Uguru, O. P., Jafry, T., Chinsinga, B., Mvula, P., Chunga, J., Zimba, L. M., Mwape, M., Nyundo, L., Mwiinga, B., & Chungu, K. (2016). The Role of Social Actors in Watewr Access in Sub-Saharan Africa: Evidence from Malawi and Zambia. *Water Resources and Rural Development*, *8*, 25–36. doi:10.1016/j.wrr.2016.08.001

Schaltegger, S., & Synnestvedt, T. (2002). The link between "green" and economic success: Environmental management as the crucial trigger between environmental and economic performance. *Journal of Environmental Management*, *65*, 339–346. PMID:12369398

Soderbaum, P. (1990, June). Neoclassical and Institutional Approaches to Environmental Economics. *Journal of Economic Issues*, *24*(2), 481–492. doi:10.1080/00213624.1990.11505047

Soderbaum, P. (1999). Values, Ideology and Politics in Ecological Economics. *Ecological Economics*, *28*(2), 161–170. doi:10.1016/S0921-8009(98)00139-6

Spash, C. L. (1999). The Development of Environmental Thinking in Economics. *Environmental Values*, *8*(4), 413–435. doi:10.3197/096327199129341897

Stavins, R. (2004). *Environmental Economics*. Harvard University.

Stern, D. I. (1997). Limits to Substitution and Irreversibility in Production and Consumption: A Neoclassical Interpretation of Ecological Economics. *Ecological Economics*, *21*(3), 197–215. doi:10.1016/S0921-8009(96)00103-6

Van den Bergh, J. C. J. M. (2007). Evolutionary thinking in environmental economics. *Journal of Evolutionary Economics*, *17*(5), 521–549. doi:10.100700191-006-0054-0

Van den Bergh, J. C. J. M., & Verbruggen, H. (1999). Spatial Sustainability, Trade and Indicators: An Evaluation of the 'Ecological Footprint'. *Ecological Economics*, *29*(1), 61–72. doi:10.1016/S0921-8009(99)00032-4

Wackernagel, M., & Rees, W. E. (1997). Perceptual and Structural Barriers to investing in Natural Capital: Economics from an Ecological Footprint Perspective. *Ecological Economics*, *20*(1), 3–24. doi:10.1016/S0921-8009(96)00077-8

Chapter 5
Water Demand Management and Household Level Metering in Low-Income Areas of Urban Africa

Marta Marson
University of Turin, Italy & Turin Centre on Emerging Economies (OEET), Italy

ABSTRACT

Increasing the level of water metering is an objective of most initiatives for the operational restructuring of African water utilities promoted by donors and development agencies from the 1990s. Water metering penetration is a common benchmarking indicator to measure the performances of water utilities. In contrast with other benchmarks and targets set for the African water sector, which remain largely unmet, water metering at household and at water point levels are quite successful. The study discusses the arguments behind the widespread acceptance of the target of 100% metering, focusing on the suitability of household level metering for low-income settlements of urban Africa. An empirical analysis shows that metering is not an effective water demand management tool for domestic consumption, probably due to the fact that average consumption is already low, and it can hardly be reduced further. The case study shows that universal metering ambitions might discourage household level connections.

INTRODUCTION

Increasing the level of water metering is an objective of most initiatives for the operational restructuring of African water utilities promoted by donors and development agencies from the '90s (Tyman and Kingdom, 2002; WSP 2008). Water metering penetration is among the most common benchmarking indicators which are adopted to measure the performances of water utilities throughout the world. Two indicators are used for this purpose: proportion of metered connection over the total number of connections and proportion of metered water over the total volume of water sold. Household level water metering is usually considered a good practice and recommended (IBNET, 2017).[1] Under an economic

DOI: 10.4018/978-1-7998-8809-3.ch005

perspective, measuring water consumption is expected to improve efficiency in the allocation of the water resource, by allowing the adoption of volumetric fees instead of fixed fees and the transmission of price signals to water users. For this reason, metering is considered a main tool for Water Demand Management (WDM) (Arlosoroff, 1998; Macy, 1999; Gumbo and Van der Zaag, 2002; Mulwafu et al., 2003), which aims at keeping the demand for water compatible with environmental scarcity. The idea is that consumption can be limited by confronting people with their actual consumption and charging them accordingly. Under a more technical perspective, metering is used to reduce Non Revenue Water (NRW), caused by leakages and illegal connections, as it makes it possible for technicians to identify the area where leakages occur and to fix the issue. Metering is not only supported by water conservation arguments, but sometimes also by an equity argument (Staddon 2010), because, through metering, everyone is charged what she actually consumes. This reflects a liberal view of equity, while the actual social outcomes of a metered system in terms of social equity might be more complex and depend on the tariff system which is used.

In contrast with other benchmarks and targets set for the African water sector, which remain largely unmet - for example access to safely managed water source in urban Sub Saharan Africa was only 50% in 2017 (JMP 2019) -, water metering at household and at water point level is quite successful. *"In Africa, water metering is surprisingly widespread, with many utilities reporting 100% metering in their service areas. (…) the average metering ratio is 75%"* (Banerjee, 2010, p. 6). The idea that metering level should approximate universal coverage is widely accepted and recommended for developing countries (Kingdom et al., 2006; WSP, 2008).

The study discusses the arguments behind the widespread acceptance of the target of 100% metering, focusing on the suitability of household level metering for low-income settlements of urban Africa. An alternative to universal metering is selective metering, which can be based on the selection, by the utility of the customers to be metered, or can be based on their voluntary enrolment in a metering program. Voluntary enrolment is prone to an adverse selection bias, as only the customers whose consumption is low are likely to save money by measuring the actual volumes, while customers with high consumption will probably prefer the flat rate (Lago and Möller-Gulland, 2012). Adverse selection is related to information asymmetries and it occurs when a product or service is selected by only a certain group of people who offer the worst return for the company and, in this case, also for the society which fails to achieve the water conservation outcomes envisaged. Selective metering can also be part of a strategy by the water utility to discriminate between customers' categories, instead of allowing self-selection. The chapter argues that poor, informal neighbourhoods of African cities are not a priority target in this sense.

The chapter shows that the universal household level metering, and particularly metering in low-income areas of African cities does not significantly contribute to reduce per capita consumption, which, in any case is already very low. Moreover, it shows that the objective of universal metering of household connections can prevent water utilities of Sub-Saharan Africa from connecting poor households in Low Income Areas, because of the high cost and complexities it implies.

In the first part of the chapter, to provide a background to the study, household level metering is discussed based on the review of available literature. Household level water metering is discussed in its consistency with the features of African low-income urban settlements and of African utilities operations. The second part of the chapter is the core of the study and it is twofold: first it presents an empirical exercise, second it describes a case study. Regression models are estimated to assess the impact of metering on average domestic consumption and on overall consumption, i.e. including all the categories of users. This is done by estimating the same model, same set of regressors, with two alternative dependent variables and by comparing the resulting coefficients. Results seem to show that metering is not an effective water demand management tool when domestic consumption is considered, probably due to the fact that consumption is, in average, already low and it can hardly be reduced further. The different impact of metering level on domestic vs. overall water consumption provides some orientation for priorities when allocating scarce resources to costly metering programs, supporting the potential of metering for industrial and non-domestic water demand management and in affluent neighbourhoods.

The case of Lilongwe Water Board (Malawi) is discussed on the basis of key informants' interviews, available literature and internal reports. The aim is to understand how metering contributed or failed to contribute to the extension of household level water supply in low-income areas. This is done by analyzing challenges in the development and implementation of efficient and effective billing and fees' collection practices. The contribution of household level metering to WDM in Lilongwe is also discussed and alternative WDM strategies are considered.

The third part of the chapter, based on the analysis of literature, the empirical evidence and the case study of Lilongwe, draws some recommendations, further research directions and concludes.

BACKGROUND

Universal water metering has seldom been criticized by practitioners and scholars and the few critical contributions available refer only to developed countries and to the recent initiative of introducing metering in the UK (Jenkis, 2006; NCC, 2002; Staddon, 2010; Zetland, 2016). Arguments against universal metering from this literature are the inability of metering to achieve the expected reduction in domestic consumption due to low price elasticity, inaccessibility of meters by consumers and failure of water bills to communicate volumetric information in an accessible way (Staddon, 2010) and the high costs of metering and their disproportionate impact on low-income groups was discussed by Jenkins (2006). The share of metering costs over operating and maintenance costs can be substantial and should be carefully considered, jointly with the opportunity cost of investing in meters. Jenkins (2006, p. 8) suggests that *"policy makers should reflect carefully on the use of metering, particularly when financial resources are subject to competing demands in relation to the need to improve the quality, access, and availability of water resources within society."*

The cost of metering devices is estimated to account for 1/10 of connection costs by the Africa Infrastructure Country Diagnostic (AICD) studies (Africon, 2008) but the total cost is much higher if operation and maintenance cost of meters is considered. The estimation of the costs of metering, including purchase, maintenance, reading and volumetric billing is not straightforward and few data are available. The water services regulator of England and Wales (OFWAT, 2004) estimated this cost to account for about 6% of an average bill and other regulators and authors estimate this cost to be much higher (UREGNI, 2007; Staddon, 2010, p. 5). The most recent developments of the metering technology, like prepaid meters and SMART metering, can also be seen as solutions to address the challenges posed by labor intensity of traditional meters systems. Prepaid metering programs are sometimes justified as a means to reduce the high costs and the complexity associated with reading and billing of traditional meters (Peters, 2011). Smart metering, which is based on ICT for the automatic transmission of data, is appreciated, particularly in developed countries, for its potential in the reduction of the labor needed by traditional metering. Some good practices that are promoted in the water sector in the developing world are here reviewed and discussed in their consistency and relevance for universal metering, namely:

community participation in water services, regulation of informal providers, and low-cost water supply solutions and quality of metering.

Community participation approaches have a long history (Kwo 1984 and 1986): self-help was the main British strategy for implementation of the new welfare oriented colonial development policy from the '50s (Page, 2003). Community participation also gained momentum after the structural adjustment in the '80s and '90s to fill the gap between the pressing needs of vulnerable strata of African societies and the minimum state. As properly described by Page (2003), community participation is now a "*development orthodoxy*". Community participation in the management of water supply and of public taps particularly was first adopted in rural areas of Africa but it is increasingly adopted also in urban low-income areas, where private participation was first considered as a preferred option to fill the gaps left by the public service, after structural adjustment. Community participation in the water sector is about involving users in the construction and operation of water supply systems. Communities can provide funds before construction, voluntary or paid workers during construction and contribute to operate and maintain the system. As part of the operation of any community run system, managers are also expected to organise the collection of fees from users (Yacoob, 1990; Yacoob and Walker, 1991). Central to the organisation of such systems is some variant of a ''community water committee'': a representative body in charge of managing water supply in a particular community (Page, 2003). Community based water management is not necessarily limited to public water points and does not necessarily imply the adoption by community entities of market instruments for the allocation of the resources managed and more cooperative solutions (Ostrom, 1990) can be adopted. Mara and Alabaster (2006) propose three solutions for group connections, namely *standpipes cooperatives*, *yard tap cooperatives* and *in house multiple taps cooperatives*. When metering is considered, Mara and Alabaster are quite radical and only provide for in house multiple taps cooperatives to be metered at a collective level, while the remaining two should not be metered at all. As specified by the authors, cooperatives are to be intended here with the meaning of consumers' cooperatives or bulk purchasers and not with the meaning of service providers, which is closer to the World Bank concept (Ruiz-Mier and Van Ginneken, 2006) and to the case of regulated alternative providers addressed in the next paragraph.

Small scale, informal water providers are important actors for water access in urban Africa. They sell water from private sources, like a water well, or sometimes resell utility waster to users outside the reach of the utility or to users who are not connected. The regulation of small-scale informal providers, and more generally the regulation of water reselling by individuals and businesses other than water utilities, are receiving increasing attention both by practitioners and scholars (Gerlach and Franceys, 2009 and 2010). The debate focuses on how to enjoy their potential to expand water access, while mitigating the risks related to their informal status (unreliable water quality, uncontrolled prices). Delegated management is built around a contractual relationship between water utilities and small-scale private operators who have the financial incentives to increase access and improve services (WSP, 2009). When metering is considered, delegated management implies moving upstream the boundaries of direct management by the central water utility of the last water meter (WSP, 2009). Better regulation of reselling can also reduce the need of metering each and every tap beyond certain, contractually identified boundaries. Unregulated reselling of utility water instead can only be tolerated in a fully metered environment. Otherwise, the few connected users could resell any volume of water while paying a flat monthly fee. Tolerance without regulation and without purposely designed devices and tariffs does not yield optimal outcomes. As noticed by Boland and Whittington (1998), the widespread adoption of increasing block tariffs (IBT) in metered environments can have regressive effects on unconnected users: the connected users who resell water to the unconnected ones tend to charge at least the rates corresponding to high consumption levels. A partially metered environment needs convenient regulatory solutions for reselling by household and small-scale providers, so that only purposely authorized, metered customers are allowed to resell and the tariff applied to them and the final price should be controlled.

Low-cost water supply solutions: the consideration of the high costs associated to universal metering at household level, is particularly important and relevant if we consider the orientation of the World Bank and donors agencies towards lower water supply standards (ladders) in order to accelerate access. This is advocated by Banerjee and Morella (2011, p. 247): *"the availability of lower-cost technologies has the potential to reduce the funding gap"*. The solution proposed focuses on efficiency savings and try to downscale the standards for water supply and sanitation. This basically means to focus on water provision through public water points instead of private household level connections. This solution falls short of health and hygiene standards that it can ensure, as it is confirmed by the WHO classification in Table 1. Current SDGs definition of water access follows very similar criteria for collection time and safety of water, but is less precise about water volumes, which are nonetheless confirmed by the volume of six cubic meters per month confirmed as lifeline consumption of households (WHO 2021). Dagdeviren and Robertson (2008, p. 4) share the view that low-cost solutions are not optimal: *"Low-cost and quick-fix solutions may be important in mitigating the adverse effects of lack of access to water in the slums."* because *"a decent solution that is sustainable over the long term is beyond the commercialised world of WSS systems"*. If water supply costs must be reduced, relaxing metering standards offers an opportunity for saving without affecting the service standards.

Table 1. Water supply ladder

Service level	Access measure	Needs met	Level of health concern
No access (quantity collected often below 5 l/c/d)	More than 1000m or 30 minutes total collection time	Consumption – cannot be assured Hygiene – not possible (unless practised at source)	Very high
Basic access (average quantity unlikely to exceed 20 l/c/d)	Between 100 and 1000m or 5 to 30 minutes total collection time	Consumption – should be assured Hygiene – handwashing and basic food hygiene possible laundry/bathing difficult to assure unless carried out at source	High
Intermediate access (average quantity about 50 l/c/d)	Water delivered through one tap on-plot (or within 100m or 5 minutes total collection time)	Consumption – assured Hygiene – all basic personal and food hygiene assured; laundry and bathing should also be assured	Low
Optimal access (average quantity 100 l/c/d and above)	Water supplied through multiple taps continuously	Consumption – all needs met Hygiene – all needs should be met	Very low

Source: Howard and Bartram 2003

The quality of metering is also important. Metering is considered a main tool to measure and address NRW problems through water balances, which allow the estimation of NRW by comparing water flows at production, at bulk meters and at users' meters, but also minimum night flows allow the estimation of NRW (Morrison 2004). However, meters accuracy is itself a cause of apparent water losses, which are differences between the volumes metered at production and the ones metered at consumption which are not due to actual water loss, but to measurement errors. Low pressure is a driver of under registration, while rationing practices and failures to provide 24 hours water services can cause the presence of air in the pipes and result in over registration. Poor maintenance and testing of meters are also among the main causes of poor metering performances. Metering errors for new and properly installed meters are estimated to be around 5% and to increase with the age of the meter and with low pressure, but actual errors in Kampala were found to be higher than 20% in 2007-2009 with 6.6% failed meters over total number of meters installed (Mutikanga et al., 2011; Mutikanga, 2012). Moreover, the practice of estimating the consumption of metered connections in the absence of records from timely readings for billing purposes is widespread in Africa (Chikasema, 2009). This practice is due to the high costs of reading and it jeopardizes the effectiveness of metering as a mean to convey information to customers. The poor metering accuracy which characterizes most African utilities is very relevant when considering the costs and benefits of metering. This was properly expressed by Mutikanga (2012, p. 103), as follows: *"when metering is inefficient, all benefits associated with metering are lost"*.

ASSESSING EFFECTIVENESS AND EFFICIENCY OF UNIVERSAL METERING

Metering and the Level of Consumption: An Empirical Exercise

It is recognized that in the context of Sub Saharan Africa consumption levels are still within, or even below, the minimum amounts that are necessary for a decent life. For example, Banerjee and Morella (2011, p. 135) recognize that there is no strong evidence of wasteful use of water in Africa, and that *"the relatively modest levels of consumption would not be further reduced by a more aggressive use of demand management tools"*.

The existence of positive externalities from water consumption, makes appropriate to compare the costs of water supply to the total benefits resulting from it, which should include both the private and the public ones. Public benefits of water (and sanitation) systems are related to public protection from communicable diseases (Budds and McGranahan, 2003), and are of particular important in developing contexts. Water supply clearly provides private benefits to the receiving household. However, if people are unwilling or unable to purchase enough water of good quality to protect their own health, and they contract diseases, then the health of others is also put at risk (Budds and McGranahan 2003). For this reason, both the user's private benefits and the related external (collective) benefits should be considered, which might challenge the priority of investing in WDM in low-income areas of urban Africa.

The ability of metering to reduce domestic consumption was assessed and compared to the ability of metering to reduce the average consumption, including all the categories of users. Regressions on panel data, with 158 annual observations from 56 utilities from 20 countries of Sub Saharan Africa, between 1997 and 2009 from the IBNET database[2] are here presented. Fixed and random effects were tested, after detecting and eliminating hard outliers[3]. Two dependent variables were used with the same regressors, in order to capture the specific features of domestic water demand against the metering level. These variables are:

- **average per capita daily consumption**: this is expressed in litres per capita per day (lpcpd), and it includes total consumption (domestic, industrial, commercial and institutional);

- **average per capita daily domestic consumption**: this is expressed in litres per capita per day and it includes only domestic water consumption, i.e. consumption by from households connections and public water points, as opposed to institutional and commercial/industrial consumption.

It is important to notice that both these figures are only estimates and that they have important limitations in their accuracy when trying to capture the real per capita consumption levels. A main challenge to the reliability of these estimations is related to the fact that in the countries considered several households use the same tap. The percentage of users served through public water points was taken as control variable because, for obvious logistic reasons, their consumption is usually lower than the one of users enjoying household connections. Per capita GDP of the country, expressed in Purchasing Power Parity, was taken as control variable because higher living standards are expected to increase water consumption. The average revenue of the water utility per cubic meter sold was taken as a proxy for the average price of water.[4] For each dependent variable, two lag structures were tested: in the first, all the variables are taken in the same year; in the second frame, in order to reduce potential endogeneity and to account for the inertia of behavioural adaptation to change, the variable about the level of metering and the one approximating water price are taken with a lag of one year. This lag structure unfortunately reduces the number of available cases so that it can be only considered as robustness check. The results are shown in Table 2.

Table 2. Effects of metering on consumption: regressions results

	Dependent variable: average per capita daily consumption, including domestic, industrial, commercial, and institutional				Dependent variable: average per capita daily domestic consumption			
	Without time lag		With time lag		Without time lag		With time lag	
	Random effects	Fixed Effects	Random effects	Fixed Effects	Random effects	Fixed Effects	Random effects	Fixed Effects
Metering level (% water metered on total water sold)	-0.70***	-0.69**			-0.39*	-0.38		
Metering level lagged one year			-0.73***	-.0.77*			-0.68**	-1.12
% of users served through public water points on the total number of users	-24.05*	-20.77	-17.45	0.07	-11.17	-3.95	-13.96	-0.13
Per capita GDP of the country in PPP	0.01***	0.01*	0.01***	0.01	0.01***	0.00	0.01***	-0.01
Average revenue per m3 (tariff)	-8.05	-4.28			-2.79	5.19		
Average revenue lagged one year			3.98	11.37**			1.59	10.73***
Constant	126.07***	128.14***	126.40***	128.87***	77.30***	84.18***	106.00***	155.31**
N obs	*158*	*158*	*109*	*109*	*158*	*158*	*109*	*109*
N groups	*56*	*56*	*34*	*34*	*56*	*56*	*34*	*34*
Wald $\chi^2(4)$	*55.54****		*34.64****		*29.55****		*13.89****	
Within R^2		*0.21*		*0.11*		*0.05*		*0.16*
F (4,55/33)		*3.07***		*3.76***		*0.71*		*11.06****

Source: author elaboration

The results show that metering has a highly significant negative coefficient when the impact on consumption including institutional and industrial is considered. On the contrary, the coefficient of the same variable is still negative, but smaller and less significant or not significant when only domestic consumption is considered as dependent variable. These results are also robust to the lagged time frame. This seems to show that metering is not an effective water demand management tool when domestic consumption is considered, probably due to the fact that levels of consumption are already low and cannot be reduced further.

The other variables have the expected sign, but they are not significant in all cases. The variable approximating water price unexpectedly shows some significant positive coefficients in the lagged models, possibly due to some reverse causality. A similar finding, in facts, was explained by Banerjee (2008) as resulting from Increasing Block Tariff structure: *"the tariff rates are near cost-recovery levels at high levels of consumption. Particularly for high-volume non-residential consumers, utility clients pay a substantially higher price per unit of consumption"*. The lagged structure of the models however should in principle reduce this effect. The lack of significance (or lower significance) of the impact of metering level on the average level of per capita domestic consumption is not an argument against metering as such, but it at least provides some orientation for priorities when allocating scarce resources to costly metering programs. At the same time the results confirm the importance of metering for industrial and non-domestic water demand management.

Countries and individual cities in the developing world are competing to attract foreign investments in fields related with the exploitation of natural resources and also in water intensive sectors. The patterns of industrial water consumption and the corresponding charges by utilities are particularly important because commercial tariffs are often below the thresholds conventionally adopted for full cost recovery levels (Banerjee et al., 2010) for both O&M and capital costs. The effort toward cost recovery tariffs for domestic consumption, did not yield proportional outcomes when commercial and industrial tariffs are considered[5]. In fact, some utilities have industrial tariff lower than the domestic ones and their average ratio was below 3, still in 2007 (source author elaboration on IBNET data).

Metering in Low-Income Areas by Lilongwe Water Board: A Case Study

The paragraph describes the successes and challenges encountered by Lilongwe Water Board (LWB), the public utility for Lilongwe city, in the effort to provide water to Low Income Areas (LIA), which are informal settlements and traditional housing areas. The analysis is based on the review of secondary sources, like reports, published studies and on semi-structured interviews carried out in November 2013 with the staff of LWB.

Lilongwe is the largest city of Malawi and its administrative capital since 1975. Lilongwe has witnessed a high urbanization rate ever since, accelerated by the relocation of all government head offices from Blantyre to Lilongwe from 2005. About 76% of this population lives in informal settlements, where water is supplied mainly through public kiosks. The water coverage was 55% in 2012 (VEI, 2013). Lilongwe is divided into 59 administrative areas, identified by a progressive number. There is not a strict correspondence between the boundaries of the low-income areas (LIAs) and the ones of the administrative areas and at least 16 administrative areas include LIAs. The households' metering level has been 100% at least since the beginning of the 21th century, as household level metering was the dominant strategy.

The case study is relevant to the present analysis because, despite donors' funding was available for household connection in LIA, LWB gave up implementing the corresponding program also due to the complexity, the high cost and the frequent failures associated with the management of such connections in informal areas. It will be argued that demanding metering standards were among the elements determining such complexity, high costs and failures. Lastly, also in Lilongwe, the opportunity to address non-domestic water demand management was not fully sized yet.

According to LWB annual report 2011-2012 (LWB 2013), meters are the 4th asset of LWB in terms of value (notes to the financial statements, p. 42) but their estimated useful life is 5 years only (p. 35), while for other assets the depreciation time horizon is much longer. The results obtained by Mutikanga for Uganda range from 3 to 13 years for the optimal replacement time (Mutikanga, 2012).

505 MKW, corresponding to around 1 €, was the monthly rent of a meter charged to users and it is also the cost of the first 5 cubic meters (6 cm is usually considered the basic household consumption), so meter rents accounts for half the average bill for low-income households, thus much higher than in the UK. Average consumption per residential connection is lower in administrative units with LIAs than in other areas. In the analysed period (2007-2013), average monthly consumption per domestic connection was 16 cubic meters in administrative areas with LIA, against 41 cubic meters units which do not include LIAs, as shown in Figure 1.

The consumption in areas without LIAs is strongly higher than the one in areas with LIAs throughout the years and it also shows a much more marked pattern of seasonal fluctuations, while LIAs consumption does not fluctuate as much, because it corresponds only to basic needs. This difference in the level of consumption is particularly important if we consider that shared connections are a reality in LIA, where the number of persons relying on each connection is much higher. By considering 8.8 persons per connection as indicated by LWB (VEI, 2013) for the whole town, it is found that in the units including LIAs people consume a daily amount of 60 litres of water. This optimistic estimate is below optimal consumption levels for the WHO (Howard and Bartram, 2003; WHO 2021), so that no evidence of wasteful use of water can be found. Conversely the consumption level in other areas is too high, despite 100% metering is practiced.

Figure 1. Average monthly consumption in administrative units including LIA and in other units. Source: LWB billing system

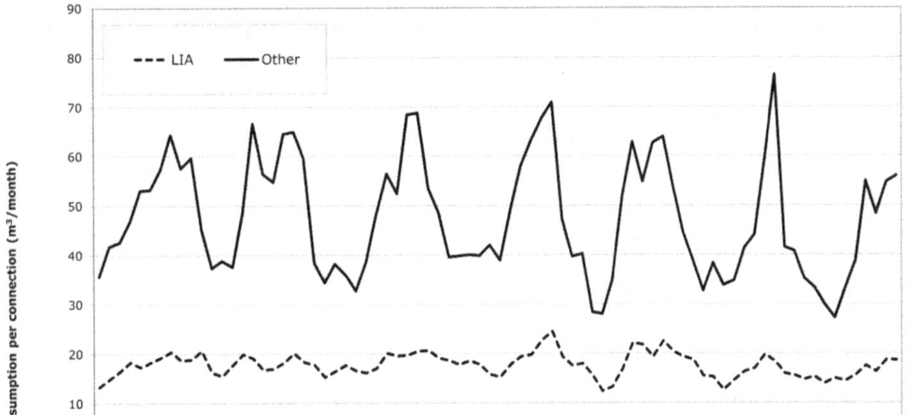

Within a financed project an Access to Water Facility was established, with a micro finance fund to finance household water connections. The facility was expected to almost double the number of household connections for low-income customers. The idea was to provide credit for the connection cost (11,000 mkw at that time), allowing the customer to pay back in monthly installments. Unfortunately, the Access to Water Facility was never implemented. According to sector stakeholders the demand for household connections in LIAs is high, but connections are too difficult and costly to manage, even when land tenure rights are there. This is due to the fact that houses are one close to the others and that it is difficult even to understand the boundaries. This has a huge impact on costs like meters' readings and billing.

The average consumption at public kiosks in Low Income Areas was 16 litres per capita per day, considering 250 persons per kiosk (VEI, 2013). This is a very low level of consumption, and it corresponds to *"no access"* in the terms of the WHO (Howard and Bartram, 2003). However, the main strategy selected by LWB and its partners to achieve the project objectives in terms of access, following the idea of low-cost solutions is based on public water kiosks for unplanned areas. The construction of 372 water kiosks was foreseen within the EIB/WF project, expected to serve more than 90,000 persons in low-income areas, with 250 persons per point. This part of the project, as opposed to the one related to household level connection, was implemented.

According to the research on LWB carried out by Chikasema (2009), the main problems related to metering practices are the lack of maintenance and replacement, the referencing system to locate the meters in the field, and the slow and unreliable flow of information from metering to billing offices. According to the author, the bill issued without a reading, based on estimations accounted every month for around 20% of the total, with some areas above 40%. He also found a direct relation between the levels of NRW and the number of meters with estimated water consumption. In the most recent years, the intervention on metering was in fact associated to the objective of reducing NRW. At first, the focus was leakages, since it was expected that at least half of the total NRW was related to illegal connections and poor functioning of meters. Since 2011, more than a quarter of existing meters were replaced and the meters removed were recalibrated and reinstalled elsewhere, but reduction of NRW achieved was far below expectations. Better results seem to be instead associated with the more recent adoption of the so called *caretaker model* and with bulk metering at the level of District Metered Areas – DMAs (Morrison 2004). This system allocates staff of different departments, both field staff and staff of local and central offices, to specific geographic areas, to increase their accountability and commitment to the performances of that area. In the experience of LWB bulk metering proved to be more strategic than household level metering to address the challenges posed by NRW rates.

Non domestic users are also being addressed by efforts by LWB to better manage water demand. LWB social media published on 21th March 2013 that *"the water demand has completely outstripped production. (…). This is mainly coming in because of developments in Kanengo Industrial site, remarkable development in City Centre and other areas. The current production capacity can no longer be increased because of the water availability in the current source (Lilongwe River)"*.

Total water demand exceeded production since 2012, and commercial and industrial consumption volumes accounted for 18% on average of the total demand, in the period between September 2012 and October 2013. In the peak period of November their share of consumption increases to 22%. The number of industrial and commercial connections is above 2,000, and area 29 alone, where big firms are concentrated, accounts for more than 20% on the total commercial and industrial consumption. Firms are from agribusiness sectors like tobacco manufacture, processing pigeon peas, soya beans, chilly and companies producing seeds varieties. Moreover, there are beverages companies and one company producing cotton. The average consumption per business in the whole city is 505 cm, per month while there are many users above 1,000 cubic meters per month. These volumes are particularly impressive if one considers that the total monthly consumption from all the water kiosks in LIA is around 50,000 cubic meters. There was a 7% increase in industrial and commercial consumption total volume between 2008 and 2012. The increase in the turnover of the manufacturing companies operating in Lilongwe fully accounts for the growth in the consumption levels, with around 59% growth in turnover in the same period (unpublished data from the Malawian Revenue Authority). However, the absolute volume remains very high, and it competes with domestic demand. This is being addressed by LWB, through pricing policy, as it is possible to see in Figure 2.

The trend of the tariff system between May 2007 and August 2013 is characterized by a huge increase of the rates for industrial and commercial consumption, while for households and kiosks the increase was lower. However, insolvency by businesses (Simbota, 2013) is partially jeopardizing the positive effect of the pricing policy.

Figure 2. Average revenue per cubic meter from different categories of customers (local currency MKW per cm). Source: LWB billing system

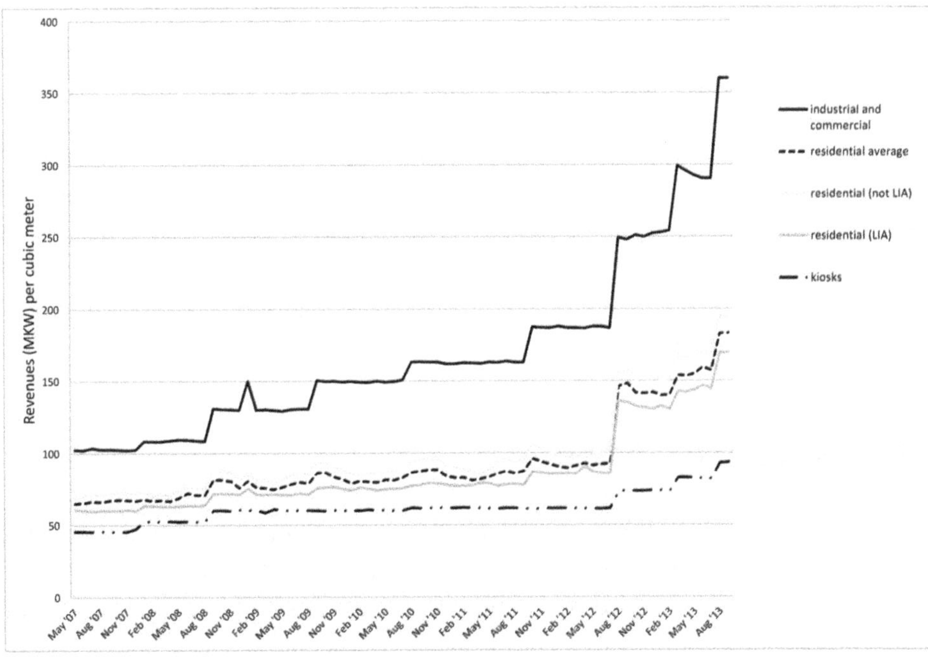

SOLUTIONS AND RECOMMENDATIONS

Unfortunately, there seem to be no quick fix solution for the problems discussed which, as it is shown in the chapter, encompass many complex and interrelated issues. The main recommendation arises from the consideration that technology is not neutral, and technologies should be appropriate (Schumacher 1973) to purse the goals associated with existing needs, in the specific context, not vice-versa adopting technologies developed for the specific conditions of western countries, like utilities financial capacity and formal urbanization patterns. Universal metering is a precise mix of technologies in this sense, but is should not be taken as the only option when considering metering options for African cities.

Most SSA cities are easily classified into better off areas and low-income areas, often corresponding to the informal settlements, usually un-served or unevenly served. This marked spatial segregation allows for the adoption of different strategies by utilities, with universal household level metering only adopted for the wealthier areas.

For low-income areas, the point is not a dual alternative weather to meter or not the water flowing to those areas, which is instead necessary for leakages and NRW control, but is where to place the last meter. Staddon (2010, p. 15) notices that *"there seems to be relatively little critical discussion of the optimal location for water meters (…) it is simply assumed that these should be located at the service inlet to each and every house in the land."*

If household level metering is reconsidered in favor of group metering, there is a chance to substantially increase the number of water points in LIA as a mean to ensure affordable and sustainable access to water to the poor. Delegated management is already a good example of how it is possible to move the threshold of the utility responsibility upward to the level of the local small- scale provider, identified as market agents or as community groups. Most solutions, however, assume that, even beyond the threshold of the last utility meter, the allocation of water will be based on the market, spreading the need to meter to each and every tap (WSP 2009). As a matter of fact, the management practices adopted in these cases are mostly market oriented, so that Page (2003) points at communities as *"agents of commodification"* referring to the role they can play in transforming water into a commodity.

Collective metering, as proposed by Mara and Alabaster (2006) might conveniently dimensioned groups of users can easily manage all the challenges that typically prevent water utilities from connecting households in low-income areas, i.e. risk of overconsumption, high turnover of dwellers, illegal reselling, arrears in the payments. Group connections can also overcome the problem pointed by Boland and Whittington (1998) of the regressive effect of increasing block tariffs in case of shared connections. This problem is in fact overcome by rescaling the ceiling of the basic consumption block in order to account for the number of users relying on the connection.

FUTURE RESEARCH DIRECTIONS

Further research could explore implications of shared meters in term of the consumption behaviour of households and their cooperation strategies. Moving the meters upstream, with different households using the same meter, poses problems of free riding and cooperation. More particularly, in the terms of the game theory, each household faces up to the decision whether to over-consume or not, and with uncertainty about the consumption of each of the other households, or, to simplify, about the average consumption of the other households (Table 3).

The incentives to free riding and opportunistic behaviours basically depend upon two variables: the penalty for overconsumption (difference between the first block tariff and the tariffs for consumption above this ceiling) and the dimension of the group (number of households using the same meters and entering a contractual relation with the water utility jointly) that is going to pay the penalty. The higher the penalty for over-consumption, the lower the incentive to over-consume and free ride. The higher the number of households, the higher the share of the overconsumption costs that each over-consuming household can externalize.

Table 3. Household consumption game

		Other households (average)	
		Normal consumption	**Over consumption**
Household	Normal consumption	Cooperation	Free riding
	Over consumption	Free riding	Tragedy of the commons

Source: author elaboration

With realistic values of these variables, in a single simultaneous game free riding cannot be avoided and it will probably result instead as the dominant strategy, because each individual is confronted with the same game structure. If all the households opt for free riding and over-consume, the result will be the lowest right cell (Table 3), corresponding to the lowest total pay off.

The tragedy of the commons however is not necessarily the case, as the game is not simultaneous, and it is not single. As a matter of fact, users can monitor the total consumption and the average consumption of their fellows over the months and adjust their behaviour accordingly. When the game is repeated because the decision takes place every month and people can learn how to cooperate, in line with Ostrom (1990).

CONCLUSION

The analysis provided some arguments against the opportunity of metering all household connections in poor areas of African cities. The main argument is related with the high cost of metering, including installing reading and billing. This cost, when compared to the benefits of metering each household, does not seem to be consistent. Complementary to that, it is argued that a service standard based on 100% metering, can prevent utilities from making household connections in poor, informal dwellings. It was pointed out that water demand management should prioritize non-domestic users, while the consumption by poor users does not need to be reduced and it is not very sensitive to metering. It was also shown that the operations of African utilities in the field of metering are not efficient and effective and this jeopardizes the potential benefits from metering. The high costs of metering and the complexity of the related practices were also considered and identified among the reasons preventing utilities from adopting a strategy of water access which is based on household level connection, regardless the widely recognized advantages of this solution.

Moreover, the lower the number of meters, the higher the accuracy that the utility can assure for their testing, functioning and reading. The study has shown that metering at household level can be too complex and costly to manage and that alternatives can be explored to reduce costs, without scaling down the supply ladder and keeping the domestic demand volumes controlled.

ACKNOWLEDGMENT

The Author wish to thank the staff of Lilongwe Water Board and for co-operation and interesting insights.

This research received no specific grant from any funding agency in the public, commercial, or not-for-profit sectors.

REFERENCES

Africon. (2008). *Unit Costs of Infrastructure Projects in Sub-Saharan Africa*. The World Bank. https://www.eu-africa-infrastructure-tf.net/attachments/library/aicd-background-paper-11-unit-costs-summary-en.pdf

Arlosoroff, S. (1998). Water demand management. *UNCHS (Habitat) Regional Conference on Sustainable Consumption Patterns in Asian Cities*. http://www.fukuoka.unhabitat.org/docs/occasional_papers/project_a/09/water-e.html

Banerjee, S., Foster, V., Ying, Y., Skilling, H., & Wodon, Q. (2010). *Cost Recovery, Equity, and Efficiency in Water Tariffs Evidence from African Utilities*. Policy Research Working Paper 5384, The World Bank.

Banerjee, S., & Morella, E. (2011). *Africa's Water and Sanitation Infrastructure. Access Affordability and Alternatives*. The World Bank. doi:10.1596/978-0-8213-8457-2

Boland, J., & Whittington, D. (1998). *The political economy of increasing block tariffs in developing countries*. Economy and Environment Program for Southeast Asia. Special Paper No. 2. https://idl-bnc-idrc.dspacedirect.org/bitstream/handle/10625/31784/118117.pdf?sequence=8&isAllowed=y

Budds, J., & McGranahan, G. (2003). *Privatization and the provision of urban water and sanitation in Africa, Asia and Latin America*. IIED.

Chikasema, G. (2009). An assessment of effective metering and billing as a water demand management tool. A case of Area 49 In Lilongwe, Malawi. In *Innovative Practices in the Water Supply and Sanitation Sector*. Sun Press.

Dagdeviren, H., & Robertson, S. A. (2008). *Reforming Without Resourcing: The Case of the Urban Water Supply in Zambia*. International Poverty Centre. https://uhra.herts.ac.uk/bitstream/handle/2299/2697/903025.pdf?sequence=1

Gerlach, E., & Franceys, R. (2009). Regulating Water Services for the poor: The case of Amman. *Geoforum*, *40*(3), 431–441. doi:10.1016/j.geoforum.2008.11.002

Gerlach, E., & Franceys, R. (2010). Regulating water services for all in developing economies. *World Development*, *38*(9), 1229–1240. doi:10.1016/j.worlddev.2010.02.006

Gumbo, B., & Van der Zaag, P. (2002). Water losses and the political constraints to demand management: the case of the City of Mutare, Zimbabwe. *Physics and Chemistry of the Earth, Parts A/B/C*, *27*(11-22), 805-813.

Howard, G., & Bartram, J. (2003). *Domestic Water Quantity, Service, Level and Health*. World Health Organization. www.who.int/water_sanitation_health/diseases/WSH03.02.pdf?ua=1

IBNET. (2017) *International Benchmarking Network Water and Sanitation Utilities Context Information & Indicators – Definitions*. IBNET. https://www.ib-net.org/toolkit/ibnet-indicators/metering-practices/

Jenkins, J. O. (2006). *Water metering: in search of a more critical debate*. University of Hertfordshire. https://uhra.herts.ac.uk/bitstream/handle/2299/3355/903361.pdf?sequence=3

Joint Monitoring Programme for Water Supply, Sanitation, and Hygiene. (2019). *Estimates on the use of water, sanitation and hygiene by region (2000-2017).* UNICEF and WHO. www.washdata.org

Kingdom, B., Liemberger, R., & Marin, P. (2006). *The Challenge of Reducing Non-Revenue Water (NRW) in Developing Countries How the Private Sector Can Help: A Look at Performance-Based Service Contracting*. Water Supply and Sanitation Sector Board Discussion Paper Series, paper n 8

Kwo, E. (1984). From our correspondent: community education and community development in Cameroon: The British Colonial experience, 1922–1961. *Community Development (Columbus, Ohio)*, *19*(4), 204–213.

Kwo, E. (1986). Community education and community development: Anglophone Cameroon experience since 1961. *Community Development (Columbus, Ohio)*, *21*(3), 61–168.

Lago, M., & Möller-Gulland, J. (2012). *WP3 EX-POST Case studies: Comparative Analysis Report*. EPI WATER. http://www.feem-project.net/epiwater/docs/epi-water_dl_3-2.pdf

Lilongwe Water Board (LWB). (2013). *Lilongwe Water Board Annual report 2011-2012*. Author.

Macy, P. (1999). *Urban water demand management in Southern Africa: the conservation potential. Sida Publications on Water Resources No. 13*. Sida.

Mara, D., & Alabaster, G. (2006). A new paradigm for low-cost water supplies and sanitation in developing countries. *Water Policy*, *10*(2), 119–129. doi:10.2166/wp.2008.034

Morrison, J. (2004). *Managing leakage by District Metered Areas: a practical approach*. Water 21. IWA. http://www.geocities.ws/kikory2004/39_Water21_5th_article_DMA.pdf

Mulwafu, W., Chipeta, C., Chavula, G., Ferguson, A., Nkhoma, B. G., & Chilima, G. (2003). Water demand management in Malawi: Problems and prospects for its promotion. *Physics and Chemistry of the Earth*, *28*(20-27), 787–796. doi:10.1016/j.pce.2003.08.003

Mutikanga, H. E. (2012). *Water Loss Management: Tools and Methods for Developing Countries.* UNESCO-IHE.

Mutikanga, H. E., Vairavamoorthy, K., Sharma, S. K., & Akita, C. S. (2011). Operational Tools for Decision Support in Leakage Control. *Water Practice & Technology*, *6*(3), wpt2011057. Advance online publication. doi:10.2166/wpt.2011.057

National Consumer Council (NCC). (2002). *Towards a sustainable water charging policy*. A Discussion Paper. NCC.

OFWAT. (2004). *Annex - Future water and sewerage charges 2005 -10 Final determinations.* https://webarchive.nationalarchives.gov.uk/20150604023659/https://www.ofwat.gov.uk/regulating/charges/ltr_rd3003_annex.pdf

Ostrom, E. (1990). *Governing the commons: the evolution of institutions for collective action*. Cambridge University Press.

Page, B. (2003). Communities as the agents of commodification: The Kumbo Water Authority in Northwest Cameroon. *Geoforum, 34*(4), 483–498. doi:10.1016/S0016-7185(03)00049-6

Peters, S. (2011). *Municipal Consumer Debt in South Africa, in Financial and Fiscal Commission.* 2012/13 Submission for the Division of Revenue Technical Report. South Africa. https://www.ffc.co.za/docs/submissions/technical_report/2011/Financial%20and%20Fiscal%20Commission%202012%20-%202013%20Technical%20Report.pdf

Ruiz-Mier, F., & van Ginneken, M. (2006). *Consumer cooperatives: An alternative institutional model for delivery of urban water supply and sanitation services*. The World Bank. https://documents1.worldbank.org/curated/en/131841468137400796/pdf/355470WN50coop1version0January02006.pdf

Schumacher, E. F. (2011). *Small is beautiful: A study of economics as if people mattered*. Random House.

Simbota, M. (2013). *LWB losing millions through non-payment bills*. MBC Malawi.

Staddon, C. (2010). *Do Water Meters Reduce Domestic Consumption? A summary of available literature*. Heednet. http://www.heednet.org/metering-defraHEEDnet.pdf

Tyman, M., & Kingdom, B. (2002). *A Water Scorecard: Setting Performance Targets for Water Utilities. Viewpoint*. The World Bank. https://openknowledge.worldbank.org/handle/10986/11351

UREGNI. (2007). *Utility Regulator Strand 1 Submission to the Water Review*. UREGNI. https://www.uregni.gov.uk/sites/uregni.gov.uk/files/media-files/Utility%20Regulator%20Strand%201%20Submission%20to%20Water%20Review%20%20September%202007.pdf

VEI. (2013). *Peri-Urban Water and Sanitation Project /Private Sector Service Contract to the Lilongwe Water Board Annual. Progress Report. Contract Year 3: November 2011 - October 2012*. VEI.

Water and Sanitation Program (WSP). (2008). *Developing Effective Billing and Collection Practices, WSP Field Note 44119 April 2008*. The World Bank. https://www.wsp.org/sites/wsp/files/publications/527200820356_SA_PIP2_May2008.pdf

Water and Sanitation Program (WSP). (2009). *Improving Water Utility Services through Delegated Management, Field Note Serving the Urban Poor*. The World Bank. https://www.wsp.org/sites/wsp/files/publications/Af-imp_through_delegated_mgmt.pdf

World Health Organization. (2021). *The measurement and monitoring of water supply, sanitation and hygiene (WASH) affordability: a missing element of monitoring of sustainable development goal (SDG) targets 6.1 and 6.2*. UNICEF and WHO. https://washdata.org/sites/default/files/2021-05/unicef-who-2021-affordability-of-wash-services-full.pdf

Yacoob, M. (1990). Community self-financing of water supply and sanitation: What are the promises and pitfalls? *Health Policy and Planning, 5*(4), 358–366. doi:10.1093/heapol/5.4.358

Yacoob, M., & Walker, J. (1991). Community management in water supply and sanitation projects-costs and implications. *Aqua AQUAAA, 40*(1).

Zetland, D. (2016). The struggle for residential water metering in England and Wales. *Water Alternatives*, *9*(1), 120–138.

KEY TERMS AND DEFINITIONS

Commercial and Industrial Water Demand: Commercial and industrial water demand include water for stores, offices, hotels, laundries, restaurants, manufacturing plants and,generally productive units.

Non-Revenue Water (NRW): The proportion of produced water that is not billed to users because it is lost due to leakages or to illegal connections.

Public Water Points: At a water point, people can draw water for their household needs like drinking water, washing (it is not meant for irrigation or industrial needs). Public taps and protected sources in rural areas are often referred to as "improved water points".

Tariff Structure: A tariff structure defines rules and rates applied to different groups of customers at different levels and timing of consumption. Tariff structure are common tools of price discrimination.

Universal Water Metering: Universal water metering corresponds to the general idea of providing all users with a water meter and it is opposed to selective metering, which can be based on the selection, by the utility of the customers and group of customers to be metered, or, in some cases, can be based on their voluntary enrolment in a metering program.

Water Access: Proportion of population using safely managed drinking water services.

Water Demand Management (WDM): Water scarcity can be addressed by increasing water supply (new infrastructure and services) and by controlling water demand. Water demand management measures are related with behaviours and include awareness campaigns, restrictions on use and, notably, water tariffs.

ENDNOTES

1 See also many African water regulators, like ESAWAS Regulators Association (https://www.esawas.org/index.php/en/publications/regional-benchmarking-of-large-wss-utilities-report?task=download.send&id=29&catid=7&m=0), WASREB (https://wasreb.go.ke/impact-reports/) and NWASCO (http://www.nwasco.org.zm/index.php/regulatory-tools/monitoring-performance-reporting).

2 The International Benchmarking Network for Water and Sanitation Utilities (IBNET) is an initiative to encourage water and sanitation utilities to compile and share a set of core cost and performance indicators. It sets forth a common set of data definitions (https://www.ib-net.org/).

3 Hard outliers are defined as the ones located outside the so called "outer fences" of the distribution, identified by multiplying the limits of first and third quartiles of the distribution by 3 times the value of the interquartile range.

4 This figure is not suitable however to get a good picture of price elasticities from coefficients of the model. For this purpose, rates charged to different users' categories and to domestic users and complete tariff structures in the common case of Increasing Block Tariffs (IBT). With IBT the rate per unit of water increases as the volume of consumption increases. Consumers face a low rate up to the first block of consumption and pay a higher price up to the limit of the second block,

and so on until the highest block of consumption. Increasing block tariff are an example of price discrimination based on quantity.

5 Under-pricing can sometimes be due to investment promotion priorities transferred to utilities to compete with other countries and cities and attract investments. Utilities can also be driven by microeconomic principles of price discrimination so that a higher price is charged to the low elasticity segment, and a lower price is charged to the high elasticity segment. Lastly, there could be a capture of regulators and utilities managers by business groups.

Chapter 6
Water Scarcity and Household Food Security:
A Case of Ulundi Local Municipality in KwaZulu-Natal, South Africa

Nomfundo Nomcebo Zulu
University of Zululand, South Africa

ABSTRACT

The study employed the post-positivist epistemology to examine the impact of water scarcity to food security in three rural areas of Ulundi Local Municipality. A stratified random sampling technique was utilised to sample 400 respondents. Data was collected through self-designed questionnaires. Correlation analysis was used to determine the relationship between the variables. The findings showed a strong positive correlation between economic and social development, water access, and scarcity. There was also a positive relationship between water access and household food security. Water access also had a strong positive relationship with public health and development. The study concludes that water access influences food security, public health, human growth, and development. It is recommended that the municipality should improve water access in order to improve food security, public health, and development in the rural communities.

1.0 INTRODUCTION

Water is an essential need for one to survive. However, water scarcity has and still is causing many problems in different countries. Adapted from White (2012:3), water scarcity can be understood as "the lack of access to acceptable quantities of water for human and environmental uses". Thus, the lack to access acceptable quantities of water has a negative impact on food security in rural areas of Ulundi Local Municipality (ULM) where agricultural activities are perceived to be responsible for poverty alleviation. Food security in rural areas is mostly secured through subsistence farming. In this case, food security is defined as the access by every person at all times to sufficient food for an active and healthy life. This

DOI: 10.4018/978-1-7998-8809-3.ch006

includes the ready availability of nutritionally adequate and safe foods (Bukenya, 2017). Food security and water scarcity has been central issues in developing countries such as South Africa and it has been prioritised by the South African government due to high numbers of people who suffer from poverty.

Water is also essential for manufacturing, agriculture, economies, and the existence of human and other life forms. According to Stratford Global Intelligence [SGI] (2014) when areas are under water stress, effects can include reduced food and livestock production, population movements, additional costs of living, starvation and other problems. The study done by International Fund for Agricultural Development [IFAD] (2012) shows that agriculture has been and still is the most pillar of survival for South Africans, more specifically in rural areas. Moreover, agriculture (farming, forestry, fisheries and livestock), is the central basis of development in rural areas through promoting employment and income. Thus, the inability to easily access water has many impacts on the socio-economic development of the country and may lead to households being food insecure (IFAD, 2012).

Therefore, water is the main driver of economic and social development while it also has a basic function in maintaining the integrity of the natural environment. Additionally, water is only one of several vital natural resources and water issues must not be considered in isolation (UN-Water, 2008). Nonetheless, sustainable management of water resources (including the provision of safe drinking water and irrigation, adequate sanitation, protection of aquatic ecosystems, and flood protection) poses enormous challenges in many parts of the world, specifically in rural areas. Despite the fact that this resource is renewable, it is stated that most of the world's population lives under conditions of water scarcity; this is referenced most specifically in developing states (Hering and Ingold, 2012).

1.1 Problem Statement and Study Justification

Water scarcity has caused multiples of problems in different countries. A study conducted by Rinkesh (2017) clarifies that there are millions of people all over the world who do not have access to water, or, if they have access, that water is unable to be used. Water crises, mostly in rural areas, are high and therefore they cause an increase in the number of households who are food insecure. ULM has water crises whereby communities cannot access water for weeks. This further affects their agricultural practice as they have to depend on rain water for their gardens. However, the constitution of the Republic of South Africa (1996) Chapter 2, Section 27 state that everyone has the right to have access to sufficient food and water; and social security, and thus, government should provide proper assistance for those who are unable to access adequate food and water (Republic of South Africa, 1996).

Thus, this emphasise that water is a basic need that should be accessed by anyone regardless of their location. However, the appropriate social assistance is not effectively provided at ULM. Most of the rural areas are still experiencing the inability to access potable water at all times. For instance, the trucks carrying water (water cans) come once a month with a limited scale of water to be provided regardless of the number of members in the household. Moreover, this study has become useful as it focuses on rural areas. Most of the studies cover the issues of water and food security at a national level and others are internationally based. Thus, this study was conducted at a local level in rural areas with the aim of filling the gap in issues of water scarcity and food security in rural areas at a local level.

1.2 Hypothesis

*1.2.1 Null hypothesis (H*o) - The null hypothesis (Ho) of this study is anticipating that "there is a statistically significant relationship between water scarcity and household food security in rural areas".

1.2.2 Alternative hypothesis (H1) - The alternative hypothesis (H1) is an inverse of the null hypothesis; it is assuming that "there is no statistically significant relationship between water scarcity and household food security in rural areas".

2.0 THEORETICAL FRAMEWORK

The study used ecological economics theory to explain how water scarcity affects food security and the people's lives. It focused on water as a human basic need, and equity in water. Additionally, the ecological economics framing of the study was mainly treating water as a natural capital. A study conducted by Costanza (1989) explains that ecological economics addresses the relationships between ecosystems and economic systems in the broadest sense, thus, stressing the importance of these relationships in order to promote a sustainable economy. Thus, the aim of ecological economics is to improve and to promote a deeper logical understanding of the complex linkages between human and natural systems, in order to make effective policies that will promote ecologically sustainable environment through equal distribution and allocation of scarce resources such as natural and social capital (Romeiro, 2012).

Economic activities depend upon the availability of water resources, including its quality (Logar, 2018). Water, elaborate Distefano and Kelly (2017), is an important factor of production contributing to economic activity across all sectors and regions of the global economy. Water scarcity may therefore go beyond having important consequences for people, society and ecological systems but may also pose a threat to economic growth of a country. Therefore, an ecological economics framework serves as a guide toward more sustainable water resources decisions in three important ways. It provides a needed theoretical review to neo-classical economic analysis. Lastly, it helps to identify the program of institutional reform that has the best chance of delivering more sustainable water resources management practices. Hence, water is, therefore, the most critical component of the economy (Lant, 2005).

Access to safe water is a fundamental human need and, therefore, a basic human right (UN Document, 2003). Nonetheless, United Nations Development Program [UNDP] (2004) expresses that formally acknowledging water as a human right and expressing the willingness to give content and effect to this right, may be a way of encouraging the international community and governments to enhance their efforts to satisfy basic human needs and to meet the Sustainable Development Goals (SDGs) as water forms a part on it. Therefore, as part of government's strategy to alleviate poverty in South Africa, a policy for the provision of free basic water (FBW) was introduced. The Department of Water Affairs and Forestry (DWAF) commenced the implementation of Phase 1 of a national free basic water strategy, in February 2001 (Department of Water Affairs and Forestry, 2002). This was to further promote equity in water access.

According to Phansalkar (2007) equity in access to and use of water and the distribution of the impacts of water resource development intervention have four elements, that is:

- Social equity: equity between different groups of people living in the same location (i.e. if nearby towns can have access to water, also rural communities must be able to have access).

- Spatial equity: equity between people living in different regions (i.e. rural people must have same benefits compared to their urban counterparts).
- Gender equity: equity between men and women in sharing labour costs, efforts to access and use water, and its benefits (i.e. giving women priority because water scarcity pose a major threat to them since they are the ones who fetch water from long distances).
- Inter-generational equity: equity in enjoyment of natural resources, including water, across generations of people (i.e. giving rural communities preference over natural resources found in their areas).

Therefore, the increase in population causes an increase in water demand. Thus, sustainable management of water resources is critical in order to balance demand and supply of water. Additionally, promoting equity in water also promotes increased agricultural and economic activities in rural areas.

2.1 Contribution to Knowledge

The study intends to add to an existing knowledge by providing valuable information on the strategies that can be improved or implemented in order to address the concern of water scarcity. The study is providing the information on the causes and challenges of water scarcity, and further assert an insight on what needs to be done in order to address such challenges. Furthermore, as stated earlier, most of the studies based on water scarcity and food security focus on other countries and continents. Therefore, this study provides an understanding on what is happening in rural areas of KwaZulu-Natal, South Africa.

3.0 METHODOLOGY

3.1 Research Philosophy and Paradigm

The study employed the post-positivist epistemology. Epistemology, defined by Moser (2015), "is the theory of knowledge, the philosophical study of the nature, origin, and scope of knowledge". Whereas, post-positivist is "a certain pluralism which balances both positivist and interpretivist approaches. It further focuses on researching issues in the context of involving experiences of the majority and announcing the result of what the majority says is acceptable" (Panhwar, Ansari, and Shah, 2017). Furthermore, post-positivism is grounded on epistemology which is a philosophical viewpoint that investigates the nature of reality, the relationship between a researcher and what can be known, and the hypothesis that changes the method of knowing and the achievement of findings (Moriarty, 2011). Thus, those active in research process attain understanding through the real life and natural setting of the area studied (Thomas, 2010).

3.2 Design

The researcher used a descriptive design linked with quantitative method. Clark and Creswell (2015) explain that quantitative research method draw conclusions for large numbers of people, uses numerical data from selected subgroup, demonstrate relationships, examine probable cause and effect, bias controlled and establish efficient data analysis. The study used questionnaires in order to collect the quantitative data, where the respondents were visited for the purpose of asking questions and to record respondents' answers. The questions were designed in two languages (IsiZulu and English) for simplicity purposes as most of the people in ULM are using IsiZulu language. Moreover, descriptive study helped the researchers to understand the concerns and causes of water scarcity and further be able to understand what strategies needed to be implemented in order for households to be food secure.

3.3 Study Population and Sampling Method

The study constituted of three rural communities under ULM with a population of 35 198, that is, Makhosini (12 015), Kwa Hhodlweni (11 050) and Mpungamhlophe (12 133). Nonetheless, the study was using quantitative research method, thus, probability sampling methods were the most suitable for this research. Probability methods, according to Clark and Creswell (2015), are based on the principles of randomness and probability theory. Thus, stratified random sampling (SRS) was utilised to sample 400 households. Shi (2015) explores that stratified random sampling is a technique that attempts to restrict the possible samples to those that are less extreme by ensuring that all parts of the population are represented in the sample in order to increase the efficiency. Therefore, the study consisted of 136 respondents from Makhosini, 126 respondents from Kwa Hhodlweni, and 138 respondents from Mpungamhlophe. Therefore, the sample size of the study was enough to cover three mentioned rural areas.

3.4 Sample Size Determination and Sampling Procedures

The sample size was determined based on an appropriate sample formula and in this instance the study was using a much simple version of sample estimation formula developed by Miller and Brewer (2003): $n=N/[1+N(\alpha)^2]$; where, n= total sample size, N = the sampling frame (35, 198), α= error margin, set at (0.05) and 1= constant value; hence; $n=35{,}198/[1+35198(0.05)^2]$. Therefore, n=35,198/[1+35,198 (0.0025)] n=35,198/[1+87.99] n=35,198/88 n=399.98 n=400. Therefore, 400 sample size was approximate for the study.

The study employed stratified random sampling method to draw the elements i.e. respondents from the target rural household populations within the target communities. A simple proportional formula {P x n/N}, where P= rural households per community, n= total sample and N= total number of rural household population, was used to select the respondents per strata. Using the above formula, the total sample for Makhosini can be determined as {12,015 x 400/35,198 = 136), Kwa Hhodlweni {11, 050 x 400/35198 = 126), and Mpugamhlophe {12,133 x 400/35198 = 138), a summation of which will be 400, which is an approximates for the study.

Similarly, the selection of the units of analysis i.e. gender representation from each community could follow a similar proportionate sampling formula pattern; {P x n/N}, where P= rural household population per each community, n= total sample per each community and N= gender classification per each stratum. Therefore, gender representation across the study areas was determined as follows; Makhosini community {male; 6002 x 136/12,015= 68, then female; 6013x136/12015=68}, for Kwa Hhondlweni community {male; 5030 X 126/11050 = 57, then female; 6020 x 126/ 11050= 69}, and finally, Mpungamhlophe community {male; 5102 X 138/ 12133 =58, and female: 7031X138/12133=80}.

3.5 Ethical Considerations, Reliability and Validity of the Study

The researchers obtained ethical clearance certificate from University of Zululand Research and Ethics Committee (UZREC) that gives a permission to go and to collect data from Ulundi Local Municipality. Secondly, the researchers consulted the traditional leaders as to also obtain the permission to perform such activity in those particular areas. Further, the researchers made it clear to the respondents that participating in the study is voluntary, and that should they for some reason want to withdraw from it, they have the right to do so at any time. In addition, respondents were promised that their privacy will be respected at all time and that everything they share will be treated as confidential. Nonetheless, the Cronbach's alpha was used in testing internal validity. Cronbach's alpha is a measure used to assess the reliability, or internal consistency, of a set of scale or test items. Thus, Cronbach's alpha, states Goforth (2015), is one way of measuring the strength of that consistency. In addition, methodologists recommend a minimum coefficient between 0.65 and 0.8; coefficients that are less than 0.5 are usually unacceptable. Therefore, the Cronbach's alpha value of 0.70 for all the items was considered generally acceptable and within the best standard practice for data reliability.

The validity of survey instruments is the level to which an instrument measures what it is supposed to measure, and performs as it is designed to perform. Hence, the validation of an instrument involves collecting and analysing data to assess the accuracy of an instrument. There are numerous statistical tests and measures to assess the validity of quantitative instruments, which generally involve pilot testing (Dewar College of Education, 2009). Therefore, for validity purposes, a pilot study was conducted. A study conducted by Rouse (2013) states that a pilot study offers a platform for a researcher to test logistics, prove value and expose deficiencies before committing huge resources or spending a considerable amount of time, energy or money on a large-scale project. The researchers pre-tested the instrument from one hundred and seven (107) respondents from B-North Section at ULM to answer the questionnaires that were going to be used for the actual project. Thus, the questionnaires indicated to be valid.

3.6 Data Analysis

For the presentation of the collected data, the study exploited Statistical Package for Social Sciences (SPSS) software as it assured that the conclusion made was reliable. The software consists of inferential and descriptive statistics, show skewness and kurtosis, measures of central tendency and measures of frequency. The study however adopted descriptive and inferential statistics, the measures of central tendency (mean and standard deviation). Manikandan (2011) state that central tendency is the statistical measure that identifies a single value as representative of an entire distribution. It aims to provide an accurate description of the entire data.

Graphs and tables were also used to provide a meaningful interpretation of the collected data. As Elston (2017) stipulates that to present data honestly and in a format that promotes meaningful interpretation, tables and graphs are more helpful. Studies identify two types of strategies that can be used to analyse data, namely: deductive data analysis strategy and inductive data analysis strategy. However, the study was focusing or rather used deductive data analysis strategy since it can be used to analyse a study based on quantitative approach. In addition, deductive approach offered the possibility to explain or to generalise research findings to a certain extent. Fabio and Blustein (2016) explore that a deductive approach is concerned with deducting conclusions from premises or suggestions.

4.0 RESULTS AND DISCUSSION

4.1 Features of the Respondents

The results showed that the majority of the respondents, at 297 (74.2%) were between 18-35 years, about 87 (21.8%) were between 36-60 years while the least number of respondents, at 16 (4%) were 60 years and above. Moreover, most of the respondents who numbered 217 (54.2%) were females while 183 (45.8%) were males. In terms of respondents aged between 18-35 years, the results further showed that the majority of the female respondents who numbered 158 (39.5%) were on average higher than their male counterparts who are 139 (34.8%). The disaggregated data revealed that, at all levels of age categories, adult females appeared to be most likely interested in water issues compared to their male counterparts. This is because females appeared to be mostly affected by water scarcity and household's food security. Females seems to be mostly engaged in domestic household activities of cooking, washing, cleaning, childcare and water haulage and therefore, turn to have direct effects in terms of water and food shortage.

Similarly, the study compared the educational attainment of the respondents and their employment status. The results revealed an unacceptably high rate of unemployment at 351 (87.8%) among the residents compared with only 49 (12.2%) who were actively employed and earned wages and salaries. This could be attributed to the lack of employable skills and low levels of tertiary educational achievement at 97 (24.2%) among the black residents. The data showed that among the illiterate respondents, about 4 (1%) were unemployed, nearly 31 (7.8%) of the respondents with primary education were unemployed while 8 (2%) were in active employment. Most of the respondents with secondary education were unemployed as they lacked appropriate qualifications for seeking decent employment. Thus, the level of respondents' tertiary education further affects their ability to attain stable employment that will help them to address the issues of poverty.

Although women appeared to be much more concerned with water and food security, comparatively, most women who were 190 in the community remained unemployed (47.5%). Moreover, about 161 (40.2%) of males were unemployed with just about 22 (5.5%) actively employed. Poverty, hunger and inequity of water access looms if equal opportunities are not created for women (who constitute the majority of the population) to have access to income earning jobs. Moreover, there were remarkable variabilities in terms of age and employment. A disaggregation of the data revealed the rate of unemployment that appeared to be heavily concentrated among the youthful populations. Approximately, 275 (68.8%) of the respondents aged between 18-35 were unemployed with some 22 (5.5%) gainfully employed. Similarly, about 63 (15.8%) respondents aged between 36-60 were unemployed while 24 (6%) of the same age group was gainfully employed. The unemployment rate among respondents who aged 60 and above stood at 13 (3.2%) with about 3 (0.8%) in active employment.

4.2 Importance of Water and Challenges of Water Scarcity

Water serves as an important catalyst for food production and growth. Hence, it plays a major role in sustaining food production while ensuring the growth in the country's economy. Therefore, water is linked with food security and it is significant for ecosystem where food security and nutrition of present and future generations depends on (FAO, 2017). Thus, the inability to access water poses a threat to food production and economic development. Nevertheless, this is in line with the study that was conducted by CONAGUA (2015) which explores that water scarcity as a challenge can cause difficulties to human and economic progress. Moreover, water is crucial in every aspect of life; hence, life is impossible without water. A study conducted by the World Food Programme [WFP] (2017) elaborates that poor access to water has a negative impact towards household's food security. Therefore, this further affects the livelihoods of the residents and poses a threat to their economic growth. Thus, in the table below, the study analysed the respondents' perceptions and values attached to water, in terms of food, livelihoods and economic growth.

Table 1. Water is important for food, livelihoods and economic growth

Response Categories	Frequency	Percent (%)
Agree	351	87.8
Strongly agree	49	12.2
Total	400	100.0

Source: Fieldwork (2019)

The results showed that most of the respondents, who were 351 (87.8%) agreed while 49 (12.2%) strongly agreed that water is important for food, livelihoods and economic growth, and thus valued water as a necessity for human survival, food production and economic activities in the community. Therefore, the results implied that all respondents see water as important for food, livelihoods and economic growth. Nevertheless, water is intrinsically part of the social and economic life of the people. Scarcity of water would, therefore, likely affect the everyday life, health, growth and development of residents in the district. Hence, a study conducted by European Commission (2002) stresses that water should be conserved in a sustainable manner, this will further allow water resources not to be diminished and thus decrease water supply.

4.3 Diminishing Water Supplies Affect Growth and Economic Prospects

Water resources should be protected in order to ensure sustainable water supply for citizens. Additionally, improved water supply and water resource management ensures a country's economic growth and contributes towards poverty eradication (Sanctuary, 2007). Thus, poor management and diminishing water supply is posing a threat to the economy. Therefore, the figure below analysed the perceptions of the respondents in the effects of diminishing water supplies on growth and economic prospects.

Therefore, the results showed that more than half to the tune of 356 (89%) of the respondents agreed, 42 (11%) strongly agreed while some 2 (2%) disagreed. The overwhelming confirmation of respondents is an indication that water has an inverse effect on household's food security, reducing the rate of growth and economic activities in the communities. This is further explored by FAO (2009) stating that although agriculture plays a major role in food security, however it is also the cause of water scarcity as it accounts for about 70% of water removals.

Additionally, this further asserts that water plays anthropogenic roles in all facets of human and inextricably linked to human biological life and it should be used efficiently and effectively in order to ensure that future generations also benefit. Nevertheless, it constitutes an essential determining factor for housing (building and construction), agricultural production, trade, commerce, recreational, traditional and customary life of the people in the community. A reduction in water allocation, therefore, implied reduction in demand-supply, which inexorably could result in competition, conflict over water and ultimately, reduce the life quality and livelihood standards of people in the area.

Figure 1. Effects of diminishing water supplies on growth and economic prospects
Source: Fieldwork (2019)

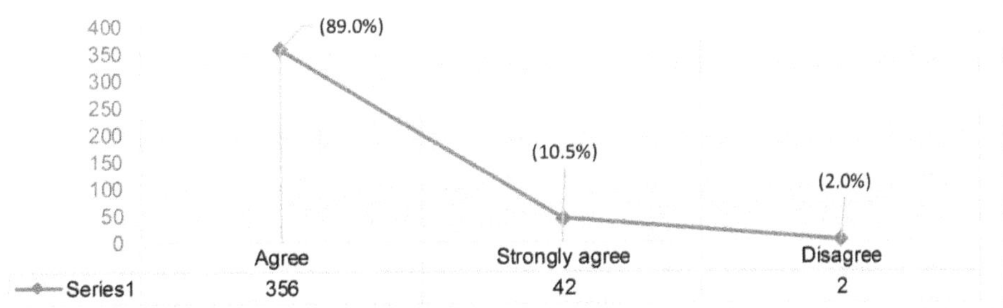

4.4 Allocation of Free Basic Water (6kl) to Households

In 2007, a nationwide community survey discovered that South Africa had made significant progress in improving access to water (Bailey, 2010). However, it should be clearly accepted that the right to water is a basic human right, and that the lack of adequate water supply and sanitation constrains opportunities and intensifies the problems of vulnerable groups particularly those who were previously disadvantaged and those living in rural areas. Thus, Figure 5.3 analysed the respondent's perception on whether the South African government is providing each household with 6kl of free basic water per month.

Figure 2. Government's allocation of free basic water
Source: Fieldwork (2019)

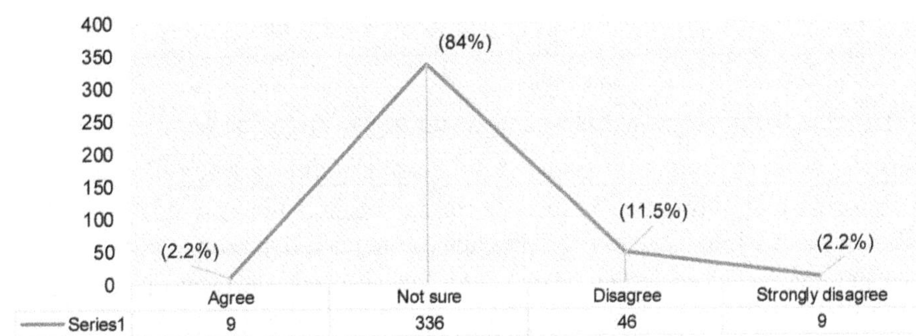

The study interrogated the respondents on their perceptions concerning their household's access to the 6kl of free basic water per month provided by the South African government since 2001. This was to enable the study to discover the effectiveness of government's policy and state commitment towards basic water supply in the community. The findings showed that most of the respondents who were 336

(84%) were not sure of their reception of or access to 6kl of monthly water supply, 46 (12%) disagreed, while 9 (2.2%) each agreed and strongly disagreed respectively with state's supply of minimum of 6kls of water to their households. Zululand District Municipality [ZDM] (2017) reports that an approved indigent/deprived household must receive fully subsidised water at a minimum of 6kl per month. However, the results meant that government either lacks the capacity for water supply or there could be leakages and free riding on water supply which present government supply from reaching targeted beneficiary communities.

4.5 Water Conservation in Response to Increasing Population

Water is a very essential resource in our everyday life. However, with an increase in population, agricultural and industrial activities, and poor rainfall, water is becoming scarce. Over the years rising populations, growing industrialisation, and expanding agriculture have pushed up the demand for water. However, efforts have been made to collect water. This has been done through building of dams and reservoirs (World Economic Forum [WEF], 2015). Hence, the study investigated the respondents' perceptions on whether conserving water is important due to a continuously increase of the population.

Figure 3. Conserving water is important due to a continuously increase of the population
Source: Fieldwork (2019)

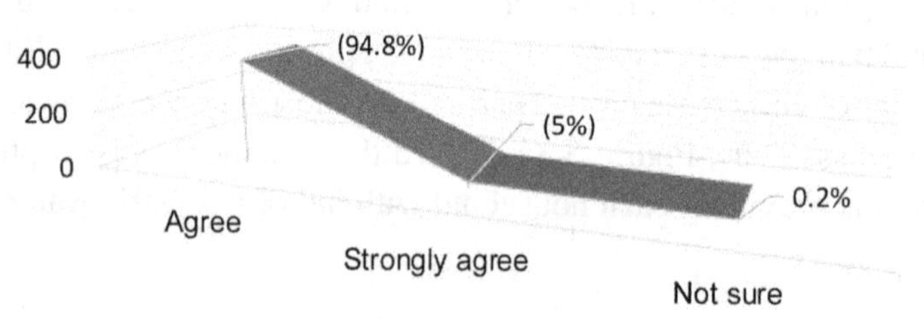

	Agree	Strongly agree	Not sure
■ Series1	379	20	1

This study found (see Figure 3) that the majority of the respondents who were 379 (94.8%) agreed and thus considered water conservation as an essential priority amidst a spiralling population, about 20 (5%) strongly agreed while only 1 (0.2%) of the respondents was not sure of the effects of population increase on water conservation. Inferably, the practice of uncontrolled human population growth has the capacity to disrupt human ecology, affect water access and consumption patterns, especially in developmental states like South Africa. Measures for water conservation are therefore no longer indisputable but must be highly prioritised to ensure human-water continuum and ecological sustainability (WEF, 2015).

4.6 The Effect of Water Used in Agriculture to Ecosystem Degradation and Its Effect to Water Scarcity

Although agriculture is seen as an approach in securing food in rural areas, a study conducted by FAO (2016) explores that agriculture is both the cause and victim of scarcity; it is the largest user of water globally and also the largest contributor to water withdrawals. Nevertheless, this will further affect water resources and other natural resources. Hence, the study analysed whether water use in agriculture affect ecosystem degradation and contribute to water scarcity.

Figure 4. Water use in agriculture affect ecosystem degradation and contribute to water scarcity
Source: Fieldwork (2019)

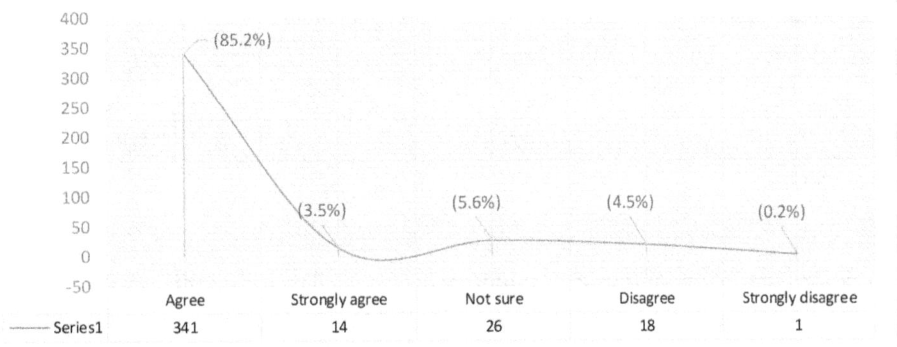

Therefore, from Figure 4, the study presents the perceptions of the respondents on the effects of water utilisation in agriculture on ecological degradation and water scarcity in the community. The results showed that more than half of the respondents who were 341 (85.2%) agreed while 14 (3.5%) strongly agreed. Approximately, 26 (6%) of the study participants were uncertain, 18 (5%) disagreed while only 1 (0.2%) strongly disagreed. Even though water was perceived to be relevant for agricultural production, nonetheless, rapid and uncontrolled utilisation of "treated water" for agriculture could likely reduce the availability of drinkable water in the area (Herselman, 2013). The results portray a perilous situation, since an overdependence on potable water for irrigation coupled with the nature and methods of farming practices could likely degrade ecological systems and pollute community drinking water sources in the community. Moreover, a polluted ground or surface water could increase the cost of local water purification, logistics /technologies and water tariffs beyond an already marginalised and peripheral community with high rates of unemployment and low income earning households.

4.7 The Impact of Water Scarcity to Household Food Security in Rural Areas

South Africa seems to be food secure at national level but the same cannot be said about households in rural areas (Hart and Aliber, 2009). Thus, the shortage of water in rural areas may further affect their ability to access food through agricultural production. Therefore, water is important for food security especially in rural areas where agriculture play a major role in food accessibility (FAO, 2017). Additionally, as stated earlier by Wheeler and Kay (2011), agriculture is central to food security and economic growth in developing countries and provides the main source of livelihood for three out of four of the world's poor. Therefore, the inability to access water at all times have a negative impact on household's food security.

Moreover, in South Africa, the research studies have shown that the role played by agriculture towards poverty reduction and achieving food security is imperative and effective (Dotse, 2016). In most rural areas, agriculture serves as a way to be food secure and provide opportunities for unemployed residents. Thus, for agricultural activities to take place, a substantial amount of water is needed. Therefore, water scarcity has a negative impact on household's food security. Hence, the study questioned the respondent's perceptions on whether water scarcity affects household's food security, specifically in rural areas.

Table 2. Water scarcity affects household's food security, specifically in rural areas

Response categories	Frequency	Percent (%)
Agree	381	95.2
Strongly agree	19	4.8
Total	400	100.0

Source: Fieldwork (2019)

Table 2 indicates that some of the respondents who were 381 (95%) and 19 (5%) respectively agreed and strongly agreed that water scarcity directly affects household's food security, specifically in rural areas where agriculture was perceived as a pillar for human survival and growth. These results implied that water is a key player in agriculture and without it performing agricultural activities is difficult. Therefore, this further implied that there is a need for a more sustainable approach to water resource management in agriculture at the global, regional and local/ household level in order to avoid depletion of water resources.

4.8 Increasing Household Food Security and Improving Farming Through Efficient and Sustainable Utilisation and Development of Water Resources in Agriculture to Reduce Poverty

Although agricultural practice is the largest contributor of water scarcity, improving its efficiency further contribute to poverty reduction in rural areas. "Improving agricultural efficiency, while sustaining and enhancing natural resources, such as water, is a crucial requirement for farmers to increase food supplies on a sustainable basis" (FAO, 2017). Therefore, the study questioned households whether in order to reduce poverty, the focus should be at increasing household food security and improving farming through efficient and sustainable utilisation and development of water resources in agriculture.

In order to reduce poverty, Table 3 indicates that about 386 (96.5%) of the respondents agreed, 2 (0,5%) were not sure while 12 (3%) strongly agreed that the focus should be on increasing household's food security and improving farming through efficient and sustainable utilisation and development of water resources in agriculture. The results implied that rural households should be assisted to specialise in the agricultural sector and thus, allowing them to be more productive and integrated in agriculture

(Christiansen, 2017). Therefore, providing such assistance will allow households to alleviate poverty not only through subsistence farming but also through commercial farming.

Table 3. Improving farming as a solution to efficient and sustainable utilisation and development of water resources in agriculture

Response categories	Frequency	Percent (%)
Agree	386	96.5
Strongly agree	12	3.0
Not sure	2	.5
Total	400	100.0

Source: Fieldwork (2019)

4.9 Quality of Drinkable Water: A Powerful Environmental Determinant of Health

Water plays a critical role in human health. It ensures the effectiveness of a person's body. A study conducted by Momentum Occupational Care South Africa (2018) reveals that there is a large amount of water covering earth's surface and most of the human body is composed of water. Thus, this proves the significant linkages between water, health and ecosystem. Additionally, lack of water hinder proper sanitation and thus, increase chances of being exposed to water related diseases such as Cholera and Diarrhea. Nonetheless, People are exposed to unsafe drinking water from rivers and lakes due to water scarcity. According to WHO (2010) drinking water quality management is the primary prevention of waterborne diseases. Hence, poor drinking water quality management is posing a threat to human health and other environmental species. However, the study analysed respondents' perception on whether the quality of drinkable water is a powerful environmental determinant of health.

Table 4. Quality of drinkable water is a powerful environmental determinant of health

Response Categories	Frequency	Percent (%)
Agree	381	95.2
Strongly agree	19	4.8
Total	400	100.0

Source: Fieldwork (2019)

The results showed that most of the respondents at 381 (95%) agreed that the quality of drinkable water is a powerful environmental determinant of health. Moreover, 19 (5%) strongly agreed that the quality of potable water plays a pivotal role in ensuring people's health. The results implied that the health of human beings strongly depends on the quality of water they drink. Therefore, inability to access clean water may result in water related diseases (such as cholera). Thus, proper governance of water should be prioritised in order to ensure human's access to clean drinkable water at all times. Nevertheless, WHO (2010) suggests that monitoring in water sector is needed to ensure suitable provision of water to people.

4.10 Women: The Victims of Water Crises

Women and girls play a critical role in household food security and water access; however, there are still inequalities that prevent women and girls to access resources such as land. Moreover, Vogelstein (2017) postulate that women become victims of water crises, and thus, the cost of water shortage and inadequate infrastructure, fall extremely on women and girls. Additionally, women and girls face challenges in responding to and adapting to climatic changes due to their lack of access to resources. Additionally, the economic cost of water crises mostly affects women in rural areas or low-income societies (Fonjong and Ngekwi, 2014). Thus, women and girls should be empowered in order to broaden agricultural development and food security (Habtezion, 2012).

Figure 5. Women are victims of water crises
Source: Fieldwork (2019)

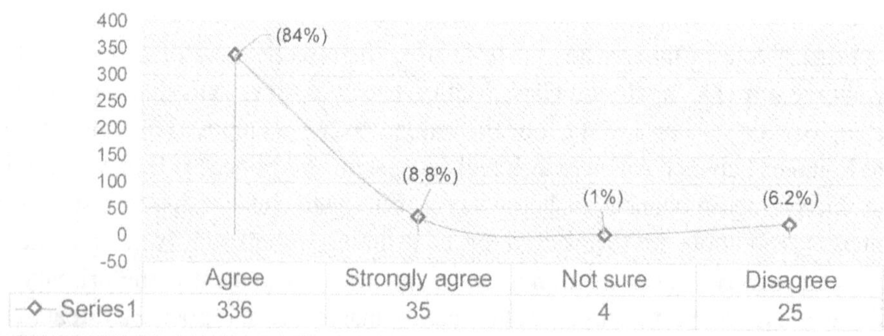

In Figure 5, the study presented respondents' perceptions on women as victims of water crises. The results implied that most of the respondents who were 336 (84%) agreed while 25 (6.2%) disagreed that women are victims of water crises. Nonetheless, 35 (9%) respondents strongly agreed and 4 (1%) were not sure. Although both males and females are affected by water scarcity, women tend to be affected more since they are the ones who usually walk long distances to fetch water and bring it back to the household. Nevertheless, even though women are the victims when it comes to water crisis, men are more dominant in when it comes to representatives in water units and women are less represented in water units especially in rural areas. Therefore, this calls for women empowerment through water use security (WRC, 2011).

4.11 Decentralisation in Achieving Partnership and Good Governance

National government has a significant role to play in ensuring water access to households. However, in most areas it has shifted its powers of water access to local government (decentralisation). Thus, decentralisation assists in achieving partnerships and good governance through transferring powers or responsibility from central government to local government with an aim to improve service delivery (Biswas,

2004, Sutherland, Scott and Hordijk, 2015). Decentralisation strategy has allowed local governments to play a leading role in water access. At ULM, Zululand District Municipality is responsible for water provision for households. This allows communities to easily access water and other services. Moreover, good local governance comprises of citizen participation, partnership among key stakeholders at a local level and partnership with the national government (United Nations Development Programmes, 2004). Hence, for service delivery to be effective, every stakeholder has a key role to play. Therefore, the study examined the respondents whether decentralisation assists in achieving partnership and good governance.

Figure 6. Decentralisation assists in achieving partnership and good governance
Source: Fieldwork (2019)

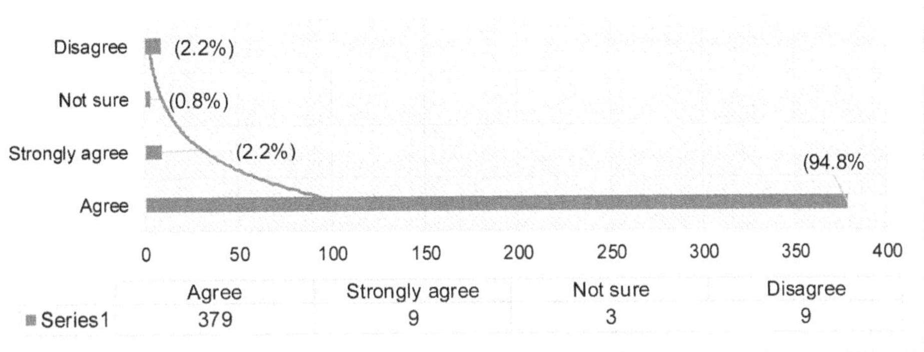

Figure 6 above presents the perceptions of the respondents on decentralisation as the way of assisting to achieve partnerships and good governance. The results found that the greatest number of respondents who were 379 (95.8%) agreed, 9 (2.2%) strongly agreed while some 3 (0.8%) were not sure and 9 (2.2%) disagreed. The results implied that local government assist local communities to develop and easily get access to services (EC, 2016). Furthermore, the response the study received provides an indication that divolving power from national government to local government assist locals to easily access services such as water. Nevertheless, local government can easily reach communities whenever they need assistance.

4.12 The Local Government Responsibility in Water Service Delivery

According to Kanyerere and Pietersen (2016), water service authorities are metropolitan municipalities, district municipalities and local authorised municipalities. Hence, the national government has decentralised it powers to local government for effective water services delivery. Nonetheless, the study questioned respondents whether water service delivery is the responsibility of local government.

The study found that most of the respondents who were 364 (91.0%) agreed that water service delivery is the responsibility of local government. Nonetheless, 11 (2.8%) strongly agreed while 2 (.5%) disagreed and they were 23 (5.8%) who were uncertain. The study showed that Zululand District Municipality is responsible for water provision at Ulundi Local Municipality. The results showed that there are members in the community who are not aware that local government is responsible for water service delivery. Thus, this means that there should more awareness to communities in terms of service delivery (EC, 2016).

Table 5. Water service delivery is the responsibility of local government

Response Categories	Frequency	Percent (%)
Agree	364	91.0
Strongly agree	11	2.8
Not sure	23	5.8
Disagree	2	.5
Total	400	100.0

Source: Fieldwork (2019)

4.13 The Role of NGOs in Providing Aid Directly to the People Who Need it Most

NGOs meet government halfway through reaching the poor with services such as food and water. Similarly, NGOs said to have natural resource conservation, land, and water management activities at heart, as they play a crucial role in improving food security and water access (World Food Programme [WFP], 2012). Thus, as stated earlier, NGOs call for the need to be treated as partners in all initiatives and at all stages of action, through active collaboration and coordination between government, CSOs, agencies, communities and other relevant institutions (Human Science Research Council, 2013). Nonetheless, they also work both independently and in conjunction with agencies from developed countries, private-sector infrastructure operators, self-help associations, and local governments in order to assist the poor (Werker and Ahmed, 2007). Hence, the study analysed the respondents' perceptions on whether NGOs are very effective in getting aid directly to the people who need it most.

Figure 7. NGOs are very effective in getting aid directly to the people who need it most
Source: Fieldwork (2019)

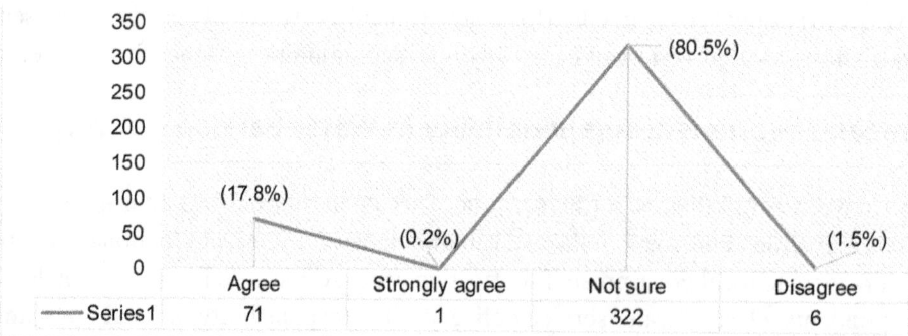

From Figure 7 above, the results showed that most of the respondents who were 322 (80.5%) were not sure whether NGOs are very effective in getting aid directly to the people. Furthermore, 71 (17.8%) agreed while only 1 (0.2%) strongly agreed and 6 (1.5%) disagreed. This showed that communities are not aware of the aid provided by NGOs. Moreover, the results implied that there should be awareness at a grass root level that provide communities with knowledge about NGOs and their role (Ramana, 2013).

4.14 Hypothesis Test Results

Table 6. Correlation Analysis between the Variables

Correlation Coefficients (Spearman's rho)			
Variables	Influence of (Water Access/Scarcity)		
	Correlation Coefficients (r)	Sig.(2-tailed)/ (p-value)	N
Economic and social development	0.454**	0.000	400
Rapidly population growth	0.619**	0.000	400
Household's food security	0.575*	0.000	400
Public health risk and development	0.758**	0.000	400

**. Correlation is significant at the 0.01 level (2-tailed).

The decision rule (region of acceptance) for hypothesis testing was based on the significance level which stated as; {accept null (Ho) at 95% confidence, with e = 0.05, if p-value $\leq$ alpha level (α) = 0.05)}. The findings in Table 2 shows a strong positive correlation between economic and social development, water access and scarcity (r = 0.454, p-value=0.00, N = 400). Based on the decision rule, we accept the null hypothesis and conclude that water access and scarcity of water influences sustainable socio-economic development. Similarly, there was evidence of a strong relationship between population growth and water access/scarcity (r = 0.619, p-value = 0.00, N = 400), and therefore, based on the decision rule for significance level, we accept the null hypothesis and conclude that rapid population density coupled with urban expansion, industrialisation and irrigation expansion, are associated with water access and scarcity in the community.

Moreover, the study found the existence of a strongly positive correlation between water access/ scarcity and household food security (r = 0.575, p-value = 0.00, N = 400). In effect, the results meant that, access/scarcity of water has the capacity to affect food production and food security specifically in rural areas where agricultural practice is seen as a pillar for survival. Hence, the study proceeded to accept the null hypothesis and rule out the alternative hypothesis. The study further discovered that water access/scarcity has a strong positive relationship with public health and development (r = 0.758, p-value = 0.00, N = 400).

5.0 CONCLUSION AND RECOMMENDATION

The study found the existence of a strongly positive correlation between water access/scarcity and household food security (r = 0.575, p-value = 0.00, N = 400). In effect, the results meant that, access/scarcity of water has the capacity to affect food production and food security specifically in rural areas where agricultural practice is seen as a pillar for survival. Hence, the study proceeded to accept the null hypothesis and rule out the alternative hypothesis. The study further discovered that water access/scarcity has a strong positive relationship with public health and development (r = 0.758, p-value = 0.00, N = 400).

Therefore, the study concludes that water access influences food security, public health, human growth and development. The results mean that a shortage or scarcity of water could pose a serious threat to food security, public health risk and development of the community. Moreover, this could result in widespread consumption of polluted surface water that could expose the population to water borne disease (pathogens, protozoa, bacteria) infections. The effect size is likely to increase among low income earning rural households, widen rural-urban development gaps, and increase public protestation and conflict among rural end-users of water.

Moreover, the evidence from the study confirmed that water scarcity is affecting household's food security. Hence, it suggests that more attention should be paid to address water issues within the municipality. Analysing the causes of water scarcity, agriculture became the most causes of water scarcity; thus, the study suggests that for agricultural purposes, communities must most of the time use rain water in order to save the drinking water. Nevertheless, water from the river can be utilised in this instance; however, the local government (ZDM) must provide each household with water for gardens.

6.0 FUTURE RESEARCH DIRECTIONS

It is important to look at water resources management in rural areas. This will help to understand the causes of water crisis in most communities.

ACKNOWLEDGMENT

The authors would like to express gratitude to study particiants, the KwaZulu Natal Department of Co-operative Governance and Traditional Affairs (COGTA), the Management Committee of the Zululand District and the late Mayor of uMlalazi Local Municipality, His Worship Councillor, Thelumoya Zulu for the insightful comments they made when we disseminated findings to key statekolders. In both COGTA and ZDM, we express our sincere gratitude to Dr Gladys Makolomakwe, Ms Nokubonga Caluza and Mr Phindithemba Manqele for organising the meetings and affording us an opportunity to present the study findings. At UNIZULU, we express gratitude to the Community Engagement Office, and the Communication and Marketing Department especially Ms Mammusa Lekoa and Ms Precious Shamase for attending and documenting these community engagement online workshops.

REFERENCES

Bailey, D. (2010). *Free Basic Services in Rural Kwazulu-Natal: The Built Environment Support Group*. https://www.ggln.org.za/media/k2/attachments/SoLG.2010-BESG.pdf

Biswas, A. K. (2004). Integrated water resources management: A reassessment. *Water International*, *29*(2), 248–256. doi:10.1080/02508060408691775

Bukenya, J. O. (2017). Determinants of Food Insecurity in Huntsville, Alabama, Metropolitan Area. *Journal of Food Distribution Research*, *48*(1), 73–80.

Christiaensen, L. (2017). *Agriculture in Africa - Telling Myths from Facts: A Synthesis.* Policy Research Working Paper No. 7979. Washington, DC: World Bank Publications.

Clark, V. L. P., & Cresswell, J. W. (2015). *Understanding Research: A Consumer's Guide.* PEARSON Publications.

CONAGUA. (2015). *Governments Contribution to Integrated Water Resource Management: The Scarcity on the Rise.* www.un.org/waterforlifedecade/waterandsustaibledevelopment2015/pdf/SP_Government_IWRM_sancho_FORMAT

Costanza, R. (1989). *What is Ecological Economics?* Elsevier Science Publications.

Department of Water Affairs and Forestry (DWAF). (2002). *Free Basic Water Implementation Strategy, Version 2.* http://www.dwaf.gov.za/Documents/FBW/FBWImplementationStrategyAug2002.pdf

Dewar College of Education. (2009). *Research rundowns: Uncomplicated Reviews of Educational Research Methods.* https://researchrundowns.wordpress.com/quantitative-methods/instrument-validity-reliability/

Distefano, T., & Kelly, S. (2017). Are we in deep water? Water scarcity and its limits to economic growth. *Journal of Ecological Economics, 142*, 130–147. doi:10.1016/j.ecolecon.2017.06.019

Dotse, L. N. A. (2016). *Exploring the relationship between water scarcity on food and nutritional security in rural households in the Nqgeleni location, Eastern Cape.* http://wiredspace.wits.ac.za/bitstream/handle/10539/21715/FINAL%20RESEARCH%20REPORT%20LAURA%20CORRECTIONS%20%282%29%20%281%29.pdf?sequence=1&isAllowed=y

EC. (2016). *Supporting decentralisation, local governance and local development through a territorial approach.* https://europa.eu

Elston, D. M. (2017). Presentation of data. *Journal of the American Academy of Dermatology.* Advance online publication. doi:10.1016/j.jaad.2017.11.009 PMID:29128450

European Commission (EC). (2002). *Water in the European Union.* European Communities Publication.

Fabio, A. D., & Blustein, D. L. (2016). *From Meaning of Working to Meaning Lives: The Challenges of Expanding Decent Work.* Frontiers Publications.

FAO. (2016). *Coping with water scarcity in agriculture: A global framework for action in a changing climate.* https://www.fao.org/3/a-i6459e.pdf

FAO. (2017). *Water for Sustainable Food and Agriculture: A report produced for the G20 Presidency of Germany.* https://www.fao.org/3/a-i7959e.pdf

FAO. (2017). *Water at a Glance The relationship between water, agriculture, food security and poverty.* https://www.fao.org/3/ap505e/ap505e.pdf

Fonjong, L. N., & Ngekwi, M. A. (2014). Challenges of Water Crisis on Women's Socio-Economic Activities in the Buea Municipality, Cameroon. *Journal of Geography and Geology, 6*(4), 122–131. doi:10.5539/jgg.v6n4p122

Food and Agriculture Organization (FAO). (2009). *Water at a Glance: The relationship between water, agriculture, food security and poverty*. https://www.fao.org/3/ap505e/ap505e.pdf

Goforth, C. (2015). *Using and Interpreting Cronbach's Alpha*. https://data.library.virginia.edu/using-and-interpreting-cronbachs-alpha/

Habtezion, S. (2012). *Gender, agriculture and food security*. https://gest.unu.edu/static/files/tm4_africa_gender-climatechange-and-food-security.pdf

Hart, T. G. B., & Aliber, M. (2009). Should subsistence agriculture be supported as a strategy to address rural food insecurity? *Agrekon*, *48*, 4.

Hering, J. G., & Ingold, K. M. (2012). Water Resources Management: What Should Be Integrated? *Journal of Science*, *336*(6086), 1234. PMID:22679083

Herselman, J. E. (2013). *Guideline for the Utilisation and Disposal of Water Treatment Residues*. http://www.wrc.org.za/wp-content/uploads/mdocs/TT%20559-13.pdf

Hordijk, M., Scott, L. M., & Sutherland, C. (2015). Resilience, transition or transformation? A comparative analysis of changing water governance systems in four southern cities. *Environment and Urbanization*, *26*(1), 130–146. doi:10.1177/0956247813519044

Human Science Research Council (HSRC). (2013). *Civil Society Organisation' Participation in Food Security Activities in South Africa*. www.hsrc.ac.za/en/research-data/ktree-doc/13815

IFAD (International Fund for Agricultural Development). (2012). *Agricultural cooperatives: paving the way for food security and rural development*. CapeNSSANews Publications

Kanyerere, T., & Pietersen, K. (2016). *Assessment of Water Service Delivery in the Municipalities of City of Tshwane, City of Cape Town and Ethekwini*. https://etd.uwc.ac.za/xmlui/bitstream/handle/11394/5332/Makhari_cl_msc_ns_2016.pdf?sequence=1&isAllowed=y

Lant, C. (2005). *Water Resources Sustainability: An Ecological Economics Approach*. North-Central Section- 39th Annual Meeting.

Logar, I. (2018). *Environmental Economics (EnvEco)*. https://www.eawag.ch/en/deparment/es/main-focus/environment-economics-enveco/

Manikandan, S. (2011). Measures of central tendency: The mean. *Journal of Pharmacology & Pharmacotherapeutics*, *2*(2), 140–142. doi:10.4103/0976-500X.81920 PMID:21772786

Miller, R., & Brewer, J. (2003). *The A-Z of social research*. Sage Publications.

Momentum Occupational Care South Africa. (2018). *Water scarcity and associated health risks*. www.ocsa.co.za

Moriarty, J. (2011). *Qualitative methods overview: improving the evidence base for adult social care practice*. London School of Economics and Political Science Publications.

Moser, P. K. (2015). *Epistemology*. https://www.researchgate.net/publication/288829852_Epistemology

Panhwar, A. H., Ansari, S., & Shah, A. A. (2017). *Post-positivism: An Effective Paradigm for Social and Educational Research.* https://www.researchgate.net/publication/317605754_Post-positivism_An_Effective_Paradigm_for_Social_and_Educational_Research

Phansalkar, S. J. (2007). Water, equity and development. *International Journal of Rural Management, 3*(1), 1–25. doi:10.1177/097300520700300101

Ramana, E. K. (2013). *The Role of NGOs in Rural Development with Special Reference to Women.* https://shodhganga.inflibnet.ac.in/bitstream/10603/18591/9/09_chapter%203.pdf

Republic of South Africa. (1996). Constitution of the Republic of South Africa, Act 108 of 1996. *Government Gazette*, 378.

Rinkesh, K. (2017). *What is Water Scarcity?* https://www.conserve-energy-future.com/causes-effects-solutions-of-water-scarcity.php#abh_about

Romeiro, A. R. (2012). Sustainable development: An ecological economics perspective. *Estudos Avançados, 26*(74).

Rouse, M. (2013). *Pilot program (pilot study).* https://searchcio.techtarget.com/definitions/pilot-program-pilot-study

Sanctuary, M. (2007). Making water a part of economic development: The economic benefits of improved water management and services. Stockholm International Water Institute (SIWI) Publications.

Shi, F. (2015). *Study on a Stratified Sampling Investigation Method for Resident Travel and the Sampling Rate.* Hindawi Publications.

Stratford Global Intelligence (SGI). (2014). *Water Scarcity: Examining Impacts Around the World.* https://cdn2.hubspot.net/hubfs/515194/B2C_content/B2CWF1/Water_Scarcity-slideshare_-final.pdf

Thomas, P. Y. (2010). *Research Methodology and Design.* https://uir.unisa.ac.za/bitstream/handle/10500/4245/05Chap%204_Research%20methodology%20and%20design.pdf

UN Documents. (2003). *The Dublin Statement on Water and Sustainable Development.* www.un-documents.net/h20-dub.htm

UN-Water. (2008). *Integrated Water Resources Management (IWRM).* www.un.org.waterforlifedecade/iwrm.shtml

United Nations Development Programme (UNDP). (2004). *The human right to water and sanitation.* https://www.un.org/waterforlifedecade/human_right_to_water.shtml

Vogelstein, R. (2017). *Women Around the World.* https://www.cfr.org/blog/women-around-world-week-75

Werker, E. D., & Ahmed, F. Z. (2007). *What Do Non-Governmental Organizations Do?* https://www.hbs.edu/faculty/Publication%20Files/08-041.pdf

Wheeler, T., & Kay, M. (2011). Food crop production, water and climate change in the developing. *Journal of Outlook on Agriculture, 39*(4), 239–244. doi:10.5367/oa.2010.0017

White, C. (2012). *Understanding water scarcity: Definitions and measurements.* Global Water Forum Publications.

World Economic Forum (WEF). (2015). *Global Risks 2015: 10th Edition.* Insight Report. https://www3.weforum.org/docs/WEF_Global_Risks_2015_Report15.pdf

World Food Programme (WFP). (2012). *Water for food security: Lessons learned from a review of water-related interventions.* https://www.odi.org/sites/odi.org.uk/files/resource-documents/11610.pdf

World Health Organisation (WHO). (2010). *Water for Health: WHO Guidelines for Drinking-water Quality.* https://www.who.int/water_sanitation_health/WHS_WWD2010_guidelines_2010_6_en.pdf?ua=1

Zululand District Municipality (ZDM). (2017). *Indigent Policy Status: Final (2017/18).* http://www.zululand.org.za/media/6137/Indigent%20Policy%20-%20Final.pdf

Chapter 7
Equitable Access to Water at Alfred Duma Local Municipality

Nokuphila Ndimande
University of Zululand, South Africa

ABSTRACT

Water plays a central role in the life of society. However, factors such as population growth, pollution, and poor allocation and distribution mechanisms place severe pressures on adequate and equitable water supply. The aim of this chapter was to look into equitable water access in the Alfred Duma Local Municipality as well as the ecological governance framework that supports water access in local areas. The chapter also looked at the position of local municipality in water access and the impact of ecological scale on water provision. Many people are still unable to exercise their constitutional right to water in Alfred Duma Local Municipality, where most women feel disempowered, marginalized, and excluded from the process of making water access decisions. This brings challenges to disadvantaged and arginalized groups socially, economically, and environmentally where vulnerable and marginalised groups have no opportunity to equitably benefit from water access.

1.0. INTRODUCTION

Water is the most precious resource on earth's surface, covering more than 70% of the planet and vital to the survival of humans, plants and animals (Hossain, 2015). Since the 1980s, water has become a prominent topic and noteworthy item globally on the political agenda. The motivation behind this is that water resources are very limited and water demand has increased due to factors such as industrialization, urbanisation, population growth, economic development, agriculture, energy and environmental security (Khadza, 2010). Biswas (2008) points out that due to the increasing population, there will be a water crisis in the future and the present trend will affect available water resources. This will affect existing water planning, management and allocation of water resources.

Equitable access to water is all about closing gaps of disparities in water access through treating vulnerable and marginalised groups differently to ensure an equitable outcome (Luh, Baum and Bartram, 2020). Global Water Partnership (GWP) (2000) contends that equitable access to water aims at

DOI: 10.4018/978-1-7998-8809-3.ch007

satisfying the needs and rights of people and people must be in the centre of water management to ensure social equity goals of access to water. Therefore, water access will be understood to be equitable when it contributes to making change within the society more equitable. Equitable access to water should be analysed within the framework of historical processes and ensuring equal opportunities in water resources management through maintaining fair play and justice in water, where society should be involved in water resources (Hamberto, 2011).

Water inequity must be measured in the context of uneven distribution of various impacts affecting poor water supply and inequality in water access can be measured based on income, household size, economic status and dwelling unit (Nastiti, Komarulzaman and Sudradjat, 2018). According to Goff and Crow (2014), the world's most disadvantaged people have not benefited from advances in equal access to water, despite the fact that it is a fundamental human right for all people to have sufficient quantities and standard of water for the sustenance of human health.

This chapter posits that human right to water must be universally recognised in addressing the equitable access to water and also consider gender dimensions, rural and urban disparities. The Sustainable Development Goals (SDGs) approved by the United Nations in 2015, especially Goal 6 of SDGs aims to ensure that by 2030 everybody has access to clean, accessible drinking water. Clifford-Holmes et al., (2017), posit that more than a third of the world's population lacks access to safe drinking water and the situation is deteriorating. The majority of those affected by inequitable access to water are those who live in rural areas. Access to water directly impacts daily life and human survival. Water access provides social power, healthy bodies, viable livelihoods, landscapes are being transformed, energy is being generated, wealth is being created, and life itself is being created (Goff and Crow, 2014). According to the Water Protocol, equal access to water must be sufficient in terms of quantity and quality, and it must be given to all members of the community, especially those from deprived backgrounds or who face social exclusion (United Nations General Assembly, 2016).

Meyer (2007) notes that rural poor who constitute just under half of the black population in South Africa has inequitable access to water and sanitation because many South Africans live in conditions with a high poverty rate. The need to correct historical inequities formed the universal human rights enshrined in the 1996 Constitution (De Visser, Cottle & Mettler, 2009). Everyone has the right to adequate water, according to section 27(1) (b) of the Constitution. This is valued in the Constitution as there are huge inequalities that exist between the rich and the poor households (Meyer, 2007).

The Free Basic Water Policy was developed in response to the unique historical situation surrounding services in South Africa, with the aim of providing six kilolitres per household of eight people per month, recognizing that everyone has the right to basic water supply and sanitation and reasonable steps must be taken to realise this right (Muller, 2011). The Water Services Act of 1997 mandates that municipalities adopt the FBW policy with at least three tariff tiers, the first of which is 0-6kl. The Alfred Duma Local Municipality created an indigent program to determine which households are eligible for free basic water. According to the indigent scheme, each household that is registered as indigent should receive a credit on their water bill for the amount of free basic water they have received, and indigents are typically classified based on their monthly household income (uThukela District Municipality, 2017). The municipality received 4500 applications from indigent households, but only 965 applications were approved due to time constraints and system problems (Alfred Duma Local Municipality, 2019).

Saner et al. (2018) maintains that the gaps in inequitable access to water are huge due to poverty and the unavailability of proper infrastructure. Equitable access to safe drinking water, survival, growth, and development are important basic necessities that are still needed to be fulfilled in many developing countries (Workman, 2019). Lange and Hassan (2006), postulate that decentralization of water management and governance by Integrated Water Resources Management (IWRM) approaches is the best way to deal with such issues. Lange and Hassan (2006) firstly suggest that decentralisation should improve all the water users and ensures more equitable allocation of the resource among competing sectors. Secondly, a more efficient use of water can be reached if local knowledge about water demand and supply is mobilized to design and implement allocation strategies.

The municipality also established the representative forum that represents the interest of the residents, at the level of the district and municipality in the participation of IDP process (UThukela District Municipality, 2019). The municipality has developed the ward committee including the water committee and representative's forum. Ward committees also encourage participation in the community to make the council aware of the needs and concerns of the community and keep people informed of the activities in the municipality and community organisations participate and contribute to issues of improving service delivery.

Sebova (2015) notes that the economic growth of many local municipalities is very low as it is associated with high unemployment, poverty, lack of water and inadequate water and lack of infrastructure. This brings hardships to local municipalities as they cannot support the provision of services to local communities. Municipalities must resolve capacity constraints by increasing the supply of local infrastructure and services, as well as structuring tariffs to limit demand growth, especially in the water sector. Sebova (2015) also suggest that small municipalities have bigger problems of efficiency and have insufficient funds for the provision of services in their local communities and thus can be determined through values of capital income and expenditure and the lower the economic efficiency of the small municipality the greater are the effects on the competitiveness of the whole economy.

According to the Alfred Duma Local Municipality (2019), the low levels of economic growth indicate that the municipality has growing levels of inequality. According to the municipality, as the wealthy get richer and the poor get poorer, inequality will continue to rise over time. Inequitable access to basic services remains one of the most visible spatial imprints of apartheid where services are provided well in urban areas whereas in rural areas there is high services' backlog in the municipality (Alfred Duma Local Municipality, 2017). The municipality is also having less economic growth and high unemployment rate in the municipality which lead to lack of economic development taking place within the local communities. The uThukela economy is largely driven by the tertiary sector with the community services (23%), finance (13%), transport (12%) and trade (14%) sectors. The institutional structure that is capable of providing water supply within the municipality is uThukela District.

UThukela District made a partnership with Umgeni Water to operate and manage bulk potable water supply to enhance service provision within the municipality and also assesses future infrastructure needs in order to enhance water services provision within uThukela District. uThukela District Municipality was the only Water Services Authority that manages both Bulk Water Services and Reticulation Services in the KZN Province (uThukela District Municipality, 2019). Maintenance and refurbishment costs of water and wastewater works are currently unaffordable by uThukela District Municipality due to financial constraints. Therefore, working in partnership with Umgeni Water Board is hoping to address water service backlogs and to ensure that there is a sustainable supply of bulk potable water within the time frame that is specified in National Development Plan (NDP) as Umgeni Water has sufficient capacity to maintain, refurbish, upgrade and build new wastewater infrastructure. Therefore, this chapter aim to investigate the ecological governance framework supporting access to water in local areas and the equitable access to water.

1.2. Background

Water access is one of the major challenges for the 21st century. Khadka (2010) notes that water is not directly addressed as a human right at international and national level. Water is essential for economic growth and provides the basis for long-term livelihoods in much of Africa, but freshwater supplies are under pressure due to overexploitation, rising demand, and continued environmental degradation (Palmer and Munnik, 2018). It should also be remembered that water shortages and inequities do not only endanger humans; they also endanger natural ecosystems and biodiversity by degrading habitats and disrupting ecosystem functions (Goff & Crow, 2014).

There is also an enormous challenge of securing sustainable equitable access to water and water crisis consequences from human decisions which results in uneven distribution of resources, benefits and cause the issue of equity to be critical to water management (Palmer and Munnik, 2018). Nkuna (2012) posit that challenges of water access, use of conservation and distribution of supply are one of the issues of water governance and decision about water are taken, by whom, at what geographical scale and who benefits. South Africa is a semi-arid country where 5 million people, mainly in rural areas, lack access to water. Hove et al. (2019), argue that many authorities lack the capacity to communicate with and respond to community needs, despite legislative and policy commitments to the right to water, cooperative governance, and public engagement. In post-apartheid, the South African government cherished water as basic human rights in the 1996 constitution. The National Water Act of 1998 and the Water Services Act of 1997 were part of a series of legislative and regulatory changes aimed at redressing segregation, promoting equal access, and assisting communities in providing services, with a focus on cooperative governance and public involvement.

According to DWAF (2008), it was against this backdrop that the Department of Water and Forestry, as policy maker and implementer, launched the water supply and sanitation programme in 1994 to meet a constitutional goal of ensuring that all South Africans have access to adequate water and a healthy climate, with an emphasis on rural areas. South Africa is facing a continuous challenge in providing equitable access to water for its people and in 1994, at the end of apartheid it shows that water and sanitation were unavailable to 41% of the black household population (15.9 million people) (DWAF, 2008).

According to Van der Mescht and Van Jaarsveld (2012), decentralisation and growth agendas shifted a variety of obligations to local levels, where various constraints arose, including a lack of revenue, financial distress, poor administrative and financial management, corruption, and public procurement exploitation. The constitution of South Africa placed water as an important tool for changing people's lives and municipalities has made a progress towards providing citizens with basic water services which is recognized to the adoption of the Free Basic Water policy (Muller, 2008).

1.3. Main Focus of the Chapter

1.3.1. Theoretical Framework and Empirical Literature

According to Pink (2016), water is the most valuable resource on the planet and the most pressing human rights issue of the 21st century. Water is one of the most important commodities and it is used basically for all, including household domestic use, industrial and agricultural production, and economic development (Oki and Quiocho (2020). The United Nations General Assembly notes that safe, clean drinking water and sanitation is a basic human right and is essential for full enjoyment of life and human rights.

By 2030, the United Nations Sustainable Development Goals (SDGs) fully encourage universal access to clean and accessible drinking water, as well as proper sanitation and hygiene. Goal 6 of the Sustainable Development Goals aims to ensure that everybody has access to clean water and sanitation (United Nations-Water, 2016).

Everyone, without exception, has the right to adequate and clean water that is suitable, physically available, and affordable for personal and domestic use under the human right to equal access to water protocol of the UN (United Nation Human Rights Council, 2014). The Human Rights Council makes disadvantaged and marginalized communities a priority and addresses their needs. Inequitable access is often hampered by social exclusion and weak governance, which is characterized by a lack of cooperation among decision-makers. Water as a human right is also endorsed by South Africa's constitutional provisions, the Free Basic Water Policy (2000) and the South African Constitution (1996), both of which value water as a right, revealing that water is a critical basic need for all citizens.

According to Muller (2008), many poor households oppose the Free Basic Water Policy because most citizens in South Africa still lack access to water in their communities. Rober (2015) proposes that municipalities be granted a position in implementing water policy, and that the municipality's failure to provide water services to the city has been described as one of the Free Basic Water Policy shortfalls. According to Koolwal and van de Walle (2013), water access is critical for lifting people out of poverty by harnessing the creative energy of women and children, enabling small businesses to thrive, and providing domestic work that is necessary for dignity and social networking among household members. In addition, Moloi (2004) notes that the lack of equitable access to water in rural communities leads to the unsustainable growth of water supply systems, a lack of economic development, and the inability to attract investments that would create jobs.

Similarly, Goswami and Bisht (2017) found that water is a critical component of long-term economic growth and sustainability. Water provides for agriculture, industries and humans, which all contributes to an increase in socio-economic development of the country as water is a source of life and prosperity. Furthermore, according to Saner et al., (2018), there are difficulties in accessing water due to factors such as population development, pollution, and inefficient allocation and distribution systems, which put severe strains on adequate and equitable water supply and sanitation. Low-income households continue to contend with competing uses of water and a lack of access to affordable drinking water (Saner et al., 2018). As a result, many countries have reprioritized the value of water allocation, putting greater emphasis on the water markets and opposes human right to water, and the increasing treatment of water as a fundamental human right.

The study used ecological economics theory to describe equitable access to water in rural communities. Ecological economics is a branch of economics that focuses on water as a fundamental human right and water equity. It emerged in the last decades of the twentieth century in response to concerns about environmental conservation and economic sustainability (Underwood, 1999). Ecological economics tackles the neoclassical economic paradigm's flaws by considering natural-world goods and services as essential components of the human economy. According to Yazdi et al., (2017), Ecological economics is an interdisciplinary theory that seeks to link natural resources with humankind. Ecological economics emphasise ecological size, water equality, women's engagement in water supply, water governance, and decentralisation. In most societies, people do not have equal use of ecological resources or the satisfaction of ecological benefits; this is more relevant in poor communities as well as racial minorities where there is fair distribution of resources.

The Global Water Partnership (2000) defines IWRM as a procedure which promotes the coordinated improvement and management of water, land and related sources, with the intention to maximize the resultant economic and social welfare in an equitable way without compromising the sustainability of vital ecosystems. IWRM is responsible for establishment of water policy and laws which use basin as the scale of management, water pricing allocation, establish water rights and participation in decision making (Giordano & Shah, 2014). Khadza (2010), posits that good governance is a key component of water management, as it strengthens a variety of aspects such as decentralization, capacity building, and enhancing monitoring and evaluation, analysis, and learning at all levels. In addition, Medeiros and Sivapalan (2020), argue that water governance is a challenge because knowledge is inaccessible and society's participation in water resource management is limited.

Di Baldassarre et al. (2019), posit that water governance is one of the primary difficulties in achieving sustainability and water management. Furthermore, governance process assists in project transparency and accountability and improves equitable access to water for the poor, especially in development countries. Women should play a central role in the supply, management, and safeguarding of water, according to IWRM's Dublin Principle 4. Women should be active in the decision-making process and implementation of water problems at all stages, according to this theory. This increases the role of women in water resource management and promotes women empowerment (Solanes and Gonzalez-Villarreal, 2008). The Sustainable Development Goals also aims to promote gender equity and women empowerment. The majority of women in rural communities have little say in decision-making, depriving them of control over water resources. It is noted that without basic literacy, people are unable to engage effectively in levels of social organization. Emler (1999) postlate that education gives people a greater inclination to speak out for change and the ability to speak up in their best interests, allowing them to control water resources. Jacobs, Rivett and Chemisto (2018) believed that incorporating the citizen's voice into decision-making processes would allow individuals to engage while also ensuring accountability and openness in the decision-making process. Furthermore, the necessity of considering local perspectives and involving people in water supply decision-making, ensuring equitable water access and environmental governance in local areas, is imperative. Women tend to be more vulnerable to equitable access to water and sanitation than men and this is because women are hindered by their social roles and positions (Figueiredo and Perkins, 2013). Harris et al., (2017), suggest that women and men also have different relationships to water access, uses, expertise, experiences, and governance. As a result, women should be involved in water resource management because they are the ones who are in charge of water provision, especially for domestic purposes.

A key process to enabling women inclusion and participation in water and sanitation delivery is to ensure the effective decentralization of decision making. According to Overton et al. (2014), decentralization entails bodies segregated from the national centre by law, in which local members are granted formal powers to rule on a variety of public issues, a political base in the community, and a small field of authority entrenched with the right to make decisions on areas within their jurisdiction. Bazaanah (2020) note that the decentralisation model should be viewed as a governance strategy that recognizes the importance of sustainable water resource governance and efficient utilization for rural communities' long-term development. Therefore, in achieving equitable access to water in rural communities, it is vital to note the relationship between the ecological economics principles, which includes water governance, women participation and decentralisation in order to promote water resource management and the protection of available water resources which will ensure equitable access to water.

1.3.2. Problem Statement of the Chapter

According to the Constitution of the Republic of South Africa, everybody has the right to adequate water. Equitable access to water aims at ensuring that all vulnerable and marginalized groups have the opportunity for meaningful participation in water and equitably benefit from water access (World Health Organisation, 2019). However, many people still cannot claim their fundamental right to water. This brings challenges to disadvantaged and marginalised groups socially, economically and environmentally. Free Basic Water Policy aims at ensuring equitable access to water through acknowledging the fundamental rights of all people to basic water and correct the historical situation and injustices of the past through providing equal access to services. The Constitution (1996), the Municipal Structures Act (1998) and the Water Services Act (1997) states that Municipalities are constitutionally mandated to ensure that water is delivered to their respective communities.

Alfred Duma Municipality have difficulties to provide equitable access to water and sanitation where many communities in the area have no access to water as a basic human right and cannot pay for services as poverty and unemployment is the main issue in rural communities (Alfred Duma Local Municipality, 2019). Hunter et al., (2010) notes that in rural communities, water is usually poor in quality, quantities and it is not easily accessible. This is because many households cannot afford the cost of minimum water supply and sanitation services, receive subsidized water and sanitation services from government. In this case, the community of Alfred Duma Local Municipality cannot pay for basic services and this led to service delivery being ineffective in a municipality where most of decision-making powers and responsibility, together with accompanying resources are inadequate and rural communities cannot express their dissatisfaction with the provision of services and local areas are unable to raise their voice in decision making.

Small municipalities are impacted by weak structures and water management systems, according to several writers (Clifford-Holmes, Palmer, De Wet and Slinger, 2016; COGTA, 2009). Hamer et al., (2017) notes that when it comes to considering the problems of equal access to water in local areas, citizen voices are rarely heard and accepted in the aspects of institutional dysfunction which leaves the picture of the challenge's incomplete faced by communities and how to provide solutions to satisfy the needs of the citizens. Therefore, it is important to involve local people in decision making processes and local experiences to be considered in order to achieve water access. The Dublin's Principles of IWRM emphasise that people should play an important role in water resource management through voicing their concerns through decision making (Solanes and Gonzalez-Villarreal, 2008). Thabethe (2011) states that the cost of providing free basic water is borne by District Municipalities and at local levels, the use of subsidies can both constrain and distort the provision of services. Therefore, the study aims to investigate equitable access to water as a basic human right and the ecological governance framework that supports the access to water in local areas. It also aims to investigate the ecological scale in the distribution of water supply in rural areas and the role of local municipality in equitable access to water.

1.3.3. Objective of the Chapter and Research Questions Investigated

- How equitable access to water is ensured in rural areas?
- What is the ecological governance framework that support the access to water in local areas?
- What is the ecological scale associated with the distribution of water supply in rural areas?
- What is the role of local municipality in equitable access to water?

1.3.4. Hypothesis of the Chapter

- **Null hypothesis (H_0):** There is no relationship between ecological scale, decentralisation, women participation in decision making and equitable water to access within the Alfred Duma Local Municipality.
- **Alternative hypothesis (H_A):** There is a relationship between ecological scale, decentralisation, women participation in decision making and equitable water access within the Alfred Duma Local Municipality

1.4. Study Area and Methodology

Study Area

The study was conducted in the village of Ekuvukeni in the Alfred Duma Local Municipality. In the South African province of KwaZulu-Natal, Alfred Duma Local Municipality is part of the uThukela District Municipality. The community of Ekuvukeni is located in the Alfred Duma Local Municipality and is defined as Ward 34.

Stats SA (2015), indicated that Ekuvukeni is a remote rural area with a population of 11018 people and 2606 households and it is governed by a Sithole-surnamed traditional tribal authority or chief. In terms of language, IsiZulu is the most widely spoken language in the region. Since a large portion of the population relies on social assistance and households often engage in subsistence farming, economic development is slow. According to Alfred Duma Local Municipality (2017), the unemployment rate in the Alfred Duma Local Municipality is 57.2%, while the youth unemployment rate is 66. 8%. The following figure 1 represents the map of the study area.

Figure 1. Map of the study area

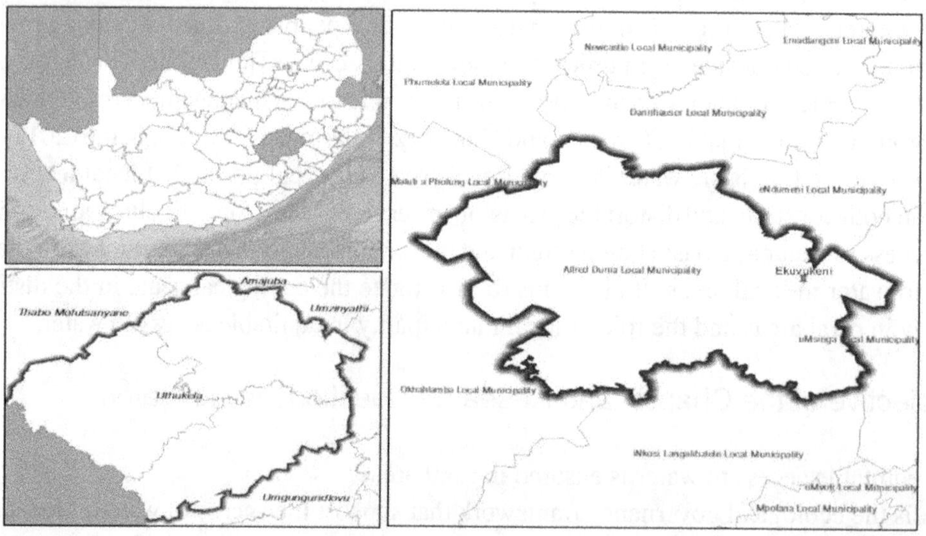

Research Philosophy and Paradigm

The study adopts the post-positivism paradigm. Post-positivism balances both positivist and interpretivist approaches. It focuses on researching issues in the context of involving experiences of the majority and announcing the result of what the majority says is acceptable (Panhwar, Ansari, and Shah, 2017). Post-positivism does not seek to validate or invalidate any research's scientific or quantitative positivism. It emphasizes a proper knowledge of the research study's directions and viewpoints by allowing the researcher to do research using numerous approaches in response to research questions (Fox, 2008). The study looked at the equitable access to water at rural communities. This was done through observing and by means of understanding human behaviour as true knowledge is based on observations and experiments. Therefore, using post-positivist philosophy it allowed the study to investigate realities taking place at Alfred Duma Local Municipality with equitable access to water services, using statistics to examine the findings.

Research Design

The researchers used an exploratory descriptive design. The exploratory descriptive design helps the researcher to realise and report some relationships among different aspects of the phenomenon under the equitable access to water. As a result, exploratory research aims to confront new issues with little or no prior research. Tewksbury (2009) posit that quantitative research is typically considered to be the more scientific approach to doing social science research than the qualitative approach. Quantitative research mainly generates statistics through the use of large survey research, using methods such as questionnaires or structured interviews. The study used questionnaires in order to collect the quantitative data and respondents were asked questions during the field visits. The questions were translated into IsiZulu which is the most dominant language in order to enable easy understanding and participation by members of the community.

Study Population and Sampling Method

Ekuvukeni has a population of 11018 and 2602 households under Alfred Duma Local Municipality. The household heads were provided with questionnaires and the study area was selected as it has problems of access to water and many service delivery protests occurs in the area due to lack of access to water. Sekaran and Bougie (2013) contend that the population is an entire group of people or the elements of interest that the researcher wishes to investigate and make up the population for a research project. The population size of Ekuvukeni is 2602 households (Stats SA, 2011). The sampling frame used in the study is the number of households.

Sample Size Determination and Sample Procedures

The sample size was calculated using the formula developed by Lohr (2010). The sample size (n) is calculated according to the formulae:

$$n = [z^2 * p * (1 - p) / e^2] / [1 + (z^2 * p * (1 - p) / e^2 * N)]$$

Where: z = 1.96, a confidence level of 95%, p = 0, 5 proportion (expressed as a decimal), N = 2606, population size, e = 0, 05, where 100-95/100 margin of error.

$n = [1.96^2 * 0.5 * (1 - 0.5) / 0.05^2] / [1 + (1.96^2 * 0.5 * (1 - 0.5) / 0.05^2 * 2606)]$
$n = [3.8416 * 0.5 * (1-0.5) / 0.0025] / [1 + (3.8416 * 0.5) * (1-0.5) / (0.0025 * 2606)]$
$n = [1.9208 * 0.5 / 0.0025] / [1 + 1.9208 * 0.5 / 6.515]$
$n = [0.9604 / 0.0025] / [1+0.9604/6.515]$
$n = [384.16] / [1+0.9604/6.515]$
$n = [384.16] / [1+0.1474]$
$n = [384.16] / [1.1474]$
$n = 384.16 / 1.1474$
$n = 384.16 / 1.147 = 334.809$
n = 335, (nearest whole number)

Therefore, a sample size of 335 respondents sufficed for the study. The study adopted the stratified random sampling method to draw respondents from a household population of 2602. Pirzadeh, Shanian, and Hamou-Lhadj (2011) posit that stratified sampling technique is generally used when the population on which sampling is applied is heterogeneous as a whole and can be divided into homogeneous sub-populations, referred to as strata. The population is separated into a desired number of partitions and then the sample elements are drawn from each stratum. The size of the sample from each stratum is kept proportional to the size of the stratum which is called proportional stratified sampling.

In this study, the researchers constructed and divided the household population in order to make a total population size of 2606. A proportional formula [P x n/N] was used to calculate the sample where P=households in the community officials, n= total sample size and N=total number of household population. Using the stratified approach, the community of Ekuvukeni was divided into three sections, namely kaGimi, KwaMngadi and kaGadeni. Using the above formula, the total sample for KaGimi was determined as {844 x 335/2606=108.4}, KwaMngadi {804 x 335/2606=103.3} and KaGadeni {888x335/2606=114.1} this calculation adds up to 335 in total. Table 1 represents the sample size distribution of the households and municipality officials.

Table 1. Sample size distribution for household and official respondent

Population strata	population	Sample fraction/ proportion (p*n/N)
kaGimi	844	108.4
KwaMngadi	804	103.3
kaGadeni	888	114.1
Municipality official respondents	70	8.9
Total	2606	335

Source: Stat SA (2011)

Data Analysis

According to Kawulich (2004) data analysis is the process of reducing large amounts of collected data and makes sense of them. The collected data was analysed through a Statistical Package for Social Sciences (SPSS). The SPSS assist the researcher to analyse data in terms of graphs, tables, charts and statistics which involve mean, median, mode and standard deviation. It also helped the researchers to analyse, transform, and produce a characteristics pattern between different variables. SPSS also assisted in generating inferential statistics and determination of relationships between the variables. Using the chi-square statistical procedure, with 95% confidence, the decision rule for the hypothesis was stated as "accept the null hypothesis if the critical value (p-value) is < 0.05 and do not accept the null hypothesis if p-value is ˃ or = 0.05". The data in the analysis is organized using numbers, tables, and graphs (Bar graphs, histograms and pie charts). Furthermore, to summarize the result, the study considered measures of central tendency which includes mean, mode and median. The researchers also used inferential statistics for data presentation in the chapter and this assisted the researchers to present data into statistics.

The thematic analysis was also used in the study. Thematic analysis assisted the researchers to determine accurately the relationship between the concepts and compare them with computer-generated data. It also allows the researchers to develop themes and provide interpretation for data. Thematic analysis helps the researchers to ensure that the themes are effectively linked to the data and provides the opportunity to code and categorise data into themes. This helps the researcher to go through all of the data and figure out what the most important significance is and what the data is trying to suggest or tell about the analysis. Mohammed (2012) posit that thematic analysis also provides rich, complex data and detailed data which is well-suited for the study and also provide a relationship between the variables and factors in order to create a reasonable and logical chain of data.

Ethical Considerations, Reliability and Validity of the Study

This study was planned and carried out in compliance with the University of Zululand's Research Ethics Policy, specifically section 6.2. (University of Zululand, 2016). The researchers gained permission from the Alfred Duma Local municipality to collect data. Furthermore, the researchers ensured that participants were provided complete information and that their involvement was voluntary. Furthermore, participants were not harmed or injured and anonymity and confidentiality were guaranteed because the information gathered was exclusively used for academic purposes. The study did a literature search and a pre-test of the research tools to verify validity and reliability. The reliability of the instruments was determined using the Cronbach Alpha. Maree (2016) defines reliability as the extent to which a measuring instrument is repeatable and consistent. A pilot study was also undertaken to ensure the validity of the study. The pilot study was undertaken at Ekuvukeni in order to test if the community households do understand the questions. The researchers pre-tested the instrument to 50 respondents which enabled a revision of grammatical and typographical errors, repeated questions and inconsistencies associated with the framing of the questionnaires. According to Sekaran and Bougie (2010), the acceptable range for Cronbanch Alpha value is 0.60 and above. Higher Alpha values shows reliability of an instrument. The Cronbach alpha for this study was 0.85 which shows that the instruments were reliable.

2.0. RESULT AND DISCUSSION

2.1. Features of the Respondents

This section presents the demographic factors that was important in the study to understand the age of respondents, gender of respondents, educational level, and household income level and employment status.

2.2. Age Categories

The research sought to know the gender of respondents as it is shown in Table 2. This was important for this study to show the presentation of respondent age.

Table 2. Age of the respondents

Age range	Frequency	Percent (%)
18 – 25	112	33.4
26 – 35	114	34.0
36 – 44	66	19.7
45 – 59	35	10.4
60 and Above	8	2.4
Total	335	100

Source: Fieldwork (2020)

From Table 2, about 33.4% of the respondents reported to be in the age group 18-25 years old, 34.0% were between 26-35 years old, 19.7% were between 36-44 years old, 10.4% between 45-59 years old, while only 2.4% of the respondents were between 60 years and above. These findings suggest that a large group of respondents who participated in the study were between 18-35 years old and were millennials. The result indicates that very young respondents are shouldering the responsibility of a household. The result also implies that 18-25 and 26-35 years are considered the most energetic with human resource capacities and potential which involves the development of water resources in rural communities. While most of the adults are not really involved in the development of water resources, because they are illiterate and their participation in access to water is complicated in planning, during the construction phase, later in operation and maintenance.

2.3. Gender and Education Status

Table 3 represents the gender and education status of the respondents. The result indicates that 63.6% of the respondents were females and 36.4% were males. This shows that females were more interested in the study than males. This is because women tend to be more important than men in water access as women are responsible for water provision in their homes and are hindered by their social roles and positions. Drawing from Principle 4 of IWRM, the findings agree that women should play a central role in the provision, management and safeguarding of water provision (Solanes & Villarreal, 2008).

The results also show that 7.5% of the females had a primary school certificate, 28.1% had high school certificate, 11.3% had diplomas, 12.2% had bachelor's degree and 4.5% had postgraduate certificates which make the total of 63.6% of females based on their educational levels.

Table 3. Gender and educational levels of the respondents

Gender		Educational qualification					Total
		Primary school certificate	High school certificate	Diploma	Bachelor's degree	Post graduate degree	
Male	N	6	55	26	28	7	122
	% of Total	1.8%	16.4%	7.8%	8.4%	2.1%	36.4%
Female	N	25	94	38	41	15	213
	% of Total	7.5%	28.1%	11.3%	12.2%	4.5%	63.6%
Total	N	31	149	64	69	22	335
	% of Total	9.3%	44.5%	19.1%	20.6%	6.6%	100.0%

Source: Fieldwork (2020)

In addition, 1.8% of male respondents had a primary school certificate, 16.4% had a high school certificate, 7.8% had diplomas, 8.4% had bachelor's degrees and 2.1% had postgraduate certificates. These findings suggest that even though there are respondents who decided to further their studies from diplomas to degree and postgraduate level, the majority of respondents have high school certificates in both males and females. It can be concluded from the above table 3 that a large number of respondents were progressive in education but they are still far away from higher education which is important today to create knowledge in society.

The results also indicate that a majority of respondents had some formal knowledge which can help in understanding aspects of water resource management. This finding agrees with Emler (1999) who suggest that without basic literacy, people are unable to engage effectively in levels of societal organization. Education offers individuals a stronger willingness to speak out for change and the ability to speak up in their best interests, allowing them to manage water resources. Therefore, educated and well-informed society promotes good participation in water resources.

2.4. Employment and Income Status

As illustrated by Table 4, the results indicated that 56% of the respondents were unemployed, 40% were employed and 4% were partly employed. According to Statistics South Africa (2015), the unemployment rate in the Alfred Duma Local Municipality is high labelled as 57.2%, while the youth unemployment rate is 66. 8%. In terms of monthly income, 63% of the respondents earn income between R0-R7200, about 21% earn income between R7201-R16500, and 15% earn income between R16501-R33400.

These findings show that the majority of 63% respondents earn between R0-R7200 because most of them are not employed. There are very few respondents (15%) that earn from R16501-R33400 on a monthly basis. Only 0.6% of respondents earn between R33401-R57400 as illustrated in table 4. In terms of employment and income the results indicate that there is still a problem of unemployment within the municipality as a high number of households are not working and they are from a low-income group. This suggests that community household cannot pay for water services as they of the lack of income. This is supported by the Alfred Duma Local Municipality (2017) that the unemployment rate in the Alfred Duma Local Municipality is 57.2%, while the youth unemployment rate is 66. 8%.

Table 4. Employment status and monthly income of the respondents

Monthly income	Employment status							
	Employed		Unemployed		Partly-employed		Total	
	n	%	n	%	n	%	n	%
R0 – R7, 200	27	8	181	54	3	1	211	63
R7, 201 – R16, 500	55	16	6	1.8	10	3	71	21
R16, 501 – R33, 400	50	15	1	0.2	0	0	51	15
R33, 401 – R57, 400	2	1	0	0	0	0	2	0.6
Total	134	40	188	56	13	4	335	100

Source: Fieldwork (2020)

2.5. Equitable Access to Water as a Basic Human Right

It can be observed in figure 2 that there is a large number of respondents who strongly disagree (39%) and disagree (28%) which indicates that there is no equitable access to water in the community. While 19% of respondents were not sure whether they have equitable access to water within their community and 13% respondents agree and only 1% of respondents strongly disagreed that they have equitable access to water. These results suggest that the Alfred Duma community of Ekuvukeni does not have equitable access to water. Inequitable access to essential services remains one of the most evident geographical fingerprints of apartheid, with services being delivered well in urban areas but a substantial service backlog in rural areas (Alfred Duma Local Municipality, 2017). This means that the rural people in the area are not provided with water which is a fundamental human right to them.

Figure 2.
Source: Fieldwork (2020)

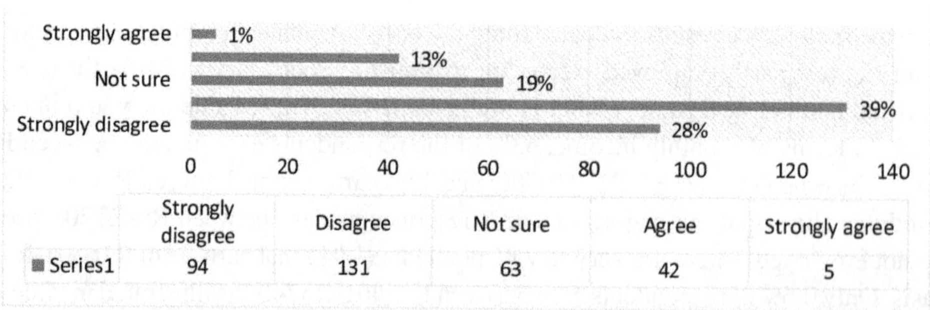

According to the Constitution of the Republic of South Africa, everyone has the right to adequate water. Many individuals, however, are still unable to exercise their fundamental right to water. Equitable water access attempts to ensure that all groups have the opportunity to participate meaningfully in water and benefit equally from water access. This is supported by Yadzi et al. (2017), who postulate that in most societies, people do not have equal access to ecological resources or enjoyment from ecological advantages, which is especially apparent in poor communities and ethnic minorities where resources are unequally distributed. According to the study's conclusions, this indicates that the community does not have equitable access to water.

2.6. Challenges of Inequitable Access to Water

From table 5, it indicates that 51.6% respondents strongly agree that there are many challenges of equitable access to water and 37.3% respondent strongly agree. Meanwhile, 10.1% respondents disagree and 0.9% respondents were not sure about challenges facing their community. The Alfred Duma Local Municipality (2019) shows that inequitable access to basic services is a problem, and while basic services are well-supplied in urban regions, the quality of services in rural areas, especially those supplied to impoverished households, remains low. Due to time constraints, the municipality received 4500 applications from poor households, but only 965 were granted indigency status. As a result, inequitable and unequal service delivery within the municipality occurs, where indigent households are being neglected. According to WHO (2019), inequitable access to water poses social, economic, and environmental difficulties to underprivileged and marginalized populations if water access does not reach all groups in order for all the groups to have the opportunity for effective participation in water and equal benefits from water access.

Table 5. Challenges of inequitable access to water in rural areas

Monthly income		Challenges of inequitable access to water and sanitation in rural areas				Total
		Disagree	Not sure	Agree	Strongly agree	
R0 – R7, 200	Count	3	28	100	80	211
	% of Total	0.9%	8.4%	29.9%	23.9%	63.0%
R7, 201 – R16, 500	Count	0	3	45	23	71
	% of Total	0.0%	0.9%	13.4%	6.9%	21.2%
R16, 501 – R33, 400	Count	0	3	27	21	51
	% of Total	0.0%	0.9%	8.1%	6.3%	15.2%
R33, 401 – R57, 400	Count	0	0	1	1	2
	% of Total	0.0%	0.0%	0.3%	0.3%	0.6%
Total	Count	3	34	173	125	335
	% of Total	0.9%	10.1%	51.6%	37.3%	100.0%

Source: Fieldwork (2020)

It can also be noted that 63% of respondents earns between R0-R7200 and 15.2% respondents who earns between R16500-R33400. According to the findings, many people are unemployed, and if they were employed, they would be able to pay for water services rather than relying on government subsidies, because rural areas perceive water as a plentiful resource that the government must provide at no cost. This depicts bad financial implications for rural communities as they cannot afford to pay for water services which bring challenges in equitable access to water in Ekuvukeni community. This is also supported by Saner et al., (2015), who postulates that low-income households continue to struggle with competing water usage and a lack of access to affordable drinking water. As a result, several governments have reprioritized the importance of water allocation, putting increased emphasis on the water sector, which is increasingly being treated as a basic human right.

2.7. Problems That Delay Water Service Provision in Rural Areas

Based on Table 6 the results indicate that there are 28.7% respondents who agree, 20.6% respondents disagreed and 20.3% respondents strongly agree that the remoteness of rural areas from urban areas brings problems to the water service provision as construction of water and waste water treatment plants and reticulation facilities such as pipes is economically unviable. It can be also observed that 19.4% respondents were not sure and 11.0% respondents strongly disagree. According to the findings, the majority of respondents agree that there are issues that cause water supply delays in rural areas. According to Alfred Duma Local Municipality (2019), there is a high rate of aging infrastructure services as a result of water theft and vandalism, which has been documented to interrupt service delivery in nearby areas. It should also be mentioned that service distribution is unequal, with more urban bias at the expense of rural areas. Because the maintenance and refurbishing expenses of water and wastewater facilities are now unaffordable for the uThukela District Municipality due to budget constraints, the municipality cooperated with uMngeni to alleviate water backlogs.

The municipality allocated R 2052000.00 to Ekuvukeni Regional Bulk Water Supply to renovate and maintain the Oliphantskoop water treatment facility, which distributes water to neighbouring areas including Ekuvukeni. There are also concerns with the water reservoir, which must be refurbished and maintained in order for the neighbourhood to receive water services. The results are supported by Mc-Conkey and Wilsenach (2010) who posit that the fundamental challenge to rural water service is that rural areas are remote from major cities, making it impossible to reticulate water to homes and making purification plant treatment economically unsustainable.

Table 6. Problems that delay water service provision in rural areas

Response categories	Frequency	Percent (%)
Strongly disagree	37	11.0
Disagree	69	20.6
Not sure	65	19.4
Agree	96	28.7
Strongly agree	68	20.3
Total	335	100

Source: Fieldwork, (2020)

2.8. Lack of Economic Development

In Table 7, the result represents the lack of economic development and non-attraction of investments to provide employment. The result indicates that 48.7% of respondents agree and 33.1% respondents strongly agree. Furthermore 4.8% of respondents disagree while 1.2% of respondents strongly disagree that the lack of economic development affects access to water. The findings reveal that many respondents see water as a significant driver of economic development, despite the fact that the Alfred Duma Local Municipality claims that the municipality has low economic growth and a high unemployment rate, resulting in a lack of economic development in the local communities. The tertiary sector, which includes community services (23 percent), finance (13 percent), transportation (12 percent), and trade (14 percent), drives the uThukela economy.

The community of Ekuvukeni is facing a number of challenges of developing themselves economically because they lack income to pay for water services due to lack of employment opportunities and lack of development in the community. The findings show that people living in rural areas have insufficient access to water, which leads to poverty. This is because water is essential for economic development, and agricultural activities in communities cannot take place owing to a lack of enough water, limiting economic growth. The findings support Moloi's (2007) conclusions that rural conditions lead to the unsustainable development of water supply systems through a lack of economic growth and the inability to attract investments to create jobs. Furthermore, according to Goswami and Bisht (2017), water is a major driver of long-term economic growth and development. Water is a source of life and prosperity for agriculture, industries, and individuals, all of which contribute to the country's socioeconomic development.

Table 7. Lack of economic development and non-attraction of investments to provide employment

Response categories	Frequency	Percent (%)
Strongly disagree	4	1.2
Disagree	16	4.8
Not sure	41	12.2
Agree	163	48.7
Strongly agree	111	33.1
Total	335	100.0

Source: Fieldwork (2020)

2.9. The Impact of Free Basic Water Policy (FBW) on Rural Communities

From Table 8, the results indicate that a total of 29.9% respondents strongly agree and 49.0% respondents agree while only 21.2% respondents were not sure about FBW. The respondent's level of education also indicates that they have knowledge about Free Basic Water Policy where about 44.5% respondents have high school certificate, 20.6% respondents have a bachelor's degree and 6.6. % respondents have postgraduate degrees. This means that respondents do have knowledge and ideas on FBWP as they have a bachelor's degree and proceeded to postgraduate studies. The findings suggest that the majority of respondents have knowledge on Free Basic Water Policy. This suggests that the free basic water policy has not been implemented well as the community still do not have access to water. To assess which households are qualified for free basic water, the municipality established an indigent program. Each home that is registered as needy should receive a credit on their water account for the amount of free basic water they have received, according to the indigent plan, and indigents are normally classed based on their monthly household income. The municipality created an indigent program to determine which households are eligible for free basic water. According to the indigent scheme, each household that is registered as indigent should receive a credit on their water bill for the amount of free basic water they have received, and indigents are typically classified based on their monthly household income (uThukela District Municipality, 2017).

The Municipality received 4500 applications and only approved 965 due to system challenges and time limitations. As the municipality approves 965 applications out of 4500, this means that only a small percentage of destitute households obtain free basic water and that rural people are still struggling in receiving adequate water supply. According to Robert (2015), rural municipalities find it challenging to adopt and implement the FBWP. The goal of the Free Basic Water policy was to ensure that no one was refused access to water because they couldn't afford it. The municipality is responsible for fully comprehending and implementing the Free Basic Water Policy. According to Muller (2008), civil society has attacked the FBW policy and its implementation for failing to reach all of the poor in areas where water is scarce, and the FBW policy is reaching more non-poor individuals than poor people. One of the shortfalls of the Free Basic Water Policy, according to Robert (2015), is the municipality's incapacity to supply water services to the community.

Table 8. Educational level and knowledge On Free Basic Water Policy

Free basic water policy provides 6Kl of water per household per month		Educational qualification					
		Primary school certificate	High school certificate	Diploma	Bachelors' degree	Post graduate degree	Total
Not sure	N	5	34	16	10	6	71
	% of Total	1.5%	10.1%	4.8%	3.0%	1.8%	21.2%
Agree	N	14	72	35	33	10	164
	% of Total	4.2%	21.5%	10.4%	9.9%	3.0%	49.0%
Strongly agree	N	12	43	13	26	6	100
	% of Total	3.6%	12.8%	3.9%	7.8%	1.8%	29.9%
Total	N	31	149	64	69	22	335
	% of Total	9.3%	44.5%	19.1%	20.6%	6.6%	100.0%

Source: Fieldwork (2020)

2.10. The Role of Women in Decision Making on Water Resource Management

As shown in figure 3, the study represents the views of respondents on decision making process based on water resource management. The result implied that half of the respondents strongly agree at 50%, while

42% respondents agree and 8% respondents were not sure about women participation in water management. Gender equality and women's empowerment are among the 5 sustainable development goals. According to the findings, half of the respondents believe that they have no say in the decision-making process in rural communities, preventing them from participating in water resource management. According to Dublin principle 4 of IWRM, which improves women's engagement in water resource management and promotes women's empowerment, women should be participating in the decision-making process and implementation at all levels of water concerns (Solanes and Gonzalez-villarreal, 2008).

Harris et al., (2017) notes that women and men frequently have different relationships to water in terms of access, uses, knowledge, experiences, and governance; as a result, women should have a role in water resource management because they are the ones who are in charge of water provision, especially for domestic requirements. Women are significant agents in providing water for their families and communities, therefore they play a vital role in water management. This empowers women and gives them the confidence to take on greater water management tasks. Many women are disenfranchised and marginalized in decision-making processes, making them more vulnerable to equitable water access.

Figure 3.
Source: Fieldwork (2020)

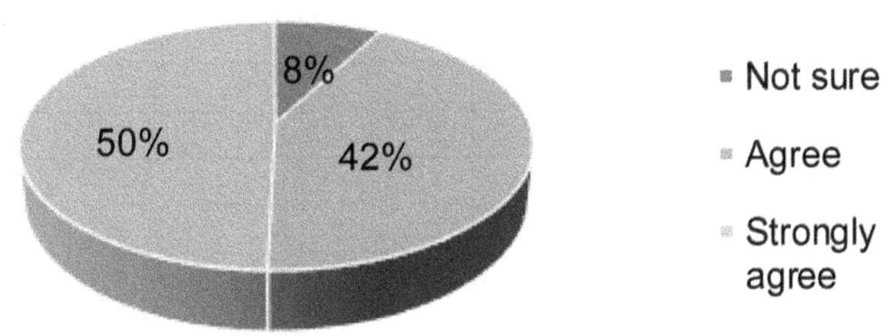

2.11. The Importance of Decentralisation in Local Areas

Table 9 represents the results on decentralisation which emphasise that local areas should participate in water supply and resource management. The results indicate that 55.2% respondents agree and 30.7% strongly agree, while 14.0% respondents were not sure about the importance of decentralisation. The result indicates that the decentralisation must be in the centre of sustainable water resource management. The municipality has developed the ward committee including the water committee and representative's forum. In the ward committee, it is where the community and community organisations participate and contribute to issues of improving service delivery. Ward committees also encourage community participation in order to inform residents about municipal activities and to make the council aware of community needs and concerns. In addition, the municipality formed a representative forum to represent the interests of citizens at the district and municipal levels in the IDP process (UThukela District Municipality, 2019). The findings imply that when it comes to water management, local regions must be recognized and considered in order to ensure that local issues are expressed and understood in order to solve the issues of water availability that communities and municipalities confront.

According to Jacobs et al., (2018), decentralization emphasizes the necessity of considering local perspectives and involving communities in decision-making as well as realistic water supply systems. People become full citizens as a result of decentralization, which improves service delivery, brings government closer to the people, and allows the government to better understand their requirements (Olum, 2014). As a result, decentralizing water supply and sanitation systems is critical for ensuring equal access to water in rural communities. The study's findings also coincide with Bazaanah's (2020) conclusion that decentralisation should be viewed as a governance pathway that recognizes the need of sustainable water resource governance and efficient utilization for rural communities' long-term development.

Table 9. Decentralisation is important for local participation in water supply

Decentralisation		Educational qualification					Total
		Primary school certificate	High school certificate	Diploma	Bachelors' degree	Post graduate degree	
Not sure	N	9	13	14	9	2	47
	% of Total	2.7%	3.9%	4.2%	2.7%	0.6%	14.0%
Agree	N	17	84	32	42	10	185
	% of Total	5.1%	25.1%	9.6%	12.5%	3.0%	55.2%
Strongly agree	N	5	52	18	18	10	103
	% of Total	1.5%	15.5%	5.4%	5.4%	3.0%	30.7%
Total	N	31	149	64	69	22	335
	% of Total	9.3%	44.5%	19.1%	20.6%	6.6%	100.0%

Source: Fieldwork, (2020)

2.12. Lack of Funds to Supports the Provision of Services

Based on Table 10, the result shows that 39.7% of the respondents strongly agree, 50.1% of the respondents agree while 0.6% of the respondents disagree. Only 9.6% of the respondents were not sure about the size of the economy which leads to inadequate funds available to support the provision of services. The findings suggest that the large number of the respondents strongly agree that the scale or size of the economy affect the provision of services in the local municipalities. According to Alfred Duma Local Municipality (2019), the municipality's budget is insufficient to support service supply. Even though the municipality receives an equitable share as well as Municipal Infrastructural Grants (MIG), they believe that this is insufficient to capacitate the municipality and carry out its duty.

The results are supported by Saner et al., (2015) who posit that many municipalities in middle and low-income countries face budgetary and institutional constraints, which typically result in low labour productivity, making it difficult to provide high-quality service. Many local governments do not have adequate financial resources to deliver services to their citizens.

Table 10. In most local municipalities the scale or size of the economy is lower, this lead to inadequate funds available to supports the provision of services

Response categories	Frequency	Percent (%)
Strongly disagree Disagree	0 2	0 0.6
Not sure	32	9.6
Agree	168	50.1
Strongly agree	133	39.7
Total	335	100.0

Source: Fieldwork, (2020)

Many local municipalities' economies are growing at a very slow pace, owing to high unemployment, poverty, a lack of water and sanitation, and a lack of infrastructure (Sebova, 2015). Local governments have difficulties as a result of their inability to support the provision of services to local communities, which has an impact on the community of Ekuvukeni's access to water for domestic use and running their small businesses. Municipalities should increase the establishment of local infrastructure and services to address capacity constraints through structuring tariffs to moderate growth in consumption especially in water. Sebova (2015) suggest that small municipalities have more efficiency problems and insufficient money for providing services in their local communities, which can be measured by capital income and spending values, and the lower the economic efficiency of the small municipality the greater are the effects on the competitiveness of the entire economy.

2.13. The Role of Municipalities in Equitable Access to Water in Rural Areas

As illustrated by Table 11, the study represents the findings based on gender and perceptions that the municipality play its role in equitable access to water in rural areas. The result implied that 28.7% of the respondents disagree and 28.7% of the respondents are not sure about the role played by the municipality in equitable access to water. Nonetheless, 19.7% of the respondents strongly disagree, 17.0% of the respondents agree and 6.0% of the respondents strongly agree. The results are also supported by the Alfred Duma Local Municipality (2019) claims that there is unequal service delivery within the municipality, with urban demands being met while rural ones are being ignored. This shows that uThukela is failing to perform its part in ensuring equal water access. Service delivery has been reported to be hampered by high rates of aging infrastructure, which are the result of water theft and vandalism, and have been observed to disrupt service delivery, as well as a lack of resources for infrastructure construction in order to undertake municipal catalytic projects (Alfred Duma Local Municipality, 2019).

Municipalities are constitutionally accountable for providing essential services to their residents, according to national legislation, policies, and directives, according to the Alfred Duma Local Municipality (2019). Although uThukela District operate as a water service authority (WSA), uThukela Water and Umgeni Water are the water service providers (WSP). They provide water services on behalf of the WSA, provided that there is a signed water services development plan (WSDP) and agreement between a municipality and Umgeni water in alignment with the norms and standards of water provision. The results indicate that there is a strong contrast between the respondents on a role played by the municipality in providing water to the community.

Table 11. Gender and perceptions that the Municipality play its role in equitable access to water

Gender		Municipality play its role in equitable access to water in rural Local					Total
		Strongly disagree	Disagree	Not sure	Agree	Strongly agree	
Male	N	19	28	35	28	12	122
	% of Total	5.7%	8.4%	10.4%	8.4%	3.6%	36.4%
Female	N	47	68	61	29	8	213
	% of Total	14.0%	20.3%	18.2%	8.7%	2.4%	63.6%
Total	N	66	96	96	57	20	335
	% of Total	19.7%	28.7%	28.7%	17.0%	6.0%	100.0%

Source: Fieldwork, (2020)

2.14. Hypothesis and Implications

The study adopted the chi-square technique in determining the associations between the variables. The chi-square was utilized because it is a nonparametric and a distribution free technique which did not strictly require normality of the sample distribution utilised by the study. With 95% confidence interval the decision rule for the hypothesis test was stated as "accept (H_0) if p-values are greater than (˃) 0.05 and do not accept (H_0) if p-values are less than or equal to (≤) 0.05".

To test the null hypothesis (H_0), the study assumed that "there is no statistically significant relationship between ecological scale, decentralisation, women participation in decision making and equitable water access within the Alfred Duma Local Municipality". The alternative hypothesis (H_A) presumed that "there is a statistically significant relationship between ecological scale, decentralisation, women participation in decision making and equitable water access within the Alfred Duma Local Municipality". As illustrated by Table 12, the Chi-square ($\chi 2$) test result revealed that the values for the variables were; ecological scale ($\chi 2$= 245.549a, df = 3, p-value = 0.032, and n=335); decentralisation ($\chi 2$= 86.281b, df =2, p-value = 0.025 and n = 335); women participation ($\chi 2$= 664.987b, df = 2, p-value = 0.01 and n = 335) and equitable water access ($\chi 2$= 205.788a, df = 3, p-value = 0.013 and n = 335), this significantly influence the ecological economics principles through realisation of decentralisation, women participation in decision making process and water access. The study decided to rule out the null hypothesis which makes the alternative hypothesis to remain valid for the study.

Therefore, from the decision rule on the region of acceptance of the null hypothesis (H_0), the study did not accept the null hypothesis (H_0) as it deviate from the findings of the study. The study rather accepted the alternative hypothesis (H_A) which states that there is a statistically significant relationship between ecological scale, decentralisation, women participation in decision making and equitable water access within the Alfred Duma Local Municipality. Hence, the study proceeds to conclude based on the hypothesis test results that in the Alfred Duma Local Municipality, the ecological scale, decentralised approach to water, women participation in decision making and access to water are significantly associated with equity and human rights to potable water and sanitation delivery among the population.

This is supported by Khadza (2010) who notes that good governance is a main pillar of water management, which improves various aspects, such as promoting decentralization, building capacity, and strengthening and monitoring evaluation, research and learning at all levels. Olum (2014) argues that the aim of decentralisation is to redesign the governmental system to promote efficiently and effective service delivery to enhance local participation and make local economies more prosperous and equitable. Women tend to be more vulnerable to equitable access to water and sanitation than men and this is because women are hindered by their social roles and positions in water access (Figueiredo and Perkins, 2013). The findings of the study also agree with the Dublin's Principle 4 of IWRM which state that women should play a central role in the provision, management and safeguarding of access to water. This principle acknowledges that women should be involved in decision making process and implementation at all levels of water issues. Therefore, accepting alternative hypothesis was valid to the study as it emphasise the relationship between the ecological scale, decentralisation, women participation and water access.

Table 12. Effects of ecological scale, decentralisation, women participation in decision making and equitable water access within the Alfred Duma Local Municipality

Test Statistics				
(χ^2) statistic	Ecological scale	Decentralisation	Women participation	Water access
Chi-Square	245.549[a]	86.281[b]	98.872[b]	205.788[a]
Df	3	2	2	3
Asymp. Sig.	0.032	0.025	0.001	0.013
a. 0 cells (0.0%) have expected frequencies less than 5. The minimum expected cell frequency is 83.8. b. 0 cells (0.0%) have expected frequencies less than 5. The minimum expected cell frequency is 111.7. Source: Fieldwork, (2020)				

3.0. SOLUTIONS AND RECOMMENDATIONS

Based on the above-mentioned major findings, the following recommendations and solutions are made in order to improve the equitable access to water in Alfred Duma Local Municipality:

- To improve unequal service delivery access to basic services at the municipality where urban areas are well provided with basic services whereas rural areas' services backlog is still very high, the following needs to be undertaken. The municipality must ensure that equitable access to water reach all the vulnerable and marginalised groups have the opportunity for meaningful participation in water and equitable benefit from water access. This can be done through increasing subsidies in providing Reconstruction Development Programme (RDP) houses as they are provided with flush toilets and house connections.
- To improve equitable access to water in local areas, equity in water should also be promoted to ensure effective water resource management. The municipality should also increase the free basic water allocation from 6kl to 9kl in line with the constitution that requires municipalities to take

reasonable legislative measures, within the available resources to achieve the progressive realisation of the right of access to sufficient water as access to water and sanitation to communities.

- The municipality need to improve the infrastructure in the community and possibly extend the water reticulation supply to the community. The current water infrastructure is old and is unable to cope with the increasing access to water and used water demands. The municipality should conduct maintenance and routine cleaning of water services and reservoirs to ensure that the supply of water is safe to prevent the outbreak of waterborne diseases.
- To improve ecological governance framework, proper governance and effective decentralisation could be helpful in ensuring that the scarce resources needed to transform communities, are utilised carefully for the equal benefit of all people in the Municipality. Therefore, improving water rights, water and sustain equitable access to water; the Municipality should develop strategies for effective ecological conservation and promote decentralisation.
- The Alfred Duma Local Municipality should ensure that the community are aware of the legislation that governs water management as it forms part of the water service delivery. Therefore, all the stakeholders, role-players and public society should have a clear understanding and make the community members aware that it is their constitutional right to access equitable access to water. The municipality should establish a public relations department as part of the IDP process or water committees per project implemented in the area.
- The Municipality should encourage the participation of women in decision making process in water resource management. This increases the role of women in water resource management and promotes women empowerment as women are the one who are responsible for water.
- A municipality should develop a proper water resource services management strategy that is all comprehensive and should be implemented. This should clearly define the roles that need to be played by the different role-players. It should include actions that need to be followed to effective ways of monitoring the implementation of the management strategy and approaches to evaluate the progress at different intervals.

4.0. FUTURE RESEARCH DIRECTIONS

This research opened doors for future research. It identified the challenges faced by municipalities in terms of equitable access to water and its impact on the service delivery plan. The following motivations for further research are identified to improve the capacity of the South African local municipalities to deal with equitable access to water challenges: Water governance approaches and decentralisation practises need to be further researched; this includes participation from local people which contributes to the effective service delivery process in communities. The research only focuses on equitable access to water in rural areas for further research, the study should compare the access to water in both rural and urban areas.

5.0. CONCLUSION

Water is a unique natural resource, since it is essential for the survival of all forms of life. Access to water and sanitation is a fundamental need for the poor, and therefore should be an essential component of efforts to alleviate poverty. The study concludes that there is a statistically significant relationship between the variables (ecological, decentralisation, women and equitable water). Therefore, from the decision rule on the region of acceptance of the null hypothesis (H_0), the study did not accept the null hypothesis (H_0), but rather accepted the alternative hypothesis (H_A) that there is a statistically significant relationship between ecological scale, decentralisation, women participation in decision making and equitable water access within the Alfred Duma Local Municipality.

Equitable access to water aims at ensuring that all groups have the opportunity for meaningful participation in water and equitable benefit from water access. The result suggests that many women feel disempowered and marginalised in the process of decision making. Women inclusion in governance must be in the centre of water management as it ensures that the needs of interest are presented in the fair and in a transparent manner. The study concludes that the FBW policy and its implementation have been criticized by civil society for failing to reach all the poor and the municipal inability to provide the community with water services has been identified as one of the short falls of Free Basic Water Policy.

The study further concludes that water is a basic human right and a backbone of the economy and also concludes that it is important for local experiences to be considered and for communities to be involved in decision making and practical arrangements for reliable water supply. This must be done through the empowerment of woman in decision making process. The ecological scale, decentralised approach to water, women participation in decision making and access to water are significantly associated with equity and human rights to potable water and sanitation delivery among the population. Therefore, to improve rights to water and sustain drinking water access, decentralisation, and encourage the participation of women in decision making process in water resource management is crucial. To improve equitable access to water in local areas, equity in water should also be promoted to ensure effective water resource management.

ACKNOWLEDGMENT

The first author would like to express her gratitude to NRF and her dissertation supervisor. This research was supported by the National Research Foundation [117267] https://orcid.org/0000-0002-2600-9738 and is hereby acknowledged. Dr Simphiwe Nojiyeza, my master's supervisor, provided invaluable supervisory support, guidance, and mentorship is also acknowledged. The authors would like to express gratitude to study particiants, the KwaZulu Natal Department of Cooperative Governance and Traditional Affairs (COGTA), the Management Committee of the Zululand District Municipality and the late Mayor of uMlalazi Local Municipality, His Worship Councillor, Thelumoya Zulu for the insightful comments they made when we disseminated findings to key statekolders. In both COGTA and ZDM, we express our sincere gratitude to Dr Gladys Makolomakwe, Ms Nokubonga Caluza and Mr Phindithemba Manqele for organising the meetings and affording us an opportunity to present the study findings. At UNIZULU, we express our gratitude to the Community Engagement Office, and the Communication and Marketing Department especially Ms Mammusa Lekoa and Ms Precious Shamase for attending and documenting these community engagement online workshops. At the time of writing, it is noted that there are further engagements planned with uThukela District Municipality, uThukela Water, uMngeni Water, the Department of Water and Sanitation and other stakeholders.

REFERENCES

Alfred Duma Local Municipality. (2019/2020). *Integrated Development Plan*. Available: www.Ihlm.org.za/annualdocuments

Baazanah, P. (2020). Political Ecology of Water Resource Governance in Ghana: Towards Sustainable Pathway for Decentralisation and Participatory Water Supply in Rural Communities of the Savannah Region. *International Journal of Political Science and Development*, *8*(6), 252–279.

Biswas, A. K., & Tortajada, C. (2008). Future Water Governance: Problems and Perspectives. *Water: Research and Development*, *2*(2), 129–139.

Biswas, K. (2007). *Water as a focus for regional development. Water Resources Management Series*. University Press.

Clifford-Holmes, J. K., Slinger, J. H., De Wet, C., & Palmer, C. G. (2017). Modelling in the "Muddled middle": A case study of water service delivery in post-apartheid South Africa. In *Social Systems Engineering: The Design of Complexity*. Wiley.

Department of Water Affairs and Forestry. (2002). *Free Basic Water*. Implementation Strategy Document version 2. www.dwaf.gov.za/annualdocuments

Department of Water Affairs and Forestry (DWAF). (2004). *National Water Resource Strategy*. Author.

Di Baldassarre, G., Sivapalan, M., Rusca, M., Cudenne, C., & Sanderson, M. R. (2019). Challenges in addressing Sustainable Development Goals. *Water Resources Journal*, *55*(23), 6325–6355. PMID:32742038

Dill, B., & Crow, B. (2014). The colonial roots of inequality: Access to water. *urban East Africa. Water International*, *39*(2), 187–200. doi:10.1080/02508060.2014.894212

Elmer, N. (1999). Politics: The Education Effects. Taylor and Francis. *LTD*, *1*(1), 251–273.

Figueiredo, P., & Perkins, P. E. (2013). Women and water management in times of climate change: Participatory and inclusive processes. *Journal of Water Resources*, *60*(3), 188–194. doi:10.1016/j.jclepro.2012.02.025

Fox, N. J. (2008). Post-positivism. In The SAGE Encyclopedia of Qualitative Methods. Sage.

Giordano, M., & Shah, T. (2014). From IWRM back to integrated water resources management. *International Journal of Water Resources Development*, *30*(3), 364–376. doi:10.1080/07900627.2013.851521

Global Water Partnership. (2000). *Integrated Water Resource Management*. Author.

Goff, M., & Crow, B. (2014). What is water equity? The unfortunate consequences of a global focus on 'drinking water'. *Water International*, *39*(2), 159–171. doi:10.1080/02508060.2014.886355

Goswami, K. B., & Bisht, P. S. (2017). The role of water resources in socio economic development. *International Journal for Research in Applied Science and Engineering Technology*, *887*(12), 2321–9653.

Hamberto, P. (2011). Social Equity and Integrated Resources Management. *Global Water Partnership*, *3*(15), 5–83.

Harris, D., Kleiber, D., Goldin, J., Darkwah, A., & Morinville, C. (2017). Intersections of gender and water: Comparative approaches to everyday gendered negotiations of water access in underserved areas of Accra, Ghana and Cape Town. *South Africa Journal of Gender Studies*, *26*(5), 561–582. doi:10.1080/09589236.2016.1150819

Hossain, M. Z. (2015). Water: The most precious resource of our life. *Global Journal of Advanced Research*, *2*(9), 1436–1445.

Hove, J., D'Ambruoso, L., Mabetha, D., Peter Byass, P., Kahn, K., Khosa, S., & Witter, S. (2019). 'Water is life': Developing community participation for clean water in rural South Africa. *BMJ Global Health*, *3*(4), 96–105. doi:10.1136/bmjgh-2018-001377 PMID:31263583

Hunter, P.R., MacDonald, A.M., & Carter, R.C. (2010). Water supply and Health. *PLoS Med, 7*(11).

Jacobs, C., Rivett, U., & Chemisto, M. (2018). Developing capacity through co0design: The case of two municipalities in rural South Africa. *Information Technology for Development*, *25*(2), 204–226. doi:10.1080/02681102.2018.1470488

Kawulich, B. B. (2004). Qualitative data analysis techniques. *Journal of Research in Education*, *14*(1), 96–113.

Khadza, A. K. (2010). The Emergence of Water as a 'Human Right' on the World Stage: Challenges and Opportunities. *International Journal of Water Resources Development*, *26*(1), 37–49. doi:10.1080/07900620903391838

Koolwal, G., & van de Walle, D. (2013). Accesses to water, Women's Work and Child Outcomes. *Economic Development and Cultural Change*, *61*(2), 369–405. doi:10.1086/668280

Lange, G., & Hassan, R. (2006), Introduction to water issues and water accounting in Southern Africa. *Economics of Water Management in Southern Africa*, 1-11.

Lohr, S. L. (2010). *Sampling design and Analysis* (2nd ed.). Cengage, Chapman/ Hall Press.

Luh, M., Baum, R., & Bartram, J. (2020). Equity in water and sanitation: Developing an index to measure progressive realisation of the human right. *International Journal of Hygiene and Environmental Health*, *216*(6), 662–667. doi:10.1016/j.ijheh.2012.12.007 PMID:23333082

Maree, K. (2016). First steps in research (2nd ed.). Van Schaik.

Medeiros, P., & Sivapalan, M. (2020). A hard-path to soft-path solutions: Slow-fast dynamics of human adaptation to droughts in a water scarce environment. *Hydrological Sciences Journal*, *15*(2), 2150–3435. doi:10.1080/02626667.2020.1770258

Meyer, W. N. (2007). *The economics of water. Water for life, sanitation for dignity*. Van Schaik Publishers.

Mohammed, I. (2012). Thematic Analysis: A Critical Review of Its Process and Evaluation. *West East Journal of Social Sciences*, *1*(1), 39–45.

Moloi, D. (2004). The Cholera Outbreak: A 2000-2002 case study of the source outbreak In Madlebe Tribal Authority Areas. UThungulu Region.

Muller, M. (2008). Free Basic Water Policy-A sustainable instruments for a sustainable future in South Africa. *Environment and Urbanization*, *20*(6), 68–80.

Nastiti, A., Komarulzaman, A., & Sudradjat, A. (2018). The prospect of the domestic water equity indicators in Indonesia: A Review. *IOP Conference Series. Earth and Environmental Science*, *106*(1), 1–6. doi:10.1088/1755-1315/106/1/012065

Nkuna, Z. W. (2012). *Water governance challenges for rural water supply: A case study of two local municipalities in South Africa* [Unpublished Masters dissertation]. Faculty of Natural and Agricultural Sciences, University of Pretoria.

Oki, T., & Quiocho, E. R. (2020). Economically challenged and water scarce: Identification of global populations most vulnerable to water crises. *International Journal of Water Resources Development*, *36*(3), 416–428. doi:10.1080/07900627.2019.1698413

Olum, Y. (2014). Decentralisation in developing countries: Preconditions for successful implementations. *Commonwealth Journal of Local Governance*, *15*(2), 23–37. doi:10.5130/cjlg.v0i0.4061

Overton, D. M., Smith, J. D., Barchiesi, S., Acreman, M. C., Stromberg, J. C., & Kirby, J. M. (2014). Implementing environmental flows in integrated water resources management and the ecosystem approach. *Hydrological Sciences Journal*, *59*(3), 860–877. doi:10.1080/02626667.2014.897408

Palmer, C. G., & Munnik, V. (2018). *Practising adaptive IWRM. Integrated Water Resources Management (IWRM) in South Africa*. Report No. 2248/1/18. Pretoria: Water Research Commission.

Panhwar, A. H., Ansari, S., & Shah, A. A. (2017). Post-positivism: An Effective Paradigm for Social and Educational Research. *Research Gate*, *11*(2), 253–256.

Pink, R. M. (2016). Water Rights in China and India: A Human Security Perspective. *Asian Affairs: An American Review*, *43*(2), 19–35. doi:10.1080/00927678.2016.1145987

Pirzdeh, H., Shanian, S., & Hamou-Lhadj, A. (2011). The concepts of stratified sampling of executing traces. Department of Electrical and Computer Engineering, Concordia University.

Republic of South Africa. (1996). *The Constitution of the Republic of South Africa Act 1996 (Act 108 of 1996)*. Government Printer.

Rober, C. (2015). *Rural Development putting the last first*. Routledge Taylor & Francis Group.

Romano, S. R. (2019). Transforming participation in water governance: The multi sectoral alliances of rural water committees and NGOs in Nicaragua. *International Journal of Water Resources Development*, *35*(3), 430–445. doi:10.1080/07900627.2017.1363722

Saner, R., Yiu, L., Filadoro, M., & Khusainova, V. (2018). Access to water in developing countries: Four options for provision and regulation in the water sector. *Asia Pacific Journal of Public Administration*, *37*(3), 193–206. doi:10.1080/23276665.2015.1075698

Sareen, S. (2017). Who Governs Local Access in Jharkhand? Mechanisms of Access to Government Services. *Forum for Development Studies*, *44*(2), 249–274. doi:10.1080/08039410.2016.1233135

Sebova, M. (2015). Impact of Municipality Size on Economic Performance: Evidence from Slovakia. *Journal of Applied Economic Sciences*, *11*(3), 1–14.

Sekaran, U., & Bougie, R. G. (2013). Research methods for business: A Skill Building Approach (6th ed.). Wiley.

Sen, A. (2001). *Development as freedoms*. Oxford.

Simukonda, K., Farmani, R., & Butler, D. (2018). Intermittent water supply systems: Causal factors, problems and solution options. *Urban Water Journal*, *15*(5), 488–500. doi:10.1080/1573062X.2018.1483522

Singh, V. P. (2017). Challenges in meeting water security and resilience. *Water International*, *42*(4), 349–359. doi:10.1080/02508060.2017.1327234

Solanes, M., & Gonzalez-Villareal, F. (2008). The Dublins Principles for water as reflected in a comparative assessment of institutional and legal arrangements for water resource management. *Global Water Partnership*, *12*(3), 9–37.

South Africa (SA). (1997). *Water Services Act 108 of 1997*. Government Printer.

Statistics South Africa. (2011). *General Household Survey*. Version 2. Author.

Sultana, F. (2009). Fluid lives: Subjectivities, Gender and Water in Rural Bangladesh. *A Journal of Feminist Geography*, *16*(4), 427-444.

Tewksbury, R. (2009). Qualitative versus Quantitative Methods: Understanding Why Qualitative Methods are Superior for Criminology and Criminal Justice. *Journal of Theoretical and Philosophical Criminology*, *1*(1), 38–58.

Thabethe, P. D. (2011). *An evaluation of water and sanitation provision by uThukela water in selected districts, Public Administration and Business Management* [Unpublished doctoral dissertation]. University of KwaZulu-Natal, South Africa.

Underwood, D. A. (1999). An Introduction to Ecological Economics. *Journal of Economic Issues*, *33*(1), 196–198. doi:10.1080/00213624.1999.11506145

United Nation General Assembly. (2016). *Transforming our world: the 2030 Agenda for Sustainable Development,* A/RES/70/1. Available at: https://www.refworld.org/docid/57b6e3e44.html

United Nations Human Rights Council (UNHRC). (2014). *Resolution on the Human Rights to Safe Drinking Water and Sanitation*. Geneva: UNHRC. A/HRC.RES/27/7.

United Nations-Water. (2015). *Eliminating Discrimination and Inequalities in Access to Water and Sanitation*. UN-Water.www.unwater.org/publications/eliminating-discrimination-inequalities-access-water-sanitation

University of Zululand. (2016). *Research proposal guide, R1 P2 DVC Research and Innovation, Senate approved on 17 April 2013, Ref: S625/13*. Author.

UThukela District Municipality. (2019/2020). *Integrated Development Plan Review*. The Office of the Municipal Manager: IDP Unit.

UThukela District Municipality. (2020). District Growth and Development Plan. Cooperative Government and Traditional Affairs (COGTA).

Van Der Mescht, J., & Van Jaarsveld, M. (2012). Addressing operations and maintenance challenges in smaller municipalities. Academic Press.

Winter, S., Barchi, F., & Dzombo, M. N. (2018). Drivers of women's sanitation practices in informal settlements in sub-Saharan Africa: A qualitative study in Mathare Valley, Kenya. *International Journal of Environmental Health Research*, *28*(6), 609–625. doi:10.1080/09603123.2018.1497778 PMID:30027750

Workman, C. L. (2019). Perceptions of drinking water cleanliness and health-seeking behaviours: A qualitative assessment of household water safety in Lesotho, Africa. *Global Public Health: An International Journal for Research, Policy and Practice*, *12*(6), 34–56. doi:10.1080/17441692.2019.1566483 PMID:30657028

World Health Organisation. (2019). A guide to equitable water safety planning: ensuring no one is left behind. Geneva: World Health Organisation, Licence CC BY-NC-SA 3.0 IGO.

Yazdi, H. S., Shakouri, B., Salehi, H., & Fashandi, A. (2017). Sustainable Development and Ecological Economics. *Energy Sources. Economic Policy*, *12*(8), 740–748.

Chapter 8
Assessment of the Realization of the Fundamental Human Right to Water at the Umalalazi Local Municipality, KwaZulu-Natal

Asanda Buliswa Mnguni
University of Zululand, South Africa

ABSTRACT

The aim of the study was to examine the municipal water resource governance and fundamental human rights to water in the UMlalazi Local Municipality of the KwaZulu-Natal Province. The study employed the post-positivist paradigm and closed ended questionnaires for collection of data. Stratified random sampling technique was used to sample 375 respondents, who were the households (head of households). Chi-square was used to determine relationships between the variables. The findings showed that most of the respondents strongly agreed that the 6kl of monthly water supply was not enough for meeting their domestic, consumptive, and hygiene needs. The municipality has a youthful population structure. Females were mostly affected by household water scarcity compared to their male counterparts. For the households, a day would be terribly difficult without this life-enabling commodity.

1. INTRODUCTION

An unhindered access to water is significant for human endurance and a technique to help break endless loop of neediness. However, for some, impartial access to consumable water involves day by day battle. Accordingly, water governance is a fundamental system for progression, poverty alleviation and fulfilment of basic rights to water supply (DWAF, 2011).

DOI: 10.4018/978-1-7998-8809-3.ch008

The issue of water access has deteriorated as of late in view of a few conditions, similar to population increase which has truly influenced non-industrial nations the most. Lack of water can dislocate drinking water supply, health, sanitation, mechanical cycles and numerous different parts of life. The 2020 UN World Water Development report distributed globally, highlighted the underfunding of water infrastructure around the world. One of the reasons underlying the investment gap in water and sanitation is that these services are seen mostly as social and in some cases ecological issue, as opposed to economic one, like energy. Understanding the economic significance of water and sanitation should provide an additional catalyst for greater investment (UN, 2020). There is a rivalry for water assets, but ensuring that the individuals who need water get, it is a good speculation, hence the need for good governance in water supply.

Water governance in UMlalazi Local Municipality is an unpredictable interaction as water 'flows' through the rapidly changing 'hydro social' pattern of the city creating rivalry for and conflict over its utilization and removal. This 'hydro social' cycle, which catches and incorporates 'the socio-political and biophysical measures that establish water arranged in a rapidly changing municipality that is being transformed by on-going urbanisation, post-apartheid restructuring, and the pro-growth and pro-poor agenda of national government (van Kopen *et al*, 2016). Decentralisation of water supply and sanitation delivery has appeared to improve responsibility, acknowledgement of multicultural patterns, and proficient and successful service delivery. However, it is hard to control the appropriate support of enormous number of waste water treatment plants, hence the danger for general health conditions increases. Besides, the cost of building the treatment and storage capacity or removal units, along with an additional network for the water and the water to be reused, expands the expense of the structure and cannot be recovered by the cost of saved water (Sutherland et al, 2015). The water strategy formulation and policy, enforcement and the proactive nature of water clients are likewise essential to more readily water and sanitation service delivery, if sustainable water management will be realised in the near future (Global Water Partnership, 2006).

Mostly in parched and semi-arid areas, access to sufficient water in regards to quality, amount, and moderateness is indispensable for economic development. However, a few difficulties block water supply in numerous parts around South Africa, particularly in rural areas. There are numerous advantages of improved water service delivery, including prevention of illnesses, improved fundamental medical care, and better sustenance, expanded access to institutions, like schools, improved water quality, expanded amount and access to water, decrease on schedule and exertion needed for water assortment, advancement of financial action, reinforcing of local area association, upgrades in lodging, and eventually, improved personal satisfaction (Mbatha, 2016).

To address the looming water crisis, the public authorities of South Africa has made a few interventions. Adoption of the Integrated Water Resources Management Policy by the South African government support of the African Water Vision (AWV) and the Framework for Action (FFA), contribution to the foundation of the African Water Facility (AWF) and execution of Rural Water Supply and Sanitation Initiative, among others, are some of the significant interventions embraced by the South African government lately (Department of Water and Affairs and Forestry, 2009). In 2001, the Department of Water Affairs and Forestry introduced a policy of providing each of the indigent households with free basic water. The overwhelming cholera outbreak in 2001 at Ngwelezane in KwaZulu-Natal Province was the chief justification for this initiative.

The departments of health, water affairs and forestry, and local government battled to keep the pandemic under control. Then again, The Phiri Case in Soweto additionally fills in as a rule to accomplishing value in sustainable services delivery. At Phiri, the community fought against the introduction of prepaid meters and took the government to court and the ruling was in favour of the community. The judge ruled that government should provide households with 50 litres of water for free as a realisation that water is a human right (Naidoo, 2005, Dugard, 2014 cited in Meyiwa *et al*, 2014). Much has changed in South Africa since 2003 in as far as the water issues are concerned. Nevertheless, the role of the state in relation to the right to water remains as argumentative as ever. The development of indigent policies by individual municipalities from the late 2000s has implied that the state is not simply ready to choose what is and what is definitely not an only dissemination of water yet in addition who is and who is not the meriting poor according to municipal service provision (Dugard, 2014).

In 2001, in the location to uMlalazi Local Municipality and COGTA, the strategy the strategy called for the allocation of 6kl of Free Basic Water to a family of 8, but there were municipalities that were still offering just 6kl, while Durban expanded their allocation to 9kl and other municipalities are now offering 12 kl .Research has shown that a small local municipality in the North West Province named Rustenburg Local Municipality is offering the Royal Bafokeng Nation unlimited quantities of free basic water, along with the Royal Bafokeng Government. As proposed in the Free Basic Water Policy, the goal of equitable realization of free basic water should be to ensure equal access to water for everyone and not only the indigent (Tabane, 2020). The Human Right to Water and Sanitation (HRWAS) is a rule that recognizes that clean drinking water and sanitation are fundamental for each individual's life. It was perceived as a basic human right by the United Nations General Assembly on 28 July 2010.The HRWAS has been perceived in global law through human right treaties, affirmations and different principles.

A few commentators have determined that the human right to water stipulated in the General Assembly Article 11.1 of the International Covenant on Economic, Social and Cultural Rights, is restrictive under worldwide law. The option to perfect, safe water and disinfection has clear roots in global basic liberties law, worldwide helpful law, and global water law. The option to water and disinfection has developed both in global practice and lawful conviction, the base of standard worldwide law. A few considerable rights in the Constitution's Bill of Rights are of significance to water. The Constitution is quite certain about water. Which is all well and good, it is vital for all life, including human existence. The stakeholders in water perceives water to be fundamental resource for everyday nourishment. In section 27 of the Bill of Rights it is expressed that: "Everybody has the privilege to receive adequate food, water and sanitation" (Republic of South Africa, 1996:11).

Gumbi (2017) states that the country's territories especially those that are under traditional authorities are described by underdevelopment and related neediness. An enormous number of individuals live under hopeless conditions, emerging from a high pace of joblessness, and poor delivery of water and sanitation services. The inability to give adequate water can mess up people lives to work well, restricting profitability and support of individual cleanliness. All in all, the privilege to adequate water isn't unequivocally perceived.

2. PROBLEM STATEMENT

Numerous studies have explored the challenges in the supply of water, because of dry spells and different glitches that represent a danger to human livelihoods and other living things. UMlalazi Local Municipality is one of the spots encountering challenges in the supply of water because of significant degrees of interest. It has been seen that the parts of water provisioning specialists at uMlalazi Local

Municipality are not sufficient for this local municipality which is predominantly rural since some of the households do not have satisfactory water either in their families or in their networks. This issue has gained attention and require action by water actors. The issue of precarious water supply in this space has for quite some time been an issue; a greater part of families doesn't have consumable water (UMlalazi Local Municipality, 2016). The public authority has made progress in the delivery of other essential services. Nonetheless, water service delivery is as yet an issue for most regions in South Africa (Carpenter *et al.*, 2014).

Numerous investigations have shown that a portion of the rural communities in South Africa actually don't have access to water services and thus would regularly depend on utilizing unprotected water sources for their needs (Basson *et al.*, 2011). South Africa is confronting various difficulties concerning water, both at the level of the resource and at the level of water services delivered by the districts. Berkowitz (2009) states that this issue is mostly critical in little districts; serving rural communities and it is not an issue for enormous regions and metropolitan territories. Hence, this plainly shows that the water supply framework needs to adapt to the local conditions as the water supply is insufficient, thus results in many people who do not get water for an extended period.

The residents of the local area are compelled to utilize water from streams and hand-burrowed wells which regularly has serious health outcomes to their wellbeing (Berkowitz, 2009). These days, individuals are expected to pay for water, a few groups cannot manage the costs of water since they do not have cash or they are confronted with poverty. The water crisis has additionally added to the spread of illnesses like diarrhea, cholera and skin disease on account of debased water that individuals utilize in their families. Rural individuals experience loss of sufficient water for domestic purposes. A dominant part of individuals in rustic zones experience the ill effects of an absence of access to clean water and inadequate water for food creation (Motoboli, 2002).

As per Motoboli (2002), mostly in oppressed zones, individuals bring water from numerous sources which they use for a few purposes as they attempt to support and keep up their livelihoods. However, these water sources are inconsistent and have a deficient water supply. The sources are either messy or they are dry, and they are additionally a long way from individuals. Accordingly, rural communities actually depend on sullied sources like streams and dams since they are not provided with more readily available water services. During the stormy seasons, the communities get more water and battles during the dry seasons. In non-industrialized nations, gathering water for domestic use in rural zones is a serious issue. At times a long way from homes, natural surface water supplies give inferior quality of water and are powerless against the impacts of dry spells or dry seasons (Makon, 2014). The King Cetshwayo District Municipality and UMhlathuze water play a vital role at UMlalazi Local Municipality, as the support base in the provision of water to their areas of jurisdiction. They both ensure that efficient, sufficient and equitable water is provided to the communities to sustain their livelihoods (Department of Water and Sanitation, 2014).

3. AIM AND OBJECTIVES OF THE STUDY

3.1 Aim of the Study

The aim of the study was to examine the governance of water in the UMlalazi Local Municipality of the KwaZulu-Natal Province in South Africa

3.2 Objectives of the Study

1. To explore the constitutional and policy frameworks adopted to support municipal water resource governance and human rights to water in the UMlalazi Local Municipality.
2. To investigate the decentralization of water management and the extent to which it promotes or hinders access to water services as a basic need.
3. To investigate cost recovery measures and subsidies provided for water service delivery in the UMlalazi Local Municipality.

3.3 Research Questions

What are the roles of the UMlalazi Local Municipality in the governance of water and realisation of the fundamental human right to water within the KwaZulu-Natal Province?

3.3.1 Specific Research Questions

1. How effective are the constitutional and policy frameworks implemented to support municipal water resource governance and fundamental human rights to water in the UMlalazi Local Municipality?
2. What is the extent to which decentralization of water management promotes or hinders access to water services as a basic need?
3. What cost recovery measures and subsidies are provided for water service delivery in the UMlalazi Local Municipality?

3.4 Hypotheses

- Null hypothesis (H_0); there is no statistically significant relationship between good governance, human rights, water contamination, women inclusion, agriculture and access to potable water at the UMlalazi Local Municipality.
- Alternative hypothesis (H_A); There is a statistically significant relationship between good governance, human rights, water contamination, women inclusion, agriculture, and access to potable water at the UMlalazi Local Municipality.

4. THEORETICAL FRAMING AND EMPIRICAL LITERATURE

In South Africa there are many pieces of legislation that aims to improve access to adequate and sufficient water. The country's water laws and policies were introduced to ensure equitable access to water for everyone. The legal framework in respect of water service delivery includes policies, laws, strategies, and approaches to deal with water stress situations and integrated water resource management (Thompson, 2006).

Therefore, it is important to put in place an effective framework to ensure that the water supply is efficient and the country's water resources are protected, used, developed, managed and controlled in a sustainable and equitable manner that will benefit everyone for the long time. The framework should include all living things, people, animals and plants (Thompson, 2006).

According to Gumbi (2017), Free Basic Water Policy was introduced by the government, aimed at providing 6, 000 liters of free water to the poor households, per month and to ensure that all households have access to a basic supply of water. He further argues that the free basic water policy is aimed at improving the quality of life of those citizens who cannot afford to pay for water and at contributing to the government's initiative to eradicate poverty in the rural areas the most that are faced with the water crisis.

The 6, 000 liters limit was set to achieve some rural and urban equity, since where water is carried, more than 50 meters or so, the usage per capita drops to around 25 liters. The World Health Organization regards this as the minimum that should be achieved universally before services are improved (Hlophe, 2004). In a similar study, ZDM (2017) reports that an approved indigent/deprived household must receive fully subsidised water at a minimum of 6kl per month.

Chapter two of the Constitution of South Africa, 1996 comprises the bill of rights which is the cornerstone of democracy in South Africa. Section 7(2) of the constitution states clearly that; the state must respect, promote and fulfil the rights of people as listed in the bill of rights. According to section 9(1), (2), (3), (4) of this constitution, no individual shall be unfairly discriminated against on the grounds of; for example, gender, sex, ethnic group or social origin, language and everyone is equal before the law (Republic of South Africa, 1996)

Section 27(1) b of the constitution states that everyone has a right to access clean sufficient water, to social assistance to people who are unable to support themselves and their dependents a healthy environment within their communities. The institutional framework functions within the legislation of the country, on the basis that the Department of Water and Sanitation (DWS) serves, as the body that formulates and implements the policy concerning water supply and sanitation programs (Republic of South Africa, 1996). In a similar study, (Machethe, 2011) also discovered that the right to health care, food, water and social security is enriched in the constitution. Subsection two of this section states categorically clear that; the state has full responsibility to see to it that these rights are realized progressively.

Motloung (2010) stipulates that the management of water, and not the scarcity of water, is the problem in many parts of the world. It is hereby realised that water governance provides a fundamental framework of water management towards the end of failure of management of water and water scarcity. Water governance recognises IWRM as a process where various stakeholders build ground rules to preserve water resources for cost effective and sustainable socio-economic development (Motloung, 2010). Governance entails a process of institutional interrelation wherein role players from government and civic society can meet.

Water is an important aspect of socio-economic development and it is associated with environmental impacts, it is central for all forms of life, economic growth, development-related activities, and environmental sustainability. Without water many sectors cannot function (social, economic, political and environmental functions). Therefore, the equitable water supply of sufficient quantity and quality is central to the growth of an economy of a particular geographical area and the improvement of public welfare (Gumbi, 2017). Contaminated water threatens the social health and physical health of the society and it is dangerous to human life (World Vision, 2010). The United States Environmental Protection Agency (2013), states that water is an important commodity for human life and all people on the planet depend upon it to survive.

Rightful access to water is important to live and is recognized as a fundamental human right. Human lives depend on and demand sufficient and safe water to be always healthy. South African law and policy state that basic water supply must be sufficient, safe, accessible and affordable (Republic of South

Africa, 1996). Hence, basic water needs to be provided continuously with a stipulated minimum rate of flow and quality to the people (Gumbi, 2017).

Poor governance lies at the heart of the world's water crisis (Rogers and Hall, 2003). Mostly, the rural municipalities depend on grants for infrastructural development, which makes it hard for them to provide sufficient water, of better quality for the people. Also, the poor planning and lack of resources under governance hinder the supply of water. How available water resources are managed is a key to achieving water security (United Nations Development Programme, 2013). However, Rogers and Hall (2003) have indicated that the ineffectiveness of water governance by states has been the hindering factor to water security.

Water plays a fundamental role in economic growth of the country and of a specific area and in people's lives. Failure to provide sufficient water primarily for irrigation and in domestic use (drinking, washing, cooking), has caused conflict over water and the effects of gender-influenced decisions and consequences on people's lives, economic growth and social change (Crow and Sultana, 2002).

The access to water in rural areas is generally more problematic, more differentiated, less secure and requires extensive expenditures of work, time and money. Rich and prominent households may have preferential conditions of access, and different sources of water, compared to those of poor households. The quality, reliability, and costs of water for a particular household will be influenced by a range of characteristics including conditions of the water source, geographic location of the household or enterprise in relation to the water source, past social investments in water infrastructure, and the social, economic, and even political position of a household or enterprise (Crow and Sultana, 2002).

Without access to clean water, women's daily lives are extremely affected; not only as individuals, but because they cannot carry out their prescribed role within their families and communities (United Nations 2005; and Souza 2008). Similarly, females become victims of water crises, since the cost of water shortage and inadequate infrastructure, fall extremely on women and girls (Vogelstein, 2017). Limited access to clean water and sanitation has an impact on both women's and men's health, physical and psychological integrity, privacy and access to education. However, this impact differs by gender, since the differences and inequalities between women and men influence how individuals respond to changes in water resources management (Gender and Water Alliance, 2006).

Grey and Sadoff's (2008); state that water has always played a central role in societies by alleviating poverty and ensuring sustainable socio-economic growth. Additionally, economic activities depend upon the availability of water resources, including its quality (Logar, 2018). Moreover, the survival of living and non-living organisms in the community is inextricably linked to the availability and access to water. Therefore, it is important for governance to play a critical role in improving water resources management and service delivery. Nevertheless, the water crisis that South Africa is experiencing do not only look at having too little water or clean water to satisfy the human needs, but also talks to managing water in an integrated manner and making it accessible to all at all times regardless of their settings (SIWI, 2017).

Moreover, contaminated water threatens the social health and physical health of the society and it is dangerous to human life (World Vision, 2010). The United States Environmental Protection Agency (2013), states that water is an important commodity for human life and all people on the planet depend upon it to survive.

The study employed ecological economics theory to explain that water should be treated as a fundamental human right. It focused on the water as a human basic need and equity in water. Additionally, the ecological economics framing of the study was mainly treating water as a natural capital. A study conducted by Costanza (1989) explains that ecological economics addresses the relationships between ecosystems and economic systems in the broadest sense, thus, emphasizing the importance of these relationships in order to promote a sustainable economy. Hence, the aim of ecological economics is to improve and to promote a deeper logical understanding of the complex linkages between human and natural systems, in order to make effective policies that will promote ecologically sustainable environment through equal distribution and allocation of scarce resources such as natural and social capital (Romeiro, 2012). Similarly, this will further promote equity in water access and sustainable management of water in an integrated manner (Stockholm International Water Institute [SIWI], 2017, Hassan, Mtsweni, Wilkinson, Weston, Mutundo, Magagula, Sithole, Farolfi, and Dinar, 2014).

5. CONTRIBUTION TO KNOWLEDGE

This study preserves that to achieve good governance in rural areas, ecologically sustainable management and use and access to water should be prioritised and water should be treated as a fundamental human right. Water governance is thus crucial for improving access to water and sanitation services. How communities choose to govern their water resources and services has profound impacts on people's livelihood and the sustainability of water resources. The study was conducted in rural areas to disclose whether there is good water governance and to see whether water is treated as a human right, with the aim of providing evidence to the government. Using quantitative methods has allowed the study to obtain information from the households who experience issues of unpredicted water supply from three different communities within UMlalazi Local Municipality. Therefore, the study adds to the existing knowledge by providing valuable information on the strategies and policies that can be improved or implemented to address the concern of good water governance and effective water supply.

6. DESCRIPTION OF THE STUDY AREA

The population dispersion in the metropolitan region is described by moderately high population densities inside metropolitan hubs and low densities in country territories. The metropolis has a total population of 213602 and the gender are spread as follow: male 44.7% and female 55.3%. There are 45 062 households as per the 2011 census report. The residential region is dominated by ancestral land, and 14 ancestral or traditional authorities exist inside the space. The town of Eshowe is of incredible recorded importance in that it is the organisation of King Cetshwayo, who was the ruler of the Zulus during the Anglo-Zulu war of 1979. Figure 1 presents the location of the UMlalazi Municipality. Eshowe Town is likewise viewed as the managerial and administration centre of the UMlalazi Local Municipality. The coastal town of Mtunzini is situated on the north coast of KwaZulu-Natal on the banks of the UMlalazi River, and borders on the uMlalazi Nature Reserve. Mtunzini is an important eco-tourism destination in South Africa (UMlalazi Local Municipality, 2016).

The city region comprises of business cultivating territories in an expansive, nonstop band from the West of Eshowe to kwaGingindlovu (also known as Gingindlovu) and the North of Mtunzini. The remainder of the space comprises of ancestral positioned land, where land is inhabited by the poor households. It is characterized by undulating topography, causing difficulties for the delivery of engineering services. The uMlalazi Local Municipality is endowed by a number of important transportation routes, such as the N2 motorway between Durban and Richards Bay, the R34 between Richards Bay/Empangeni and Nkwaleni Valley to the north of Eshowe, and the R66 from the N2 motorway to KwaGingindlovu, Eshowe, Melmoth, Ulundi and Vryheid (Umlalazi Local Municipality, 2016).

Figure 1. Location of the UMlalazi Municipality
Source: AfriGis (2019).

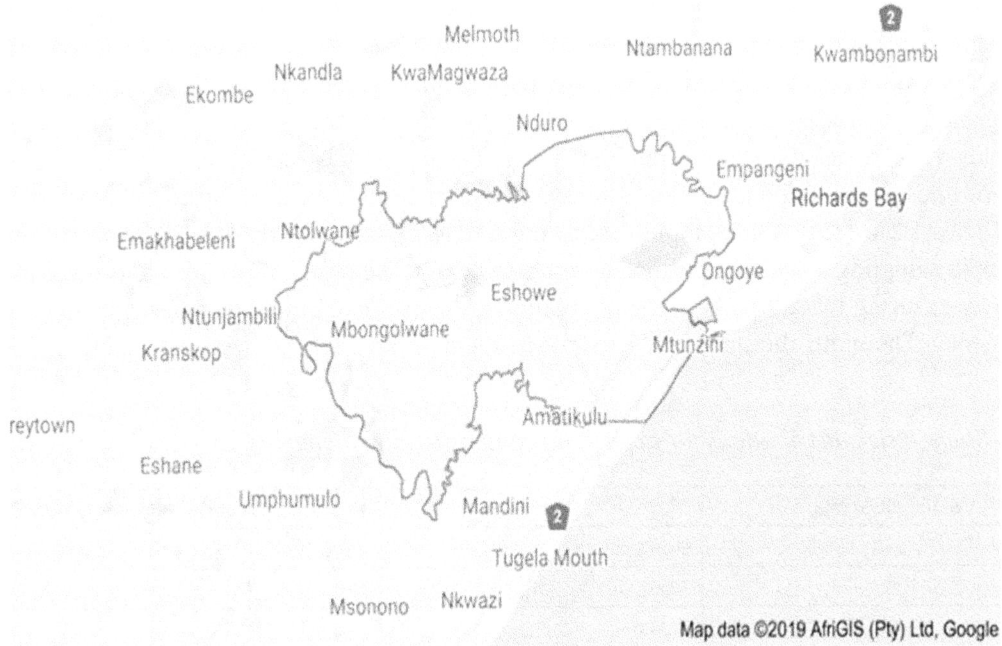

7. METHODOLOGY

7.1 Research Philosophy and Paradigm

The study used the post-positivist paradigm. Moser (2015:1) states that, epistemology is "the theory of knowledge, the philosophical study of the nature, origin, and scope of knowledge". Whereas, post-positivist is "a certain pluralism which balances both positivist and Interpretivists approaches. It further focuses on researching issues in the context of involving experiences of the majority and announcing the result of what the majority says is acceptable" (Panhwar, Ansari, and Shah, 2017). Therefore, using post-positivist philosophy allowed the study to analyse realities taking place at UMlalazi Local Municipality with a concern of water governance and water as a human right.

7.2 Design

The researcher used a descriptive design interrelated with quantitative method. Clark and Creswell (2015) explain that quantitative research method draws conclusions from large numbers of people, uses numerical data from selected subgroup, demonstrates relationships, examines probable cause and effect, bias controlled and establish efficient data analysis. The study used questionnaires to collect the quantitative data. The questions were designed in IsiZulu and English for simplicity purposes as most of the people in UMlalazi Local Municipality are using IsiZulu language. Moreover, descriptive study helped the researcher to understand the governance of water and human right to water at the Municipality and be able to understand the strategies needed to improve the water governance, to ensure access to water.

7.3 Study Population and Sampling Method

For this study, the population consists of uMlalazi Local municipality households and officials. The study primarily focused on the households. The study population covered three sections namely Ndlangubo, KwaMqhwakazi and Mlalazi. The study focused on households (heads of households) because many households do not receive a regular supply of water. The problem is escalating at an alarming rate as a number of households in the area increases rapidly. Nevertheless, the study used quantitative research method. Thus, stratified random sampling was employed to 375 households. Shi (2015) argues that stratified random sampling is a technique that attempts to restrict the possible samples to those that are less extreme by ensuring that all parts of the population are represented in the sample in order to increase the efficiency. Therefore, the study consisted the residents from all the above three mentioned sections.

Table 1. Sample size distributions for households and official respondents

Community strata	Population	Sample proportion (p*n/N)
Ndlangubo	9050	226.25
Mlalazi	3666	91.65
KwaMaqhakazi	2224	55.6
Official respondents from the municipality	60	1.5
Total	15000	375

Source: Fieldwork (2019).

7.4 Sample Size Determination and Sampling Procedures

The sample was determine based on the appropriate sample formula

The sample size equation is provided below:

Confidence level (a) 95%

Margin of Error (e) 5%

Population (p) 50%

Population size (N) (optimal) 15000

The sample size (n) is calculated according to the formula:

$$n = [z^2 * p * (1 - p) / e^2] / [1 + (z * p * (1 - p) / e^2 * N)]$$

$$n = [z^2 * p * (1 - p) / e^2] / [1 + (z * p * (1 - p) / e^2 * N)],$$

where z = 1.96, p = 0.5, N = 15000, confidence 95%, hence, e = 100-95/100 = 0.05

Hence,

n = [1.96^2 * 0.5 * (1 - 0.5) / 0.05^2] / [1 + (1.96^2 * 0.5) * (1 - 0.5) / (0.05^2 * 15000)].
n= [3.8416 * 0.5 * (1 -0.5) / 0.0025] / [1 + (3.8416 * 0.5) * (1 - 0.5) / (0.0025 * 15000)].
n = [1.9208 * 0.5 / 0.0025] / [1 + 1.9208 * 0.5 / 37.5].
n = [0.9604/ 0.0025] / [1 + 0.9604 / 37.5].
n = [384.16] / [1+ 0.9604 / 37.5].
n = [384.16] / [1+ 0.0256].
n = [384.16] / [1.02561.0262]
n=384.16/1.0256, hence, n= 374.570
n= 375 (nearest whole number)

7.5 Ethical Considerations, Reliability and Validity of the Study

The researcher obtained ethical clearance certificate from University of Zululand Research Ethics Committee (UZREC) which gave permission for collection of the data from UMlalazi Local Municipality. The researcher consulted the traditional leaders to obtain permission to collect the data within their traditional jurisdictions. Furthermore, the researcher ensured that participants operate from informed consent. Respondents were told the truth about the intention of the study and were not deceived into participating in the study. The researcher made it clear that participation in the study was a voluntary basis so that they may pull out anytime they wish. There was no violation of privacy and respondents were assured of their anonymity, the Cronbach's Alpha was used in testing the internal validity of the instruments. Cronbach's Alpha is a measure used to assess the reliability, or internal consistency of a set of scale or test items. Thus, Goforth (2015) states that Cronbach's Alpha, is one way of measuring the strength of the consistency. In addition, researchers recommend a minimum coefficient between 0.65 and 0.8. Coefficients that are less than 0.5 are usually unacceptable. Therefore, the Cronbach's Alpha value of 0.70 for all the items was considered generally acceptable and within the best standard practice for data reliability.

Validity refers to the extent to which a question or instrument measures what it is supposed to measure (Nxumalo, 2007). The validity of this research study is provided by the strong chain of evidence flowing through the study. Therefore, for validity purposes, pilot study was undertaken. A pilot study is a mini-version of a full-scale study done in preparation of the complete study. Pilot study is also referred to a feasibility study. It can be treated as a specific pre-testing of research instruments, including questionnaires or interview schedules (Polit, *et al.*, & Baker in Nursing Standard, 2002; Van Teijlinge & Hundley, 2001). The researcher distributed questionnaires to 38 participants, which is 10% of the total sample size

7.6 Data Analysis

For the presentation of the collected data, the study used the Statistical Package for Social Sciences (SPSS) software as it assured that the conclusions made were reliable. The software consists of inferential and descriptive statistics in order to provide an accurate description of the entire data. The study used the graphs and tables to provide meaningful interpretation of the collected data. This software is one of the most popular statistical packages which can perform highly complex data operation and analysis with simple instructions. The study used descriptive SPSS because exporting survey data to SPSS's proprietary format makes the process of pulling, and analyzing data clean and easy. The chi-square (χ^2) test statistic was applied in determining relationships between the variables and for testing the hypothesis. With 95%

confidence level and error margin (e) =0.05, the decision rule for the hypothesis test was stated that; accept the null hypothesis (Ho) if the critical values (p-values) are greater than (˃) the alpha level (α) 0.05 and do not accept the null hypothesis if the critical values (p-values) are less than or equal to (≤) the alpha level (α) of 0.05. To summarise data, the study used bar graphs, histograms, and pie charts in presentation of the results (Van Teijlingen & Hundley, 2001).

8.0 RESULTS AND DISCUSSIONS

8.1 Features of the Respondents

In this section, the study presents and discusses the results on the background characteristics of the respondents in order to reveal the general socio-demographic context of the study area. The themes and variables covered are gender and age categorization as well as the racial elements of the respondents. From Table 2; the findings showed that all respondents interviewed were eligible adults and with ages ranging between 25-75 years. Most of the respondents (41.6%) aged between 25-35 years, followed by those between 35-45 years (30.7%), 45-55 years (18.9%), 55-65-yearSgroup (6.7%) and those 65-75 years constitute (2.1%) of the respondents. The findings reveal a predominantly youthful population structure of 25-45 years, which constitutes (72.3%) of the population within the study area. This population group is considered the most energetic with human resource capacities and potentials which could be harnessed locally to promote the development of water resources in the rural communities.

Table 2. Gender and age of the respondents

Gender	Age range (in years)										Total	
	25-35		35 -45		45-55		55-65		65-75			
	n	%	n	%	n	%	n	%	N	%	N	%
Male	88	23.5	46	12.3	40	10.7	14	3.7	6	1.6	194	51.7
Female	68	18.1	69	18.4	31	8.3	11	2.9	2	0.5	181	48.3
Total	156	41.6	115	30.7	71	18.9	25	6.7	8	2.1	375	100

Source: Fieldwork (2019), n = 375

In terms of gender, Table 1 revealed a slightly higher rate of males (51.7%) compared with females which constituted (48.3%) of the population. The findings suggest possibilities for male domination in decision making on the provision of critical public and environmental services including water and sanitation in the Municipality. However, adult females appeared to be most likely interested in water issues compared to their male counterparts. This is because females appeared to be mostly affected by water scarcity and household's food security. Similarly, females become victims of water crises, since the cost of water shortage and inadequate infrastructure, fall extremely on women and girls (Vogelstein, 2017). Additionally, females seem to be mostly engaged in domestic household activities of cooking, washing, cleaning, and childcare and water haulage and therefore, turn to have direct effects in terms of water and food shortage.

In addition, the study interrogated the respondents on their racial elements. This was to understand the opinions of the respondents on race whose livelihoods are mostly impacted in terms of water scarcity. From Table 2, the findings reveal that the study area is predominantly constituted by black populations (97.9%) as compared with Coloured resident population (2.1%). The results convey to the forefront issues of racialisation, segmentation and segregation against native black populations on matters of water governance. Since blacks constitute the majority of the populations, scarcity of water in the area is likely to affect their standards of living and livelihoods, particularly poor households and indigenous blacks who rely on water for agricultural activities.

8.2 Free Basic Water Policy (FBWP)

According to Gumbi (2017), Free Basic Water Policy was introduced by the government, and it aimed at providing 6, 000 litres of free water to the poor households, per month and to ensure that all households have access to a basic supply of water. The free basic water policy is aimed at improving the quality of life of those citizens who cannot afford to pay for water and at contributing to the government's initiative to eradicate poverty in the rural areas to the most that are faced with the water crisis. Thus, Figure 2 analysed the respondent's perception on whether the South African government is providing each household with 6kl of free basic water per month.

Figure 2. Free Basic Water Policy is aimed at improving the quality of life of those citizens who cannot afford
Source: Fieldwork (2019), n = 375

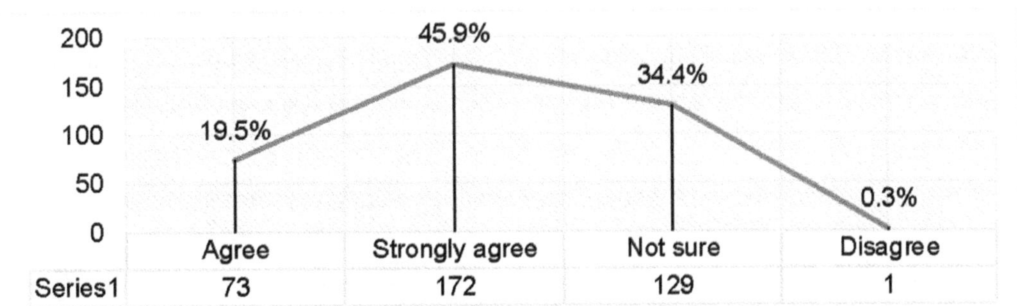

From figure 2, the results showed that most of the respondents (45.9%) strongly agreed on their perceptions of FBWP at aiming to improve the quality of life of those citizens who cannot afford to pay for water and at contributing to the government's initiative to eradicate poverty in the rural areas especially the most that are faced with the water crisis. About 34.4% were not sure, 19.5% agreed and 0.3% disagreed. Free basic services were part of the government's response to debates about social welfare policy. Free basic water is a National Government Policy but it can only be implemented by the local government. In a similar study, ZDM (2017) reports that an approved indigent/deprived household must receive fully subsidised water at a minimum of 6kl per month. The finding affirms that the policy should target the poor and make sure that all the households have access to water supply; access within 200m of the house and the assurance of supply.

Figure 3. South Africans should have access to safe and affordable basic water and sanitation
Source: Fieldwork (2019), n = 375

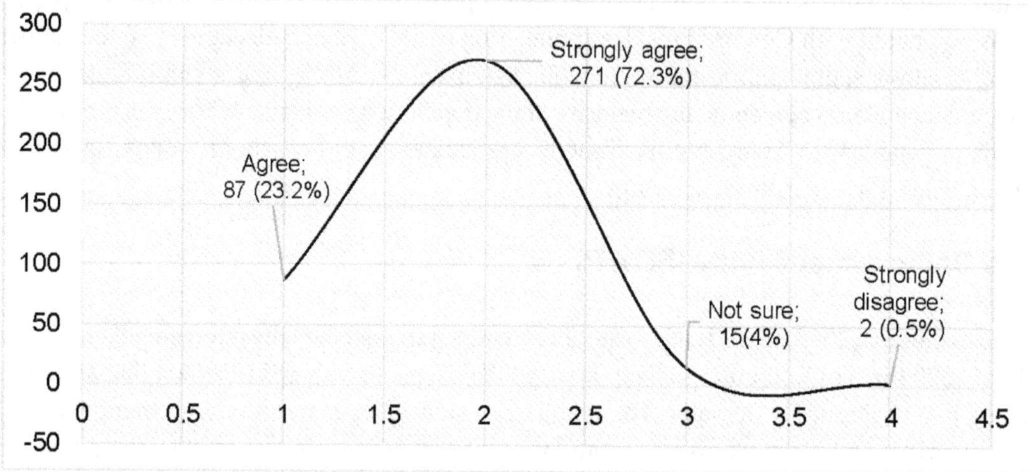

From Figure 3, the study showed that most of the respondents (72.3%) strongly agreed that South Africans should have access to clean water and sanitation, (23.2%) agreed, (4%) were not sure, and (0.5%) strongly disagreed. In a similar study, (Machethe, 2011) also discovered that the right to health care, food, water and social security is enriched in the constitution. Section 27(1) (b) unequivocally states that; everyone has the right to access to sufficient food and water. Subsection two of this section states categorically clear that; the state has full responsibility to see to it that these rights are realized progressively. The results implied that without water there is no life. The water required for each person or domestic use must be safe, therefore free from micro-organisms, chemical substances and radiological hazards that constitute a threat to a person's health. Hence, water facilities and services must be affordable to all.

Table 3. Water and sanitation services should be equitable, affordable, and sustainable and gender sensitive

Response categories	Frequency	Percent (%)
Agree	67	17.9
Strongly agree	283	75.5
Not sure	19	5.1
Disagree	5	1.3
Strongly disagree	1	0.3
Total	375	100

Source: Fieldwork (2019), n = 375

As depicted by Table 2, the study showed that most of the respondents (75.5%) strongly agreed that water and sanitation services should be equitable, affordable, and sustainable and gender sensitive, (17.9%) agreed, (5.1%) were not sure, (1.3%) disagreed, and (0.3%) strongly disagreed. The findings of the study agree with the constitution adopted in 1996 that guarantees the right to water for every citizen, stating that 'everyone has the right to have access to sufficient food and water and social security, and thus, government should provide proper assistance for those who are unable to access adequate food and water' (Republic of South Africa, 1996).

8.3 The Institutional Framework

In this section the study addresses the first objective which is aimed at; to explore the constitutional framework adopted to support human rights in South Africa. This is because the 1996 South African constitution states that everyone has a right to access clean water, to social assistance to people who are unable to support themselves and their dependents. The institutional framework functions within the legislation of the country, on the basis that the Department of Water and Sanitation (DWS) serves, as the body that formulates and implements the policy concerning water supply and sanitation programs. Section 27 of the Constitution ensures that all South Africans have access to clean, sufficient water and a healthy environment within their communities (The Republic of South Africa, 1996). Hence, the study analysed whether everyone has a right to access clean, sufficient water and a healthy environment.

Figure 4. Everyone has a right to access clean, sufficient water and a healthy environment
Source: Fieldwork (2019), n = 375

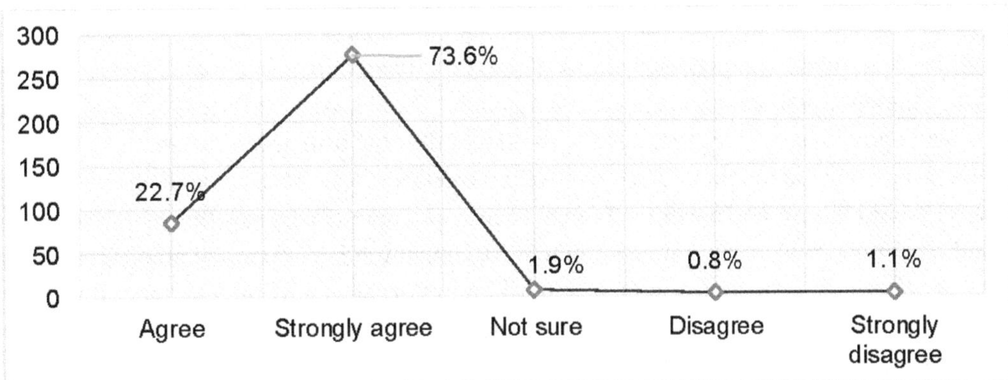

As depicted by Figure 4, the findings showed that most respondents (73.6%) strongly agreed, (27.7%) agrees, (1.9%) was not sure, (1.1%) strongly disagreed and (0.8%) disagreed that everyone has a right to access clean, sufficient water and a healthy environment. The results prove that people are aware that water is important to their everyday lives and they should have access to it. In a similar study, (Machethe, 2011) also discovered that the Constitution of the Republic of South Africa, 1996 section 27(1) (b) states that; everyone has the right to access to sufficient food and water. Moreover, everyone has a right to access clean water, to social assistance to people who are unable to support themselves and their dependents.

8.4 Water Governance

In this section, the study seeks explores the constitutional framework adopted to support human right to water in South Africa. This section further argues that governance play a critical role in improving water resources management and service delivery. Hence, without proper management of water resources, current and future generations might be affected in terms of water access.

Figure 5. Water governance provides fundamental framework for water management
Source: Fieldwork (2019), n = 375

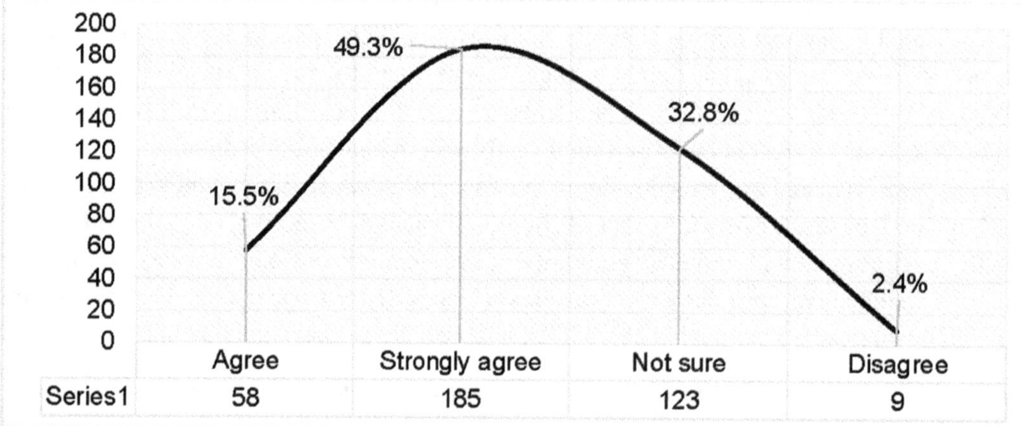

From Figure 5, the results showed that 49.3% of the respondents strongly agreed that water governance provides a fundamental framework for water management, 32.8% were not sure, 15.5% agreed and 2.4% disagreed. Water governance is responsible to provide for the core principles, mechanisms and processes that are necessary to enable municipalities to move progressively towards the social and economic uplift-ment of local communities, and ensure universal access to essential services that are affordable to all.

The findings of the study agree with Motloung (2010) that water governance provides a fundamental framework of water management towards the end of failure of management of water and water scarcity. Water governance recognises IWRM as a process where various stakeholders build ground rules to pre-serve water resources for cost effective and sustainable socio-economic development (Motloung, 2010).

8.5 Importance of Water in the Municipality

Water is an important aspect of socio-economic development and it is associated with environmental impacts. Water is central for all forms of life, since it is central for economic growth, development-related activities, and promotes environmental sustainability. Water is a key driver of sustainable growth and development, water is a source of life and prosperity (Goswami and Bisht, 2017). Hence, water provides for agriculture, industries and humans, which all contributes to an increase in socio-economic develop-ment of the country. Without water, there will be no sustainable livelihood and socio-economic develop-ment. Hence, this study questioned the respondents' perception on water as a key driver for economic and social development. The aim was to analyses whether the respondents agree to the statement that water plays a major role in socio-economic development.

Figure 6. Without water many sectors cannot function
Source: Fieldwork (2019), n = 375

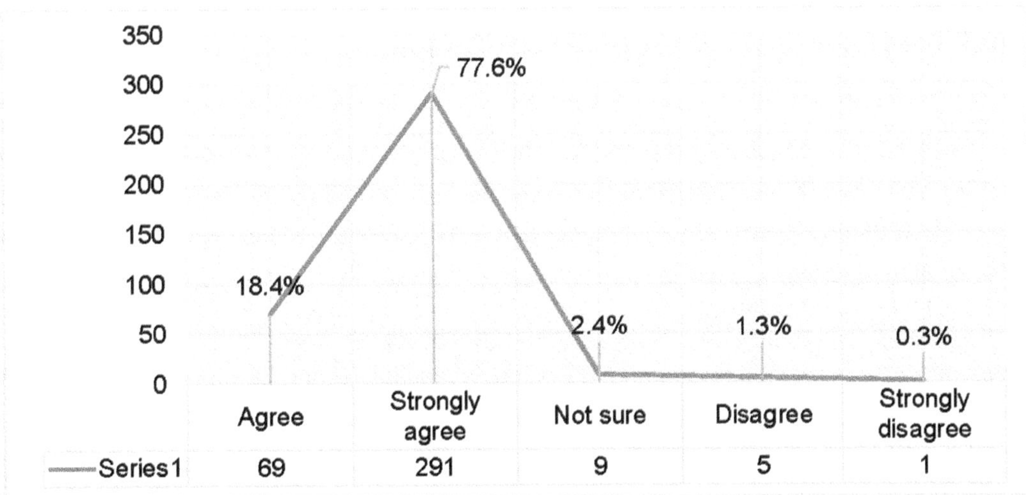

From Figure 6, the results showed that most of the respondents strongly agreed that without water many sectors cannot function (social, economic, political and environmental functions). About 18.4%of the respondents agreed, 2.4% were not sure, 1.3% disagreed, and 0.3% strongly disagreed. In a similar study, (Gumbi, 2017) also discovered that without water many sectors cannot function; therefore, the equitable water supply of sufficient quantity and quality is central to the growth of an economy of a particular geographical area and the improvement of public welfare.

Table 4. Water is an important aspect of socio-economic development

Response categories	Frequency	Percent
Agree	78	20.8
Strongly agree	285	76.0
Not sure	10	2.7
Disagree	2	0.5
Total	375	100

Source: Fieldwork (2019), n = 375

From Table 4, the study showed that most of the respondents (76.0%) strongly agreed that water is an important aspect of socio-economic development. Again, (20.8% agreed, 2.7% were not sure, 0.5% disagreed. The results implied that respondents regard water as a key driver for economic and social development. The findings affirm Grey and Sadoff's (2008) assertion that water has always played a central role in societies by alleviating poverty and ensuring sustainable socio-economic growth. Additionally, economic activities depend upon the availability of water resources, including its quality (Logar, 2018).

Moreover, the survival of living and non-living organisms in the community is inextricably linked to the availability and access to water.

8.6 Problems Hindering Sufficient Water Supply

Limited access to water and sanitation is a risk to health, dignity, and ability to engage in occupations. Water scarcity is today a global phenomenon that affects close to 2.8 billion people worldwide, especially in developing countries (Mbua, 2013). Both the availability and quality of water are crucial. A study conducted by Masuku *et al.*, (2017) found that in South African rural communities, poor infrastructural conditions, including limited storage facilities, transportation and access to new technology pose a threat in achieving socio-economic development for rural households. Hence, the inability to access water and being exposed to poor infrastructural conditions further pose a serious threat to rural areas

Table 5. Poor governance lies at the heart of the world's water crisis

Response categories	Frequency	Percent (%)
Agree	82	21.9
Strongly agree	258	68.8
Not sure	27	7.2
Disagree	6	1.6
Strongly disagree	2	0.5
Total	375	100

Source: Fieldwork (2019), n = 375

From Table 5, the study showed that most of the respondents (68.8%) strongly agreed that poor governance lies at the heart of the world's water crisis. Besides, 21.9% agreed, 7.2% were not sure, 1.6% disagreed, and 0.5% strongly disagreed that poor governance lies at the heart of the world's water crisis. The results implied that without proper management of water resources, current and future generations might be affected in terms of water access. The findings of the study agree with (SIWI, 2017) who states that, it is important for governance to play a critical role in improving water resources management and service delivery. Nevertheless, the water crisis that South Africa is experiencing do not only look at having too little water or clean water to satisfy the human needs, but also talks to managing water in an integrated manner and making it accessible to all at all times regardless of their settings.

9. HYPOTHESIS TEST RESULTS

In this section, the study provides answers and discusses the test outcome for the hypothesis. The null hypothesis (H_0) for the study was that "there is no statistically significant relationship between good governance, human rights, water contamination, women inclusion, agriculture and access to potable water at the UMlalazi Local Municipality" while the alternative hypothesis (H_A) assumed that "there is a statistically significant relationship between good governance, human rights, water contamination, women inclusion, agriculture, and access to potable water at the UMlalazi Local Municipality". The study adopted the chi-square technique in determining the relationships between the variables. As a nonparametric tool, the chi-square is a distribution free technique which did not strictly require normality of the sample distribution. Moreover, since the sample size was reasonably large (n = 375) and the sampling was randomly determined, the study assumed the possibility for sample error will be effectively neutralised. With 95% confidence interval and error margin of (e) = 0.05, the decision rule for the hypothesis test was stated as "accept H_0 if p-value < 0.05 and do not accept H_0 if p-value > or = 0.05".

Table 6. Relationship between good governance, human rights, water contamination, women inclusion, agriculture and access to potable water at the UMlalazi Local Municipality

Variables	Observed (OB) and Expected (EX) scores								n
	Agree		Strongly agree		Not sure		Disagree		
	OB	EX	OB	EX	OB	EX	OB	EX	
Good governance	58	93.8	185	93.8	123	93.8	9	93.8	375
Human rights	86	75.0	265	75.0	19	75.0	4	75.0	375
Water contamination	59	93.8	306	93.8	9	93.8	1	93.8	375
Women inclusion	85	75.0	269	75.0	10	75.0	12	75.0	375
Agriculture	56	93.8	295	93.8	23	93.8	1	93.8	375
Access to potable water	87	93.8	271	93.8	15	93.8	2	93.8	375

a. 0 cells (0.0%) have expected (EX) frequencies less than 5.

In Table 6, the findings showed that the least expected scores was (75.0) and the highest (93.8), meaning none of the expected scores were less than 5, a condition which implied that the analysis did not violate the conditions associated with chi-square test. From Table 7, the finds showed the chi-square values for the variables were; access to potable water (chi-square = 491.549a, df = 3, p-value = 0.00 and n= 375); good governance (chi-square = 188.189a, df =3, p-value = 0.00 and n = 375); human rights to water (chi-square = 664.987b, df = 3, p-value = 0.00 and n = 375), water contamination (chi-square = 661.789a, df = 3, p-value = 0.00 and n = 375), women inclusion in planning and decision making (chi-square = 686.853b, df=3, p-value = 0.00 and n = 375) and agriculture (chi-square = 592.371a, df=3, p-value = 0.00 and n = 375).

The findings indicate statistically significant relationships between the variables. Therefore, from the decision rule on the region of acceptance of the null hypothesis (H_0), since the p-values of the calculated chi-square statistics were less than the critical value of 0.05, the study proceeded to rule out the null hypothesis (H_0) in favour of the alternative hypothesis (H_A). The study concludes based on the hypothesis test outcome that in the UMlalazi Local Municipality, good governance, human rights to water, contamination of water, women inclusion in decision making and agricultural activities in the area are significantly associated with access to potable water and sanitation delivery among the population.

Table 7. Effects of good governance, human rights, water contamination, women inclusion and agriculture on access to potable water in the UMlalazi Local Municipality

Chi-Square Test Statistics						
Variables	**Access to potable water**	**Good governance**	**Human rights**	**Water contamination**	**Women inclusion**	**Agriculture.**
Chi-Square	491.549[a]	188.189[a]	664.987[b]	661.789[a]	686.853[b]	592.371[a]
Df	3	3	3	3	3	3
Asymp. Sig.	0.000	0.000	0.000	0.000	0.000	0.000
N	375	375	375	375	375	375

a. 0 cells (0.0%) have expected frequencies less than 5. The minimum expected cell frequency is 93.8.
b. 0 cells (0.0%) have expected frequencies less than 5. The minimum expected cell frequency is 75.0.

Hence, to improve and sustain drinking water access, the Municipality should strengthen measures to promote good water governance, improve rights of localities to water, control the rate of water pollution, expand access and inclusion of women in planning and streamline activities for ecologically friendly agricultural activities in the communities. The findings of the hypothesis agree with Gumbi (2017) who states that rightful access to water is important to live and is recognized as a fundamental human right. Human lives depend on and demand sufficient and safe water to be always healthy. South African law and policy state that basic water supply must be sufficient, safe, accessible and affordable. Basic water needs to be provided continuously with a stipulated minimum rate of flow and quality to the households. Moreover, contaminated water threatens the social health and physical health of the society and it is dangerous to human life (World Vision, 2010). The United States Environmental Protection Agency (2013), states that water is an important commodity for human life and all people on the planet depend upon it to survive.

10. CONCLUSION AND RECOMMENDATIONS

The study concludes that there is a relationship between good governance, human rights, water contamination, agriculture and access to water. It can be presumed that access or scarcity of water has the capacity to affect good governance, human rights to water, and agricultural purposes. The results mean that access or scarcity of water could pose a serious threat on people livelihoods, public health risk and development of the community. Moreover, this could result in widespread consumption of polluted surface water that could expose the population to water born disease. The evidence from the findings confirmed that some residents still suffer from water scarcity, as a result of poor governance.

The study concludes based on the hypothesis test outcome that in the UMlalazi Local Municipality, good governance, human rights to water, contamination of water, women inclusion in decision making and agricultural activities in the area are significantly associated with access to potable water and sanitation delivery among the population.

Recommendations include the following:

- The study suggests that each household be provided with metered water inside their homes and tanks and taps to avoid long distances to fetch water.
- Secure a sustainable funding source and a stronger mandate for coordinating intergovernmental organizations.
- Promote greater collaboration among organizations engaged in water governance to build understanding and coordinate action.
- Explore and develop guidelines and principles to govern non-governmental processes, thereby ensuring that they are legitimate, accountable, and have relationships with existing public governance structures.
- The Municipality should strengthen measures to promote good water governance, improve rights of localities to water, control the rate of water pollution, expand access and inclusion of women in planning and streamline activities for ecologically friendly agricultural activities in the communities to improve and sustain drinking water access

ACKNOWLEDGMENT

The authors would like to express gratitude to study particiants, the KwaZulu Natal Department of Cooperative Governance and Traditional Affairs (COGTA), the Management Committee of the Zululand District and the late Mayor of uMlalazi Local Municipality, His Worship Councillor, Thelumoya Zulu for the insightful comments they made when we disseminated findings to key statekolders. In both COGTA and ZDM, we express our sincere gratitude to Dr Gladys Makolomakwe, Ms Nokubonga Caluza and Mr Phindithemba Manqele for organising the meetings and affording us an opportunity to present the study findings. At UNIZULU, we express gratitude to the Community Engagement Office, and the Communication and Marketing Department especially Ms Mammusa Lekoa and Ms Precious Shamase for attending and documenting these community engagement online workshops.

REFERENCES

Ashton, P. J., Hardwick, D., & Breen, C. M. (2008). Changes in water availability and demand within South Africa's shared river basins as determinants of regional social-ecological resilience. In M. J. Burns & A. B. V. Weaver (Eds.), *Advancing Sustainability Science in South Africa*. African Sun Media.

Castro, J. E. (1992). *El conflicto por el agua en México. Los casos de Tuxtla Gutiérrez, Chiapas y Ciudad Juárez, Chihuahua, 1986-1991* (Unpublished thesis). Latin American Faculty of Social Sciences (FLACSO).

Costanza, R. (1989). *What is Ecological Economics?* Elsevier Science Publications. doi:10.1016/0921-8009(89)90020-7

Creswell, J. W. (2015). *Research design: Qualitative, quantitative, and mixed method approaches*. Sage publications.

Crow, B., & Sultana, F. (2002). Gender, Class, and Access to Water: Three Cases in a Poor and Crowded Delta. *Society & Natural Resources*, *15*(8), 8. doi:10.1080/08941920290069308

DEAT (Department of Environmental Affairs and Tourism). (2004). *A National Climate Change Response Strategy for South Africa*. Author.

Goforth, C. (2015). *Using and Interpreting Cronbach's Alpha*. https://data.library.virginia.edu/using-and-interpreting-cronbachs-alpha

Goswami, K. B., & Bisht, D. (2017). The Role of Water Resources in Socio-Economic Development. *International Journal for Research in Applied Science and Engineering Technology*, *5*(12), 1669–1674.

Grey, D., & Sadoff, C. (2008). *Water for Growth and Development: A Theme Document of the 4th World Water Forum*. Commission National del Agua Publications.

Gumbi, N. (2017). *Challenges of water supply management* (Doctoral dissertation). University of Limpopo.

International Water Management Institute (IWMI). (2007). *Water for Food, Water for Life: A Comprehensive Assessment of Water Management in Agriculture*. Earthscan and International Water Management Institute.

Logar, I. (2018). *Environmental Economics (EnvEco)*. https://www.eawag.ch/en/department/es/main-focus/environment-economics-enveco

Machethe, E. M. (2011). *The causes and impact of water shortage on the households of Ga-Kgapane Township in the Limpopo Province (Research dissertation)*. University of Limpopo.

Masuku, M., Selepe, M., & Ngcobo, N. (2017). Status of Household Food Security in Rural Areas at uThungulu District, Kwa-Zulu Natal, South Africa. *African Journal of Hospitality, Tourism and Leisure*, *6*(2), 19–38.

Mbua, L. (2013). Water supply in Buea, Cameroon: Analysis and the possibility of rainwater harvesting to stabilize the water demand. University of Germany.

Motloung, S. (2010). SIBU and the crisis of water service delivery in Sannieshof, North West Province. *TD: The Journal for Trans disciplinary Research in Southern Africa, 6*(1), 25-56.

Muller, M, (n.d.). *Inter-basin Water Sharing to Achieve Water Security- A South African Perspective*. Department of Water Affairs and Forestry

Nxumalo, Z. F. (2007). *The role of continuous assessment in primary school* (Doctoral dissertation).

Pilot, Van Teijlingen, & Hundley. (2001). Research methodology. *Journal of Quantitative Approach, 12*(2), 12-17.

Republic of South Africa. (1996). *Constitution of the Republic of South Africa*. Republic of South Africa.

Rogers, P., & Hall, A. (2003). *Effective water governance*. Global Water Partnership. http://www.gwp.org/global/toolbox/publications/ background papers effective water governance.pdf

Romeiro, A. R. (2012). Sustainable development: An ecological economics perspective. *Estudos Avançados*, *26*(74), 78–99.

Shi, F. (2015). *Study on a Stratified Sampling Investigation Method for Resident Travel and the Sampling Rate. Europe*. Hindawi Publications.

Singh, N., Åström, K., Hydén, H., & Wickenberg, P. (2008). Gender And Water From A Human Rights Perspective: The Role Of Context In Translating International Norms Into Local Action. *Rural Society*, *18*(3), 185–193. doi:10.5172/rsj.351.18.3.185

Stats, S.A. (2011). *Formal Census*. Statistics South Africa.

Stockholm International Water Institute (SIWI). (2017). *Governance of water and waste: A key to sustainable development?* https://www.worldwaterweek.org/event/7450-governance-of-water-and-waste-a-key-to-sustainable-development

Thompson, H. (2006). *Water law: A practical approach to resource management & the provision of services*. Juta & Co Ltd.

United Nations. (2005). *Valuing water resources. Module 10*. http://www.cap.net.org/node/1551

United Nations Development Programme. (2007a). *User's guide on assessing water governance*. UNDP.

Van Teijlingen, E.R., & Hundley, V. (2001). *The importance of pilot studies*. Academic Press.

Vogelstein, R. (2017). *Women around the World*. https://www.cfr.org/blog/women-around-world-week-75

Zululand District Municipality (ZDM). (2017). *Indigent Policy Status: Final (2017/18)*. http://www.zululand.org.za/media/6137/Indigent%20Policy%20-%20Final.pdf

Chapter 9
Durban Community's Experiences of Class-Based Urinary Diversion Toilets

Innocent Simphiwe Nojiyeza
https://orcid.org/0000-0002-9948-2147
University of Zululand, South Africa

ABSTRACT

The aim of this chapter is to look at the integrated water resources management (IWRM)-related institutional shift in governance of water and sanitation in Durban and the extent to which class is considered in providing rural inhabitants with facilities they regard as sub-standard. The key finding is that rural communities remain uncomfortable with nearly 90,000 UD toilets constructed by eThekwini Municipality, despite current efforts being made to relieve them of the most hated chores of emptying vaults when they fill up. In a study conducted in rural areas of Durban through interviews with 70 households and focus group discussions, communities continue to reject UDs on the grounds of lack of equity, proximity to areas with flush toilets, lack of communication, and lack of sustainable institutional mechanisms associated with this project.

INTRODUCTION

Promoters of IWRM view decentralisation of water and sanitation services to local government (Mehta et al, 2016) and related cost saving mechanisms (Sutherland et al, 2014 and van Koppen, et al, 2016) as innovative approaches (Sutherland et al, 2014) to provision of sanitation solutions. A geographic scale which is socially reproduced and has class connotations is often overlooked in IWRM related multi-scaling and rescaling of sanitation provision which may result in contradictions and conflicts on hydrological boundaries (Mehta et al, 2016, Pelling, 1999, Walker, 2005). In Durban the provision of dry sanitation to poor rural households which manifested in spatial differentiation whereby households residing beyond the waterborne sanitation belt (Sutherland, et al, 2014) were served with facilities different from those given to their urban counterparts residing within the waterborne sanitation belt. In

DOI: 10.4018/978-1-7998-8809-3.ch009

a country which experienced centuries of colonialism and apartheid where different race groups were served with different levels of amenities, the dawn of democracy in 1994 saw the continuation of further discrimination with a class connotation (Swyngedouw and Heynen, 2003).

The dominant narratives (Bryant and Goodman, 2004) of decentralisation promoted by the National Water Act of 1998 (Bourblanc, 2012, Bourblanc, 2015) which is based on the Dublin principles of IWRM (Mehta et al, 2016) together with the intention of the Durban local state to save costs of rolling out water infrastructure to previously under-served communities (Swyngedouw and Heynen, 2003, Escobar, 1998) is responsible for sanitation conflicts and contradictions (Walker, 2003, Hellberg, 2014).

These contradictions are reflected in two sides of the Durban rural sanitation story. The one side of the story is told by households that are unhappy with urine diversion (UD) toilets, and the other side of the story is told by the councillors who authorised the implementation and the municipal officials that actually carried out the implementation of such a controversial and dialectical project. The latter are supported by the fact that eThekwini Water and Sanitation received many water and sanitation awards for being one of the best performing municipalities and not only in South Africa, but the whole of Africa. As early as 2002 eThekwini Water and Sanitation Department was bestowed a National Geographic Award for Sanitation, followed by the 2003 Dubai International Award for Best Practices (for sewage disposal education), and in the same year also awarded the 2003 South African National 'Excellence in Innovation' Award, and the Ford Foundation's 2003 Impumelelo Award. EThekwini Water and Sanitation was and continue to be one of the recipients of the 'best municipal delivery award' and 'Blue Drop' awards from the regulator, the Department of Water Affairs. One of the sponsors of eThekwini Water and Sanitation, Bill and Melinda Gates Foundation (2010) wrote in a blog, that Durban's Neil Macleod 'has been a leader in thinking through how to improve sanitation for the poor in Durban'.

Whilst communities and councillors continue to be unhappy, eThekwini Municipality's Water and Sanitation Department continue to get accolades from international organisations. The latest award being the 2014 Stockholm Industry Water Award which indicates that the world at large have high regards for eThekwini Water and Sanitation. Councillors, who authorised the introduction of UD toilets, find themselves under immense pressure from communities to oppose the scaling up of the project whilst officials on the other hand continue with the implementation. Officials did, however, change critical features of the UD system, that show the limits of the IWRM full cost-recovery approach: one crucial U-turn was in May 2014 when the municipal council approved a $3.5 million project in which city employees would empty the UD containers (instead of that being the household's responsibility). How can the recipients, councillors and officials have such different positions regarding 90 000 toilets built around the municipality?

A related question concerns the class nature of the sanitation delivery in rural areas which is perceived by households as discriminatory on the basis of class as peasants rather than racial apartheid which was practiced by many National Party Administrations since 1948. To these households, racial-based apartheid has subsequently been replaced by class apartheid and geographical discrimination, through which petit bourgeoisie of all races continues to flush their toilets and forget whilst their peasant counterparts continue to face their human waste in their daily lives. The 'sanitation belt' within the Durban municipality is the geographical manifestation of the class-biased sanitation policy.

Durban is a case study of sanitation implementation that is funded directly by rate payers and not the international agencies. The local government and not the Department of Water and Sanitation re-established in 2014, Department of Human Settlements (responsible for sanitation since 2009) and KZN Department of Cooperative Governance and Traditional Affairs, is responsible for the conceptualisation and implementation of UD toilet project. Durban is a case study of a water and sanitation authority that is managing water and sanitation on the basis of a fully-fledged local authority in line with IWRM principle of decentralisation. Durban is also a case study of a local authority with democratically elected councillors who played a role in the choice of the sanitation technology and crafting of the institutional mechanisms that influenced the roll-out of the project. The Durban sanitation story is a good example of what technocrats that met in Dublin in 1992 did not anticipate; a municipality providing a subsidy to all its households, but discrimination based on class resulted in the project being rejected. In this article, we will present an overview of the UD toilet technology, pointing out how it contributed to class antagonism within the city. I also argue that lack of equity, poor communication, institutional focus and closeness of UD toilets to areas provided with other technologies and lack of responsiveness to local issues contributed to the rejection of UD toilets by households.

URINE DIVERSION TECHNOLOGY AND CLASS ANTAGONISM

With the extension of its boundaries in 2000, eThekwini Water was faced with the challenge of rolling out sanitation to households beyond the waterborne edge (Sutherland, et al, 2014), as well as households that showed a strong dislike of Ventilated Improved Pit Latrines (VIPs). Facing budget constraints and keeping the cost and future scarcity of water in mind, eThekwini Water and Sanitation did not consider low-flush septic tanks or biogas digester as an option (Hawksworth, et al, 2006). eThekwini Water and Sanitation embraced UD toilets as an 'affordable option' that showed a vision toward the future. Flush toilets, with full sewerage service, along with flush toilets leading to household septic tanks are reserved for higher-income groups, with VIPs and UDs predominantly served to peri-urban households and rural peasants. UDs are a type of dry sanitation that diverts the urine into the sandy soil, leaving the solid waste to dry out and decompose into human compost. Ash, dry soil or sawdust is sprinkled over the faeces after defecation by members of the rural households. The ash absorbs the moisture and also controls odours and flies. The toilets are built to be rotated between two chambers, once one is filled, the other is used and the first decomposes. The Durban UDs were not designed to harvest urine which is soaked way in the yard or garden (Austin, 2006: 1, Duncker, Matsebe and Austin, 2006:5, Lutchiminarayan, 2007:15, Hawskworth, et. al., 2006).

The advantages of the UD toilets according to those that promote them and disadvantages according to those that are either critical or opposed is that:

- The entire structure can be built above ground, there is thus no need for expensive digging and lining of pits;
- Urine is diverted, no water is used for flushing; and
- The volume of the processing vault is fairly small, as it is emptied periodically (Esrey, et al, 1998, Duncker, Matsebe and Austin, 2006:5).

The Water Research Commission conducted a study in 2006 that raised the issue of UD acceptance as well as building standards used. In a study conducted by Duncker, Matsebe and Austin (2006) in Northern Cape, Eastern Cape, North West and KwaZulu Natal (Umnini and eHlanzeni) the sources of dissatisfaction were highlighted:

The UD was not the only toilet found in the yards of the respondents. Some households also used pit toilets and flush toilets. Some people who once experienced the use of flush toilets will under no circumstances accept UDs, with all the inconvenience associated with it. Pit toilets were still in use by especially children under the age of ten and by those households that did not use the UDs. Children under the age of 10 did not use UDs because the seat was usually too high, with fears expressed that they might fall into the vault. Proponents of IWRM and Community Led Total Sanitation advocate for consultation with communities for any type of technology to be accepted. In the case of UDs, no matter eThekwini officials' claims to the contrary, many people were ill informed about the device. This affected their perceptions regarding the handling of human waste, as the vaults were designed to be emptied by households. The perceptions of people around handling of human excreta are negative, in no small part because of a very high disease burden, including Durban's status as the world's city with the highest HIV positive population, and hence a high diarrhoea prevalence. The difficulty in keeping excrement dry, with high levels of diarrhoea and a very humid environment from November through March meant that many people resisted the task of emptying the UD's vaults. In most of Durban, the vaults were shallow, requiring regular emptying. Instructions to dig holes and plant fruit trees were not well communicated, or accepted.

Even in wealthy Durban, UDs were chosen because the full operating and maintenance costs are borne by households and no longer the municipality. Yet it soon became evident that removing and replacing the covering slab, when emptying the vault or unsealing the covering slab, are difficult tasks. If households wished to get additional support regarding equipment failure e.g. doors which regularly needed replacing, it was difficult to locate the community-based sanitation committees that could assist in replacing slabs, doors, vaults and other spare parts. Though such sanitation committees were encouraged by the municipality and allied NGOs such as the Mvula Trust, they did not survive for a long time. Moreover, in a bid to cut operating costs, environmental health officers and community health workers were only considered as part of the project during the construction phase. Another complicating factor was that in many cases, the UDs construction was not executed properly. For all these reasons, a substantial proportion of households in some of the case communities decided not to use or maintain the toilet. Some toilets were not used as toilets at all but as storage places for building material or other valuables because the toilet was the only place that could be locked. Some toilets were also used as animal pens or hen houses.

Even in previous studies conducted, it was established by Roma, et al, (2013:307) that out of 17448 households surveyed in Durban 1465 (8.4%) households converted their UDs into flush toilets. Those that did not afford to convert their UDs into flush toilets continued to use their un-improved pit latrines. Roma et al (2013:307) found 2243 (14%) of Durban households that chose to use un-improved pit latrines instead of using the 'ecologically friendly' UDs. Despite concerns raised by various households in Durban, 12760 (80%) households in Cliffdale, Masuku, Mnamatha and Ngonweni easily adopted the UDs because they were not close to any community that use other types of sanitation facilities. Whilst other communities convert their UDs into flush toilets, there are those (1117or 7%) in areas such as Ngonoma, Wood Glen Rural, Mahlabathini, Msunduzi, Bux Farm and Denge who never used UDs, despite being allocated to them (Roma, et al, 2013:308).

This set of problems is not unique to Durban. In sites across South Africa, Duncker (2005:2) recorded a variety of similar problems: poor design and construction, lack of community participation in project design, resistance to handling human waste, the perception that urine is harmful to plants, the sense that UD toilets are a sub-standard technology compared to flush toilets, and the designers' failure to take into account the different sanitation needs of men and women.

Can the UD system be reformed? Proponents of UD toilet technology (Guness, et al, 2004:4, Mnguni, et al, 2008:1) acknowledge that buried UD waste has a potential negative impact on the quality of soil and groundwater pollution but indicated that such effects could be mitigated through secondary treatment and sanitisation of human waste before any reuse is recommended. Hawsworth, et al., (2006) argue that introduction of UD toilets reduces the chance of either surface or groundwater becoming polluted. There are also new technical fixes being developed to address odours, especially when faecal matter does not dry properly. Austin (2006:8) suggests a process of augmenting vaults to ensure one storage facility keeps faeces for a period of 12 months to allow pathogens to die off in order to be safe for handling by households, an advance on the drying of faecal material in a pit during the filling phase (Brouckaert, Ridgway and Buckley, 2004). This is one of the standing concerns of households regarding the safety of handling waste when they empty toilets when both vaults are full. Although garden rakes, spades and gloves are provided, the danger of being contaminated with pathogens is always an issue (Austin, 2006:1, Mnguni et. al, 2008:1).

Since the early stages of UD experimentation, grievances about UDs remain unresolved, the toilets have become increasingly controversial, with strong supporters and strong resistance. Lutchminarayan (2007) conducted a study of UDs in rural areas including eMzinyathi, Mtamuntengayo, and Sawpits, as well as areas in which UD toilets were not installed, such as Ogunjini, Adams Mission and Bux Farm.

- Only 16% of households with UD toilets never used them, whilst 77% of households stated that their UDs did not smell, 82% reporting no flies, 79% reporting that toilet pedestals are always clean (Lutchminarayan, 2007:68).
  - In the same study 23.6% of households surveyed at eMzinyathi reported that their UDs smell, with 23% of Mtamuntengayo and Sawpits reporting similar grievances (Lutchminarayan, 2007:70).
  - At eMzinyathi 1% of households have flush toilets with sewerage facilities and, 2.6% flush with septic tanks, with 94.4% having no flush capacity: 3.9% ventilated improved pit latrines (VIP); 25.2% pit latrines without ventilation; 3.2% open defecation: and 64.5% UDs. At Sawpits, 1.1% flush with sewerage facilities, 1.7 have septic tanks, 52.2% of households use UDs, 8.9% use VIPs, 25.6% use pit latrines without ventilation, 1.1% use buckets and 8.9% openly defecate (Lutchminarayan, 2007:71).
  - At Mntamuntengayo, the figures are: 48.9% UDs, 1% VIPs, 9.4% pit latrines without ventilation, 4.1% open defecation, and nobody surveyed had either sewerage facilities or septic tanks.
  - At Ogunjini 2.2% flushed with sewerage facilities and, 25.3% with septic tanks, while 50.6% had pit latrines without ventilation, 1.1% buckets, 20% open defecation (UDs were not yet installed) (Lutchminarayan, 2007:71).
  - At Bux Farm, 1.2% of households had septic tanks, 8.4% had VIPs, 45.8% had pit latrines without ventilation, 44.6% openly defecated and no UDs were installed.
  - At Adams Mission, 11.5% of households had septic tanks, 65.6% pit latrines without ventilation, 6.4% with buckets, 16.6% openly defecating and no one was served with UDs (Lutchminarayan, 2007:68).

Utilisation rates varied. In Mtamuntengayo 55% of the households with children under the age of 12 years used their UDs, whilst 41% resorted to open defecation. Children under the age of 5 and those under the age of 15 have a high risk of contracting intestinal infections and helminth infestations, including cholera, typhoid and paratyphoid fevers, dysentery, hookworm, schistosomiasis and filariasis, because of their weak immune system (Mnguni, et. al, 2008:2). This finding contrast earlier findings by Ducker, Matsebe and Austin (2006:98) that most children at eHlanzeni and eMnini did not use UDs. In both studies, very few households reported that they participated in hygiene awareness programmes (Lutchminarayan, 2007:70, Duncker, Matsebe and Austin, 2006:97). Duncker, Matsebe and Austin (2006:97) concluded that training focused on the operation and maintenance of the UDs but without reference to health and hygiene promotion. Such training would have included washing hands with soap after using a toilet, before cooking, when handling food, before eating and feeding babies and availability of hand washing facilities. Ironically, Lutchminarayan (2007:81) concluded that areas with UDs show better hygiene standards of toilets and hygiene practices than those without.

Hawksworth, et al, (2008) tested 124 households at Mtamuntengayo and Sawpits and concluded that 90% of the households tested positive in ascaris and protozoan infections: these infections were suspected to be caused by open defecation in areas that are served with UDs. But, those who used their UDs were also still vulnerable to protozoan and ascaris infection. Hawksworth et al., (2008) concluded that the high level of infected households in rural areas is a cause for concern. Moilwa and Wilkinson (2006) concluded in a study conducted at Augrabies, Northern Cape that although health and hygiene promotion form part of water and sanitation projects, very little behavioural change is observed after the completion of the project cycle. Continuation of health and hygiene messages beyond the construction phase and vault-emptying phase could assist in alleviating low hygiene behavioural pattern such as washing hands with soap after emptying a UD, washing hands with soap after defecation and before handling food (Moilwa and Wikinson, 2006:6).

In a study conducted at Zwelibomvu in Durban West (Mnguni, et. al., 2008) it was established that no hand washing facilities were provided, 42% of the toilets observed had flies, 84% of the waste burial sites were accessible to humans and had a very high risk of human exposure to residual faecal matter. Moilwa and Wilkinson (2006:4) found that two thirds of the households washed their hands with soap only, 28% washed their hands with both water and soap and 3% washed hands with water and soil after emptying toilet vaults. Health and hygiene practices were never applied after an intensive programme of promoting such an initiative. Mnguni, et al., (2008) recommended that there is a need for provision of more health and hygiene education focussing on handling waste during excavation of UD vaults and burial of such in a pit that is about 250cm followed by planting a tree as a sign of identification. Guness, et al, (2008) concluded that health related bacteria have a potential to re-grow either during storage or after the burial of human waste.

In 2014, eThekwini Water and Sanitation agreed to re-visit some of the outstanding concerns of the communities regarding vault emptying. The municipality agreed to empty all UDs in Durban, but poor workmanship was still not going to be addressed except in pilot houses chosen in 2012. The focus on emptying UDs will only be on those households that used their facilities. The decision was taken after studies conducted by both the municipality and UKZN revealed that acceptance of the UDs could improve if emptying facilities can be provided and paid for by the municipality (Roma et al, 2013). In December 2012 uMzinyathi, eSikebheni and Lahlumlenze areas were identified as pilot areas chosen to repair doors and other areas in which there is evidence of poor workmanship. In houses that were occupied by people with disabilities, the municipality will provide facilities that are big enough for people using wheel chairs to utilise. Some of the households identified in this study were going to be beneficiaries of the repairs. All 30 households that were chosen to participate in the pilot phase of refurbishing toilets benefitted through broken doors being refurbished, new vaults being installed and faecal matter being removed from both vaults (Interview with community member and council staff on 20 October 2012). Schools that initially used UDs and VIPs were upgraded to flush toilets linked to septic tanks as part of the project funded by the Department of Human Settlements of which eThekwini Water is a service provider for KZN province (Interview with a council official on 20 October 2013).

By June 2014 the emptying of all UDs by the municipal staff or companies contracted by the municipality in a similar manner as the VIPs commenced after the approval of such an initiative was granted by eThekwini Human Settlement Committee chaired by Councillor Nigel Gumede at the end of May 2014. eThekwini Municipality confirmed in their recent media statement after council committees gave approval of this move that:

"About R39 million has been set aside to empty 90 000 UDs that could pose a health risk for Durban residents if left full. The city will implement a two-year programme to empty these specialised toilets in a safe and controlled way. The toilets operate without water and have a divider so that the user, with little effort, can divert the urine away from the faeces. The city's head of water and sanitation, Neil Macleod, in a report tabled before the human settlement and infrastructure committee yesterday, said the approach thus far had been that the rural householder was responsible for emptying the toilets. Macleod said in peri-urban and some urban areas, the city had recently provided a pit emptying service to residents, with various types of toilets. "This has led to conflict with the original policy of rural homeowners being responsible for their UDs," he said. According to Macleod, research had shown that although rural residents were happy to use the toilets, there was dissatisfaction about having to empty them. "Research has also shown some degree of health risk in the emptying process," he said. The programme will provide free emptying services to alleviate health risks and the burden on householders. "Some residents in these communities are becoming more affluent and are installing higher levels of service in their houses," said Macleod. This would result in a gradual decrease in the number of UDs as residents no longer qualified for the free service (Mpume Madlala, Natal Mercury, 04 June 2014)"

This is regarded as the step in the right direction in terms of addressing some of the concerns of the communities although emptying is a drop in an ocean compared to other grievances dealt with in the next sections of this article.

METHODOLOGY

In Durban, 70 participants were drawn from various rural areas. The study adopted qualitative research in seven wards located in South, North and West Durban where UDs were built. The study was conducted at KwaNgcolosi and KwaDinabakubo sections of Ward 9, Embo and KwaNgcolosi areas of Ward 8, Umnini, Magcino, eHlanzeni and Danganya areas of Wards 98 and 99 respectively, Zamani B Section of Ward 91, Mpushini section of Ward 100, Ngonweni section of Ward 3 (see the map below). These areas are very close to other wards where UDs were built, so they were likely to share common concerns. Interviews were conducted with councillors, project facilitators, members of various technical management committees, members of various project steering committees, ward committees as well as at least 10 households selected randomly from each of the above mentioned areas. Interviews were guided by a wide range of questions and discussion was open ended to avoid bias and allow informants to give shape to the information provided. The spirit of enquiry that informed this research was one of wanting people to have a sustainable sanitation that meets their needs.

Figure 1. Map of Rural Durban (Source: Ethekwini Municipality, 2010

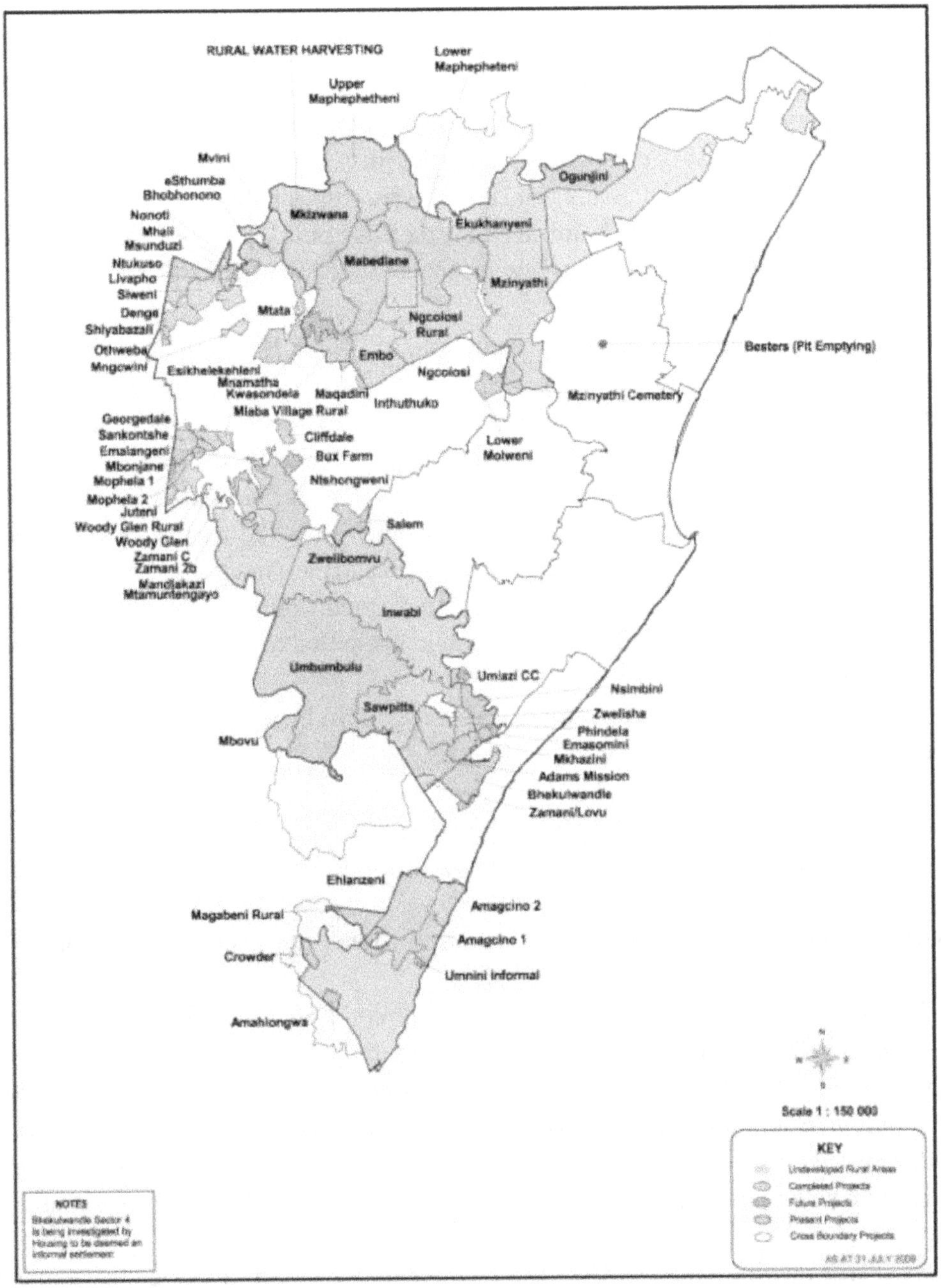

FINDINGS

This article will review four issues that arose repeatedly in interviews, namely issues of equity, communication, sensitivity to local contexts, and institutional focus.

Lack of Equity Associated With UD Roll-Out

Many of the areas where UDs were built are in close proximity, sometimes literally next door, to other areas with flush toilets. Hawksworth, et al., (2006) highlighted that more affluent communities have waterborne sewerage perceived as a prestigious system that all communities aspire to have. In some cases the areas with flush toilets were RDP houses and others were well-established middle class households. At KwaNgcolosi in various sections of Wards 8 and 9, there are public facilities such as schools, clinics and a traditional court that have flush toilets and those that have VIPs. The trend at KwaNgcolosi is as follows:

Bongimpilo Creche situated at Mshazi, does not have a UD, but is served through VIPs built by an NGO called Valley Trust that operated in this area since the dawn of democracy in South Africa. It was established that VIPs were no longer used, since the institution was provided with flush toilets by donor agencies. Ndlokolo Primary School, and Hlahlindlela High School utilise flush toilets and the Valley Trust toilets are either not utilised at all or only utilised if there are mechanical water cut-offs in the area. Sihlangusabasha and Labora Primary Schools are still using Valley Trust VIPs for both educators and pupils. The other public spaces visited were the Local Clinic and the Traditional Court. Both institutions are having fully functional flush toilets. eKuphikeni Traditional Court which operated in the 80s as a Department of Home Affairs offices was the first public institution to have improved sanitation facilities, followed by the, local clinic.

Mrs N, a community participant from eMshazi, KwaNgcolosi laments that eThekwini Municipality is not taking them seriously as residents:

"They could have asked us to dig our own pits that are deep like the Valley Trust toilets. We are provided with spades, rakes and shovels that we could use to empty toilets once both the chambers are full".

Her arguments are supported by Austin (2006:1) who wrote extensively about low level of UD acceptance in most parts of South Africa.

"Emptying toilets is not acceptable, that is why we have not done so. We don't want these toilets, and we are wondering why they built them. We have water in the dam (referring to Inanda dam whose construction in 1984 led to the displacement of 1080 households from Maphephetheni, KwaNgcolosi, KwaNyuswa and Amaqadi traditional areas) and why are we not using water from the dam to flush. Our schools and traditional court have flush toilets and why we do not have them." These are some of the questions that remain un-answered by some of the many members of 70 households that participated in the study.

The views shared by Mrs N and her Ngcolosi community were also supported by Councillor TC, a former DA Ward Councillor responsible for Ward 8 and also former eThekwini Member of the Executive Committee, which includes areas such as Ngcolosi, Waterfall, Hill Crest, Kwanogwaza, Molweni, Langerfontein and Wushwini. He acknowledged that people resisted the introduction of UDs and there is still a challenge in accepting them. Putting human waste in their gardens, selling of urine, empting toilets when both chambers are full are some of the issues that people raise consistently. The unhappiness in the community is also highlighted in studies conducted in other parts of South Africa by Austin, (2006:4). TC stated that:

"When UDs were proposed, there was a lot of resistance, people were unhappy with putting waste in their gardens. The reason for that is that it is not socially acceptable to put human waste in the garden. The other reason for resistance is that rural communities were annoyed by the fact that they are expected to accept second best facilities compared to their urban counterparts. People base their argument on the fact that If flush toilets are provided in urban areas, why not in rural areas?"

He further stated that UDs are not dignified and there is enough water in Durban for everyone to flush.

"People want dignity when it comes to toilets and people are not lazy to build their own toilets. If the city can provide people with material, they can build their own toilets. There is enough water to flush toilets in our city. People can flush using their tanks in their homes, e.g. one flush after everybody urinated could save water. For 45 years people were not given anything, people feel they are entitled to get something from the government."

Ngcolosi, Umnini, Amaqadi and Embo residents reject the proposed move by the city to purchase urine that could be used as fertilizer. They are happy with the use of cow dung and found this move as an insult because they are black, Zulu and poor they are subjected to conditions where they could even sell their urine for a living. Facilitators from Kwandeni, Wood Glen and Zamani also rejected the sale of urine because there are no facilities to harvest urine and it is socially unacceptable, making communities to accept UDs was difficult and to add harvesting of urine in the equation could result in facilitators being killed in most communities. Umbumbulu community cited no problems in the harvesting and sale of urine, but raised a concern about failure of the City to cater for people with disabilities, especially those that are wheel chair bound.

A similar trend is also observed at Embo which is situated next to an affluent White suburb of Botha's Hill where most of the residents have flush toilets whereas Embo residents have UDs. At KwaDinabakubo there are households with RDP houses in which flush toilets are available, especially in Upper Molweni which includes suburbs and Lower Molweni where there are either UDs, VIPs or traditional latrines. Molweni is situated next to Waterfall, Cresthouse, Valley of Thousand Hills and Hill Crest which are affluent White suburbs with flush toilets and modern day amenities. Other areas that constitute Ward 9 are Wushwini, Umgababa, Lower Langerfontein, Madimeni, Roman Catholic 1 and 2 and KwaNgcolosi which are mainly serviced with UDs. In the South Coast, the residents of Magcino and Magabheni Township enjoy flush toilets whereas other rural areas within Luthuli Traditional Authority such as Danganya, eHlanzeni and Umnini have UDs. The local schools such as Mnganuwakhe Secondary, Power Primary and others are served with flush toilets and this implies that the infrastructure to build flush toilets for everyone is available and it is the political will to do so that lacks. The Municipality argued that the reason to build UDs was the fact that such areas were beyond the 'waterborne sanitation edge' where technically, it is difficult to install sewerage facilities, but the findings of the study point out that the 'waterborne sanitation edge' is not only artificial, but human induced class antagonism.

In Ward 59, there is also a big divide between the services rendered to Amaotana and Etiyeni areas that still have UDs and their neighbours at Trennace Park that are served with flush toilets. Councillor

TG noted that a toilet war has degenerated into a Zulu-Indian war because the predominantly Indian areas of Phoenix and Trennace Park are provided with flush toilets whereas the predominantly Zulu areas of eTiyeni and Amaotana are provided with UDs. The other areas such as Ohlange and Bhambayi have flush toilets within their premises included as part of the RDP houses. In the Durban West Region, the provision of UDs in areas such as Zamani, Wood Glen and Kwandeni which are situated next to Mpumalanga Township which is served with flush toilets precipitated resistance from communities in accepting them. What complicated matters is that certain sections of Zamani, i.e. IA, 2A were upgraded and provided with flush toilets whilst Sections 2B and 2C remains with UDs which prove that the 'waterborne sanitation edge' has class connotation amongst the black bourgeoisie.

The proximity of Mpushini section of eMbumbulu to areas such as Adams Mission and eFolweni where communities enjoy flush toilets is often cited as causes of great concern by communities. There is a general feeling that UD construction is not sensitive to the needs of people with disabilities, elderly, children, women, and therefore lack equity. Former Councillor M from Sankontshe area near Mpumalanga Township had an experience to share about how UD construction discriminates against the elderly.

"I have a UD in my household at Sankotshe area of Ward 5, it is not user friendly especially during the wet seasons therefore my family resorted to use our previous traditional pit latrine. Factors that motivated us not to use the UDs is that it is not suitable for old people who may not have the energy to climb on the stairs and it is not suitable for people with disabilities using wheel chairs. My parents are both wheel-chair-bound and they are unable to climb the stairs each time they visit the toilet. There should always be somebody that goes with them each time they visit the toilet. As communities, we were not given choices about which facilities we preferred, and UDs were imposed".

Facilitators that were responsible for health and hygiene education in the nearby Zamani agree with Councillor M that *"people with disabilities were not catered for, at Wood Glen for example, to the extent that toilets with stairs were constructed in houses with people living with disabilities. In certain areas this was avoided."* Some of the perceptions that users of UDs grapple with is that they are sub-standard, poorly constructed, gender insensitive, their construction lacks participatory and community acceptance (Duncker, 2009). This contrasts findings by Moilwa and Wilkinson (2006:5) that vault emptying does not have a stereotypical gender bias and both sexes are exposed to microbial risk from vault emptying.

Another Councillor M who resides in Ward 98 and has a traditional home at uMbumbulu cited her inability to use a UD because they are not designed for females.

"When trying to use a UD in my Umbumbulu rural home, I could not because the urinal is designed for men and not for us females. It is situated very high and only a standing man is able to use it."

Mrs. NDN, the official of Lizalise Palliate Care Centre situated at Mpushini area of Umbumbulu, highlighted the plight of people with disabilities as follows:

"My husband who is wheel-chair-bound is not using a UD, because it is small and stairs were constructed and it is not possible to drive in a toilet using a wheel chair. When I requested a bigger toilet they simple built another one without stairs and the size is still the same and the wheel chair could still not be driven in and out of the toilet. Children are also unable to use it as the urinals are situated far up. There are flush toilets at Mpushini Primary School which is few metres away from the Centre. People

that complained about poorly constructed toilets were insulted by construction workers. If you complain, they would swear at you. People with disabilities were called by names when they raised concerns about the type of sanitation facilities that were constructed and this promoted discrimination. For us it was a quadruple discrimination, we live with disabilities, we are Zulu, we are poor and we live in rural areas. Even people that live with disabilities that frequently visit the Palliate Care Centre find it difficult to spend time here because we do not have sanitation facilities that are suitable to their disability situation".

Former Project Steering Committee member for Wards 100, 96 and 93 which catered for Sobonakhona Traditional Authority, Mr. SVD also came to a similar conclusion that eThekwini municipality use UDs to discriminate some of its citizens:

"eThekwini Municipality introduced UDs in our areas because we are black, it is discriminatory and looking down upon us simple because we are black. We are not provided with good facilities, simple because we are black. Our IsiZulu speaking mayor and councillors are just like coats placed in a garden to intimidate rabbits and prevent them from eating crops, if rabbits could be made aware that they are intimidated by hanged coats they could simple eat everything in the garden. When coming to constructing poor facilities in black areas, eThekwini Municipality is just the same as the previous apartheid government, they are just whites in black skins"

Mrs N who resides in Zamani B agrees with SVD that *"the municipality cannot build such toilets in areas where Whites live, they did this because we are black, Zulu and poor."* Other Zamani residents Mrs M, N, Miss H, Mrs. M, K and N agree with Mrs. N regarding discrimination.

"No explanation was provided regarding why the municipality decided to construct UDs in this area. We thought it was a temporal measure whilst RDP houses were under construction. When building RDP houses, flush toilets and not UDs are constructed. RDP houses were promised, but they were never delivered. Only Sections 4 and 6 have RDP houses and we were excluded. We were excluded because the former KZN Member of the Executive Committee for Rural Development, Environment and Agriculture, Meshach Hadebe regarded this area as a zoo and therefore was not suitable for further development. In every meeting that Hadebe attended he referred to us as animals that live in a zoo that do not deserve modern day amenities such as flush toilets".

Former MEC Hadebe was demoted by the ruling ANC to the position of the KZN Deputy Speaker of the Legislature after May 2014 general elections for making inappropriate comments when addressing an election rally that voters who do not vote for the ANC will lose their pensions and child support grants. This was interpreted as using child support grant and pension benefits to buy votes in a similar manner that the Inkatha Freedom Party leader Mangosuthu Buthelezi used state institutions to maintain his party's hegemony in KZN from 1994 to 2004.

The local headman of Ngonweni area under AmaQadi Traditional Authority, induna N highlighted that UDs are used as a tool to discriminate Amazulu and rural communities. The town planners and promoters of such projects live in affluent suburbs of Umhlanga Rocks and design sub-standard facilities for rural areas in which they do not even live in.

"Our citizenship, democratic and constitutional right to water and sanitation and dignity is violated by the municipality, when it comes to services to rural black communities, they argue that they have no money, but they have money for 2010 soccer stadiums, speed humps, swimming pools and other amenities they construct exclusively in urban areas. We live next to a very big dam here, but we are worse off when it comes to water and sanitation. This government is showing no respect for us. Instead of giving us flush toilets, they are asking us to collect our urine and sell it to them. If selling urine is good, why are they not doing it in urban areas, if UD toilets are good why they do not build them in urban areas?"

These are a number of questions that Induna N and his Ngonweni subjects believe were never resolved by eThekwini Municipality. To a group of people already sensitive to the vestiges of inequality under apartheid, this proximity raised questions immediately. How can they have waterborne sewerage and we cannot? There is also another argument that the pinch of spending valuable financial resources on water for flush toilets was also not readily apparent in places where people still have unlimited access to water. Some households did their own "upgrade" to flush toilets, with septic systems, built by a local plumber. Another key finding related to lack of equity is lack of communication with beneficiaries of the project.

Communication

Many problems can be traced back to either poor communication with households or a lack of communication. Councillors, claim that communities were consulted adequately before UDs were installed whereas communities claim that eThekwini used a top-down approach (Roma, et al, 2010:589). A former municipal councillor, NN that participated in the study highlighted the following issues regarding consultations:

"I was part of the decision to install the UDs in which the anticipated improvement in terms of health and costs were highlighted by the officials after they conducted studies. What impressed us is that there are no emptying costs as faecal matter dry and decompose easily. The UD is good because, no flies enter the toilet facility; no solid waste disposal budget would be set aside to empty toilets since the communities would do this on their own. I accompanied the officials when they went to present this project to the community. Community representatives were part of the decision-making processes. A draft Integrated Development Plan (IDP) was presented to communities and the Council. This was subsequently followed by convening Imbizo IDP Forum in which councillors, communities and officials extensively participated. The people were given an opportunity to highlight the needs of their ward. The IDP responded to the needs of the community. Water and sanitation formed part of the IDP although it was not number 1 priority."

Similar reasons for the choice of UDs were cited by Austin (2006:1). The above statement is supported by other councillors that claim that they saw the value of UDs in other areas and fully consulted their communities before the implementation. Another councillor MP claim that the suggestions came from communities because of the un-hygienic nature of pit latrines without ventilation that most households used before the introduction of the UDs. Communities were given an option to dig pits for VIPs and when they resisted, they were then given an option of UDs. The type of toilets that were to be constructed were initially built in demonstration houses and facilitators that were drawn from such communities visited each household to explain toilet usage, maintenance procedures including health and hygiene promotion.

Another councillor TC claimed that: *"the Infrastructure Committee did a research about sanitation options suitable for rural areas. Various affected communities were consulted about the proposed plans. The consultations were extensive"*. The health and hygiene motives as well as consultations are also cited in works of Austin, (2006:1) as either inhibiting or facilitating factors.

At Kwangcolosi, households claimed that there were no meetings convened in which the municipality presented UDs and other options in which they were expected to choose from. At least all Ward councillors that participated in the study claim that there were various meetings in which communities were consulted. There are also party representative councillors that support such a claim. Members of PSC, facilitators and TMC are divided on the issue of consultation. There are those that believe that consultation took place and there are equally those that claim that consultation did not take place. Lack of consultation, feature strongly in the reasons for the rejection of UDs by households. At Umgababa, Umnini, Magcino, eHlanzeni and Danganye most households thought that the introduction of the UDs was still the continuation of the flush toilet project that was still supposed to be rolled out from Magcino to other areas. There are households that have both flush toilets and UDs and they claim that UDs were installed whilst they were away either in the nearby Amanzimtoti or Durban.

The material used to construct such toilets was of poor quality and approval by building inspectors was guaranteed. Toilets were poorly constructed to the extent that most of them broke down during the construction phase. There are no avenues that communities could use to complain about poor workmanship, because most of the construction camps are either relocated or were combined with other municipal wards as a result of the municipal demarcation process. Some councillors also support the allegation made by communities that consultation did not take place prior to the installation of the UDs. Councillor TB concluded that: *"There is a lack of community ownership which resulted in structures such as PSC, TMC and facilitation teams becoming dysfunctional."* The lack of options and the imposition of UDs were repeated again and again. This is not surprising since eThekwini says it was in no position to consult on options since no other option existed within its budget. Although there might have been fear of opening a can of worms, this approach ultimately backfired. Being consulted is important to people, if only to tell them about the limitations and make them realise that there is only one option (if that is the case).

Communication was the responsibility of the Institutional Support Developers (ISDs), which eThekwini holds only (5 out of 24?) did a proper job. ISDs were appointed by the municipality, on the recommendation/ selection by ward committees. The agenda of the users are always not incorporated in project design and implementation (Roma et. al. 2010:590, Duncker, 2009:1). There is a gap in communication between officials and politicians, for an example, members of the community claim that there is R400 connection fee that is charged by the municipality when UDs are installed and the officials together with many councillors were claiming that there are no charges levied as the project is part of poverty alleviation initiatives. Some communities acknowledge that health and hygiene promotion was part of the project, but indicated that it was mainly a once off intervention that only took place during the construction phase. It was mainly an operational matter whereby the facilitators explained to households the procedures they need to follow when emptying the toilets using rakes and spades they were provided with, pouring sand in a pit during each and every toilet visit. Washing hands with soap after defecation did not form part of the protocol due to the fact that toilet facilities were situated far away from the grey tanks and taps that were provided. The cups that the municipality provided in each of the households was for pouring sand into pits so that the faecal matter could decompose and this was not meant for washing hands. Communities were advised not to bring water closer to the toilets because washing toilets with water could bring about a smell and decomposition could take longer (Interview with a council official in 2011).

For communities of eMagcino, Umnini, Danganya and eHlanzeni, there was no health and hygiene promotion that was provided. Some households discovered that UDs were even constructed without their consent and they did not request them. This highlights lack of communication between the municipality and the communities. Households also claim that the criteria used to appoint facilitators, TMC and PSC members was active participation in various structures of the ruling ANC. Those that are not active members were excluded. This claim was repeated again and again at Kwangcolosi, Ngonweni, Mpushini and Zamani. Facilitators that were responsible for training at Kwandeni, Wood Glen and Zamani although they highlight that the choice of facilitators was the prerogative of the ward councillor, acknowledged that they provided comprehensive health and hygiene education which included hand washing, waste disposal, locating toilets away from houses and "one home one garden" campaign. Another facilitator that was also based at Zamani B praised the quality of health and hygiene education that was provided:

"The health and hygiene education that households were provided with includes hand washing with soap, keeping rubbish such as disposable napkins that are not decomposable out of toilet pits, cleaning and operating toilets as well as general maintenance of UDs. Facilitators conducted a door to door campaign educating households. Health and hygiene education was provided before construction, during construction and during the handover stage. Every 6 out of 10 households handle UDs properly. Education provided was good to our best of knowledge."

The good education that facilitators claim to have provided is disputed by various households. Mrs N from Zamani B states:

"No information was provided, we just saw people constructing toilets calling it development. We did not participate in deciding what type of development is a priority for us. It was just top-down approach. I was not happy about the failure of the local councillor and the municipality to convene a community meeting, also the failure on the part of the municipality to give us various choices and options. We would have preferred to start by installation of electricity before toilets"

Another group of households, also from Zamani that included M, K and N households noted that instead of health and hygiene promotion:

"We were given a bucket, spade, and a rake including basic education on how to use it. We were told that if you do not accept a toilet, you will not get an RDP house. No meeting was convened to explain anything. Those that were responsible for health and hygiene education simple left their pamphlets with anyone they found at home during their visit. All the things they said when they gave us tools to clean toilets are not at all happening. They said that if the toilet is full, one could empty it with ease, unfortunately it is not easy as the toilet smells very bad. They also indicated that it is up to you to decide what to do with the faeces after emptying the toilet. We do not have electricity, water and RDP houses in this area, despite having accepted UDs as a condition to get RDP houses. Our toilets are full, the municipality promised us that they will empty the toilets in each and every time after one year or six to twelve months, but to date they have not done so. People that provided health and hygiene education has since left the project. In the community meeting we held a year ago, we raised the issue of toilets, but up until

now nothing has happened. In every meeting the issue of toilets was raised consistently. The pits are shallow and in large families toilets fill up easily."

M, N and H households claim that:

"The health and hygiene education provided was not good at all. We were only provided with equipment that we could use to empty toilets and basic skills such as use of toilet paper only, washing hands after defecation was not part of the education provided. The emphasis was on how to empty them. Other people still use newspapers instead of toilet paper, because of poor health and hygiene education that was provided. No explanation was provided on how to clean the toilet seats. We were told not to put water inside the toilet pit, but there are instances where girls are menstruating and blood stains drop inside the pit and get mixed with faeces."

At Mpushini area of eMbumbulu households claim that "no health and hygiene education was provided besides toilet usage protocols." One member of a household at Mpushini claimed that:

"The facilitators that handed over toilets were very impatient and handover in most instances did not even last for more than 5 minutes. No visitations took place after the handover, no explanation was provided and the focus was on ensuring that households sign to authorise that their toilets were completed. Children pour faeces in urinals and that results in most latrines blocking even if one use detergents the smell refuse to go away. Children sometimes pour sand in urinals and the whole system blocks and there is no one to help as the site office relocated to an unknown place".

Even at Ngonweni, households were not impressed about the standard of health and hygiene education that the municipality claimed to have provided:

"We were given user guides such as things that should not be put in toilet pits such as newspapers and that was provided as part of a door to door campaign," commented a 50 year old Mrs M. Another commentator had a different view that the standard of health and hygiene education was very poor, to the extent that it was just the distribution of spades, cups, rakes and jugs that people should use when they defecate and when they empty toilets. This was just a handover process and it last for few minutes and it should not be called health and hygiene education, comments Mr N. People were not necessarily saying that they cannot pay for water, but they need explanation and engagement.

Sensitivity to Local Contexts and Institutional Focus

Specific areas need attention to different issues. Many times this is what allows for the human element to be prioritised within a bureaucratic approach. For example, a story of no wheelchair access could have been prioritized during the construction phase of the project. There was no need to provide households living with disabilities with two UDs, but a toilet that is accessible even if someone is using a wheel chair. The fact that areas served with UDs were in close proximity with those with flush toilets highlighted lack of sensitivity to local dynamics. The municipality could have utilized the services of Durban Solid Waste to ensure that everyone is getting access to black plastic bags for disposal of napkins and other solid wastes. The Durban Botanical Gardens could have assisted with providing trees to plant as suggested by proponents of UDs. The municipality could have explored emptying toilets in a similar manner that they are using with VIPs, even if it is for a pilot phase. Although emptying UDs is budgeted for in the 2014-2015 financial year as indicated in the previous section of this article, it is important for the municipality to address all the issues identified in this study as well as those identified by Roma et al (2013).

As much as 14319 (90%) of households received information about how to use UDs and 1073 (6.7%) did not receive such information (Roma et al, 2013:308), it is clear that UD usage information is not equivalent to health and hygiene promotion. The job creation that resulted in councillors approving UD project did not yield good results. Even Roma et al (2013:308) acknowledged that only 2196 (13.7) households observed an increased job creation as a result of UD project whereas 13254 (83.3%) believe that UDs did not contribute towards job creation. Other challenges such as smell from toilets, lack of privacy from doors not closing properly, poor material and workmanship for construction and urine pipes that were not properly constructed (Roma et al, 2013:308) are problems that emptying UDs by municipal officials will not be able to address. The fact that 11130 (70%) of the households were not satisfied with their UDs (Roma et al, 2013:308) implies that even if the municipality take over the emptying, they will still be dissatisfied because it is not lack of emptying that is problematic but a combination of many issues dealt with above and in the next sections of this article. For instance, eThekwini Water could have acted on allegations of water pollution issues reported in Joint Services Board (JSB) constructed boreholes.

There are two institutional tendencies that flavours eThekwini's engagement around UDs. The first is that eThekwini is driven by technical people with engineering backgrounds, and lacks grounding in basic community development skills. It is not simply a lack of human resources in this area, but recognition that these skills are actually skills and ones that need to be prioritised and valued alongside the technical rather than treated as an after-thought that existing project managers or consultants can take care of (Roma et. al, .2010:589). Secondly, officials are ultimately accountable to the Council, which means that they must "toe the line" and attend to top-down dynamics. This places them in a somewhat invidious situation, which at very least bounds what they can and cannot do. The institutional mechanisms and institutions that were established to promote the introduction of water and sanitation projects within eThekwini Municipal jurisdiction were PSC, TMC, clerk of works, project managers (PM), ISDs and health and hygiene education facilitators. The ISD consultants were appointed by the eThekwini Municipal officials in accordance with procurement policies. PSC, TMC and facilitators were appointed by ward councillors in consultation with communities and their ward committee. There were also project sites where building material was kept, also where members of the communities that had technical queries could go and request either a member of a PSC, TMC or facilitators to resolve.

The experiences of how the various institutions established to implement the UD project varied from community to community and councillor to councillor that participated in the study. In certain areas, such as Zamani, Kwandeni, Wood Glen and other areas next to Mpumalanga Township, Embo and KwaNgcolosi facilitators, TMC, PSC and ward committee members are still visible and operational whereas in areas such as Mphushini and other eMbumbulu areas, uMnini and Ngonweni institutions collapsed and project sites are either relocated or shut down and communities no longer have any mechanisms for further recourse if they have issues to raise. Even in areas where such institutions still exist, the perception of ordinary people about the usefulness of such structures varies, depending on how close they are with decision-making individuals. Some of the people indicated that they were not involved in the election of such institutions and cited political interference and ANC hegemony that inhibited ownership and inclusiveness of such structures.

In Ward 8 which includes areas such as KwaNgcolosi, Molweni, Hillcrest, KwaNogwaza and Langerfontein, the DA Ward Councillor at that time worked hand in hand with ANC's party representative councillor NN and IFP's party representative Councillor M in identifying members of the PSC, TMC, facilitators and contractors that procured various services in various stages of the project cycle. Each party appointed at least 3 service providers including facilitators. Councillor NN pointed out that the number of facilitators increased to about 40 in various stages of the project cycle. PSC, TMC and Ward Committee members were paid allowances for each of the meetings that they attended. Both councillors TC and NN agree that white residents did not participate in the election of the ward committee and therefore it is one of the few such committees that was not under the chairmanship of the ward councillor TC, but chaired by a member of the community. TC cited busloads of ANC members to meetings as the reason why the few whites that attended the meeting declined to be nominated, whereas NN cited cocooning culture amongst whites as the reason they preferred to work through Ratepayer's Association's other than Ward Committees. Councillor M cited politicization of development by the ANC as the prime reason why the institutions created are not inclusive and even the attempt by residents of Lower Molweni to seek recourse through Human Rights Commission and Public Protector for violation of their water and sanitation rights were unsuccessful, because ANC members supported the UD implementation whereas other residents were opposed.

At uMnini, Ngonweni, Mshazi, Embo, Zamani, and Mpushini, ANC hegemony was cited as the reason why communities were excluded in the election of PSC, TMC and appointment of facilitators. In all these areas former ward councillors SX, ZM, MP, NN and TM appointed ANC activist exclusively. Traditional leaders and other community members that are not associated with branch executive committee members of the ANC were excluded. In all these areas members of the community alleged that active members of the ANC are served with flush toilets and those that are not associated with the ruling party are served with UDs. At KwaNgcolosi community participants highlighted that PSC, TMC and facilitators are dysfunctional. They further alleged that only ANC members participated in meetings, including the view that an inkosi Bhekisisa Bhengu that is an ANC Member of the KZN Provincial Parliament and the chairperson of the environment portfolio committee member and former Councillor MP prevented people without ANC membership cards from participating in meetings and having a say in their development. Councillor MP disputed that political interference by the ANC is the reason for the dysfunctionality of the institutions and under-utilisation of the sanitation facilities. He cited misuse of the facilities as the main reason for failure and emphasised democratic participation in the election of PSC, TMC and appointment of facilitators.

"The community members met to elect facilitators, PSC members as well as the TMC. Facilitators conducted research in the area, collected data about living conditions in each household. Households with eight members were allocated one UD and those with more than eight were allocated an additional UD. The data collection was conducted in various stages of the project cycle and it was coupled with provision of health and hygiene education. Households were informed about the type of toilet facilities they were about to get, they were further provided with spades, rakes and cups they could use to fill sand in the pit to allow faecal matter to decompose. UDs were designed in such a manner that faecal matter would be kept dry and decompose naturally once sand is poured every after defecation has taken place. Toilets are becoming full before their full one year life cycle because of misuse and because some families are more than the stipulated number. Initially households were supposed to dig their pits, unfortunately most households refused. Misuse of the toilet facilities includes mixing faecal matter and urine and other

liquid substances. What complicated the matter further is that most of the houses belong to land lords who do not live in the area and let out their houses to tenants that are not complying with toilet usage protocols. Vandalisms, broken doors and mixing of urine and faecal matter are very common especially in houses occupied by tenants."

Councillor MP further added that the project created employment opportunities, skills transfer and promoted technical engagement between communities and municipal officials. He elaborates on how this happened:

"The Health and Hygiene Education ISDs were identified by officials through an open tender process. Once consultants were appointed, they were given further staff compliment drawn from communities in which they were going to operate. People with a minimum grade 12 qualification were identified and elected by their people from their areas. ISDs are accountable and contractually obliged to present high quality health and hygiene education. The process of working with ISDs was a good experience as it created employment opportunities in the area, skills were transferred from ISDs to local people that resulted in many of our youth securing full time employment with consultancies. Members of the community played an important role in selecting members of the TMC as well as PSC. Apart from members of the community, clerks of work, engineer, ISDs and project manager participates in PSC. Ordinary people managed to share their views with PSC, TMC, and engineer, clerk of work and project managers."

The above view is supported by Councillors NN, NNI, TC and various people that served as members of the PSC, TMC and facilitators. One PSC member from eHlanzeni disputed that institutions created functioned well as they were supposed to. He stated that even building inspectors approved toilets that were falling apart during the construction phase. Such poor workmanship that was condoned by PSC, TMC and facilitators resulted in the death of Mr. M in the South Coast when an approved toilet collapsed whilst he was inside defecating. This view was supported by Councillor M who stated that:

"The quality of the materials that were used to construct the toilets was of poor standard. Doors and corrugated iron roof that they used were of poor quality, broke down during the construction phase and were not replaced. There is no evidence of vandalism taking place in certain areas, and it points to the problem of poor quality of material used to construct the toilets that is responsible for the breakdown."

For SVD the problem with PSC, TMC and facilitators was beyond just poor quality of material that neither of the structures created to monitor this nor councillors, raised concerns about, but even job creation benefited only members of the ANC. What complicated matters at Mpushini is that Councillor TM left the area and settled in the affluent Amanzimtoti and his assistant, M alienated members of the community when he accumulated and appointed people on the basis of their party affiliation. Members of Sobonakhona Traditional Council were also alienated by TM and M's actions. SVD resigned from the PSC because the clerk of works, Mrs B was no longer inviting him to PSC meetings because he was not part of Councillor TM's inner political circle. A member of Sobonakhona Traditional Council, Mrs DN highlighted that political infighting amongst members of the ANC, resulted in difficulties for the council to work with PSC, TMC and councillor TM. The internal conflicts destroyed TMC, PSC and facilitators were resigning from the project. Another complicating factor was conflicts between Councillor TM and the IFP party representative Councillor PM. In meetings in which the traditional council invited councillors, a lot of the time was spent sorting out their fights rather than reporting on the progress regarding the project. Even when the project site was shut down, no explanation was provided, only members of the ANC knew why such a move took place.

Political in-fighting amongst ANC members was also cited as the reason why the residents of Wood Glen demolished their UDs which resulted in the halt of the project which then resumed after the in-fighting was over. At Ngonweni the site office relocated to Matata where the then ward councillor ZM resides. All PSC, TMC, contractors and facilitators were active members of the ANC and no-one came from Ngonweni. They were either from Matikwe, Umzinyathi and Matata which are regarded as the strongholds of the ANC. AmaQadi Traditional Authority and Qadi Development Committee were some of the structures that were excluded when TMC and PSC members were appointed. In a ward committee, Ngonweni is not represented and therefore job creation and other benefits that were accrued from the project did not benefit anyone in this area.

At Zamani, KwaNdeni, Wood Glen and surrounding areas, TMC, PSC and facilitators are still vibrant and they were left with few toilets to handover before they focus on the water provision project. All 9 members of the PSC and 5 facilitators responsible for Zamani B are still active and the site office with clerk of works that visit the site 3 times a week, project manager and ISD are still active in the implementation of the project. The ISD, PSC, TMC meets once a month to receive progress reports and PSC members report back to the community that elected them. Any member of the public that have concerns is welcome to visit the project office to meet with the ISD, TMC and or facilitators. The facilitators are under the management of the supervisor and all operate from the project office. In Wards 8 and 91 about 700 UDs and 608 were constructed respectively, strong institutional mechanisms were cited as the reason for such significant strides. In both these wards, site offices still exists and structures such as PSC, TMC and facilitators are still active. Communities of both these wards are still not happy despite strong institutional mechanisms. Community participants at Zamani B complained about emptying their toilets, carrying of a bucket of sand each time they defecate, smell from toilets, shallow pits and poor material used which result in doors being broken and political appointment of such structures as reasons for their lack of acceptance of these facilities. Other reasons for lack of acceptance even if there are strong institutional mechanisms, are perceived discrimination, provision of dry sanitation in areas with flush toilets and lack of municipal support in emptying toilets.

CONCLUSION

There is no doubt that the implementation of UDs in the rural areas of Durban is still associated with class-based discrimination. The households regard the provision of such toilets as lowering their dignity as people living with disabilities, women, and those that are not part of the ruling elites. Collective action in the form of demolition of UDs at Wood Glen are some of the experiences that households reported about. Poor institutional mechanisms, cost saving and community management of such complicated facilities are some of the challenges that households remains grappling with. As much as the communities and councillors are telling different sanitation stories, perhaps if the municipality can provide a (free) service to empty UDs by developing local contractors or wider service, there may be a possibility for such facilities to be accepted as proposed in the Water Services Development Plan presented by Neil Macleod in May 2014. This move needs to be coupled with addressing all other issues highlighted in this article as well as other studies.

In zoning the allocation of sanitation facilities and health and hygiene promotion, it is imperative for the municipality to urgently attend to equity issues by taking steps to move middle class areas toward dry sanitation or lower flush toilets, through changes in regulation and/or incentives otherwise flush toilets are perceived as a privilege of 'petit bourgeoisie' and UDs as a 'curse' for being part of the 'class' of poor rural peasants. It is also imperative for eThekwini Water and Sanitation to attend to outstanding technical issues raised by households such as hand washing, repair of facilities and implement a new round of health and hygiene education. There is also a need to consistently ensure that there are mechanisms for households to raise issues or get attention to their queries relatively easily, e.g. site offices, household visits by health and hygiene workers and align programmes with community-led total sanitation principles that are in line with water- shed management.

REFERENCES

Austin, L. M. (2006). *Operational safety of urine diversion toilets in Durban, South Africa*. In *32nd WEDC International Conference*, Colombo, Sri Lanka.

Bourblanc, M. (2012). Transforming Water Resources Management in South Africa. *'Catchment Management Agencies' and the ideal of Democratic Development Journal of International Development, 24*, 637-648.

Bourblanc, M. (2015). The South African 'Ecological Reserve', A Travelling Concept. *Politikon, 42*(2), 275-292.

Brouckaert, C. J., Ridgeway, T., & Buckley, C. A. (2006). *A mathematical analysis of the drying of faeces in an urine diversion toilet- sensitivity analysis*. Water Research Commission.

Brown, J. C., & Purcell, M. (2005). There's nothing inherent about scale: Political ecology, the local trap, and the politics of development in the Brazilian Amazon. *Geoforum, 36*(5), 607–626. doi:10.1016/j.geoforum.2004.09.001

Bryant, R. L., & Bailey, S. (1997). *Third World political ecology*. Routledge.

Bryant, R. L., & Goodman, M. K. (2004). Consuming narratives: The political ecology of alternative consumption. *Transactions of the Institute of British Geographers, 29*(3), 344–366. doi:10.1111/j.0020-2754.2004.00333.x

Duncker, L. C. (2008). Prejudices and attitude change to dry toilets in South Africa CSIR. Academic Press.

Duncker, L. C., Matsebe, G. N., & Austin, L. M. (2006). Use and Acceptance of Urine- Diversion Sanitation Systems in South Africa. Water Research Commission.

Enzensberger, H.M. (1990). A critique of political ecology. *Kursbuch, 33*, 3-31.

Escobar, A. (1998). Whose knowledge, whose nature? Biodiversity, conservation, and the political ecology of social movements. *Journal of Political Ecology, 5*, 53–82.

Esrey, S., Gough, J., Rapaport, D., Sawyer, R., Simpson-Hebert, M., Vargus, J., & Winblad, U. (1998). *Ecological Sanitation. Swedish International Development Corporation*. SIDA.

Flores, A., Buckley, C., & Fenner, R. (2008). *Selecting Wastewater Systems for Sustainability in Developing Countries.* In *11th International Conference on Urban drainage*, Edinburgh, UK.

Guness, M., Jackson, S., Rodda, N., Smith, M., Trotter, D., Buckley, C., & Macleod, N. (2010). *Impact of Buried Urine Diversion Waste on Environmental Quality and Plant Growth.* Water Research Commission.

Guness, M., Pillay, S., Rodda, N., Smith, M., Buckley, C., & Macleod, N. (2009). *Quality of Leacheate from Burried Urine Diversion Toilet Waste.* Water Research Commission.

Hawksworth, D., Archer, C., Hanssen, A., Tronnberg, L., Lutchminarayan, R., Knight, S., Smith, M., & Rodda, N. (2008). Occurrence of Helminths and Protozoan Parasites (Giardia and Cryptosporidium) in the Filing Vaults of Urine Diversion Toilets in the eThekwini Municipal Area (Durban, South Africa). *WISA Conference*, Sun City.

Hawksworth, D., Archer, C., Rodda, N., Smith, M., Appleton, C., & Buckley, C. (2009). *An Improved Method for the Recovery of Ascaris Ova from Solid waste from Urine Diversion Toilets.* Water Research Commission.

Hellberg, S. (2014). Water, life and politics: Exploring the contested case of eThekwini municipality through a governmentality lens. *Geoforum, 56*, 226–236. doi:10.1016/j.geoforum.2014.02.004

Heynen, N., Perkins, H. A., & Roy, P. (2006). The political ecology of uneven urban green space: The impact of political economy on race and ethnicity in producing environmental inequality in Milwaukee. *Urban Affairs Review, 42*(1), 3–25. doi:10.1177/1078087406290729

Le Billion, P. (2000). The political ecology of transition in Cambodia 1989-1999: War, peace and forest exploitation. *Development and Change, 31*(4), 785–805. doi:10.1111/1467-7660.00177

Lutchminarayan, R. D. (2007). Sanitation, Water and Hygiene in eThekwini Municipality, Durban, South Africa: Baseline Cross-Sectional Study. UKZN.

McCarthy, J. (2002). First World political ecology: Lessons from the Wise Use movement. *Environment & Planning, 34*(7), 1281–1302. doi:10.1068/a3526

Mehta, L., Movik, S., Bolding, A., Derman, B., & Manzungu, E. (2016). *Introduction to the Special Issue- Flows and Practices: The Politics of Integrated Water Resources Management (IWRM) in Southern Africa Water Alternatives.* Academic Press.

Mnguni, M., Ndlovu, S., Gounden, T., Pfaff, W., Rodda, N., & Buckley, C. (2008). Health Risk Assessment of the Operation and Maintenance of Urine Diversion Toilets in eThekwini Municipality. *WISA Conference*, Sun City.

Mnkeni, P. N. S., & Austin, L. M. (2009). Fertiliser Value of Human Manure from pilot urine-diversion toilets. *Water S.A., 35*(1), 134–138.

Moilwa, N., & Wilkinson, M. (2006). The Effect of Hygiene Communication on Emptying of Urine Diversion Toilets. *32nd WEDC International Conference*, Colombo, Sri Lanka.

Murray, A., & Buckley, C. (2010). *Designing Reuse-Oriented Sanitation Infrastructure: The design for service planning approach.* UKZN.

Neil Adger, W. (2001). Advancing a political ecology of global environmental discourses. *Development and Change*, *32*(4), 681–715. doi:10.1111/1467-7660.00222

Pelling, M. (1999). The political ecology of flood hazard in urban Guyana. *Geoforum*, *30*(3), 249–261. doi:10.1016/S0016-7185(99)00015-9

Pillay, S., & Buckley, C. (2005, October). *LCA Studies of Water Delivery Systems*. Water Research Commission.

Roma, E., Buckley, C., Jefferson, B., & Jeffrey, P. (2010). Assessing user's experience of shared sanitation facilities: A case study of community ablution blocks in Durban, South Africa. *Water S.A.*, *36*(5).

Roma, E., Philp, K., Buckley, C., Xulu, S., & Scott, D. (2013). User perceptions of urine diversion dehydration toilets: Experiences from a cross-sectional study in eThekwini Municipality. *Water S.A.*, *39*(2), 305–312. doi:10.4314/wsa.v39i2.15

Schroeder, R. A. (1993). Shady practice and the political ecology of resource stabilization in Gambian garden / orchards. *Economic Geography*, *69*(4), 349–365. doi:10.2307/143594

Sutherland, S., Hordijk, M., Lewis, B., Meyer, C., & Buthelezi, S. (2014). Water and sanitation provision in eThekwini Municipality: A spatially differentiated approach. *Environment and Urbanization*, *26*(2), 469–488. doi:10.1177/0956247814544871

Swyngedouw, E., & Heynen, N. (2003). Urban political ecology, justice and the politics of scale. *Antipode*, *1*(5), 899–918. doi:10.1111/j.1467-8330.2003.00364.x

Van Koppen, B., Tarimo, A. K. P. R., van Eeden, A., Manzungu, E., & Sumani, P. M. (2016). Winners and Losers of IWRM in Tanzania. *Water Alternatives*, *9*(3), 588–607.

Von Munch, E., & Mayumbelo, K. M. K. (2007, October). Methodology to compare costs of sanitation options for low- income peri-urban areas in Lusaka, Zambia. *Water S.A.*, *33*(5). doi:10.4314/wsa.v33i5.184017

Walker, N. J., Still, D. A., Ravenscroft, P., Hazelton, D., King, W., Erskine, S., & Eardley, R. (2010). *Financial Sustainability of Sanitation Services in South Africa*. Water Research Commission.

Walker, P., & Fortmann, L. (2003). Whose landscape? A political ecology of the 'exurban' Sierra. *Cultural Geographies*, *10*(4), 469–491. doi:10.1191/1474474003eu285oa

Walker, P. A. (2003). Reconsidering 'regional' political ecologies: Toward a political ecology of the rural American West. *Progress in Human Geography*, *27*(1), 7–24. doi:10.1191/0309132503ph410oa

Walker, P. A. (2005). Political ecology: Where is the ecology? (2005). *Progress in Human Geography*, *29*(1), 73–82. doi:10.1191/0309132505ph530pr

Walker, P. A. (2006). Political ecology: Where is the policy? *Progress in Human Geography*, *30*(3), 382–395. doi:10.1191/0309132506ph613pr

Winblad, U., & Simpson-Hebert, M. (Eds.). (2004). *Ecological Sanitation revised and enlarged edition* (2nd ed.). Stockholm Environment Institute.

Chapter 10
The Political Ecology of the Decentralized Water Management in Zimbabwe:
Theory and Empirical Evidence From Sanyati Catchment Area

Winmore Kusena
https://orcid.org/0000-0002-9948-2147
Midlands State University, Zimbabwe

ABSTRACT

The chapter assesses the notion of local scale and decentralization that emanates from the IWRM principles. Evaluation of the benefits of decentralization was done through the political ecology lens that critically examines fairness and power struggles across spaces. Sanyati catchment was used to draw empirical evidence in light of the theoretical expectations of decentralization towards catchment water security. Qualitative approach was used to collect data. Interviews were the main sources of data collection. The findings showed that decentralization has failed to produce the desired results compared to what is assumed in the dominant narrative that highly esteems the decentralization management approach. The chapter showed that what brings results are not local scales and suggests that probably fair and transparent resource distribution and allocation at any scale may bring about water protection that does not trigger the tragedy of the commons.

INTRODUCTION

Integrated Water Resources Management (IWRM) is popular in Zimbabwe for the establishment of catchments and sub-catchment councils (decentralization) and other water reforms (Mapedza et al. 2016; Conyers 2003). Decentralization has led to a process where power and institutions are taken to the local people. The concept of a local scale in decentralization emanated from the, Dublin principles

DOI: 10.4018/978-1-7998-8809-3.ch010

which assume that any development agenda that is taken to the local scale inevitably yields desired results compared to activities spearheaded from exterior scales which are perceived to be rather distant (Purcell 2002). This was done with the assumption that conservation efforts would be fruitful, and this has been a dominant narrative that the Government of Zimbabwe adopted since the inception of IWRM in 1998. However, Brown and Purcell (2005) are of the notion that there is absolutely nothing inherent about local scale since scale is a social construct whose configurations can be changed from time to time. Scales always require a more explicit understanding of particular scalar conformations instead of generalisations. In preference to questioning the aspect of the effectiveness of scale from a theoretical and methodological perspective only, the study prototypes Sanyati catchment configurations to assess the results of the existing dominant narrative that assumes the efficacy of decentralization to catchments and sub-catchments. Sanyati catchment case was used to ground truth the dominant narratives related to human interaction and water resources, focusing on the pros and cons of the involvement of the local scale in water governance. This is especially more pronounced when channels of communication available prohibit or undermine networking in the scalar arrangements (Swyngedouw, 1997). However, many in political ecology have since identified this conception of geographical scale (Brown and Purcell, 2005; Green 2016) and are not comfortable with it, arguing that scale is just a social construct with no direct implications on the success of any project agenda (Mohan and Stokke, 2000; Herring, 2001; Myers, 2002). Nonetheless, the IWRM concepts of decentralization and local scale greatly influenced the Zimbabwean water management approach.

The question that prompted this study from a political ecology perspective is, if decentralization and local scale are a panacea to catchment water ills, why then do catchments continue to face a plethora of challenges years after the introduction of the highly esteemed IWRM principles (Manzungu, 2002), or is it time to outrightly change the scalar configurations. The political ecology of water management is seemingly well understood, and the dynamics are well articulated (Hellum and Derman, 2002). However, results in catchments seem to be in the contrary to what is anticipated from decentralization. The discourse of local scale has its own tenets and dominant narratives. There is a strong assumption that when decisions concerning resources are crafted by the local scale they yield socially and environmentally impartial resolutions (Bassett and Zueli, 2000). However, this paper seeks to establish whether the tenets of decentralization are applicable to Sanyati catchment and its sub-catchment councils. This brings about the spatial aspect as a pertinent field of enquiry because generalisation concerning local scale may be creating grey areas of untold successes or failures in the decentralization discourse due to different social constructs. This focuses on how the local scale is politicised and whether the outcomes are yielding intended outcomes or maybe there are struggles within the local scales themselves bearing in mind that scale has no permanent qualities (Swyngedouw 2004; McCarthy 2005; Paulson et al 2005; Smith, 1993), scale qualities are bound to change over time and through political power struggles and the existence of varying agendas in struggles. What might have worked a decade ago may not be used as a contemporary standard of measure.

In this contested terrain, Escobar (2001) is of the view that local scale brings about desirable outcomes in the management of natural resources. However, Brown and Purcell (2005) argue that nothing is intrinsically desirable about a local scale. Schroeder, (1999) also uncovered through findings from National Environmental Action Plans in Africa the detrimental effects of local scale management, which in this case is represented by the decentralized water management to catchments and sub-catchment councils in Zimbabwe. Zimmerer (2002) challenges the perceived challenges of local scale and is convinced that these must be verified through empirical investigations that are area specific. This actually supports the conception that local scale itself does not lead to any social and ecological outcome; rather the political interests and agenda that produce scalar arrangements do (Brown and Purcell, 2005). If anything, Zimmerer, (2002) is of the view that local containment poses threats to the environmental and socio-economic sustainability.

This article examined the tenets of decentralization; with an attempt to ascertain the narratives and presumptions put forward to support the efficacy of decentralization to Sanyati catchment. Although the study is not comparing Sanyati case to any other scale, the presumed advantages of decentralization are interrogated through political ecology theory. We sought to understand what is really being yielded though local scale in Zimbabwe using Sanyati catchment or there could be a need for new management approaches.

THE DECENTRALIZATION CONCEPT IN ZIMBABWEAN WATER SECTOR

IWRM in Zimbabwe is based on the concepts of social equity, economic efficiency and environmental sustainability (GWP 2000; Biswas 2005). This paper seeks to establish whether equal access for all users (particularly marginalized and poorer user groups) to an adequate quantity and quality of water necessary to sustain human wellbeing has been achieved in Sanyati catchment. GWP (2006) states that the right of all users to the benefits gained from the use of water need to be considered and the financial benefits generated from the use of water for economic purposes. Decentralization is a Sustainable Development aspect that emphasizes on bringing all stakeholders to the table to set policy and make sound, balanced decisions in response to specific water challenges faced (Abell and Thieme, 2009). However, it has to be noted that some political ecologists are skeptical about the benefits of sustainable development and the effectiveness of geographical scale in the management of resources (Weale, 1992; Adams 1990; Hecht and Cockburn, 1992) although no feasible alternative blueprint has been suggested so far.

Nonetheless, the existing response to water challenges range from model-based methods that quantitatively describe causal relationships and interactions among system components (Jakeman and Letcher, 2003; Parker et al., 2002) to participatory ways that favour decision-maker and stakeholder participation in the assessment and decision-making process. Zimbabwe has effectively operationalized the two models to come up with a hybrid approach that is scientific and participatory in nature under IWRM.

Zimbabwe has been implementing and attempting to align water reforms to IWRM since 1999 with a number of water reform measures taking place in order to comply with the Dublin Principles. These include the formation of ecologically based water management boundaries; creation of catchment councils and sub-catchment councils (Derman and Manzungu, 2016). All this has been unfolding primarily for the purpose of bringing real participation to the black majority and small holder farmers who had been side-lined during the 1976 Water Act era (Manzungu and Derman, 2016).

Hellum and Derman (2002) allude to the fact that all small water users' interests were not taken into consideration. Their concerns were in most cases dismissed at meetings as the structures then favoured large users, mainly because they were well resourced in terms of knowledge, skills and finances (Pilosof, 2012). This concurs with the GoZ (1994) Land Tenure Commission Report which alludes to the fact that water was mainly allocated to white farmers without meaningful share to the communal areas (Swatuk, 2002). Those were the anomalies that decentralization sought to address. That was the case then, but the question remains whether this has been achieved to date. However, besides knowing the perceived merits of the participatory approach through decentralization and devolution of power to catchments and

Sub-catchment councils, little has been reported concerning the results yielded from the decentralization and local people's participation in Zimbabwe, particularly at catchment level. Reports have mainly been at national and even regional scale (Manzungu and Derman, 2016; Mehta, 2016; van Koppen et al 2016). In most of these reviews Zimbabwe emerged as a country with one of the most progressive water reform processes within the Southern African region, but this was an analysis mainly based on statements of intents from catchment plans and not the real translation of the plans into intended outcomes (Mapedza and Geheb, 2010).

Besides reports on Manyame, Sanyati and Mazowe's incapacities to participate due to donor withdrawal (Derman and Manzungu, 2016) not much has been understood about the involvement of the local people and what has been achieved so far amid the glaring political and economic challenges the country is seized with. If it is really about financial incapacity, there is a need to understand if the predicament has had similar effect within the local scale of Sanyati catchment. How the water reforms played out in Zimbabwe is a function of unequal power dynamics amongst the stakeholders (Mapedza and Geheb 2010). This resonates well with Arnstein's (1969) Ladder of Citizen Participation which states that sometimes grass root or local scale participation is only manipulative and therapeutic without real citizen control. In some cases, consultation and information flow takes place but the powerful make final decisions; whilst they deliberately proffer consultation and informing as real participation.

Study Area

The Sanyati catchment is one of the seven catchments of Zimbabwe (Figure 1). The catchment lies between latitudes 160 161 and 190 301 south and longitudes 270 451 to 310 161 east. The Sanyati river system covers areas drained by the Kwekwe, Zivagwe, Munyati, Mupfure, Sengwa, Ume and Nyaodza rivers. There are 15 hydrological sub-zones, with areas varying from 2527 to 11427 km^2. These hydrological sub-zones form the ten (10) sub-catchments, namely: Sengwa, Nyaodza, Zivagwe, Upper Munyati, Muzvezve, Lower Munyati, Upper Mupfure, Upper Middle Mupfure, Lower Middle Mupfure and Lower Mupfure. The catchment constitutes a combination of four natural climatic regions and each region has its specific climate characteristics. These regions range from 2 to 5.

Sanyati catchment has the highest concentration of urban centres in the country. There are three cities namely Kwekwe, Gweru and Kadoma and four towns Chegutu, Redcliff, Kariba and Gokwe. The catchment is also home to semi-urban centres like Chivhu, Mvuma, Beatrice ad Mubaira. On the one hand, its cities' economies are based on formal employment, large and small businesses, informal trading, cross border trading and vending. On the other hand, the communal lands and farms side is based on commercial and semi commercial farming, subsistence farming, small-scale mining and small-scale businesses.

Data Collection and Analysis

The study employed a qualitative approach to assess the effectiveness of decentralization using Sanyati catchment as the local scale. Interviews with Sanyati catchment hydrologist, sub-catchment representatives and purposively sampled water users from the ten sub-catchments were the main sources of data. ZINWA agreements and permits database was instrumental in identifying water users in different sectors and their participation in water management issues.

Figure 1. shows Sanyati catchment zones, rivers and location in relation to other catchments

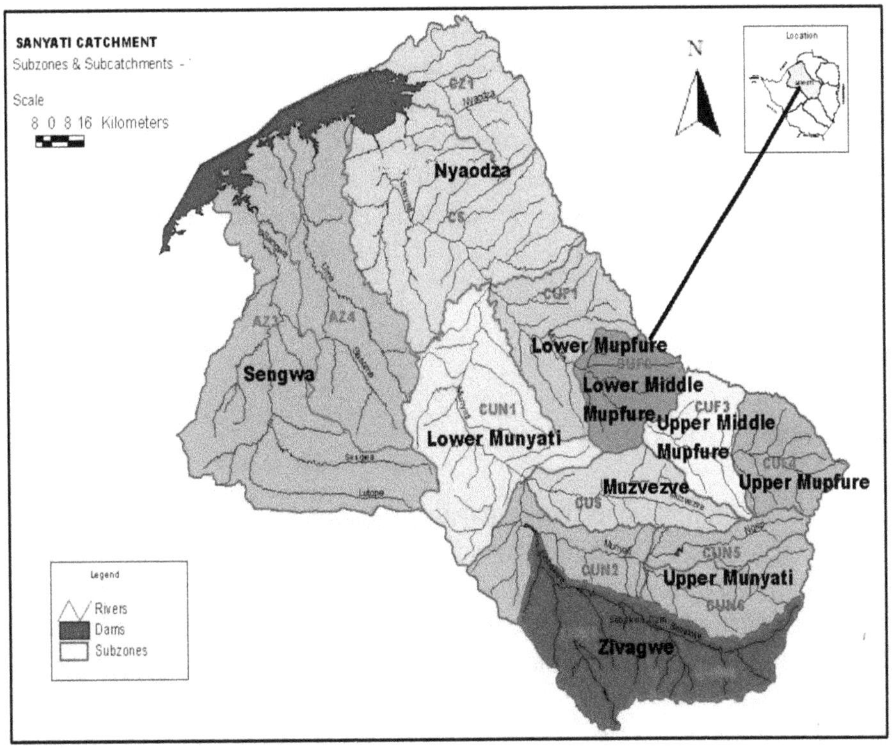

Table 1. Major rivers and sub-catchments in Sanyati catchment

Sub-catchment	Catchment Area (km2)	Major River
Lower Mupfure	4 062	Mupfure
Lower Middle Mupfure	2 666	Mupfure
Upper Middle Mupfure	2 527	Mupfure
Upper Mupfure	2 871	Mupfure
Lower Munyati	7 297	Munyati
Upper Munyati	7 776	Munyati and Ngezi
Zivagwe	6 896	Sebakwe
Muzvezve	3 271	Muzvezve
Nyaodza	14 028	Sanyati and Nyaodza
Sengwa	18 660	Sengwa and Ume
TOTAL AREA	**70 054**	

The researchers used a computerised file for Councillors and representatives in the Sub-catchment Committee and Zimbabwe National Water Authority (ZINWA) maintained by ZINWA Sanyati catchment. The file contained names of all Sub-catchment Councillors, Sub-catchment Council employees and

ZINWA employees, their contact numbers, physical address, email addresses as well as supplementary demographic information pertaining to occupation, position held in the council together with tenure. Therefore, Sub-catchment chairpersons, secretaries, and treasurers and water permit supervisors, who were responsible for the day-to-day functioning of Sub-catchment Councils were interviewed.

Purposively selected ZINWA employees responsible for coordinating the Sub-catchment council affairs as well as providing secretariat services to the Sub-catchment Councils were part of the sample. Nvivo 12 Pro was then used for interview data coding and analysis. Frequency query was done in Nvivo in order to create codes and themes for findings presentation and discussion. These themes were extracted from interview transcripts texts and notes obtained from the field.

RESULTS AND DISCUSSION

Intended Decentralization Gains

The 1998 Water Act set up a decentralized water management structure, based on seven catchment councils which were drawn according to ecological boundaries. The catchment councils have sub-catchments which are meant to bring water management to users, both small and large scale.

According to the Water (Sub-catchment Councils) Regulations, 2000; Sanyati catchments mainly through their 10 Sub-catchment councils would be responsible for the following functions:

1. Regulating and supervising the issuance of permits in order to maintain user discipline
2. Monitoring water flows and use in accordance with the prescribed allocations
3. Ensuring the availability of water measurement equipment
4. Promote catchment protection from degradation in accordance with the Water Regulations, 2000
5. Issuance of permits for wastewater discharge
6. Establishing feedback channels on water issues within the catchment or sub-catchment
7. Data collection and participation in planning.
8. Collecting rates, fees and levies
9. Allocating water within the framework of the catchment plan
10. Monitor groundwater levels and abstraction
11. Monitoring water quality.

Sanyati catchment and its sub-catchment councils were to discharge their duties in consultation with the stakeholders of the river system. The objective was to involve every stakeholder in the catchment and at the same time inculcate a sense of ownership and protection of the catchment. This participatory approach was designed to eliminate or at least minimise conflicts within the confines of the catchment. Nevertheless, Sanyati sub-catchment councils were facing challenges in all the functions they were supposed to effectively perform.

Sub-catchments are in principle ideal in facilitating water management at a smaller scale, with representatives from various water users being involved. The decentralization process that took place in Sanyati is the one that was meant to champion local people's participation through devolution of power to sub-catchments. Sub-catchments are supposed to raise funds, make decisions concerning allocation and protection of water. However, the question is the level of autonomous of these institutions. There is an argument that decentralization can take several forms, part of which can be mere delegation of power or tokenism. Just the transfer of government decision-making and administrative authority and/ or responsibility for carefully spelt out tasks to institutions and organizations that are either under government's indirect control or semi-autonomous. This involves delegating some authority and decision-making powers to local officials, but Central Government retains the right to overturn local decisions and can at any time take these powers back (Mapedza and Geheb, 2010). However, decentralization is supported because it is thought to increase efficiency and to reach populations who might not see benefits otherwise if they were carried out by a Central Government on their behalf (Derman, et al, 2001).

The Reality in Sanyati Catchment

In theory, the decentralization venture has brilliant ideas as afore chronicled. However, it was imperative to assess the level of the success of decentralization in water resources management. In terms of decentralization of institutions, the catchment did a sterling job. Sanyati catchment has 10 sub-catchments. This clearly shows a stride in the positive direction towards bringing water management to the people and redressing the anomalies conveniently used in the past to prejudice the black majority under the 1976 Water Act (GoZ, 2000). Nonetheless, in terms of results, achievement of the intended gains of decentralization, a closer look at specific key area of reference was done in order to measure the efficacy of the whole process in Sanyati catchment. The assessed areas included issues to do with representation, equity, consultation, user discipline and water permit regulation and supervision.

Representation of Stakeholders in Sub-Catchment Councils

The essence of decentralization is, in this case, the inculcation of grass root representation in water management. Full stakeholder representation was enabled in Sanyati catchment. Councillors represent their wards in communal lands and issues related to water abuse were amicably addressed. During the last quarter of 2017, proliferation of illegal gold mining activities at the Ngondoma dam led to the collapse of riverbanks. However, through stakeholders based in the areas, the issue was quickly reported and addressed. This concurs with Nemarundwe and Kodzanayi, (2001) who are of the narrative that the formation of multi-stakeholder institutions in the management of natural resources such as water, is basically a shift from the centralised and state driven natural resources management regimes of the colonial period and the immediate post-independence years, towards decentralized and mainly community-based management strategy that produce favourable results for both humans and the environment at large.

However, the identified success cannot be over-celebrated as the strategy is not devoid of shortfalls. The study identified cases of under-representation in the same catchment and these were more pronounced than the successes. It was noted that representation was sometimes just therapeutic and ineffective. Chosen representatives in most cases were not as useful because they failed to do follow ups of cases raised by their constituencies. Stakeholder representatives tendered apologies a few hours or minutes before meetings and some would be represented by new members who were attending the meeting for the first time. This presented lack of continuity on matters raised in previous meetings. In some instances, there would be no communication and representation at all. Despite financial incapacity to attend meetings, some of the sub-catchment representatives lacked the urge to work with the people, for the people and the environment at large. Constant absenteeism from the monthly meetings by representatives shows that some stakeholders are treating the issue of water management lightly. The main reason that was attributed to that behaviour was the fact that some representatives were just promptly seconded into office without the requisite training and knowledge. However, in order to ensure continuity, the chairperson and the vice-chairperson of each SCC in the Sanyati catchment area automatically became members of the Sanyati CC as required at law.

Representation has been predominantly mired by lack of adequate financial resources. Most of the full council meetings are underfunded and some are cancelled due to resource strains. Financial resources constitute the lifeline of stakeholder institutions such as CCs and SCCs. The financial resources at the disposal of the respective stakeholder institutions indicate the potential to run its own affairs. This implies that a group that is seized with more financial challenges may also have compromised representation. To worsen the situation, representatives are mainly interested in their travel and subsistence allowances. Some of them do not even know what they are supposed to be doing in Council. The moment SCC funds are depleted and cannot support the travel and subsistence allowances anymore, would mark the end of meetings during that season. ZINWA ended up doing everything through catchment councillors; that is from monitoring of water use, allocation and data collection.

Equity Among Stakeholder Groups

Lack of finance stifles people's confidence and paves a way to relatively rich constituency domination. Financially powerful stakeholders always dominate the process, while protecting their own interests, the same reasons that led to decentralization as a process that would bring equality. Unfortunately, Sanyati catchment has very few permit holders and funding has always been a problem. Regrettably, ZINWA is also financially incapacitated, rendering the whole devolution process a futile exercise as the same scenario that existed during the colonial era is regenerating because of poverty among the stakeholders and the institutions.

Apparently, the issue of varying agendas is not black and white man incongruence only, but involves diverse social classes whether black or white. The polarity within the same race needs to be addressed if ever equity is to be achieved, there is politics within the very local scale, even in sub-catchments. Racial reforms have not yet so far yielded any significant success in Sanyati in terms of equity and development. As long as people are placed in the same setting, with different financial, social and above all, political backgrounds, effective checks and balances are needed for equity to be reached, lest the power struggles leave the least esteemed (particularly financially and politically) in the consortium more vulnerable than before. It was observed that the government was controlled until 2009 by a single dominant party (ZANU PF). The party set the rules of the 'political game' and was able to control the agenda (Derman and Manzungu, 2016). However, when it comes to the current domain of power and the political authority, the same party has regained close to total control again, compromising democracy and equity in the catchment.

The same goes for female representation in Sanyati catchment, most of the time the quorum is male dominated. The catchment council does not have a gender policy neither does it have a gender focal person. The system does not give preference to the women and girl child as well as marginalised groups in society. This observation violates the IWRM principle number three that underscores the involvement of women. The finding concurs with Manzungu (2002) stating that women in Southern Africa have remained in parenthesis in the implementation of IWRM. The Sanyati catchment and national-levels politics undermine women participation and economic development, respectively. This inevitably incubates struggles that may be actually influencing some negative changes in Sanyati catchment in terms of water conservation.

Catchment Protection From Degradation

ZINWA being a parastatal board under the Minister of Environment, Water and Climate, advocates for the protection of water bodies and any other form of wetlands. Efforts to manage the environment are fruitless, contamination are present in Sanyati catchment and pollution problems are dealt with as early as possible. Awareness campaigns conducted during early 2018 in Ngondoma and Muzvezve enlightened people about the importance of conservation. The proper functioning of 87 water supply stations (which, 15 have conventional treatment plants and the 72 have borehole sources) is positive testimony to catchment protection.

However, weirs in some parts of the catchment continue to be vandalized. There is antagonism in other communities who have a feeling that when a dam is to be introduced, they may be displaced like what happened in the Tokwe Mukorsi scenario (Marungwara 2014; Tarisayi 2018). In this case, the conservation of the water resources in sub-catchments is at the mercy of perceptions of the users, due to what they have seen happening in the past. For example, the changes in the politics of Zimbabwe have seen people being disadvantaged regardless of safety promises prior to elections. Glaring examples were *Operation Murambatsvina* (ZLHR 2005) and land reform programme that kept allocating and displacing people. Ngondoma gold panners were so hostile to both the police and the ZINWA staff, not to mention sub-catchment representatives. Environmental protection in Sanyati was compromised by mining activities as a result of limited livelihood alternatives and lack of certainty and security as a result of abrupt changes in policies by the central government. Water users felt helpless and stewardship spirit was being eroded due to inconsistent policy changes.

Groundwater protection is not guaranteed in the catchment. Giving its hidden nature, ground water protection from contamination through landfill disposal sites and leakage from storage tanks makes it difficult to protect. Especially when the council rarely conducts and draft comprehensive environmental impact assessments to ensure that pollution prevention measures are in place.

Siltation was high in Sengwa, Nyaodza and Ume River Systems due to sand abstractions, gold mining and river bank cultivation. The same people who are supposed to be protecting their environment are the ones degrading it. The rivers actually need urgent de-siltation to safeguard the water resources. However, no action has been taken by the CC and SCCs to that effect.

Water Monitoring (Flows, Quality, and Ground Water Levels)

Accessibility of most gauging stations is challenging, moreover there is inadequate staff to operate these gauging stations. The SCCs largely lack the technical capacity to fully perform water monitoring and data collection. As necessary as technical manpower is, Sanyati SCCs fail to attract the expertise required for data collection and monitoring because they cannot afford them.

Permit Regulation and Supervision

Zibagwe is the SCC with the most active people who apply for permits and Sengwa constitutes the least, and this is happening in the same catchment. Due to a systematic permit regulation system, the requirements are always the same and results in smooth application and granting of permits. Forms for application, requirements and terms are specified in the SI 206 of 2001. However, regardless of a clearly laid out procedure of acquiring permits which applies to all at a local scale, some sub-catchments are not producing the same results purported in the dominant narrative that acclaims decentralization as a panacea to resources management.

Borehole drilling companies such as Pote drilling services (T/A Makanaka investments) and Toramvura are mandated to only drill a borehole once they have seen an authority to drill or sink a commercial well (in form of GW3) to ensure an organized way of obtaining information about boreholes. Every aspect of water management is stated in the S1 206 of 2001 regarding use, validity period and rates to be paid which makes it easier for both the clients and the catchment.

However, people rarely apply for permits and other drillers do not follow the law, thereby drilling without a legal documentation. The study revealed that despite decentralization process since the early 2000, most water users in Sanyati do not know the difference between the Authority to drill or sink a commercial well (GW3) and the temporary abstraction permit (GW7A). There is a need for education and awareness efforts in the catchment. To make matters worse, Sanyati catchment lacks follow ups on unregistered boreholes and expired permits. The Dublin principle that regards water as an economic resource has not been fully understood in the Sanyati catchment as exhibited by some of the questions posed by the noncompliant water users. The majority of water supply schemes are not making any profit and economic use of water is limited.

For most of the users, water is a social and infinite good that is God-given. The questions and sentiments that revealed their ignorance to the finite and economic attributes of water were as follows:

- Why do I have to pay for water, when it is God-given?
- Why do I have to pay for raw water which is not treated?
- Why place such a price for the water that I pump without any financial assistance from anyone including ZINWA?
- Is ZINWA not a public organization, why not freely provide us with raw water and drill boreholes for us to irrigate?
- We must pay less money for dams in resettled farms taken over by ZINWA than for dams constructed by ZINWA.
- Why should I have to pay ZINWA for my private borehole/dam yet ZINWA contributed nothing?
- Do you know that you are derailing/sabotaging the land reform programme by charging water yet we get other inputs for free?
- Why do we have the same people representing us every time?

CONCLUSION AND RECOMMENDATIONS

The article engaged the interplay of the broad concepts of political ecology. Sanyati catchment exhibited complications instigated from both social-environmental interactions and a politicised environment. Decentralization to local scale is associated with effective environmental problem solving. If done properly, decentralization to lower tiers has a record of producing resounding successes. However, the study findings revealed evidence contrary to expectations. Local scale would be expected to yield the same positive results but that has not been witnessed in Sanyati catchment. For some SCC, the tenets of the scalar concepts have been proven whilst in most they have been totally refuted. Despite availability of a local playing field, both the environment and water users are under threat. The study suggests that what brings about change in terms of equity, effective distribution, utilisation and protection of the catchment is not theoretical ceding of power but full support and feedback is the backbone of successful water reforms and management. In terms of the participation dynamics, not much has changed from the colonial administration that favoured resourced people at the expense of the poor. Water reform success cannot be an issue of regime change only but capacity building and development within catchments. Taking water management to local scale without the requisite resources will only remain a brilliant process in terms of structure but dismally failing on the ground. Besides the mere excitement that management has been taken to local scale, results in terms of catchment protection, distribution of water and representation still leave a lot to be desired at catchment level. Decentralization does not seem to be yielding its intended benefits due to lack of financial resources, technical knowhow, people's attitudes and the national political terrain that is seasonally populist. Of interest was the fact that not only did the issue of scale bring challenges of disregard and downplay, the local people themselves also created perception problems that made water and catchment management difficult. Given the results, there is nothing desirable about local scale if not accompanied by necessary enabling structures and attitudes.

It is therefore recommended that:

- The government provides financial resources to SCC and CC to turn the rhetoric into reality. Clearly the major cause of absenteeism is lack of funding. Representatives should be remunerated in order for them to be motivated. The incentives could be factored in the permit fees.
- To allay water users' fears, they should be continually conscientized about the advantages of decentralization. Their fears, scepticism and ignorance must be cleared through education, honesty, consistency and equal sharing of resources and benefits.
- Representatives be chosen from willing people and they should rotate without favouritism to harness confidence in the whole system.
- There is a need for capacity development in the area of data collection and management. This could be done through trainings, where SCC representatives are exposed to flow measurements and data management lessons. However, knowledge transfer mechanisms should be put in place to avoid knowledge monopoly.

REFERENCES

Abel, R., & Thieme, M. (2009). Integrated Water Resources Management and Integrated River Basin Management. Encyclopedia of inland waters.

Adams, W. W. (1990). *Green Development: Environment and Sustainability in the Third World*. Routledge.

Arnstein, S. R. (1969). A Ladder Of Citizen Participation. *Journal of the American Institute of Planners*, *35*(4), 216–224. doi:10.1080/01944366908977225

Bassett, T. J., & Zueli, K. B. (2000). Environmental discourses and the Ivorian Savanna. *Annals of the Association of American Geographers*, *90*(1), 67–95. doi:10.1111/0004-5608.00184

Biswas, A. K. (2008). Integrated Water Resource Management: Is it Working? *Water: Research and Development, 24*(1), 5–22.

Brown, J. C., & Purcell, M. (2005). There's nothing inherent about scale: Political ecology, the local trap, and the politics of development in the Brazilian Amazon. *Geoforum, 36*(5), 607–624. doi:10.1016/j.geoforum.2004.09.001

Conyers, D. (2003). Decentralization in Zimbabwe: A Local Perspective. *Public Administration and Development, 23*(1), 115–124. doi:10.1002/pad.265

Derman, B., Gonese, F., & Ferguson, A. (2001). *Decentralization, Devolution and Development: Reflections on the Water Reform Process in Zimbabwe*. Centre for Applied Social Sciences, University of Zimbabwe.

Derman, B., & Manzungu, E. (2016). The complex politics of water and power in Zimbabwe: IWRM in the Catchment Councils of Manyame, Mazowe and Sanyati (1993-2001). *Water Alternatives, 9*(3), 513–530.

Escobar, A. (2001). Culture sits in places: Reflections on globalism and subaltern strategies of localization. *Political Geography, 20*(2), 139–124. doi:10.1016/S0962-6298(00)00064-0

GoZ. (1994). *Report of the Commission of Inquiry into Appropriate Agricultural Land Tenure Systems. Three volumes*. Government of Zimbabwe.

GoZ (Government of Zimbabwe). (2000). *Water resources management*. The Government Printers.

Green, K. E. (2016). A political ecology of scaling: Struggles over power, land and authority. *Geoforum, 74*, 88–97. doi:10.1016/j.geoforum.2016.05.007

GWP. (2000). *Integrated water resources management (Background Papers No 4). Technical Advisory Committee (TAC)*. Global Water Partnership.

GWP. (2006). *Gender mainstreaming: An essential component of sustainable water management. Policy Brief No. 3*. Global Water Partnership.

Hecht, S. B., & Cockburn, A. (1990). *Fate of the Forest: Developers, Destroyers, and Defenders of the Amazon*. Harper Collins.

Hellum, A., & Derman, B. (2005). Negotiating water rights in the context of a new political and legal landscape in Zimbabwe. In F. von Benda-Beckmann, K. von Benda-Beckmann, & A. Griffiths (Eds.), *Mobile People, Mobile Law: Expanding Legal Relations in a Contracting World* (pp. 177–198). Ashgate.

Herring, R. (2001). Authority and scale in political ecology: some cautions on localism. In L. Buck (Ed.), *Biological Diversity: Balancing Interests Through ACM*. CRC Press. doi:10.1201/9781420042597.ch10

Jakeman, J., & Letcher, R. A. (2003). Integrated assessment and modelling. Features, principles and examples for catchment management. *Environmental Modelling & Software, 181*(6), 491–505. doi:10.1016/S1364-8152(03)00024-0

Jessop, B. (2000). The crisis of the national spatio-temporal fix and the tendential ecological dominance of globalization capitalism. *International Journal of Urban and Regional Research, 24*(2), 323–360. doi:10.1111/1468-2427.00251

Manzungu, E. (2002). More than a headcount: Towards stakeholder representation in catchment management in South Africa and Zimbabwe. *Physics and Chemistry of the Earth, 27*(11-22), 927–933. doi:10.1016/S1474-7065(02)00095-5

Manzungu, E., & Derman, B. (2016). Surges and ebbs: National politics and international influence in the formulation and implementation of IWRM in Zimbabwe. *Water Alternatives, 9*(3), 493–512.

Mapedza, E., & Geheb, K. (2010). Power dynamics and water reform in the Zimbabwean context: Implications for the poor. *Water Policy, 12*(4), 517–527. doi:10.2166/wp.2010.141

Mapedza, E., Manzungu, E., Rosen, T., Ncube, P., & Van Koppen, B. (2016). Decentralized water governance in Zimbabwe: disorder within order. Water Resources and Rural Development. doi:10.1016/j.wrr.2016.05.00

Marugwara, W. (2014). *The plight of internally displaced communities: case of Tokwe Mukosi*. http://hdl.handle.net/11408/595

McCarthy, J. (2005). Scale, sovereignty, and strategy in environmental governance. *Antipode, 37*(4), 731–753. doi:10.1111/j.0066-4812.2005.00523.x

Mehta, L. (2014). Water and human development. *World Development, 59*, 59–69. doi:10.1016/j.worlddev.2013.12.018

Mohan, G., & Stokke, K. (2000). Participatory development and empowerment: The dangers of localism. *Third World Quarterly, 21*(2), 247–268. doi:10.1080/01436590050004346

Myers. (2002). Local communities and the new environmental planning: A case study from Zanzibar. *AREA, 34*(2), 149-159.

Nemarwandwe, N., & Kodzanayi, W. (2001). *Institutional arrangements for water resources use: A case study from Southern Zimbabwe*. Institute of Environmental Studies-University of Zimbabwe.

Parker, P., Letcher, R., Jakeman, A., Beck, M. B. G., Harris, R. M., Argent, M., Hare, C., Pahl-Wostl, A., Voinov, M., Janssen, P., Sullivan, M., Scoccimarro, A., Friend, M., Sonnenshein, D., Barker, L., Matejicek, D., Odulaja, P., Deadman, K., Lim, G., ... Bin, S. (2002). Progress in integrated assessment and modelling. *Environmental Modelling & Software, 17*(3), 209–217. doi:10.1016/S1364-8152(01)00059-7

Paulson, S., Gezon, L., & Watts, M. (2005). *Politics, ecologies, genealogies*. Rutgers University Press.

Pilosoff, R. (2012). *The unbearable whiteness of being: Farmers' voices from Zimbabwe*. Weaver Press.

Purcell, M. (2002). The state regulation and global restructuring: Reasserting the political in the political economy. *Review of International Political Economy, 9*(2), 284–318. doi:10.1080/09692290110126128

Schroeder, R. A. (1999). *Shady practices: agroforestry and gender politics in The Gambia*. University of California Press.

Smith, N. (1993). Homeless/Global: Scaling places. In Mapping the futures: Local cultures global change. Routledge.

Swatuk, L. (2002). Water reforms in Zimbabwe: Some observations based on the Save catchment experiences and suggestions for ways forward. In E. Manzungu (Ed.), *The processes and dynamics of catchment management in Zimbabwe* (pp. 45–68). Save Africa Trust.

Swyngedouw, E. (1997). *Neither global nor local, Glocalization and the politics of scale*. Guilford press.

Swyngedouw, E. (2004). Scaled geographies: nature, place, and the politics of scale. In Scale and geographic enquiry: Nature, society, and method. Blackwell. doi:10.1002/9780470999141.ch7

Tarisayi, K. (2018). Traditional leadership and the Tokwe-Mukosi induced displacements: Finding the missing link. *Journal of Disaster Risk Studies*, *10*(1), 1–22. PMID:30473759

van Koppen, B., Tarimo, A. K. P. R., van Eeden, A., Manzungu, E., & Sumuni, P. M. (2016). Winners and losers of IWRM in Tanzania. *Water Alternatives*, *9*(3), 588–607.

Weale, A. (1992). *The new politics of pollution*. Manchester University Press.

Zimmerer, K. (2000). Rescaling irrigation in Latin America: The cultural images and political ecology of water resources. *Ecumene*, *7*(2), 150–175. doi:10.1177/096746080000700202

ZLHR. (2005). *Report on the Impact of Operation Murambatsvina. The Case of Hatcliffe Extension*. Researched and compiled by Department of Sociology, University of Zimbabwe Zimbabwe Lawyers for Human Rights Sponsored by Zimbabwe Defence and Aid Fund. http://archive.kubatana.net/docs/urbdev/zlhr_uz_impact_murambatsvina_hatcliffe_0506.pdf

Chapter 11
An Analysis of Sector-Based Water Supply and Demand:
Mapping the Possible Water Development and Management Ways for Sanyati Catchment, Zimbabwe

Winmore Kusena
https://orcid.org/0000-0002-9948-2147
Midlands State University, Zimbabwe

Thomas Debwe
Midlands State University, Zimbabwe

ABSTRACT

This chapter sought to assess sector-based current water use levels in Sanyati catchment, assess a possibility of new ways of managing the catchment water using inter-basin transfer systems, and assess future water demands projection using WEAP model. The study was both quantitative and qualitative in nature. Interviews and observations were the main primary data collection instruments. Findings showed that the main sectors and institutions that utilised water in the catchment were mining, agriculture, manufacturing industry, and local authorities. Local authorities and farming sectors were the major users of water. WEAP model results showed a steady increase in demand for water as a result of both population growth and agricultural activities water demand. The study mapped possible inter-basin transfer routes and recommends water harvesting as part of development and management in Sanyati catchment to address the problem of water shortage.

INTRODUCTION

Global consumption of water is doubling every 20 years, more than twice the rate of human population growth (United Nations 2011). If the current trend persists, by 2025 the demand for fresh water

DOI: 10.4018/978-1-7998-8809-3.ch011

is expected to rise by 56 percent more than is currently available. Water as a natural resource is scarce and requires integrated approaches in its management. The supply of water is limited but the demand for water is increasing rapidly on a global scale. The imbalance between water supply and demand is expected to greatly expand as water demand inevitably increases. With respect to total water demand, agricultural demand is the largest globally whereas demand in the municipal and industrial sectors is increasing significantly (Shiklomanov, 2000). Almost 70% of the extracted fresh water is utilized for crop-raising activities globally (FAO, 2013), and more water is required in order to produce more food in the future. It is believed that if water is utilized effectively in agriculture, it will not be a bottleneck for future food production (De Fraiture and Wichelns, 2010). The demand for municipal and industrial uses is expected to increase rapidly in the near future due to burgeoning urbanization and industrialization. More specifically, the increases between 2000 and 2050 will mainly come from manufacturing (400%), electricity (144%), and domestic uses (127%) (Leflaive et al., 2012),

Water scarcity has been a critical issue for many regions and has caused numerous water conflicts. In Southern Africa, water is key to winning the battle against poverty and its scarcity could be a limiting factor to growth. Southern Africa's water sector is already being affected by unsustainable water and land use practices, over pumping, flooding, watershed degradation, wetland loss, and proliferation of aquatic weeds, sedimentation, and climate variability. No socio- economic development can take place without water. Zimbabwe's water resources are indeed limited and scarce. The situation is worsened by the occurrence of droughts and the increasing demand associated with population growth and a developing economy. Zimbabwe is an atypical sub-tropical country with one rainy season (November to March). The country's average rainfall is 657 mm/annum and varies spatially from the eastern highlands (1,000mm) to low lying areas such as valleys (400 mm) in the southern part of the country. Specifically, agriculture normally requires a sizable amount of water during the crop growing season, and much less water during the other months of the year. The improvement in water use efficiency and productivity is widely considered as the best solution to ensure that future water demand does not exceed water availability, and this requires special attention on the demand side.

Kjeldsen etal (2009) argue that in the last 20 years Zimbabwe has experienced periods of severe droughts, increasing the pressure on available freshwater resources and enhanced the occurrence of conflicts between water users. Unreliability of precipitation for the past decades in Zimbabwe has seen many problems arising in water supply for both rural and urban users (Chigumira and Mujere 2015). The Sanyati catchment runs through four different natural regions (1-4) with varying average rainfall amounts and hence the water demand levels are different and must be managed accordingly. The Sanyati catchment receives its rainfall in the upper reaches of the catchment, rainfall decreasing gradually towards the west. Surface water availability follows the same trend with the upper catchment having almost perennial flow. Rivers such as Munyati and Mupfure in Sanyati Catchment are perennial but in dry years these rivers dry up during August to November (Kjeldsen etal 2009).

Rivers in the lower catchment experience flow during the rainy season and dry up in early winter. Surface water decline in Zimbabwe will most likely affect development in the affected catchments as it provides most of the domestic water in rural areas, supports poverty reduction through irrigation and reliance on it is likely to increase as rainfall becomes more variable and demand for water becomes greater (Chikodzi 2013). The agricultural sector (which dominates the Sanyati catchment area) plays a very important role in both economic and social terms and therefore, sufficient water supply plays a crucial role in everyday life (Nhapi 2015). Climate change will invariably affect water withdrawals especially against the predicted decreasing precipitation trend (1mm to almost 15 mm per decade) and the 2°C and 3°C increase in temperature which will also mean higher crop evaporative demand (Gumbo 2006). According to Makurira and Viriri (2017), they argue that sub-catchment council offices are under-funded and, this in turn, affects other important activities such as monitoring. To manage water resources effectively, current and future water demand and use for all sectors should be estimated as accurately as possible. Therefore, this study seeks to identify surface water sources and users in the catchment; establish current uses of water levels per activity within the catchment; map the possible inter basin transfer routes in the catchment; and forecast future water demands for major water users in the catchment for the next 10 years.

Study Area

Sanyati catchment is one of the seven catchments of Zimbabwe (Figure 1). The catchment lies between latitudes 16^0 16^1 and 19^0 30^1 south and longitudes 27^0 45^1 to 31^0 16^1 east. Sanyati river system covers areas drained by Kwekwe, Zivagwe, Munyati, Mupfure, Sengwa, Ume and Nyaodza rivers. There are 15 hydrological subzones with areas varying from 2527 to 11427km^2. These hydrological sub-zones from the ten (10) sub-catchments namely: Sengwa, Nyaodza, Zivagwe, Upper Munyati, Muzvezve, Lower Munyati, Upper Mupfure, Upper Middle Mupfure, Lower Middle Mupfure and Lower Mupfure.

The catchment as constitutes a combination of four natural climatic regions and each region has its specific climate characteristics. These are regions ranged from 2 to 5 as shown in table 1:

Table 1.

Region	Rainfall
Region II	750-1000mm
Region III	650-800mm
Region IV	450-650mm
Region V	Below 650mm

The topography of the Sanyati catchment is sloping towards the North West direction as evidenced by the direction of rivers that flow into the Kariba dam. The catchment constitutes an upland area with an altitude of above 1200m, covering approximately 21527km^2 of the catchment. The areas like Mashaba Mountains in upper Munyati sub-catchment form much of the watershed and the source of Sanyati River. The upper reaches of the catchment are generally flat and gently undulating until it intersects the Great Dyke. The middle part of the catchment has an altitude of above 900m extending north East and Northwest. The topography is generally flat and undulating with ridges and hills that are as follows; Mafungautsi plateau with an altitude of 1192m in Lower Munyati sub-catchment, its main land use is for recreational parks and forestland. The other plateaus are Kwavanyangau (1181) and Chikurukwe (1385m) in Sengwa and Nyaodza sub catchment respectively and the land is classified as communal land. The low-lying areas are at an altitude of more than 300m. Ridges of 1000m and above are found in Nyaodza, Ume, Sanyati and Sengwa river basins. Basically these are the areas along the Zambezi valley and have a ridged terrain with isolated hills for example Chizarira hills (1433m) in Sengwa sub-catchment, Mapongola (1100m) and Whamira (1336) hills in Sengwa and Nyaodza sub-catchment and the land use is for national parks and wildlife management. Charara and Chirisa are classified as safari areas in Nyaodza and Sengwa sub-catchment (Sanyati River System Outline Plan, 2009).

Figure 1. Map Showing the Study Area

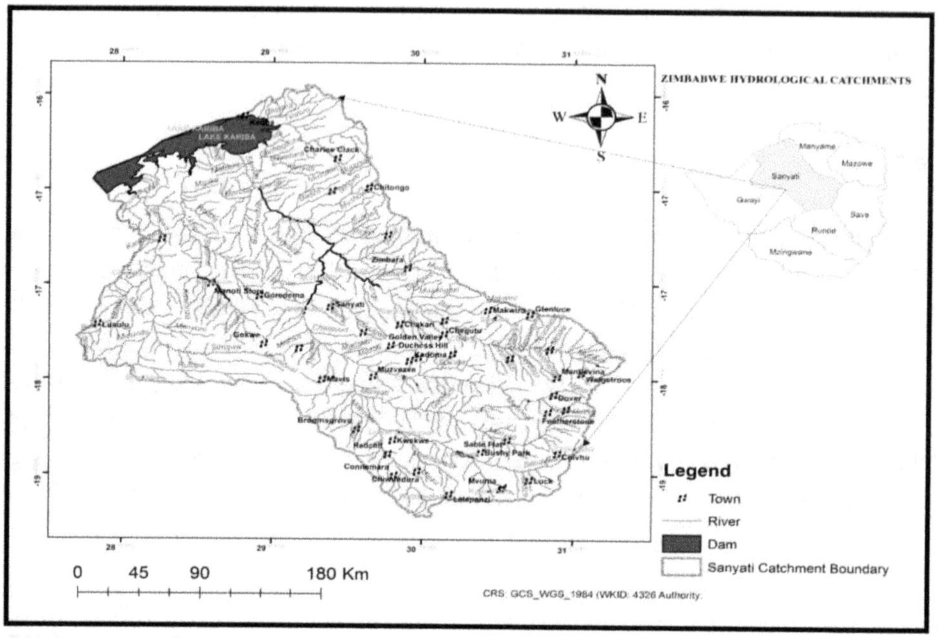

METHODOLOGY

A case study approach was used for this study because it enables researchers to closely examine data within a specified context. Hence, Sanyati Catchment was purposively chosen as a case study for assessing water demand levels and uses because it presents a cross-section of the whole country in terms of climatic characteristics. A case study was preferred because it gives an intensive description and analysis of the current water demand situation and the concerns for future water requirements. The study targeted key informants from the ZINNWA and Sanyati Catchment Council. Through interviews, the key informants provided insight on issues to do with water demand levels. These key people included Catchment hydrologists, inspector, and planning engineer. The Sanyati raw water users in ZINWA agreements and permits database and Agritex personnel were also part of the target population (Irrigation Engineer). The reason for targeting the Agritex personnel was to do with the expert knowledge on irrigation rates.

This study used all identified raw water users in the catchment as per ZINWA water agreements and permits to determine clearly the levels of usage of water per activity. However, for forecasting water requirements, Muzvezve Sub-Catchment was chosen as a reference study area within the catchment. Purposive sampling based on availability of flow data was considered among other factors such as river shape and size. Therefore in terms of monthly flow data consistence, Muzvezve River in Muzvezve Sub-Catchment was used for 10 year future water demand forecasting using WEAP model.

WEAP model is computer based software which stands for Water Evaluation and Planning System was used to project future water requirements amongst the major users within Sanyati Catchment using a single sub-catchment for simulation purposes. The model uses scenario based analysis for simulating reference years. Various models can be used to forecast different economic variables on short, mid and long term basis. Long-term projections are usually done at global level, regional levels. Secondary data for simulation was obtained from River System Outline Plan (RSOP) and ZINWA annual reports from the Hydrology and Raw Water Engineering department. Monthly flow data were obtained from the hydrological department, consumption rates and annual total volumes of water demand per sector was obtained from Sanyati Catchment Raw Water Engineering department

Annual rainfall data was obtained from the CHIRPS online data portal on the website http://chg.geog.ucsb.edu/data/chirps/ and exported in ArcMap 10.5. DEM was obtained from https://gisgeography.com/free-global-dem-data-sources/ for analysis with a spatial resolution of 1 arc-second (30 meters). This high-resolution terrain data set was opted for due to its level of accuracy. The obtained slope information was in form of raster and it was later put into a GIS environment for processing and analysis. Population statistics for Kadoma obtained from the 2012 census together with the associated population growth rates (ZIMSTATS 2012) information was used in the modelling of water demand for Kadoma City as a major demand site.

Data from the secondary sources was mainly analyzed through Microsoft excel while content analysis was used for key informant interviews and field observations results. ArcGIS 10.5 software was used for mapping possible IBTS routes for Sanyati basing on slopes and rainfall. However, the model shortfall is that it did not include other factors such as settlement and geology for IBTS mapping.

The WEAP model was run under two scenarios which projected irrigation increase and population projections within the catchment based on arithmetic growth rate of 2.01%. WEAP model for water allocation was run using C87 gauge station flow data from ZINWA. For the best results in running the model, the climatic data, soil, geology and other parameters are needed to closely imitate the reality of the Sanyati Catchment.

RESULTS AND DISCUSSIONS

Water Sources in the Catchment

The main sources of raw water in Sanyati Catchment consist of rivers and dams. Rivers draining Sanyati are both perennial and non-perennial. Perennial rivers are mainly found on the upper reaches of the catchment whilst it is the opposite in western side of the study area. One of the reasons causing this is that Sanyati catchment receives its highest rainfall in the upper reaches of the catchment, rainfall decreasing gradually towards the west. Sanyati catchment has three major rivers which are namely Sengwa, Ume and Sanyati and these feed the Zambezi Trans-boundary Rivers. However, other small rivers also drain the catchment. For example in total there are 6680 basins in the study area supplying its different consumers. Table 2 shows the major rivers in the catchment.

The major rivers shown in table 2 replenish dams that happen to be the major sources of water for most water users in the catchment. These dams also provide socio-economic development benefits to the catchment water users. Table 3 shows the dams that are found in Sanyati catchment and their capacities

Table 2. Major rivers in Sanyati catchment

Sub-catchment	Catchment Area (km2)	Major River
Lower Mupfure	4 062	Mupfure
Lower Middle Mupfure	2 666	Mupfure
Upper Middle Mupfure	2 527	Mupfure
Upper Mupfure	2 871	Mupfure
Lower Munyati	7 297	Munyati
Upper Munyati	7 776	Munyati and Ngezi
Zivagwe	6 896	Sebakwe
Muzvezve	3 271	Muzvezve
Nyaodza	14 028	Sanyati and Nyaodza
Sengwa	18 660	Sengwa and Ume
TOTAL AREA	**70 054**	

Table 3. Major Dams in Sanyati Catchment

Name of dam	Full Capacity (10^6m^3)	Level (m)	Volume as at 31 December 2017 (10^6m^3)	Volume as at 31 January (10^6m^3)	% full	Change (10^6m^3)	% Change
Cactus Poort	3	100.1	2.067	3.086	102.87	1.019	33.97
Chikomba	5.46	98.99	3.974	4.179	76.54	0.205	3.75
Claw	65.46	105.12	45.598	58.923	90.01	13.325	20.36
Clifton	10.87	98.24	7.153	7.565	69.60	0.412	3.79
L/Sebakwe	6.99	99.94	6.858	6.877	98.38	0.019	0.27
Dziva	2.88	100.01	2.868	2.886	100.21	0.018	0.63
Mamina	11.36	98.06	6.846	7.044	62.01	0.198	1.74
Ngezi	22.63	97.5	10.591	11.452	50.61	0.861	3.80
Ngondoma	7.49	99.93	3.029	7.339	97.98	4.310	57.54
Nyamafufu	9.11	98.95	7.075	6.947	76.26	-0.128	-1.41
Padre's pool	2.99	96.99	1.464	1.448	48.43	-0.016	-0.54
Sebakwe	265.73	96.05	193.571	186.278	70.10	-7.293	-2.74
Suri Suri	9.97	95.4	2.654	2.467	24.74	-0.187	-1.88
White Waters	4.9	99.13	3.792	3.705	75.61	-0.087	-1.78
Mahusekwa	2.992	99.2	2.232	2.51	83.89	0.278	9.29

An interview with ZINWA River Inspector pointed out that these dams drive the livelihoods of many water users in the catchment. This is mainly because all water users benefit in terms of water supply from these dams. The interview also revealed that some water users applied for permission to build their own dams in order to meet demand. Some of the dams included Upper Mupfure OX-West Dam, Donnington dam (5.0 $10^6 m^3$), Selusi dam (13.5 $10^6 m^3$) and Burgan dam (5.0 $10^6 m^3$)

Sanyati Water Users

ZINWA database on raw water users in the catchment reveal six categories of water users. The categories of users are Agriculture A2, Agriculture A1, Agriculture Communal, Industry, Local Authority and Mining. Agriculture A2 consists of large commercial farms mainly focusing on food cash crops such as maize and wheat, these are 320 hectares on average. Local Authorities were also a major water user in the catchment; and these refer to different municipalities that draw water from the catchment major dams. Table 4 shows number of raw water clients/users in the catchment by category.

Table 4. Updated Raw Water clients by Category for 2018-2019 water year

Category	Number of Clients
Industry	18
Mining	16
Local Authorities	8
Agriculture A2	237
Agriculture A1	46
Agriculture Communal	11
Total	**237**

Current Levels of Water Use Per Activity

Findings from ZINWA raw water database on current levels of use per activity gives different volumes used on an annual basis for each category. Local Authorities use on average up to 42.4% of the total volume of water used annually. An interview with ZINWA River Inspector indicated that, the volumes used by Local Authorities is high due to a large consumer base. The other factor is to do with some industrial activities present in Local Authorities consumer base which use treated water. As such, these industries cause higher water demand for local authorities. Some of the industries that rely on treated water include Delta Beverages, Dendairy and ZIMASCO.

Agriculture combined (A1, A2 and Agriculture Communal) as one activity has an estimated consumption rate of 39% of the total volume used in 2018-2019 water year. Water use in agriculture was mainly for irrigation purposes which extend throughout the year. Findings from key interviews alluded to the fact that more water in agriculture is used during the winter season especially by farmers engaged in wheat production. An estimated 6,5ML/ha is used during the winter season whilst early summer it is 3ML/ha and an estimated 2,5ML/ha during the summer supplementary. This is supported by FAO (2016) which states that during dry months, irrigation requires more water compared to other months of the year. According to Ministry of Lands and Agriculture (2017), only 37% of the country receives rainfall considered adequate for agriculture. Irrigation is of vital significance due to uneven distribution of rainfall and seasonality in the catchment (Buyukcangaz et al, 2017).

Mining had the lowest consumption rate in the catchment of 2.1% compared to other activities on an annual basis. This could have been motivated by the fact that miners are charged more per Megalitres ($50/ML) and the activity generally uses less water. Mining entities such as Prucible Mine engaged in water reuse since mineral processing is not depended so much on the quality of water. Findings from key informants revealed that, miners are charged more due to the cost associated with controlling pollution.

Manufacturing Industry as an activity has an average of 16% water use level for the water year 2018-2019. The water use in industry as a sector varies according to scale of operations. For example volumes of water used at Delta beverages are relatively high compared to the use of water at ZIMASCO. The factor responsible for this is different activities which require water are more at Delta compared to those which use water at ZIMASCO.

Table 5. shows the different water use levels per activity in Sanyati Catchment

Category (Activity)	Annual Water usage volumes(ML)	Percent in Use
Mining	4 090.56	2.1
Industry	32 859.00	16.9
Agriculture A1	1 412.05	0.7
Agriculture A2	72 188.84	37.1
Agriculture communal	1 534.25	0.8
Local Authority	82 500.00	42.4
TOTAL	**194 584.70**	**100**

Trend Analysis of Water Supply Levels Per Activity

Gathered evidence from ZINWA annual reports database shows that there are changes to supply levels of water for the past 3 years in the catchment. Volumes of raw water abstraction have been changing in all the sectors due to a number of reasons. Researchers were able to get information for three consecutive years (2016-2019) in order to establish the trends in water supply levels per activity in the catchment area.

Table 6. Changes in water demand levels per activity in Sanyati catchment

Category/ Major Users	Annual Volumes (ML)			Percentage in Use		
	2016-2017	*2017-2018*	*2018-2019*	**2016-17**	**2017-18**	**2018-19**
Mining	3 823.76	13 403.00	32,859.00	2.7	2.2	2.1
Industry	11 545	3 467	4 090.56	8.2	8.3	16.9
Local Authorities	84 300	85 300.20	82,500.00	59.6	52.9	42.4
Agriculture A2	37 721.00	53 552.41	72,188.84	26.7	33.2	37.1
Agriculture A1	781.25	1 716.10	1,412.05	0.6	1.1	0.7
Agriculture C	3 229.80	3 507.50	1,534.25	2.3	2.2	0.8
Totals	**141 386.81**	**161,225.13**	**194,584.70**	**100**	**100**	**100**

Table 6 is an illustration of water supply levels per activity in the catchment area since the water year 2016-2017. For major water users such as Agriculture A2, Local Authority there is evidence of high demand as indicated by the rise in respective percentages. However, according to ZINWA Sanyati Catchment engineer technician, the increase in levels of water by some sectors since the mentioned years could be attributed to supply and demand gap. Users made use of more water as it was availed. However, the changes may also be attributed to inefficient water use, expansion of activities and population growth (Madebwe and Madebwe, 2013).

Possible Inter-Basin Transfer Routes in Sanyati Catchment

The possible Inter-basin transfer routes shown on Figure 2 was generated on the basis of slope, flow accumulation and flow direction. As such four classes of slope determined made four basins with notable variations in elevation. More water accumulates and flows basin 1 compared to other basins. Justification of uneven distribution comes after calculating the basin area. However, from visual analysis one can conclude that basin 1 is bigger and hence water that accumulates here especially in Sanyati River is more than that which is accumulating in other basins. Again this fact can be attributed to the natural region in which basin 1 lies in respect to levels of rainfall received. To the users of water in Sanyati catchment it means that the distribution of water is uneven hence some of the strategies to curb this problem has been countered with construction of dams. These dams serve as key water resources to the various users with the major uses being agriculture, mining and industry

Figure 2. Possible IBTS Routes

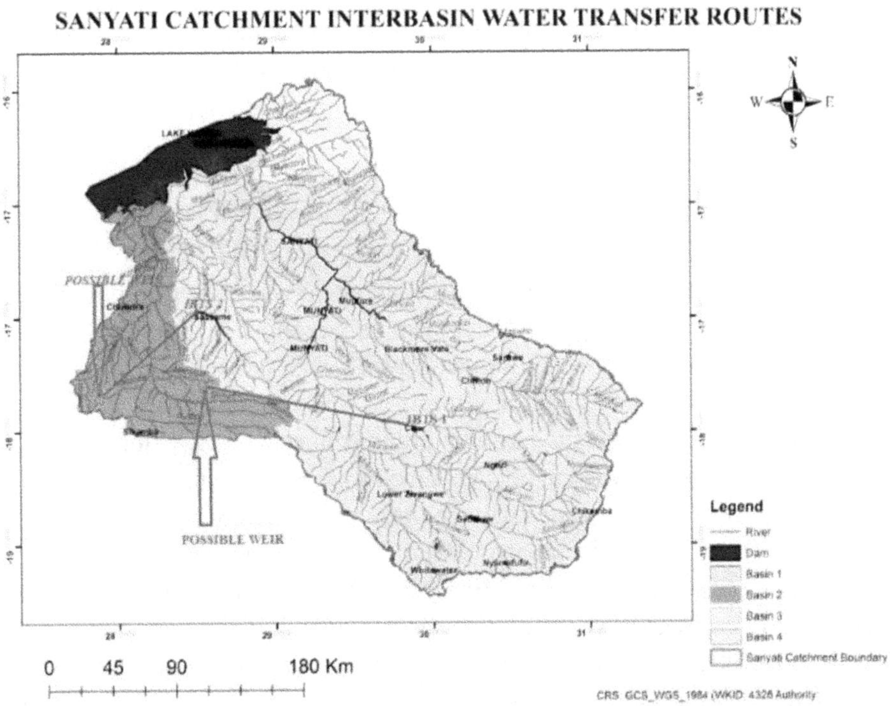

The availability and distribution of the rainfall in Sanyati catchment can prejudice water users in drier localities. The average amount of precipitation need not necessarily be a constraint to successfully carrying out an agricultural or water resources operation (Mazvimavi e.tal 2005). Therefore, IBTS serve s very important projects that address issues of water shortages in the catchment.

Figure 3. Spatial Distribution of Rainfall in Sanyati Catchment

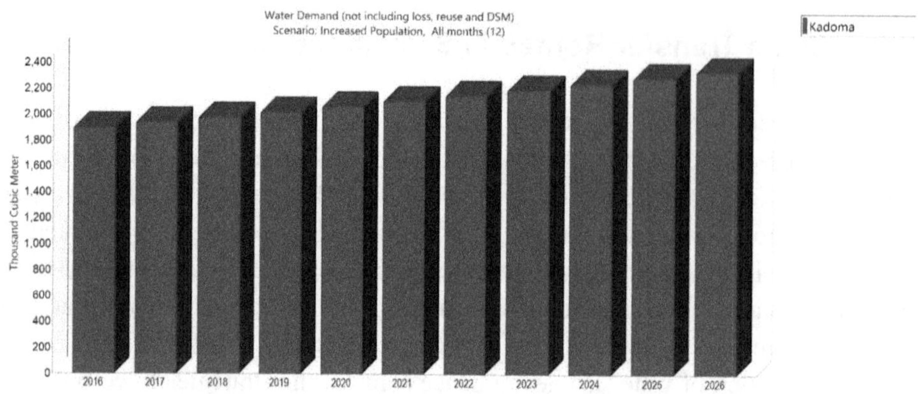

Figure 3 map shows that more precipitation is received on the upper reaches of the catchment with figures ranging from 750mm and above. The rainfall totals gradually decrease as one moves towards the western side of the catchment. These areas receive on an annual basis rainfall total ranging from 500mm and below. The difference in the amount of rainfall received in the catchment has a bearing on water availability. It means that areas in Sanyati Catchment do not have equal volumes of water since rainfall is unevenly distributed. This distribution in rainfall further explains the different climatic conditions which affect the catchment which are varied when one moves from the upper reaches of the catchment to the low-lying areas.

Future Water Requirements Simulation

A look into the future water requirements for Sanyati Catchment was done using the Water Evaluation and Planning (WEAP) using Muzvezve Sub-Catchment as a reference case. The flow data for Muzvezve river was run in the model using the current data available of 2015/2016 from ZINWA especially C87 gauge station. In terms of water abstraction, the city of Kadoma draws its water from this dam and hence Kadoma City is a major demand site. Agriculture in Muzvezve is a major water user and draws significant amounts of water annually from the dam hence a demand site in the WEAP model. The two demand sites were of concern to the researcher since these are the major water users as far as abstraction of raw water is concerned in the catchment as a whole. Simulation was done under two scenarios of population increase and irrigation expansion (future A and B). Projections indicated a steady increase in water demand levels in the next 10 years for both scenarios

The annual activity for the urban area Kadoma is based on population totals acquired from the country's 2012 census which indicated the city having an estimated population of 61 397 people. The estimated population growth rate was calculated at 2.05% per year and hence formed the basis for future water requirements. Agriculture future water demand was projected assuming that irrigation will increase by 2% annually against the background of increase in clients over the past years. Figure 3 shows the first projection in relation to Kadoma city as a demand site and how increase in population also mean an increase in the demand for water the projected years. The findings show a link between number of people and future demands. However, this could be overcome through water conservation.

Figure 3 clearly shows a steady rise in the water demand levels in the next 10 years with current accounts of 2016 for Kadoma City. It means that when populations increase the per capita demand also increases and hence water demand levels rise due to the effect of population. If population continues to grow at this rate over the next 10 years it means that, Kadoma City will require more water from the source with time. Figure 4.9 shows the level of increase or change in water demand levels for Kadoma City in the next 10 years. It depicts a sharp rise which will occur when population is on the rise until the year 2026

The same trend was exhibited when the model was run on the basis of increase in hectares under irrigation in the catchment. Therefore, it goes on to show that water demand is increasing despite unreliable precipitation which is a major input to surface water used in the catchment.

According to FAO (2016) future demand of water is set to increase due to increase in competition amongst major users around the world. As urban areas expand and populations increase over time, it means there is significant amount on the available water resources. This statement has been supported by the scenario population increase in the WEAP model for Kadoma city. There is evidence of cubic meters of water used annually as population increase at an estimated growth rate of 2.1% until the year 2026. In the case of Muzvezve sub-catchment, against the successful running of the model it means that if the current trends on head flow continue until 2026, there will be increased pressure on the source of the precious liquid which might affect other activities such as agriculture, mining and industry in the study area as a whole.

CONCLUSION

The study revealed pertinent issues that exist in the study area concerning water demand levels and future water requirements. The catchment transverse different ecological regions making the catchment an area of possible IBTS provided a proper operational criterion is established. The catchment also needs close monitoring of water developments such as dam construction because some farmers got permission to build dam and reservoirs in their premises. Demand for water is on the rise for both agriculture and local authorities. Therefore, it is of paramount importance that the raw water available in the catchment area is managed in an efficient way in order to satisfy the growing needs of water users. The rate of increase in the demand of water is closely related to the growing number of users in the past years.

Agriculture and urban withdrawals have the highest percentage in consumption of water compared to mining and industry as other users. Urban withdrawals are slightly above agriculture and that has been attributed to the wide base of users at their disposal.

The different natural regions in the catchment area have an impact on available water resources in the catchment when considering the effect of rainfall totals annually. It means that some rivers as major sources of water to users are not perennial. The different climatic regions within the catchment heavily influence the dominating type of socio-economic activities existing in a particular Sub-Catchment or area within the study area.

The future water requirements in the catchment modelled using the WEAP model in Muzvezve Sub-Catchment points to an increase in water demand levels. The scenario of population increase modelled in the model projected a significant increase in water demand for Kadoma City considering a 2.01% growth rate. It means therefore that with the expected increase in population numbers in the city, it is undoubtedly obvious that levels of water demand from the Claw dam will increase significantly as well. The increase in the number of users is the main reason for the rise of future water requirements in the near future as clearly modelled.

RECOMMENDATIONS

Taking into consideration the findings obtained from this research, the following suggestions were given:

- Sanyati Catchment Council needs to invest in serious water conservation strategies and education in order achieve efficient water use and void an obvious demand pattern that rises simultaneously with population increase.
- ZINWA and Sanyati Catchment Council should consider using the powerful WEAP model in managing its water resources as far as changing water demand is concerned by simulating storage and supply capacities of major sources of water in the study area.
- Sub-Catchments need to improve data collection within the catchment boundaries such as stream flows, evaporation rates, monthly variations in water usage timeously to enable the use of current data in water demand analysis.
- Water harvesting is also a possible management approach particularly for the western drier part of the catchment.

REFERENCES

Armstrong, J. W., & Graham, M. C. (1972). Comparative Study of Methods for Long-Range Market Forecasting. *Management Science, 19*(2), 211–221. doi:10.1287/mnsc.19.2.211

Attanasi, E.D. Close E. R., & Lopez, M.A. (1975). *Techniques for Water Demand Analysis and Forecasting: A Case Study of Puerto Rico.* Academic Press.

Banerjee, A., Chaudhury, S., Singh, D. K., Banerjee, I., Mahato, A. K., & Haldar, S. (2007). *Statistics Without Tears.* Inputs for Sample Size Calculations.

Carson, E., & Cobelli, E. (2001). *Modelling Methodology for Physiology and Medicine.* Academic Press.

Chikozho, C. (2005). *Policy and institutional dimensions of integrated river basin management: Broadening stakeholder participatory processes in the Inkomati River Basin of South Africa and the Pangani River Basin of Tanzania.* Commons Southern Africa Occasional Paper Series No. 12. Pretoria: Centre for Applied Social Sciences/Programme for Land and Agrarian Studies/ International Water Management Institute.

De Fraiture, C., & Wichelns, D. (2010). Satisfying Future Water Demands for Agriculture. *Agricultural Water Management, 97*(4), 502–511. doi:10.1016/j.agwat.2009.08.008

FAO. (2013). *World Food and Agriculture*. Food and Agriculture Organization Statistical Yearbook.

Finn, J., & Jacobson, M. (2008). Just Practice: A Social Justice Approach to Social Work. Peosta, IL: Eddie Bowers Publishing.

Global Water Partnership (GWP). (2000). Integrated Water Resources Management. Technical Advisory Committee, Background Paper, no. 4.

Groves, R. M., & Heeringa, S. (2006). *Responsive Design for Household Surveys:* Tools for actively controlling survey errors and costs. *Journal of Royal Statististics Soceiety*, 439–57.

Hancock, B. (1998). *Trent Focus for Research and Development in Primary Health Care: An Introduction to Qualitative Research.* Trent Focus.

Henry, J. G., & Heinke, G. W. (1989). *Water Use in North American Cities: Environmental Science and Engineering*. Prentice-Hall.

Howard, K. W. F. (1997). Impacts of urban development on ground water. In Environmental Geology of urban areas. Geological Association of Canada.

Hunt, N., & Tyrrell, S. (2001). *Stratified Sampling*. Webpage at Coventry University.

Kjeldsen, T. R., Lundorf, A., & Rosbjerg, D. (1999). Barriers to sustainable water resources management: A Zimbabwean case study. *Hydrological Sciences Journal, 44*(4), 529–539. doi:10.1080/02626669909492250

Moriarty, P., Butterworth, J., & Batchelor, C. (2004). *Integrated water resources management and the domestic water and sanitation sub-sector.* IRC International Water and Sanitation Centre.

Mulder, M. W. (2005). *A political policy analysis of the integrated water resources management approach in South Africa's water policy (1998 - 2001)* (Master's dissertation). University of Pretoria.

Musingafi, M., & Chando, C. (2013). The Legal Framework in the Governance of Potable Water Supply In Zimbabwe: A Global, Regional And National Overview of The IWRM Paradigm. *Journal of Law, Policy, and Globalization.*

Ogula, P. A. (2005). *Research Methods.* CUEA Publications.

Schulze, R. E. (2007). Coefficient of Variation of Annual Precipitation: South African Atlas of Climatology and Agro hydrology. Water Research Commission, Pretoria, RSA, WRC Report 1489/1/06, Section 6.3.

Shiklomanov, I. A. (2000). Appraisal and Assessment of World Water Resources. *Water International, 25*(1), 11–32. doi:10.1080/02508060008686794

Tevera, D., & Crush, J. (2003). *The new brain drain from Zimbabwe. Southern African Migration Project*. SAMP.

United Nations Development Programme. (2002). *Zimbabwe Land Reform and Resettlement Assessment and Suggested Framework for the Future.* Interim Mission Report.

Van Der Zaag, P. (1999). Water allocation Principles in Catchment Areas. In Water for Agriculture in Zimbabwe: Policy and Management Options for the Smallholder Sector. University of Zimbabwe.

Windham, C. (2008). *Description of Target and Sample Populations for Research Methods Sanyati River System Outline Plan (RSOP) 2009*. Academic Press.

Chapter 12
Addressing Energy Poverty Through Ecological Governance of Solar Home Systems in South Africa

Edzisani Ellen Netshiozwi
https://orcid.org/0000-0002-9948-2147
Management College of Southern Africa, South Africa

ABSTRACT

South Africa has long recognised the need to eliminate energy poverty, and significant progress has been made since 1996. With the recent global outcry about the use of sustainable and clean energy sources for human and ecological development, it has become essential for countries to upscale the use of non-grid electrification in order to effectively and sustainably eradicate energy poverty. South Africa implements different non-grid electrification programmes which focuses on ensuring universal access to clean energy by all, with one of these initiatives being the Solar Home Systems Programme. This study aims to contribute to academic discussion and knowledge about the role that solar home systems can play in bringing change on the current status of the energy poverty eradication initiatives which negatively affect the environment and the people. The study used a qualitative approach in assessing how the Solar Home Systems Programme contributes to energy poverty eradication and environmental preservation if managed in line with ecological governance principles. Semi-structured interviews were conducted with a total of 40 households in Limpopo and officials from government departments in the province as well as a service provider responsible for the roll-out and maintenance of solar home systems in Limpopo. Framed within the ecological economics theory which advocates for ecological and environmental solutions that take into consideration the future generation, the study established that the solar home systems can eradicate energy poverty and contribute to the protection of the environment if managed in line with ecological governance principles. The study further revealed that the Solar Home Systems Programme in South Africa failed due to lack of proper governance systems that provide solutions for the future as the programme was short-term focused and lacked proper ecological governance systems.

DOI: 10.4018/978-1-7998-8809-3.ch012

In order for the solar home systems to contribute to energy poverty eradication and environmental protection, the study recommends a total overhaul of the programme which includes ensuring that the provision of SHSs is not viewed as a temporary measure that is only meant for poor rural households but a long-term and sustainable initiative.

1. INTRODUCTION

Access to energy is central to meeting the economic needs of the country and basic social needs for households. South Africa's experience of inequality in energy provision during the apartheid period was characterised by differentiation in the provision of infrastructure (both formal dwellings and electricity grid) and a visible separation between the lower, middle and upper classes, with the benefits focused on the minority (Azimoh, 2016). In order to address this inequality, one of the country's goals entrenched in the energy policy is to reach universal access to households' electricity. Consequently, this has led to an increase in energy consumption over the years as more households receive electricity connections (Wentzel, 2005). Meanwhile, the increase in the electrification rate (i.e. the number of households electrified), has resulted in the country's further experience of an increase in the emissions of greenhouse gasses. This is because more than 90% of electricity in South Africa is generated through coal burning (Eskom, 2018) which emits greenhouse gasses. As a result, the country has been rated the 12th largest Green House Gas (GHG) emitter globally and the most carbon-intensive developing economy in the world (Parr, Swilling and Henry, 2018). This challenge can be addressed through the use of low carbon-intensive energy sources. Accessibility of energy through SHSs, therefore, provides equal opportunities for households to receive clean and reliable energy that is affordable if they are managed in line with ecological governance principles (Azimoh, 2016). *Ecological governance is not only about addressing energy poverty for social benefits, but it also focuses on ensuring sustainable and affordable access to energy and addresses ecological and environmental effects of conventional sources of energy. This is the main argument of ecological economists and proponents of renewable and clean energy technologies.*

According to Feron (2016), the notion of social justice is the main driver to energy access, because it determines the fairness or inequality between different groups, with the poor being the most affected by the inequality. To address the energy, divide for rural households, the South African Government established the SHSs programme which provides basic electricity for lighting, charging phones, watching a black and white television set and listening to the radio. The argument is that energy access through SHSs is easy as it is affordable as well as subsidised through the government's Free Basic Alternative Electricity Policy (FBAE)which reduces the cost that consumers would pay to access the technology and for daily operations (Department of Minerals and Energy, 2007). Furthermore, access to solar energy in South Africa is encouraged as the country has a daily solar irradiation averaged between 5.7kWh/m^2 per day and 8.0kWh/m^2 per day which is higher (Centre for Renewable and Sustainable Energy Studies, 2014), thereby making solar energy effective and favourable.

As postulated by Methvin and Philipp (2017), the majority of South Africans who lack access to grid electricity are poor households who live in remote areas where it is claimed to be not viable for the grid to be installed. In these areas, solar energy, which is part of the Non-Grid Electrification Programme, becomes the most viable source that can easily be accessed by poor households. The use of SHSs must however, not only focus on the motives of affordability for the grid-based electricity by both government and households. Instead, meeting the households' energy demand as part of universal access must take into consideration sustainability and contribution towards addressing the global climate change phenomena. This is also supported by the United Nations Environmental Programme (2011) which argues that the use of coal or fossil fuels in generating electricity in order to meet universal access undermines the ecosystem as it causes degradation and countries cannot continue to depend on coal to generate electricity.

As of 2017, households' connections to the grid in South Africa totalled 84.4 percent, with only 75.9 percent of households using electricity for cooking (Statistics South Africa, 2018). In rural areas such as Limpopo, there was only 60.2 percent of households who used electricity for cooking as they prefer using alternatives such as firewood due to affordability issues (Statistics South Africa, 2018). To redress these worrying trends, the SHSs programme was funded through the Renewable Energy Fund of South Africa (REFSA), providing a subsidy of R1 500 per solar home system. However, the programme was discontinued before any systems were financed as the REFSA model was identified as uneconomical as the government could not sustain the provision of the subsidy (Clark, 2005). The REFSA programme was however brought back into life by involving the private sector through the concession programme. The concession programme was meant to bring on board private investors in providing non-grid electricity and was implemented in the KwaZulu Natal, Eastern Cape and Limpopo provinces (Department of Energy, 2012).

SHSs are standalone photovoltaic systems which are suitable for off-grid use. The systems provide 12 voltages of direct current which serve as energy sources for low power appliances such as lights, radios and small TV's for an estimated time of three to five hours a day. Additionally, due to them being off-grid and stand-alone, SHSs can be more scalable compared to micro-grids. This is because rejection of the system by one household would not affect the other household who supports the system, which differs from how micro-grids operate. The programme uses a fee-for-service model which requires that households pay a monthly fee that would cover maintenance and costs for replacing the batteries as and when required (Department of Energy, 2012). The systems contain four main components which are: a 50WP photovoltaic (PV) panel which equates to approximately 250Wh/day, a charge controller, a battery (105Amp-hour), wiring and outlets for small appliances and four (4) energy efficient compact fluorescent lights (Department of Energy, 2012).

Nevertheless, the programme has experienced challenges which led to the non-achievement of the set targets. The study focused on assessing how ecological governance could contribute to the success of the programme and whether the programme is and would contribute to the eradication of energy poverty. Ecological governance as suggested by Samimi, Ahmadpour, and Ghaderi. (2012) can be defined as a system that focuses on institutional filters which facilitate a balance between anthropogenic factors and biophysical processes. This suggests that initiatives require effective management and enabling legal and regulatory structures. Ecological governance in the electricity sector requires a broad and transdisciplinary perspective that can integrate across the natural and social sciences, including the process of putting relevant ecological policies into place (Costanza, 1989)

2. LITERATURE REVIEW AND THEORETICAL FRAMING

The challenge of ecological economics requires a rethinking of the major problems of the past that were temporarily solved by economic growth. Owing to the fact that the price of electricity has been increasing, there could be a greater motive to shift to renewable energy which has been applauded as affordable. According to Visagie (2010), the demand for electricity is fairly inelastic to the changes in prices in the short-term, although there are longer-term impacts. Moreover, electricity demand is sensitive to short-term changes in income and commodity prices. Ecological governance is primarily about how ecological goals such as ecological conservation, that is coupled with sustainable development can be reached. It also focuses on how decisions that are ecologically friendly are made. In order to achieve good ecological governance, it is crucial that participation of stakeholders, access to information, adequate funding, transparency and accountability are considered from the planning and design stage of the initiatives (Woolley, 2014).

There is no universal definition of energy poverty as different countries measure energy poverty differently. However, in many countries including South Africa, energy poverty is measured through different approaches which include affordability (i.e. the share of household income used on basic energy sources), perception of households on the adequate amount of energy they need and lastly, the connectivity or electrification rate (Culver, 2017; Ismail and Khembo, 2015; Pye and Dobbins, 2015; Grevisse and Brynart, 2011; Fahmy, 2011). The approaches as mentioned above as per the definition imply that energy poverty cannot only be measured through the availability of connections or electrification rate but through affordability and the capacity that is available to meet household energy needs, thereby contributing to dignified living conditions for households. Moreover, in addressing energy poverty, the energy sources or technologies to be employed must take into consideration the reliability and sustainability of the energy provided by the technologies. These cannot be achieved if the energy poverty eradication initiatives are not managed in line with ecological governance principles.

The above definition of energy poverty shows that putting together programmes or interventions, building technologies, establishing new institutions and providing funding cannot be the only essential elements in addressing energy poverty and achieving universal access to electricity. Ecological governance, which entails proper systems and processes for the sustainable use and management of these interventions, plays a pivotal role in ensuring their success, reliability and sustainability. Ecological governance embeds ecological values in all levels of decision making and actions which are characterised by different principles, including purpose, design, process and relationships (Robertson and Choi, 2010). This notion is also supported by the South African National Development Plan (Vision 2030), which states that "*the energy system of the country must be supported by effective policies, institutions, governance systems, regulations and competitive markets*" (National Planning Commission, 2012:163).

Ecological economics is aimed at addressing the interdependence and co-evolution of natural ecosystems and human economies over time. What is vital about ecological economics is the fact that it recognises that practical solutions to serious environmental and social problems need new interdisciplinary ways that focuses on the relationship between economic, social and ecological systems (Costanza, 1989). *Some of the government's initiatives to eliminate energy poverty and ensuring affordability in South Africa were the introduction of the Free Basic Electricity (FBE) and Free Basic Alternative Electricity (FBAE) which are meant to provide a subsidy to indigent households. A study conducted by Kasangana, Makonese and Mbonane (2018), found that there are gaps in the provision of FBE and FBAE. The authors contend that there is a need to review the 50 kWh per month of the FBAE as this can hardly meet the energy needs of each household. The 50 kWh was determined in 2003, and a lot has evolved in terms of energy needs since the FBE policy was initiated. The discussion of FBE and FBAE in this study is addressing one of the critical elements affecting energy poverty, which is affordability.*

The gap in FBE implementation is evident in the energy sector statistics which show that in the year 2017, there were 11.6 million households who accessed electricity from the grid countrywide, of which 2.5 million households received free basic electricity from municipalities (Statistics South Africa, 2018). South Africa had a total number of 16.92 million households and 3.56 million indigent households nationwide during this period, meaning that 31 percent of households still did not have access to electricity, while 30 percent indigent households did not have access to FBE. This gap can be addressed through ecological governance in the form of institutional arrangements and improved capacity at local government levels as well as public awareness of the FBE and FBAE policies and their benefits. As mentioned in the introduction of this paper, ecological governance is not only concerned about eradicating energy poverty for social benefits, but it also focuses on ensuring sustainable and affordable access to energy as well as addressing ecological and environmental effects of conventional sources of energy. Providing financial support such as FBE and FBAE does not necessarily address energy poverty in isolation as how the initiatives are managed is crucial.

Ecological economics posits that the economic policy need not to only focus on efficiency, but also consider the equity and sustainability of energy usage (Costanza, 1991). In ecological economics, models of economic behaviour include consumption and production in the broader sense, which entails their ecological, social and ethical dimensions, as well as their market consequences. Drawing from the experience of developing countries that have made good progress in implementing non-grid electrification initiatives, Bangladesh is one of the countries that have successfully implemented the SHSs programme as part of renewable energy plans. Since 2002, the country installed over 3.5 million SHSs in rural villages which have generated clean electricity equivalent to approximately 141 megawatts. The programme has changed the livelihoods of over 16 million people living in areas where the grid could not be accessed while saving approximately 200,000 tons of fossil fuel per year. This is evidence that if governance systems are well in place, renewable energy programmes in developing countries can also be successful in eradicating energy poverty. These programmes do not only change the livelihood of the people, as they also have ecological benefits as a result of reduced use of fossil fuel and coal-generated electricity. The challenge that was experienced with the implementation of the programme in Bangladesh was in relation to the entry of unregulated commercial providers of lower-quality SHSs and components, putting pressure on prices (Marro and Bertsch, 2015).

In African Countries that implemented the SHSs programme such as Zambia, the positive contribution of the technology included increased pupils' performance at schools in addition to other social benefits. The improvement in schools' performance was as a result of changes in study routines as children were able to study at night. Other benefits of the SHSs in Zambia included improved night-time security, access to information through phones and entertainment through radio and TV. Many off-grid projects in developing countries have reportedly not succeeded due to technical, regulatory and institutional challenges. Institutional challenges, high upfront payment and lack of long-term maintenance plan have been reported as major barriers to rural development in many developing countries (Azimoh, 2016).

Sustainability puts emphasis on the interaction between the social, ecological, environmental and social systems (Sutton, 1998). The success and sustainability of SHSs initiatives would, therefore, be measured by their contribution to the social, ecological, environmental and economic factors. This is important in that the economy is fuelled by energy, which requires that the energy system is sustainable enough to contribute the economic wellbeing of countries, taking into consideration the ecological, social and environmental factors.

The argument to use renewable energy is on the premise that it has the ability to provide for present energy needs without depleting the resources of the future generation. The sustainability provides for a balance between theory and practice by integrating the social responses with problems associated with the environment and culture. Ecological integrity and diversity of biological factors are at the centre of the ecological model, which is informed by the economic, ecological and political factors. In implementing programmes that are meant to ensure ecological and environmental sustainability, Dernbach and Mintz (2011) postulate that there is a greater motive for ecological governance. Ecological sustainability is brought into this study due to its significance in explaining how the biosphere has to meet the current generation's needs without affecting the ability to meet the needs of future generations. The argument for the use of renewable energy technologies such as SHSs is on the premise that they have the ability to provide for present energy needs without depleting the resources for the future generation.

A study conducted by Kabir, Kim and Szulejko (2017) found that SHSs have significant health and environmental benefits in rural areas as they have a minimal installed carbon footprint. Furthermore, solar energy is an almost inexhaustible supply in the foreseeable future, which implies that SHSs have more positive environmental impacts than non-renewable energy technologies. It is argued that a solar home system can replace the use of kerosene lamps and candles, which emit substantial amounts of fine particulates, carbon monoxide (CO), nitric oxides (NOx) and sulphur dioxide (SO2), which adversely affect air quality.

Pollutants from using candles and kerosene lamps may affect the health of households by causing health risks such as impaired lung function, increased infectious illness (including tuberculosis), asthma and cancer risks. SHSs have been found to result in less pollution and reduce the inconvenience associated with the use of kerosene lamps (Kabir et al., 2017).

The advantages of using renewable and clean energy technologies include reducing the depletion of natural resources, greenhouse gases and toxic gas emissions as well as required transmission lines of the electricity grids. The theory of ecological economics, which embraces modernisation, supports the use of renewable energy sources such as solar energy technology. Schelly (2015:60) posits that using solar energy technologies "*is a step in the modernisation of capitalist production processes, geared towards ecological sustainability for the sake of both profit and industrial longevity*". If managed well, SHSs can contribute to climate change mitigation and adaptation targets. Literature has shown that the net carbon emission offset is estimated to be approximately six tons of CO2 over the 20-year life of one PV system. That is, it can easily internalise the externalities caused by coal-generated electricity (Kabir et al., 2017). One other advantage of the SHSs is that they are re-usable, which is why they are considered renewable and clean. As SHSs are removed from households due to grid encroachment (when the grid is installed in the areas), it is important that they are disposed of in a manner that would not negatively impact the environment. Therefore, the de-installed systems are re-used in other non-grid areas as informed by the non-grid electrification policy of the country.

3. METHODOLOGY

A qualitative research approach was utilised to collect data from key informants (households) and major government stakeholders involved in the SHSs programme, either at a policy or implementation level. A qualitative research approach is used to answer questions that relate to the experience, the meaning and perspective of the participants on a certain subject, most often from the standpoint of the participant (Hammarberg, Kirkman and de Lacey,2016). The study draws from the findings of the face-to-face interviews that were conducted with households from two communities in Limpopo Province (i.e. Hlatlaganya and Ha-Tshikonelo), officials from the Department of Energy (DoE), and municipalities (i.e. Polokwane Local Municipality and the Thulamela Local Municipality). Furthermore, interviews were conducted with Solar Vision, the SHSs concession company in Limpopo and an official from the Limpopo Department of Economic Development, Environment and Tourism. The interviews were conducted using semi-structured interview questions which were designed by the researcher specifically for the study. The findings from the fieldwork/ interviews are also complemented by information from available published reports and literature on climate change, renewable energy and electricity generation from solar energy sources. This study basically answers the question "*how does ecological governance of solar energy initiatives contribute to the eradication of energy poverty*".

The study uses a critical theory paradigm which aims to challenge the status quo in the energy sector relating to the use of alternative energy and how the initiatives are designed, planned and implemented. The critical theory explains what is wrong with the current social reality, identifies the actors to change it and provides clear norms for criticism and achievable, practical goals for social transformation (Thomas, 2010). The study aims to contribute to an academic discussion about the preconditions for the success and sustainability for the SHSs and how they can contribute to eliminating or addressing energy poverty while also bringing change on the current status of the effects of current dominating electricity generation sources in South Africa. The critical theory assists in identifying the challenges with the planning, design and implementation of the SHSs programme in South Africa and also provides an understanding of the status quo on the initiative by combining theory and practice. This study, therefore, contributes to both the social, economic and ecological change that would evolve in terms of theory and policy development.

The criterion sampling method was utilised in selecting the population and sample for households. In each of the two case study areas, 20 households were selected according to the criteria that they use a SHS. This was on the basis that those who use this technology are able to provide rich information regarding their experiences and perceptions. Criterion sampling is a type of purposive sampling that allows for the selection of cases which are information-rich, which becomes effective when those who are interviewed understand the processes and weaknesses of the systems (Suri, 2011). As part of the criteria set, interviews were conducted with one adult per household regardless of gender. In addition to the households' interviews, five (5) officials were selected using criterion purposive sampling and were interviewed. The purposive criterion sampling, in this case, allowed the researcher to identify individuals who are knowledgeable and have experience about renewable energy initiatives, including the roles of different institutions.

The establishment of rigor was ensured through triangulation by interviewing a wide range of informants, and also reviewing supporting documents and reports. This ensured that findings were more dependable when they were confirmed from several independent sources, thereby enhancing rigor. Reflexivity was applied to achieve trustworthiness by using a field diary and materials to reflect on how the research was being conducted as well as reflecting on own actions, feelings and conflicts experienced during the study. Reflexivity was applied continuously to ensure that the researcher reflects on the values, pre-conceptions and behaviour of participants which may have an effect on the interpretation of responses. Reflexivity, which is basically a way of making the interview process open and transparent is important because qualitative studies have some level of subjectivity as the interpretation of the participants' behaviour and the collected data is influenced by the values, beliefs, experience and the researcher's interest (Jootun, McGhee, and Marland, 2009).

The approach to analyse data was inductive, which required reading of field material based on the interviews conducted and listening to the recorded tapes of the interviews and transcribing them. The inductive data analysis method is aimed at allowing the research findings to occur from the frequent and dominant themes that are inherent in the collected raw data. The inductive data analysis, therefore, ensures that extensive and different raw data can be summarised into a brief format as well as establishing the clear links between the research objectives and the summary of findings that were derived from the raw data (Thomas, 2003).

As part of ethical consideration, informed consent was ensured by informing all participants from the onset about the aim of the study and the fact that it was for academic purposes. Participants were also encouraged to use the language that they were comfortable with and they signed informed consent forms which outlined the objectives of the study and any implications for participation. The forms highlighted voluntary participation in the study, the no incentive rule to all the informants as well as their right to withdraw from participation. Participants were also assured of anonymity. Where anonymity could not be guaranteed, for instance in the case of policy makers and implementers (service providers), relevant official's permission was sought to disclose informants' positions or the use of pseudonyms.

In analysing data, the raw data from the interviews was reorganised through re-checking the interview guide as well as identifying and differentiating between the questions that the study seeks to answer and those that were simply included in the interview guide as important, but not essential. The ideas and concepts were organised from the data by finding meaning in the language that was used by participants. This ensured an understanding of perceptions, attitudes and feelings of participants about renewable energy initiatives. The way in which informants express themselves through words can reflect their attitudes or behaviours (O'Connor and Gibson, 2003). This was crucial, as household perceptions about renewable energy were one of the themes of the study, which made it important to reflect on the attitudes and behaviours of interviewed households. After finding meaning from the language, all the ideas and concepts were organised into different categories. The categories used focused on experience, feelings, opinions, knowledge and inputs.

The study ensured that each of the response categories has one or more associated themes that offered a deeper meaning to the data in order to build overarching themes in the data. Different categories were then collapsed under main overarching themes. Possible and plausible explanations for findings were then found by providing a summary of findings in line with the themes. This included the review of whether the findings are related to expectations of the study based on the literature reviewed. In addition, major surprises in the findings were checked through assessing whether there were any differences or similarities regarding what was stated in the literature. This required reading of the field diary and listening to the recorded interviews several times, to avoid eliminating important aspects of the responses. Lastly, the information was organised into a final report by summarising the research findings including the implications of the findings, strategies and areas of future research that were identified. In presenting the research findings, pseudonyms are used.

Study Area Description

Limpopo is the fifth largest Province in South Africa, with a population of 5.8 million as at 2016 (Statistics South Africa, 2016). The Province had 1.6 million households by 2016, which are shared by different districts (i.e. Vhembe with 382 346 households, Capricorn with 378 272 households, Mopani with 338 385 households, Greater Sekhukhune with 290 489 households and Waterberg with 211 452 households).

As provided by data published by Statistics South Africa (2016), Limpopo is one of the provinces with the highest proportion of households living in formal dwellings (88.9 percent or 1, 4 million) and the lowest number of informal dwellings (4.8 percent or 77 371) in the country. The proportion of households with access to electricity has increased from 39.2 percent in 1996 to 93 percent in 2016, which shows that the majority of households in the Province have access to electricity (Statistics South Africa, 2016).

The Province has a high potential to implement successful solar energy projects due to its high solar radiation levels. According to the Limpopo Department of Economic Development, Environment and Tourism (2013), priorities for the Province in terms of renewable energy include the production of electricity through concentrated solar plants and production of electricity through off-grid sources. Statistics South Africa (2015) has also indicated that Limpopo has the lowest rate of electricity use/ consumption for cooking in the country. Wood is the main source of energy for cooking and heating (>38 percent) with paraffin and candles used mainly for lighting.

4. RESULTS AND DISCUSSION

The results of the study are presented in line with the different themes that are linked to ecological governance which are access and households' perceptions, affordability, willingness to pay, social benefits, reliability, sustainability and environmental benefits. These are also in line with the definition of energy poverty as presented in the literature review.

4.1. Access

Accessibility is driven by the notion of social justice, which determines the equity and disparity between those who stay in urban and rural areas (Feron, 2016) The main aim of assessing the accessibility of SHSs is to ensure that there are equal opportunities to receive clean and reliable energy, which the SHSs programme aims to address by providing access to energy where the grid is not viable or not economical. The SHSs programme is rolled-out through a concession programme, which uses private companies to install, manage and maintain the systems in the concession areas. The SHSs concession programme has to date failed to reach its original targets of 50 000 units in the Limpopo concession area (the national expectation/ set targets was to install 300 000 units nationally by 2025).

The study found that access to energy through SHSs is affected by different factors such as theft of the components of the system and tempering, grid encroachment by Eskom without proper planning in the concession area and household perceptions regarding the system. The number of households who access electricity through SHSs has not grown as expected due to the DoE not continuing with the tendering processes. These challenges can only be addressed through embedding ecological governance in the planning, design and implementation of the programme. In the Limpopo Province, only a total of 20 000 SHSs were installed by 2017 across the province, of which only 6,600 were still operational due to grid encroachment. These statistics imply that since 2002, the SHSs programme in Limpopo provided access to electricity to 20 000 households who previously did not have access to electricity, which is 40 percent of the intended target of 50 000. The programme has, therefore, failed to effectively contribute to improving access to electricity to the indigent as intended in terms of the government's goals.

The above statistics were also confirmed by Patrick, from Solar Vision who contends that access to energy through SHSs is affected by grid encroachment and this negatively affect the country's goal of increasing the use of non-grid technologies to address energy poverty. Patrick said:

"Despite the numbers that were planned for installation at the inception of the programme not being met, the number of SHSs that we manage continue to decline due to grid encroachment. Even though grid encroachment means that households' electricity access through SHSs is being replaced by grid-based electricity, the purpose to increase deployment of non-grid electrification programmes that are meant to ensure universal access to affordable electricity using clean energy technologies is defeated".

Discussions with Patrick also revealed that access to SHSs is a challenge in the concession areas that are closest to the grid or subject to grid encroachment, hence the planned projects have been less successful.

Public perception regarding the implementation of sustainability programmes is crucial in ensuring success. According to Patrick, household perception about access to electricity through non-grid sources such as SHSs is still problematic. When responding to the question on what the SHSs programme implementation challenges are, and how those challenges can be addressed, Patrick said:

"Some of the households in rural areas do not consider themselves as having access to electricity if they are connected through non-grid sources as they regard non-grid electrification as inferior. They only consider themselves as having access to electricity if they are connected through the grid, which therefore requires awareness for households to understand the benefits of non-grid electrification. Households also need to understand that non-grid electrification is not only for the poor so that they may accept the technology".

One other challenge regarding access to energy is the definition of "*access*" as used by the government. The Polokwane Local Municipality in its Integrated Development Plan and State of Energy report refer to households with access to electricity only as those who are connected through the grid (Polokwane Local Municipality, 2018). This promotes the perception that electrification will only reach 100 percent when all households have been electrified through the grid, which raises a question of the municipality's support for renewable energy. John, an official from the Polokwane Local Municipality, was asked about the municipality's definition of universal access to electricity. He responded as follows:

"Even though the Department of Energy (DoE) defines universal access based on access through both on-the-grid and off-the-grid sources, the Polokwane Local Municipality counts universal access based on the grid-based connections. The off-the-grid solutions are temporal, and they get removed as soon as the grid is installed. Therefore, when reporting on households with access to electricity, the municipality refers to those who have access through the grid".

He, however, added that:

"Although there is grid encroachment and access is defined based on grid-based connections, households in rural areas still prefer using conventional sources of energy such as firewood and cow dung for cooking and heating. Households in many areas within the Polokwane Local Municipality refer to

the grid-based electricity as "Mabone", which means lights. They only use the electricity for lights and charging phones, while still using other sources for all households' energy needs".

Access to electricity through SHSs in the Limpopo Province has also been affected by theft and tempering of the systems which has been highlighted as a major challenge by both households and Solar Vision. Patrick said:

"The other major challenge facing the SHSs programme is the theft and tempering of the systems. The systems are being tampered with, and components such as batteries are sold for other uses and replacing them is expensive. This becomes a problem to Solar Vision as we have to replace the batteries or panels time and again".

Households have also indicated that their systems get stolen, which means that they will not have access to electricity until the stolen components are replaced or fixed by Solar Vision. Mashudu from Hlatlaganya said:

"People steal the panels and batteries, which affects access until Solar Vision comes and replace the stolen components. We are often faced with the challenge of not having access to electricity due to the stolen components".

Similar to Mashudu's statement, Mark from Ha-Tshikonelo said:

"The main problem with SHSs is that people do not understand the implications of stealing the components and how it impacts us. My system has been stolen twice, and this requires that the service provider looks at improving the security features of the system. During the time when the components are stolen, we do not have access to electricity. That is why we sometimes feel that getting the Eskom electricity is better".

The above findings show that there are different interpretations on the definition of access to electricity. The way in which energy poverty is defined determines how the delivery of those solar energy initiatives is done.

4.2. Affordability of SHSs by Households

Affordability is an important element which contributes to the success and sustainability of non-grid electrification initiatives, as they are usually targeted for rural households who are poor. High initial investment costs have the potential to make these initiatives unsuccessful. The study found that SHSs are affordable to households as they are expected to pay only an initial capital cost of R100 for the system to be installed. Furthermore, the government provides a subsidy for the monthly maintenance costs as part of the FBAE, which requires households to only pay a monthly fee of R30 (Solar Vision, 2017).

The funding model that is utilised for SHSs is favourable for the households in poor rural areas. Despite the advantage of the government paying 80 percent of the total capital cost for each SHS and the concession company covering the remaining 20 percent, there is a potential risk that households may not take care of the system as they do not have a sense of ownership. This is because the SHS equipment once installed, does not belong to the household, but the DoE and the concession company. Households have also expressed their satisfaction with the cost of SHSs and have indicated that they are affordable as they only pay the initial cost of R100 and a monthly fee of R30. They contend that compared to electricity that is connected from the grid (which they call Eskom electricity as famously known in the rural villages of Limpopo), SHSs are cheaper and affordable. Maria, a resident of Hlatlaganya, said:

"SHSs are affordable as they cost less. I am able to use basic electricity at a cost of R30 a month, which I can afford to pay. I think that the Eskom electricity would be expensive for me to afford that is why I am happy with the solar home system".

In agreement with Maria, Takalani, who is also a resident in Hlatlaganya said:

"The solar home system provides us with cheap electricity as we only have to pay a monthly service fee of R30. This helps us to access electricity at an affordable price".

In addition, Tshifhiwa from Ha-Tshikonelo also said:

"The SHSs are affordable compared to the Eskom electricity as we are not required to pay a lot of money at the end of the month".

Majority of the households have however indicated that even though SHSs are affordable, they are unable to cook or use other appliances such as the refrigerator. In responding to a question on whether SHSs are affordable, Thendo said, yes, *but I cannot switch the fridge on.*

Consistent with Thendo's response, Michael from Hlatlaganya said:

"I agree that SHSs are cheaper and affordable as I only pay R30 a month. My problem is the fact that the electricity that I can access from the system is limited and I cannot use other appliances like a fridge and stove".

The funding model that is utilised for SHSs is favourable for the households in poor rural areas as it reduces the cost of the systems making them to be affordable by all. The funding model includes the capital subsidy which covers the total capital cost of the SHSs at 80 per cent, while the service provider/ concession company only cover 20 percent of the capital cost for each system. This subsidy, however, implies that the SHSs equipment is co-owned by the DoE and the Service Provider. The programme receives a direct allocation of approximately R101 million (2015/16 financial year allocation) from the national budget. The advantage of this model is that it eliminates a capital cost to the customer, who is often poor or cannot afford to pay the capital cost which can affect the successful roll-out of the programme.

The model is beneficial to households, although it can have adverse effects to the service provider. For example, the Distribution Board (DB) box of the system cost R1500, meaning that it could be costly for the service provider to always fix or replace them as they get stolen or breakdown. Patrick indicated the following:

"Even though the DoE provides for the capital cost of the system, it becomes solely the responsibility of the service provider to replace the stolen or faulty components. With the theft and tempering that happens, Solar Vision bears more financial risks".

He added that:

"However, the capital subsidy is a good model as it would have been difficult for households to afford to pay this subsidy which would have negatively affected the roll-out. The concern is the fact that households do not care about the safety of the systems because they know that they do not have to pay for them should they need replacement".

The SHSs also use the fee-for-service model in addition to the capital subsidy. Before the system can be installed, households are required to pay a once off payment of R100. Once the system has been installed and operational, households are therefore expected to pay a monthly fee of R30 (after the FBAE subsidy) to the service provider for the services provided as well as maintenance. The main benefit of the fee for service model is that, in the instance of a faulty system or stolen components, households do not have to worry about the cost of fixing or maintaining it. The service provider is therefore responsible for maintaining the SHSs within the concession area as and when required, and to bear such costs. Solar Vision has indicated that the only revenue they receive is from the monthly service fee. Patrick said:

"The fee-for-service model allows us to receive some revenue even though it is not sufficient. To us, the programme is no longer about making money but supporting the Government's agenda on renewable energy as we do not make enough to even sustain the company sufficiently. Furthermore, investors withdrew their investments due to challenges that we experienced regarding the delays by the DoE to approve the projects which leave us with little revenue".

Sharon, a resident from Ha-Tshikonelo, said:

"I am happy with the R30 fee payment that I have to make on a monthly basis as this is not too much and I know my solar home system will be fixed for free should anything happen to it".

In addition, Joel from Hlatlaganya said:

"The money that I pay is not too much as it helps to make sure that Solar Vision can come and fix my system if it is not working properly. I understand why I have to pay the monthly fee as it was explained to me before the system was installed".

In determining the affordability of SHSs, the World Bank has conducted a comparison between the cost of SHSs and grid electricity as of 2015, which shows that there is a cost advantage of approximately 43 percent for SHSs in contrast to grid electrification. Grid connection would cost R171 per month, compared to a total monthly cost of R98 for a solar home system (before the FBE subsidy) which shows that electricity from SHS is more affordable than electricity from the grid. This conclusion is in agreement with the findings of this study. The detailed comparison is provided in the table below.

The municipalities also agree that the FBE and FBAE are good initiatives in ensuring affordability to clean energy technologies. John explained that FBE and FBAE is a good initiative as it ensures access and affordability to energy for all. He said:

"The FBE and FBAE policies are good initiatives as many households in rural areas are unable to afford the cost of electricity. SHSs would have been expensive if the municipalities were not providing the FBE amount". Due to the FBE and FBAE policies, everyone regardless of their income can afford to access basic electricity".

Table 1. Estimated Cost Comparison of Grid versus Off Grid Electricity Connections (2015)

Description	Grid Electricity	SHS	Variances
Capital Cost, excluding wiring, lightbulbs, plugs and DB board	R25,000	R5,340	The cost of SHS is 21 percent of the cost of a grid installation
Monthly Network Charge	R85	N/A	Grid network charge for SHSs
Monthly kWh	50	12-15	The FBE is compared with the output of a SHS on current specifications
Value to Customer on a R0.63/kWh Tariff	R31.50	R7.60 - R9.45	SHSs on current specs provides less value than the off-grid equivalent
Monthly cost to Service Provider	R171 (Based upon consumption of 100kWh/month in an urban setting)	R98	Increased capacity or specifications would not increase the solar home system's monthly fee to a great extent as the fees are not based on the size of the installation but on the running costs of a set number of units

Source: World Bank (2015)

Paul from the Thulamela Local Municipality added that:

"As the municipality, we pride ourselves in providing FBE and FBAE to ensure equity in access to electricity by all. The impact of these subsidies might not be visible in the short term, but they are there".

The main finding relating to the affordability of SHSs is therefore that, SHSs are affordable on the basis that they are highly subsidised by the government through the initial capital subsidy and the FBE/FBAE that is provided by municipalities. What is important to note is the fact that FBE and FBAE are provided in line with the indigent policies, meaning that only those who meet the criteria of the FBE and FBAE will receive these subsidies. The main question that remains is how the government would make sure that SHSs can also be affordable to those who are not covered by FBE and FBAE in considering ecological sustainability.

4.3. Willingness to Pay

The study found that households in both study areas are willing to pay for SHSs if they had to choose between the grid and non-grid electricity, only if the energy generated by SHSs was reliable and sustainable enough. The interpretation of this finding is that households are currently supporting the SHSs because they do not have another option as grid-based electricity is not available. This perception is however a result of households not knowing the full benefits of off-the-grid initiatives as they only view them as a temporal way to access electricity. The main challenge was the perception inculcated to households that solar energy, especially non-grid based is inferior. The perceptions about solar energy are based on the lack of community awareness on the benefits and use of these technologies. The SHSs proved to have failed to meet the expectations of households, which gives households an impression that other sources of energy are better than solar energy. Households in both study areas indicated that they were willing to pay for SHSs even though they would have preferred that the technologies provided more capacity than they were currently providing.

4.4. Social Benefits of SHSs in Limpopo

Literature suggests that SHSs improve the livelihood of households as it provides them with access to electricity which they previously did not have (Azimoh, 2016; Wlokas, 2015; Prasad, 2016). The study found that SHSs provides social benefits with no doubt in many ways. Households in Limpopo indicated that the SHS's programme had changed their lives as they now have access to lights and communication devices, which they previously lacked. Another positive social impact of SHSs is in relation to improved conditions as children are now able to study sufficiently at night which is better than when there were no SHSs.

Households have also confirmed these claims as they have indicated that SHSs have changed their lives in many ways. Tshifhiwa, a resident in Hlatlaganya alluded that:

"SHSs provided me and my family with many benefits which include: being able to listen to the radio and being kept abreast with the news, charging phones and most importantly, not using candles as we previously did which posed safety risks as they have the potential to cause the fire".

John said, *"this programme has brought major benefits to the poor. We once realised in one of the villages which fall under the municipality that households were paying a fee to charge phones in a house that had electricity. This was not good as everyone deserves to get free basic electricity. Since the programme was rolled out in that area, households have indicated that their lives have changed as a result of the programme".*

These highlighted benefits suggest that the systems have had an overall positive impact on the improvement of the lives of who have them. The question, however, remains on whether the programme has achieved its intended goals as set by the government, especially in terms of providing quality and affordable electricity to households.

Even though households have complained in both study areas about the theft of their systems or some of the parts, the systems have improved the safety aspects in the areas. Before the systems were installed, households indicated that there was a high crime rate as the areas were very dark and thieves would easily access houses to steal without being noticed. This however changed after the systems were installed due to the presence of outdoor light. Joyce, who is a resident of Hlatlaganya said:

"Since the installation of the SHSs, we no longer have the fear of being attacked because it is no longer too dark. Before the SHSs were installed, this area was not safe at all".

Another positive impact of SHSs highlighted by households was related to education conditions. It was indicated that children are now able to study sufficiently at night compared to when there were no SHSs. Portia, a resident from Ha-Tshikonelo, said:

"My children can now do school work in the evening as we have lights. Previously, they could only do their homework in the afternoon as the light from the candles was not sufficient. This has contributed to their improved performance at schools".

4.5. Reliability of SHSs

Crucial to ensuring sustainability of the SHSs programme is ensuring that the electricity that is generated from the system is reliable. That is, households must be able to access electricity at all times without disruptions. Electricity generated from SHSs in Limpopo should be reliable based on the solar radiation levels in the province. Limited reliability of the SHSs in Limpopo was mainly attributed to the loss of parts due to theft as well as low energy due to cold weather conditions. According to interviewed households, even though the SHSs provide them with access to electricity, availability depends on the weather which raises reliability questions. Households have indicated that if it rains for more than two (2) days, they go without electricity as the system DoE not provide for capacity to store excess generated energy.

In responding to the question relating to the reliability of SHSs, Portia said:

SHS is problematic when it rains for longer periods as we have to go back to using candles and paraffin.

Paul also added that:

Even though SHSs provide households with electricity that they previously did not have, its reliability becomes a problem. When it rains for more than two days, electricity is interrupted due to low solar radiation levels on those days. This makes it difficult for households to rely on the systems for electricity access as they often have to go back to conventional sources. It also changes the perception of households about the system, making them want to be connected through the grid because they think grid-based electricity is better than SHSs because of its reliability.

As a result of reliability challenges identified, there is a need for improved technology that will allow for saving/ storage of excess electricity, which can be used in the periods when the solar radiation is low.

4.5.1 Sustainability of the SHSs Initiative

To improve the sustainability of the SHSs programme, it is important that the service providers/ concessionaires are effectively supported. Many concessionaires that are involved in the SHSs programme are not generating enough revenue, and this can have an impact on the sustainability of the programme if they run out of business. The way in which the programme is implemented has already led to international investors pulling out their investment as the programme has not yet materialised. The original target size of the concessions, 50,000 units, was set to make each concession attractive to external investors. It was estimated that with a customer base of 50 000, the service providers would generate income of approximately R29 million per annum which is collected from the monthly service fee paid by both municipalities as FBE and households' monthly fee (World Bank, 2015).

According to Solar Vision, government's support to the SHSs programme is minimal which affect sustainability and success of the programme. The government's failure to deliver the expected 50,000 in subsidised installations per concession as initially planned has created problems with the sustainability of the programme. The reduction of operating systems due to grid encroachment, coupled with failure to approve new projects by the DoE affects the sustainability of the programme. In responding to the question on what could be done to ensure sustainability and success of the programme, Patrick said:

A customer base of at least 10 000 would be desirable. With the current customer base of 6 600, solar vision might force to go out of business which may create a collapse of the programme in Limpopo. The sustainability and success of the programme also depend on proper maintenance of the systems so that they can properly function. If the programme is not financially viable, maintenance of the systems will remain a challenge.

Solar Vision suggests that the government's commitment to delivering on the planned targets for the programme would improve its sustainability.

To ensure sustainability, the DoE must ensure that the programme is managed differently so that the targets that were set at inception can be achieved. Said Patrick.

Furthermore, awareness of the programme is needed for households to understand the benefits of the systems, as currently they consider the systems as inferior and unsustainable. Financial viability of solar concessions without the capital subsidy for new installations and the operational subsidy is not possible, hence the delays in approving new projects by the DoE is a challenge. In addition, the future payment of the monthly operational subsidy by poor households is also doubtful, and this risk is entirely shifted to the concession companies. These need to be addressed to avoid the whole solar home system programme to collapse. Patrick said:

For us to continue operating effectively for sustainability, we depend on the monthly payments from both the municipalities and households. Sometimes households are unable to pay the monthly fee which affects us as the concession company, and continuous non-payment would lead to a collapse of solar vision and the programme.

Grid encroachment has been identified as one of the significant challenges in the solar home system programme which affect sustainability. Grid encroachment refers to instances where grid infrastructure has reached areas where non-grid infrastructure (solar home system) has been provided to households, meaning that households will now be connected to the grid. The main challenge is a lack of integration of the solar home system programme with grid electrification planning, as this affects the concessions because they usually do not know when the grid will be installed in areas where they have a solar home system. Patrick said:

Our biggest problem as Solar Vision is still grid encroachment by Eskom. Just when we think that we have a good number of systems to manage, then the grid comes and we are forced to remove the systems. However, because there are no new projects approved when the grid is installed in the areas where there are systems, we do not have anywhere to re-install those systems.

Some of the criteria for allocation of non-grid service provision is that off-grid systems should not be installed within two (2) kilometres from a gridline and in areas where grid electricity will be installed in the next three (3) years (DoE, 2012). Patrick indicated that:

Since there is a lack of transparency in the grid electrification roll-out plans, several concessionaires have experienced the unexpected arrival of the grid in their concession areas which puts them out of business.

Households also indicated their preference to grid electricity as they consider it as sustainable compared to non-grid electricity. The service level of the subsidised SHSs is limited and the SHS technology itself is not very flexible with limitations in its application, which affects sustainability.

4.6.1 Environmental Benefits of SHSs (Contribution to the Climate Change Agenda)

The solar home system technology still requires improvements for it to result in notable benefits to the environment. This is on the basis that the current technology is unable to meet all the energy needs of households, who tend to continue relying on fuels that can still emit GHGs. For example, households still continue to use firewood or paraffin to cook and heat water, as the SHSs cannot provide for enough capacity to cover all energy needs in a household. The study did not assess the exact amount of greenhouse gas emissions that were avoided as a result of using SHSs in the study areas as this was beyond the scope of the study.

One other area of concern with the solar home system's contribution to the reduction of GHG emissions is the fact that the systems are replaced as soon as it is possible to provide grid-based electricity in the areas. The result of the replacement due to grid encroachment means that SHSs only curb the emissions of GHGs in the short term. This is in contrast with ecological economics which requires ecologically sustainable initiatives.

5. CONCLUSION AND RECOMMENDATIONS

The study found that SHSs can eradicate energy poverty if managed in line with ecological governance. It is clear that the implementation of the SHSs programme using the concession model has not resulted in the programme being a success, thereby negatively affecting the energy eradiation goal. The failure of the programme was due to lack of proper ecological governance systems, which includes amongst others; weak institutional arrangements, gaps in the legislative framework, lack of integrated planning and no participatory decision making at the local level. Sustainability challenges relating to the programme was attributable mainly to the delays in the tendering processes which affect the service providers and limited capacity of the technology which only cater for a few appliances and lights which many households identified as a problem. Most of the SHSs' users were satisfied overall with the solar home system itself and the fact that it provided them with electricity which they could not access before, although they would be happier if it provided them with more electricity capacity.

5.1 Recommendations/ proposed solutions

To ensure sustainability of the programme, the provision of SHSs must not be viewed as a temporal measure that is only meant for poor rural households. The programme must be incorporated into the broader electrification programme of the country, with the main aim of increasing electricity access through renewable energy sources. Ecological economics is for initiatives that are meant to provide longer term and sustainable solutions to ecological and environmental problems which this initiative failed to achieve. Also, the programme did not achieve the internalisation of externalities associated with coal-based electricity generation, due to the fact that as soon as Eskom is ready and able to install the grid in the concession areas, coal-based electricity is rolled out to replace these cleaner energy technologies.

The application of the theory of ecological economics in implementing the solar homes systems programme is a pre-condition for the success and sustainability of these initiatives. This is on the basis that the implementation of initiatives that are meant to internalise negative externalities must provide long term solutions that will benefit the future generation. For SHSs programme to be a success and sustainable, there is a crucial need to put in place ecological governance systems. This relates to ensuring easy access, affordability, community participation in decision making, proper institutional arrangements and legislative framework as well as ensuring that the impact on climate change is considered. For the SHSs programme to effectively contribute to the eradication of energy poverty in South Africa, the following recommendations are made:

- **Improved technology to ensure reliability and sustainability of SHSs**

The specifications and capabilities of the technology that is being utilised for SHSs were developed 18 years ago when the programme was introduced. The improvement in the technology should include increasing the capacity or coverage of the systems so that households can use more appliances such as a fridge and a stove. The increase in the capacity would ensure that households do not continue using other non-clean fuels such as firewood, candles and paraffin which continues to emit greenhouse gasses. The improvement in the technology may include providing batteries that have high capacity to last for more days, to ensure that should it rain, households can still have access to electricity for a longer period using the stored energy.

With the prices of solar energy technology showing a declining trend, there is a need for more research on what technologies can be utilised to power houses and businesses without using the grid/ coal-based electricity. South Africa is well endowed with good solar radiation and own generation must be on an increasing trend.

- **Improved institutional arrangements for the planning and implementation of solar energy initiatives**

In the current institutional arrangements, the DoE plays the majority of the roles in the planning, implementation and monitoring of the SHSs programme. This has its own challenges as the department also has other priorities, which may lead to the programme being overlooked.

This study, therefore, proposes that a polycentric institutional model is utilised for the provision of non-grid electrification such as the use of solar home systems across the country. A polycentric institutional model allows for multiple actors and mechanisms to share powers in dealing with energy problems at different spheres (Feron, 2016). Using the polycentric institutional model in non-grid electrification programmes would allow for the provincial government to have powers on planning and setting targets for the SHSs to be installed in the province. It would also contribute to the enhancement of the capacity of the SHS per household (i.e. how much electricity can each household access, how many appliances and how many lights) as well as the delivery model (such as how households would be subsidised as well as how tendering processes and concession programme would be.

The polycentric model would also bring an advantage for the programme to become a local initiative where there can be an exchange of knowledge and the control of measures by local representatives who know the area and its dynamics well, including the needs of the people. Another benefit of localising the initiatives would be the local facilitation of the programme implementation which would ensure easy acceptance as locals would easily identify with initiatives that are being led by the people that they know and live with. With this model, it can be easy for the national sphere of government to achieve its goals as each province would work towards ensuring that there is an improvement at its sphere, therefore, contributing towards the achievement of national goals. The provincial COGTA can be the best department to manage the programme at a provincial sphere, which would also make it easy for the municipalities to implement the programme based on its authority and mandate.

- **Improved planning integration between DoE, municipalities and Eskom**

Lack of coordination between institutional stakeholders, ad-hoc tenders and implementation resulted in challenges for the SHSs. The study discovered that lack of integration in planning has failed the SHSs programme as this has led to institutions not agreeing on off-grid areas. This has negatively affected the concession companies as they experienced wastage of investment in SHSs as the grid was sometimes extended into areas where SHSs are installed, and customers suddenly switching to grid supply. Proper and integrated planning of the programme would ensure success and improve the sustainability of the programme.

- **Improved safety features for SHSs to reduce the level of theft**

The theft of the systems or some parts has been identified as a major challenge in both study areas, which negatively affect both households (in terms of access) and the concession companies (replacement costs). Theft of the systems and components affect the reliability and sustainability of the technologies, requiring focused attention to protect them. A concern regarding the theft of the system was said to be attributable to the fact that households are aware that the equipment is not under their ownership and would be replaced without them paying any additional cost; hence they do not make any effort to keep the systems safe. To reduce the risk of theft, households must own the systems which would generate a sense of responsibility in them towards the systems installed in their homes.

- **More education and awareness on the benefits of solar energy**

To reduce negative perceptions about the characteristics, reliability and utilisation of the SHSs, it is essential that education and awareness are provided to households. Information flow must be improved, which requires clarity on the roles and responsibilities of different role players. Currently, more responsibilities for awareness are given to the concession companies, which communicate with community leaders. The government must take responsibility in creating awareness, through the IDP consultation processes. As required in terms of the non-grid electrification policy guideline, households must be trained on how the system operates and how to conduct minor troubleshooting. Households must also be educated about the use of renewable and clean energy technologies and how SHSs contribute so that they can understand the different benefits of using SHSs. As it stands, they understand the systems as a temporary measure which cannot provide them with enough electricity that they require for daily use.

- **Scaling-up the roll-out of solar energy as intended**

For the SHSs programme to have a positive impact on universal access to electricity and climate change mitigation, it is vital that the roll-out is scaled-up as it was intended. Installing SHSs in small scales may affect the realisation of the intended change. Small scale roll-out also has a negative impact on the concessioners, who as a result, struggle] financially. This may lead to them pulling out of the programme. It was intended that each of the successful bidders would install 50,000 SHSs during the first five years of the programme. Scaling up would enable the economies of scale required for the businesses and the programme to succeed and be sustained. The Limpopo concession company for SHSs currently has challenges as it only has a customer base of 6 600 (as of November 2017). A customer base of 10 000 at a minimum is desirable for the company to make profits.

- **Improved storage capacity for SHSs**

One of the major challenges raised with regards to solar energy in South Africa is the storage capacity. Solar energy would be more sustainable if it can be stored for use even when the weather is not favourable. Current arguments regarding the use of solar energy are that it cannot be used on its own as it depends on the radiation levels, meaning that when there are cold weathers, energy generation will not be sufficient. This can be resolved by ensuring that there is enough storage capacity, not only for bigger projects but also for small scale generation or electricity generated for own use. Innovation is crucial so that there can be portable storage technologies.

- **Long-term initiatives**

Ecological economics advocates for initiatives that are focused on providing long-term solutions. The solar energy initiatives that must be promoted are the ones that provide sustainable energy to both households and businesses, with long-term benefits. Non-grid electrification initiatives such as solar homes systems are temporary and do not provide long term solutions to ecological and environmental challenges. The study proposes for the non-grid electrification programmes to be designed in a way that they will provide long-term and sustainable energy. Thus, the SHSs programme must not be implemented as a temporary measure as it is currently the case. This relates to ecological rationale planning and decision making, which must take into consideration an innovative way of making the technology to be more sustainable.

CONCLUSION

The study established that ecological governance, in the form of an institutional framework, can ensure the success and sustainability of solar home energy initiatives. These conclusions contribute to discussions in renewable energy by providing useful knowledge on how ecological governance can be utilised in ensuring effective planning and implementation of solar home systems initiatives in South Africa and how it links to the theory of ecological economics. The focus of ecological economics was mainly on the understanding of how and why people make decisions that have consequences for the natural environment, and how institutions and policies can be changed to bring environmental impacts more balanced with human desires and needs of the ecosystem itself.

Ecological economics motivates for ecological relational governance to be a permanent feature of a system of governance that seeks to preserve the ecosystem. The causes of failure of renewable energy programmes which consequently impact on the country's ability to meet its climate change mitigation targets were brought to the fore, in order to address them. Prioritisation of capacity building as a long-term goal is, therefore, critical for enhancing the sustainability of off-grid PV systems. The price of solar technology influences the deployment of solar energy. This study, therefore, concludes that the success and sustainability of solar energy initiatives are dependent on ecological governance.

REFERENCES

Azimoh, C. L. (2016). *Sustainability and development impacts of off-grid electrification in developing countries: An assessment of South Africa's rural electrification program*. Mälardalen University.

Centre for Renewable & Sustainable Energy Studies. (2014). *New solar resource maps for South Africa*. Author.

Clark, A. (2005). *Resource-based Technology Innovation in South Africa: Innovations in South Africa's Off-grid Concession Programme, Employment Growth & Development Initiative, Human Sciences Resource Council*. HSRC.

Costanza, R. (1989). What is ecological economics? Coastal and Environmental Policy Program. *Ecological Economics*, *1*(1), 1–7. doi:10.1016/0921-8009(89)90020-7

Costanza, R. (1991). *Ecological Economics: The Science and Management of Sustainability*. Columbia University Press.

Department of Energy. (2012). *A survey of energy-related behaviour and perceptions in South Africa: The Residential Sector*. Department of Energy.

Department of Energy. (2012). *Non-grid household electrification policy guideline*. Department of Energy.

Department of Minerals and Energy. (2007). *Integrated national electrification programme (INEP) plan to achieve universal access to electricity by 2012 masterplan / business plan*. Department of Minerals and Energy.

Dernbach, J. C., & Mintz, J. A. (2011). Environmental Laws and Sustainability: An Introduction. *Sustainability*, *3*(3), 531–540. doi:10.3390u3030531

Eskom. (2018). *Integrated report 2018.* Retrieved from http://www.eskom.co.za/IR2018/Documents/Eskom2018IntegratedReport.pdf

Feron, S. (2016). Sustainability of Off-Grid Photovoltaic Systems for Rural Electrification in Developing countries: A Review. *Sustainability, MDPI, 8*(1326), 1–26. doi:10.3390u8121326

Grevisse, F., & Brynart, M. (2011). Energy Poverty in Europe: Towards a more global understanding. *ECEEE 2011 Summer Study. Energy Efficiency First: The Foundation of a Low-carbon Society.*

Hammarberg, K., Kirkman, M., & de Lacey, S. (2016). Qualitative research methods: When to use them and how to judge them. *Human Reproduction (Oxford, England), 31*(3), 498–501. doi:10.1093/humrep/dev334 PMID:26759142

Ismail, Z., & Khembo, P. (2015). Determinants of energy poverty in South Africa. *Journal of Energy in Southern Africa, 26*(3), 66–78. doi:10.17159/2413-3051/2015/v26i3a2130

Jootun, D., McGhee, G., & Marland, G. (2009). Reflexivity: Promoting rigour in qualitative research. *Nursing Standard, 23*(23), 42–46.

Kabir, E., Kim, K., & Szulejko, J. E. (2017). Social Impacts of Solar Home Systems in Rural Areas: A Case Study in Bangladesh. *Energies, 10*(1615), 1–12. doi:10.3390/en10101615

Kasangana, K. K., Makonese, T., & Mbonane, T. P. (2018). Dissemination of Free Basic Electricity in Low-income Settlements. *Conference Paper*. 10.23919/DUE.2018.8384380

Limpopo Department of Economic Development, Environment and Tourism. (2013). *Limpopo Green economy plan including provincial climate change response.* Retrieved from https://www.environment.gov.za/sites/default/files/docs/limpopogreen_economyplan.pdf

Marro, P., & Bertsch, N. (2015). *Energy a Success in Bangladesh: Getting the Business Model Right. Mandaluyong City*. Asian Development Bank.

Methvin, T., & Philipp, F. (2017). *Energy Provision at the Base of the Pyramid: Are there viable business models to serve South Africa's low-income communities?* Impact Amplifier. Retrieved from http://www.impactamplifier.co.za/wp-content/uploads/2017/02/BOP_Final.pdf

National Planning Commission. (2012). *The South African National Development Plan (Vision 2030).* National Planning Commission.

Parr, B., Swilling, M., & Henry, D. (2018). *The Paris Agreement and South Africa's just transition.* Melbourne Sustainable Society Institute. Retrieved from www.sustainable.unimelb.edu.au

Pye, S., & Dobbins, A. (2015). *Energy poverty and vulnerable consumers in the energy sector across the EU: Analysis of policies and measures.* Insight Energy. Retrieved from http://www.innoenergy.com/wp-content/uploads/2016/03/INSIGHT_E_PR2_Energy-Poverty-Main-Report.pdf

Robertson, P. J., & Choi, T. (2010). Ecological Governance: Organizing Principles for an Emerging Era. *Future of Public Administration in 2020, 70*(S1), 89–99.

Schelly, C. (2015). Framework for Understanding and promoting Solar Energy Technologies. *MDPI Resources, 4*(1), 55–69. doi:10.3390/resources4010055

Statistics South Africa. (2015). *Electricity redistribution: Which councils are likely to feel the pinch?* Author.

Statistics South Africa. (2016). *Limpopo Community Survey 2016 results*. Retrieved from http://www.statssa.gov.za/?p=7988

Statistics South Africa. (2018). *General Household Survey: 2017*. Statistics South Africa.

Thomas, D. R. (2003). *A general inductive approach for qualitative data analysis*. Retrieved from https://www.researchgate.net/profile/David_Thomas11/publication/263769109_Thomas_2003_General_Inductive_Analysis_-Original_web_version/links/0a85e53bdc04f64786000000/Thomas-2003-General-Inductive-Analysis-Original-web- version.pdf?origin=publication_detail

Thomas, P. Y. (2010). *Towards developing a web-based blended learning environment at the University of Botswana* (D.Ed. Thesis). University of South Africa.

United Nations Environment Programme. (2011). *Towards a Green Economy: Pathways to Sustainable Development and Poverty Eradication - A Synthesis for Policy Makers*. Retrieved from www.unep.org/greeneconomy

Visagie, C. (2010). Medium Term Scenarios for Electricity Supply/Demand Balance. In *The Sustainable Energy Seminar*. Stellenbosch: AMEU.

Wentzel, M. (2005). Achieving universal access to electricity in South Africa. *Energize*, 12-15.

Woolley, O. (2014). Contents. In Ecological Governance: Reappraising Law's Role in Protecting Ecosystem Functionality (pp. v-viii). Cambridge University Press.

World Bank. (2015). *Evaluation of Rural Electrification Concessions in sub-Saharan Africa: Detailed Case Study: The South African Solar Home System (SHS)*. Concessions.

Chapter 13
The Scope and Potential for Mini-Grid Power Systems Based on Biomass Waste for Remote Areas in Sub-Saharan Africa:
Off grid

Gratitude Charis
Botswana International University of Science and Technology, Botswana

Tafadzwa Nkhoma
Manicalland State University of Applied Science, Zimbabwe

Gwiranai Danha
Botswana International University of Science and Technology, Botswana

ABSTRACT

Sub-Saharan Africa (SSA) has been experiencing an energy crisis. Socio-economic balances depend on access to clean, convenient, and dependable energy. This is critical for remote areas which are off the national grid, necessitating the installation of renewable energy sources such as bioenergy plants. These plants could valorize waste using combustion and gasification or biogas plants. The challenge is to produce a competitive levelized cost of electricity (LCOE). Nations like Germany and Sweden have successfully launched these. SSA can benchmark from these and valorize its biomass wastes. Key issues to consider would be cost-effective supply chains, sustainable harvest rates, after sales support, and good regulatory frameworks. This study was mostly a desktop review with a few field study observations. It was concluded that the stoker fired boiler and landfill gas 'biomass only' technologies would have the least capital costs, although gasification and anaerobic digestion are also competitive in terms of LCOE.

DOI: 10.4018/978-1-7998-8809-3.ch013

INTRODUCTION

The scope for off grid clean energy has been increasing in sub-Saharan Africa (SSA), encouraged by policy incentives, donor support and derisking mechanisms. The major constraints have been project risks associated with such utility-scale power projects and a general lack of local finance for expensive power purchase agreements with independent power producers (IPP) in this developing region. For example, emerging markets commissioned ~107GW of clean energy in 2018 compared to 0.87GW in SA due to the fact that such energy systems are more costly to commission and operate, beyond what the region can afford. Regardless, there are indications that investments in renewables in SSA have increased by at least 70%, with a large number of photovoltaic (PV) projects lined up for markets that previously had little grid-scale solar (Bloomberg NEF, 2020). The investment growth has been made sustainable due to rapidly declining technology costs, making renewable power systems competitive against current fossil fuels. Moreover the multilateral-backed support programs have helped make available the necessary finance and diffuse or spread the risk. Fig 1 shows the variation in the capacity and fractional share of various renewables with time.

Figure 1. Capacity additions in sub-Saharan Africa. Source: (Bloomberg NEF, 2020)

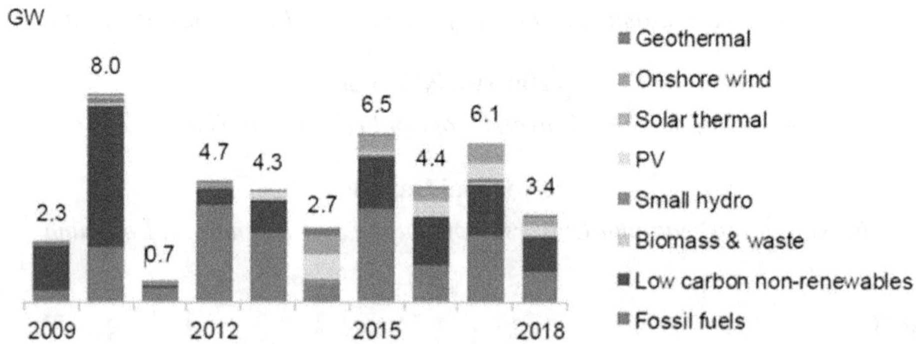

Indeed small hydro, solar and wind have been at the forefront of capacity additions in SSA (Fig 1) while biomass capacity has been fluctuating. On the contrary, South Africa, an important player in both the conventional and renewable energy, has shown a great interest in bioenergy and waste to energy (WtE). This nation planned that for the 2019-2030 period, they would have increased the share of renewables from a current ~30% to ~70% with a biomass and waste allocation of 14.5% (Bloomberg NEF, 2020) .

This projection in South Africa emulates developments in nations such as Germany, Sweden, Finland and the United States (US) that have succeeded in using biomass and its waste for bioenergy production. For SSA, which has a generous supply of biomass and waste, these fractions can be channelled to off-grid power production to improve energy access and enhance socio-economic development. This chapter explores pilot, demonstrated and undemonstrated cases above Technology Readiness Level (TRL) five for off-grid power production using products or by-products of biomass wastes. Hybrid solutions for such technologies are also discussed. The key objective is to establish the gap and opportunity for the application of such technologies in Africa. The gap and opportunity analysis will include the feasibility, ease of uptake and versatility of application of these technologies within the context of SSA. Although the focus is SSA, reference is occasionally made to Africa at large where statistics for the continent rather than the region are available and are an approximate mirror. The other objective is to identify any limiting factors such as policies and regulatory issues and provide recommendations in line with best practices or high value research.

Figure 2. South Africa's Integrated Resource Plan for 2019-30 period. Source: (Bloomberg NEF, 2020)

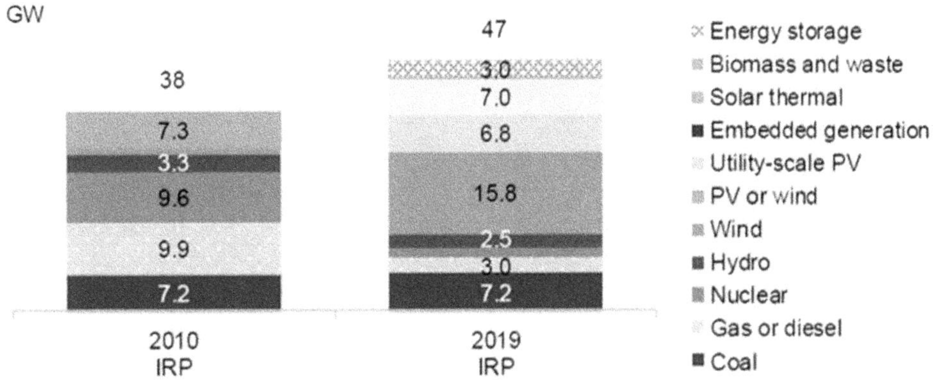

The research was mostly conducted through a desktop review using keywords such as 'bioelectricity', 'bioenergy', 'biomass inventory in SSA', 'biomass and CHP', 'off-grid' and 'mini-grid' were used in conjunction with the names of key organizations that report on such projects. Such organizations include the International Energy Agency (IEA), the International Renewable Energy Agency (IRENA) and the United Nations Environment Programme (UNEP). A number of journals were also explored using the same keywords. Case studies from other regions were noted for nations that have been at the forefront of bioenergy initiatives such as Germany, Sweden, Finland and the United States (US) for benchmarking purposes. The most suitable technologies that could be adopted for SSA were then selected and evaluated qualitatively in terms of feasibility, ease of uptake and versatility of application. Ease of uptake was further evaluated through costs of installation per KWh of electricity produced, since SSA is a developing region that could have challenges in raising high investment capital. Field observations were also made and captured in images for some forestry waste in two SSA countries, Zimbabwe and Botswana.

BACKGROUND

According to IEA (2020), SSA is one of the most affected regions, globally, by energy poverty. Although the region accommodates 13% of the global population, it only accounts for 4% of the world's energy demand. Energy demand is set to rise quickly given the large and growing population. Presently, energy supply unreliability has actually been one of the factors impeding economic growth with most nations relying on expensive and polluting sources (IRENA, 2020). Figure 3 shows the number of people without access to electricity within SSA compared to the rest of the world. Though there was a larger proportion of people without access to electricity in places other than SSA back in 2005 and 2010 the number had drastically reduced, while that in SSA increased. The ability to tap into renewable and available alternatives such as biomass could help the region ameliorate the energy poverty and achieve higher economic, social and environmental objectives.

Figure 3. Number of people without access to electricity Source: (Bloomberg NEF, 2020)

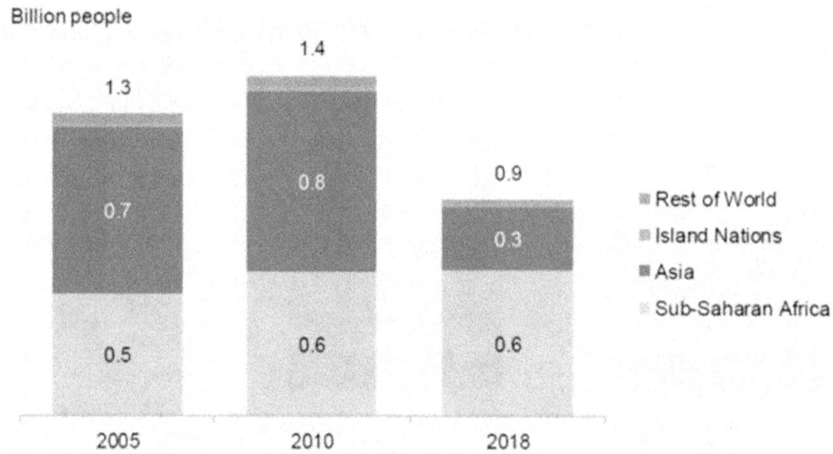

The International Energy Agency, (2015)'s report for that year reported that at least 620 million people in SSA had no access to electricity then, with around 753 million relying on traditional wood-based biomass for heating and cooking on open fire stoves that are inefficient and hazardous to respiratory health. Table 1 shows the population and percentage of people relying on the traditional use of biomass within various regions.

Table 1. Population relying on biomass from various regions (UNEP, 2017)

Region	Population relying on traditional use of biomass (million)	Percentage of people relying on traditional use of biomass
World	2722	38
Africa	754	68
North Africa	1	0.037
Sub-Sahara Africa	753	80

Evidently, SSA has a higher fraction of people depending on the traditional use of biomass than the rest of Africa. North Africa, in comparison, has managed to provide electricity access and other modern, cleaner alternatives, reducing the fraction of the population dependent on traditional biomass use to <1%. Nyika et al., (2020) have a slightly higher fraction (81%) for the SSA population that depend on traditional biomass use as fuel for for economic, household and cooking activities, probably because their study was based on latter years. The regular use of solid biomass is linked to 3.8 million premature deaths annually, primarily among women and children who mostly do the cooking (International Energy Agency, 2015). The unmonitored firewood collection is also associated with soil and forest degradation and competes for time that could have been used for other necessary social, income earning and educative activities (Nyika et al., 2020). There is a need to address such energy poverty through the provision of cleaner, alternative energy sources. The starting point could be to convert the traditional use of biomass to modern processed forms that are cleaner and offer more convenience. Avenues that do not lead to deforestation can be pursued by using waste biomass like invasive/encroacher species, agricultural and forestry residues, with socio-economic and ecological benefits, especially for remote areas that are off the national grid. The ideal outlook is to migrate from traditional usage of biomass and increase the share of *clean methods, especially for mini-grid power production* (International Energy Agency, 2015).

Waste biomass, in particular, is a plenteous resource that can be converted to energy and fuels through conventional and advanced conversion methods such as densification, incineration, biodigestion, gasification and pyrolysis (IRENA, 2016). Carbonization (slow pyrolysis) + densification would provide cleaner solid fuels for both domestic and industrial use, while a centralised incineration plant for combined heat and power (CHP) with adequate exhaust gas treatment would reduce the exposure of a larger population to health hazards. The uptake of gasification and pyrolysis advanced conversion methods has been warming up quickly and it is set to experience growth since they have a greater energy recovery efficiencies from feedstocks and as their techno-economic cases improve (Blimpo & Cosgrove-Davies, 2019). The ubiquitous nature of biomass caters for the remote areas where a sustainable biomass supply chain is available to justify the commissioning of a viable mini-grid WtE facility employing any of these conversion methods. Valorizing biomass waste residues for off-grid power generation has considerable potential to promote the socio-economic development of remote and emerging regions such as Sub-Saharan Africa (IEA, 2010). Such residues have been seriously considered for second generation biofuels because they allow migration from biomass food sources and avoid land use change (LUC) that can be an issue with some cases of energy crops. Still, sustainable measures have to be applied in utilizing these wastes for bioenergy- including using sustainable harvest rates and economic supply chain systems (Searle & Bitnere, 2017).

Of the two off-grid options- stand alone and mini-grid (MG) systems- biomass, WtE facilities would be suitable for the latter. A mini-grid system is a power network that covers an isolated community of between 40 and 40,000 customers, where a part or all of the power produced is injected into a localized distribution grid. The customers should be far from main grid, meet a certain minimum load demand and the population and household cluster densities should be high enough. It generally cheaper to install these MG systems in remote, sparsely populated areas than to extend the transmission and distribution system to many remote areas since those places have a low demand for electricity. Figure 4 shows the distinction between various power systems that can be used for remote or rural communities.

The MG generally comprises a power generator with or without storage, a distribution network, a service drop, customer installation and maintenance services. Currently, the most used power generator in Africa is the diesel set, mostly for stand-alone power systems. For MGs, the power generators include fuel oil, hydro, wind, solar and hybrid generators.

Other Considerations for Sustainable Off-Grid Bioenergy

Essentially, sustainable biomass to bioenergy projects will have to cover supply chain capacity and optimization, land use change (LUC) sustainable harvest rates and displacement effects for alternative utilization of biomass feedstocks. Alternative renewable sources can also be considered, especially for possible hybrid options to offset the possibility of using one renewable source. The capital expenditure

(CAPEX) and operating expenditure (OPEX) balances also have to be considered. The obvious advantage of biomass, liquid fuel and hydro generators is that they can be run continually with a buffer storage to cater for variations in the demand and load duration curve. Other renewable resources such as wind and solar are dependent on weather elements and hybrid solutions with biomass WtE systems can help smooth the demand-supply curves (Berahap, 2019).

Figure 4. South Africa's Integrated Resource Plan for 2019-30 period

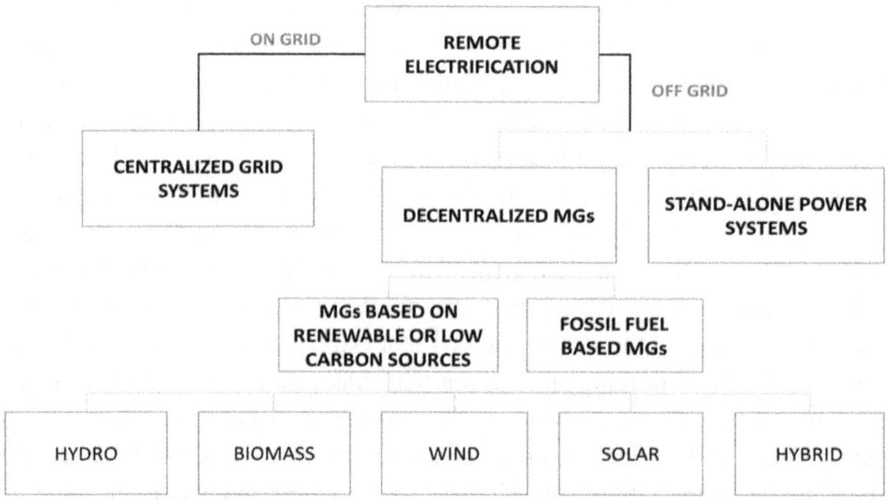

CURRENT STATE: BIOMASS AVAILABILITY, USES AND BEST PRACTICES

The Biomass Waste Inventory in SSA: Supply Chain Opportunities and Constraints

SSA and other developing countries have a large inventory of unutilized biomass residues and municipal solid waste (MSW) fractions due to expansive agricultural and forestry practices, growing populations and improved lifestyle qualities resulting in higher MSW generation (Charis et al., 2019a). The IEA studies on global bioenergy potentials show that SSA is grossly underutilizing its potential to produce modern **bioenergy and biofuels** from lignocellulosic biomass comprising high lignin, cellulose and hemicellulose. Africa only accounts for <0.1% of global biofuels despite having ~20% of the worldwide biomass (IEA, 2010). This resource includes encroacher species in savannah lands, invasive species in wetlands and water bodies and unutilized forestry and agricultural residues. The availability of such vast unutilized feedstocks gives a good justification for their conversion to off-grid power. However, techno-economic issues such as the levelized cost of energy (LCOE) then have to be considered so that the power is competitive against fossils and alternative renewables (Bloomberg NEF, 2020).

It is equally important to ensure that the supply chain for the particular biomass is regular and economically viable, implying there should be a sufficient store within a certain threshold radius. The International Renewable Energy Agency (IRENA) mentioned that, for bio-energy and biofuel industry to be more competitive than fossil sources, progressive innovation in conversion technologies, supply chain models, market development and policy support is required (IRENA, 2016). The cost for the feedstock supply significantly affects production costs, contributing about 40-70% even where feedstocks such as forest residues are cheap. This is because of the low energy density of biomass when compared to fossil fuels, which increases the costs of storage, handling and transportation of an energy unit. This makes it critical to optimize supply chain logistics and minimize costs. The seasonal and uncertain nature of some biomass feedstock supplies and variability of feedstock locations further complicate bioenergy projects (Iakovou et al., 2010).

The ensuing subsections explore the inventories of various biomass forms in SSA:

Forestry Based Wastes vs Wood Fuel and Charcoal

Africa's forest cover of 650 million hectares is approximately 17% of the global forestry area, with a potential sustainable energy supply of 9.4 PJ/year (Nyika et al., 2020). The major forest product is round wood with 237 million tonnes produced per annum compared to 15, 52, and 207 million tons of charcoal, fuel wood and industrial products (e.g. paper) respectively. Fig 5 shows the percentage shares of firewood and charcoal production in various parts of Africa. In this case, SSA comprises Western, Southern, Eastern and Central Africa. For both solid fuels, 85% of them were for domestic use while 1-3% were taken up by industry and the rest were for other uses.

Figure 5. Regional shares of firewood (left) and charcoal production (right) in 2013. Source (UNEP, 2017)

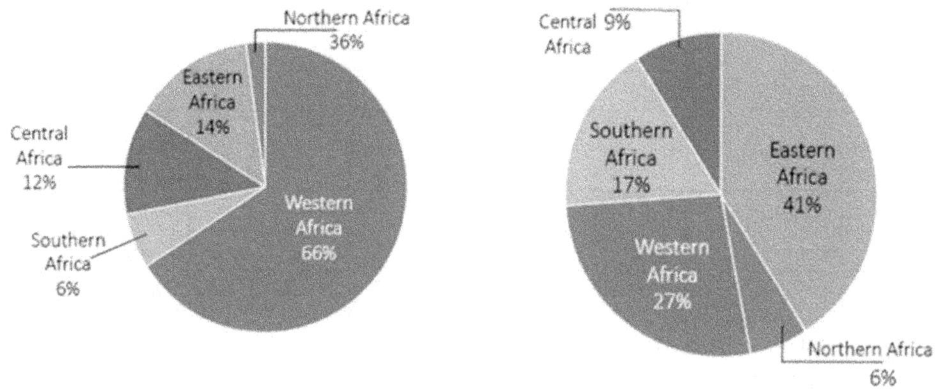

Indeed, as illustrated in Table 1, Northern Africa has a limited production and use of firewood. It however has a good percentage of charcoal production relative to its population, compared to SSA. This is a significant observation since charcoal is a cleaner fuel and it shows how SSA is trailing behind with regards the promotion of modern, cleaner alternatives of utilizing biomass.

On the other hand, forestry harvests and downstream processing activities generate wood wastes such as lops, tops, shavings, off-cuts, saw dust and chips, which have been estimated to be just above 0.4 billion tonnes per annum (Cudjoe et al., 2015). Most developing nations have not been processing this waste since this requires considerable technical knowhow and capital investment. The Food and Agricultural Organization (FAO) estimates that approximately 54% of harvested logs in Africa remain unutilized (Cudjoe et al., 2015). In Ghana, Kumasi only 30% of round wood is recovered and the residual bulk waste (especially sawdust) is dumped or incinerated (Ogunwusi, 2014). Nigeria and Zimbabwe have also experienced similar cases of waste disposal from sawmills, because downstream industries like engineered wood products, pulp and paper are either too few or non-existent in some regions, to take up all the waste generated (Arimoro et al., 2006).

Agricultural Residues

Estimates on Africa's collective biomass potential by 2020 are 13,900PJ/yr for energy crops (both woody and herbaceous) up to 5,400PJ/yr forestry biomass (plantations) and up to 5,254PJ/yr of residues (forestry & agriculture) and organic waste (Stecher et al., 2013). The bulk of the agricultural and forestry residues in Southern Africa and Africa at large are largely unexploited. Table 2 further illustrates the agricultural residues that have a large potential for bioenergy exploitation in various regions of Africa.

Table 2. Biomass from various agricultural residues in Africa

Residue type	Thousands of tonnes					Total
	Central	Eastern	Northern	Southern	Western	
Maize husks	9,543	2,514	14,161	21,130	32,974	130,321
Maize stalks	954	5,251	1,416	2,130	3,297	13,032
Maize cobs	1,303	7,168	1,933	2,884	4,501	17,789
Millet stalks	1,616	3,749	22	58	13,405	18,850
Rice straw	1,673	18,247	12,502	6	25,623	58,051
Rice husks	203	2,210	1,514	1	3,103	7,031
Sorghum straw	3,247	11,365	1,497	374	19,883	36,366
Wheat straw	36	5,260	23,833	2,976	139	32,243
Cassava stalk	2,248	1,777	-	-	4,716	8,740
Cassava peelings	9,064	7,164	-	-	19,016	35,244
Cotton stalks	533	2,027	606	83	2,671	5,919
Oil palm empty fruit branches	744	35	-	-	3,196	3,975
Oil palm shells	210	10	-	-	903	1,123
Oil palm fibre	453	21	-	-	1,945	2,420
Sugarcane	1,057	6,833	3,279	4,360	1,161	16,691
Total	32,884	123,630	60,763	33,985	136,534	387,797

Source: (Cudjoe et al., 2015)

Evidently, maize agricultural residues are the most plenteous at ~144 million tonnes produced per annum within SSA. These are followed by rice waste produced at a rate of ~51million tonnes per annum. The total amount of agricultural residues within the SSA region in 2015 was just over 327million tonnes per annum, with Western and Eastern sub-regions contributing the most, in that order.

Encroaching and Invasive Species

Bush encroachment in savannah ecosystems such as found in Botswana, Namibia and South Africa has been a major problem that reduces grazing land productivity, livestock productivity, groundwater availability and mars aesthetic outlook in residential spaces (Charis et al., 2019c). Drought hardy woody plants like *Acacia tortilis, A. mellifera, A. erubescens, Dichrostachys cinerea* and *Terminalia sericea* have progressively increased in cover and density, encroaching rangelands, cities, towns and villages. In Namibia alone, 26-30million hectares of land are affected, causing loss of income of up to US$48,4million per year in stunted livestock productivity (Harper-Simmonds et al., 2012). Bush control activities in Namibia have presented socio-economic and ecological benefits through increasing livestock productivity, adding to the renewable energy mix, industrial diversification, employment creation and economic growth. Commercial solutions that have been working well include heat production for cement kilns, electrical power generation, charcoal production, while other initiatives like gasification, pyrolysis and engineering wood products have been tabled for consideration (DECOSA, 2015). Botswana and South Africa had not yet started valorizing this waste resource significantly and the starting point would be to be quantify it. Currently, debushed encroachers are burnt on site or transported to central place for burning using hired companies. Fig 6 shows pictures of debushed *acacia tortilis*, an encroacher species in Botswana, where debushing in such residential spaces has to be carried out frequently in different areas.

There are also a number of invasive species in SSA in the form of shrubs, water plants and trees. South Africa alone has approximately 8.7 million tonnes per annum (tpa) of invasive species (Pradhan & Mbohwa, 2014).

Figure 6. Debushed acacia tortilis shrubs in Palapye, Botswana: Fresh (left) and dry (right).

MSW

By 2012, approximately 125 million tpa of MSW was generated in Africa with ~81 million tonnes (65%) from SSA. This is expected to grow to 244 million tpa by 2025. Only ~55% (68 million tonnes) of this is collected and consequently, almost half of all MSW remains within cities and towns or ends up in sidewalks, open fields, storm water drains and rivers. Only 4% of the collected waste on average, is recycled and less than that is converted to energy (UNEP, 2018).

Biogas Feedstocks

Biogas results from the microbial degradation of organic matter in anaerobic conditions. It is generally derived from broken down derivatives of lignocellulosics such as sewage sludge, food waste and animal dung or any such feedstocks rich in carbohydrates, proteins and lipids. Biogas can be harvested from landfill sites, wastewater plants, breweries and specially designed biodigesters that can utilize a defined range of feedstocks (UNEP, 2017). Across SSA, suitable substrates that have been demonstrated for anaerobic digestion include aquatic plants like water lettuce and water hyacinth; agricultural wastes/ residues such as cow and piggery dung, cassava leaves and municipal wastes including sewage sludge, processing wastes, MSW and industrial wastes.

Other Considerations on Supply Capacity of Various Biomass Forms

SSA can benefit from the abundance of biomass in it, especially forestry and residues that do not effect LUC. When such residues are used within threshold distances from the source, it is even possible to guarantee carbon neutrality. It is important to maintain sustainable harvesting rates for such residues to ensure that the soil organic content is not severely depleted and policies have to begin covering such essential issues (Searle & Bitnere, 2017). When considering the lifecycle impact of utilizing the residues, one must also factor in that carbon neutrality and economic feasibility is guaranteed when they are used within certain maximum distances from the source.

Global Best Cases on the Use of Biomass for Off-Grid Power

In mapping the potential of using biomass for mini grid power supply, it is useful to take a cue from global and regional best practices. These studies inform on what is technically feasible, while enabling objective comparisons of the techno-economic and supply chain workspace of areas where such bioenergy projects have been commissioned to the SSA cases. Four global best cases, including Germany, Sweden, Finland and the US were selected.

Germany

In Germany, the use of biomass for energy generation plays a considerable role in the energy mix. This European Union (EU) nation has made significant strides in developing biomass off-grid power generation with working technology as well as many planned and pilot plants in Africa. In 2016 an estimated 43.8 TWh of electricity was produced from biomass, equivalent to 23% of the total renewable electricity production. Of these biomass off-grid power plants, 25.1% were based on solid biomass, 73.9% were biogas and biomethane plants with 1% being liquid biomass based plants. The main technologies that were used for power production from solid biomass include steam cycles, organic Rankine cycles and gasification systems. Policy incentives in Germany have promoted continual biomass off-grid electricity production through initiatives such as fixed feed-in tariff paid per kWh. From 2012 there has been significant effort by the German government to improve policies towards improving the regulatory framework that promotes biomass off-grid power generation (Purkus et al., 2018).

In 2009, a German company (BME GmbH) designed a High Performance, Temperature Controlled (HPTC) biogas plant, having a two stage digestion system. The first plant BME GmbH installed made use of a diesel generator, replacing 80% of the diesel with biogas. Most of these plants in Germany are set up near slaughterhouses which are the main source of the digester's feedstock. With a 2.5 year payback period and installation cost of US$36,584, these digesters are said to be profitable in remote areas, to replace expensive electricity generation by diesel generators, even without a feed-in tariff (Dimpl, 2010).

Sweden

At the time of the first oil crisis in 1973, Sweden was 80% dependent on oil, however, it has transitioned to substitute oil with bioenergy, especially in the district heating and CHP sectors. Currently, bioenergy constitutes more than one third of all the energy used in Sweden (Amiandamhen et al., 2020) and biomass is the fourth largest source of electricity (IEA, 2018). The biomass raw materials used include energy crops, agricultural residues, forestry, stumps and marine biomass with the largest contributors to bioelectricity being undensified wood (55,059 GWh) and black liquor (44,181 GWh) . The use of bioenergy in Sweden has increased from 40 TW/h in the 1970s to around 140 TW/h. In 2009, bioenergy was more than the use of electricity from hydropower and nuclear power summed together. Tremendous growth of the bioenergy sector in Sweden is largely attributed to political support and the use of strong general incentives such as the Swedish carbon dioxide tax, green electricity certificates, tax exemption for biofuels for transport as well as direct investment support (IEA, 2018).

Finland

Finland has vast forest resources and sustainable wood production that supports industrial output, energy supply, economic growth and its social well-being. Finland has highly efficient combined heat and power (CHP) plants where some use the direct combustion technology and others the more modern technology of fluidized bed boilers or circulating fluidized bed technology. Gasification technology works for a wide range of low-quality forest residues. Advanced CHP plants come with greater flexibility in timing and sourcing feedstock collection which translates to lower costs due to their ability to use a wide range of biomass from forests (IRENA, 2018).

Finland has three operational plants from which lessons can be drawn from a technical and policy making view point. All the three plants produce heat and electricity. Two plants are owned and operated by the power utility, Fortum, and one is a bio-mill in central Finland owned by Metsä Fibre. The Metsä Fibre plants operates on 100% renewable energy and produces enough electricity for its use as well for other neighbouring towns and industries. This plant is much more energy efficient than the typical pulp mills fuelled by oil. The Fortum plant in Järvenpää, southern Finland, uses 99.5% biomass fuel and has a 96.5% efficiency(IRENA, 2018) . It is a high efficiency, multi-fuel CHP plant. These high efficiencies can be attributed to the flue gas condenser technology employed in the plant. The plant has a fuel flexibility that allows it to alternatively use farmyard manure up to 30%. The third plant is in eastern Finland, in Joensuu. The plant produces pyrolized fuel at high efficiency from forest residue and sawdust. The plant makes use of the fluidized bed boiler as a heat source for pyrolysis, the coke and uncondensed gases from pyrolysis are used for heat and electricity generation.

A key challenge from the Finland case study is the consistency of feedstock supply from forest biomass. This challenge is influenced by the seasonal nature of feedstock as well as the availability of enough quantity in a single location. The seasonal nature of the feedstock may be addressed by converting the biomass to gas or pyrolysis for storage to be used when the season does not support enough feedstock production.

United States (US)

The US has been exemplary, especially in WtE power plants. In 2018 biomass supplied about 46% of all renewable energy, constituting 4.6% of the total energy consumption in the nation (Campbell et al., 2018). Although this is a smaller fractional share compared to Sweden, it is significant in terms of population supported by the bioenergy. In that year, the US managed to produce 14 billion kWh of electricity from the combustion of 29.5 million tons of MSW in 68 power plants (Nyika et al., 2020). A part of the steam from these plants was also used for heating. The biomass fraction in the MSW was >60% of the combustible waste and accounted more than 50% of the power produced. The rest of the combustible waste comprised non-biomass materials such as plastics. Many of the US states have also set goals bordering around the utilization of woody biomass and MSW (Campbell et al., 2018). In 2004, close to 1,000 wood-fired plants were already operational in the US, with over 100 fluidized bed (FB) boilers installed, while the rest were the conventional pile burners, stoker fired boiler and suspension type boiler types (Bauen et al., 2004). are operating or planned for operation.

Current State on Application of Biomass For Off-Grid Power in Sub-Saharan Africa

Presently the use of biomass for off-grid power in SSA has a huge potential for implementation. Biomass currently supplies about a third of Africa's industrial-based heat, which could be diverted to electricity production (UNEP, 2017). This potential is anchored on the continued growth of the African economy, huge biomass reserves as well general increase in population and urban migration. The benefits include industrial growth, entrepreneurship, and support for increased prosperity. However before implementation can be recorded as a success story, there are milestones of concern that have to be addressed and planned for such as downstream value chain management in supply of the biomass, financing of planned and pilot plants, pollution control and monitoring with respect to the effluent from the plants, as well as adaptation of the technology for easy on-site operation. SSA has off-grid power plant case studies from which the technology, support policies and sustainability can be improved.

As discovered in earlier sections, biomass has always been used as a source of energy in SSA. However, its exploitation for the production of cleaner bioelectricity has not been practiced substantially to constitute a significant portion of the energy mix. Electricity generation from solar systems has experienced greater success as compared to biomass off grid generation. Though biomass off-grid generation is not popular, there are plants that have been run and some still being run in SSA with enough success to record and learn from Kenya, Tanzania, Zimbabwe, Mauritius and South Africa.

Kenya

In Kenya, success has been recorded at the Kilfi plant which has been in operation since 2007, the biogas plant is well adapted to farm operations. The plant makes use of agricultural waste, sisal and cow manure to produce electricity and heat as end products (Dimpl, 2010). The Kilfi biogas plant is the biggest in Kenya with an installation capacity of 150 kW_e and digester volume of 750 cubic meters. This project was financed through a tripartite Public Private Partnership (PPP) involving two German companies. The production cost of electricity in this set up was US$0.20/kWh compared to grid electricity costing US$0.18 -0.22/kWh. The success of this project lies in the training of locals to operate the plant, close liaison with government authorities for import clearances and regulation by energy authorities as well as manufacturing of pipes, wiring and civil works locally to reduce the project cost.

With the support of the German government again, Kenya set up two biomass off-grid power plants though these realized partial success. One made use of slaughterhouse waste to cater for refrigeration needs while the other processed vegetable waste and were meant to power water pumping. For both cases, generation costs remain very competitive at US$0.17/kWh. Both plants have been performing well from a techno-economic viewpoint (Dimpl, 2010).

Tanzania

Though famous in the US, Asia (especially India) and Europe, anaerobic digestion has faced a slow uptake in SSA against demonstrated potential of many substrates in the subcontinent. For instance, in 2005, the number of biogas digesters in Africa were insignificant compared to other continents, until the new African initiative launched in 2007, aimed at installing biogas digesters in at least two million households by 2020. By 2010, the number of biogas digesters had more than doubled and Tanzania had installed about 4000 units. However, only 60% of the installed plants were functional while the remaining ones performed below satisfaction or failed due to planning and construction errors, poor community awareness, poor maintenance culture, a misconception of the technology's benefits and lack of technical know-how by end-users (Dahunsi et al., 2018).

Since 1983 over 1000 medium sized biomass off grid power plants have been installed in Tanzania with support from the German government. A pilot project managed and partially financed by United Nations Industrial Development Organization (UNIDO) showed that sisal residue contained an essential substrate for anaerobic digestion for generation biogas, electricity and bio-fertilizer. These projects died a premature death as they did not realize the second and third phases of implementation due to poor adaptability of the plant design to sisal operations. Another factor that was perceived as a drawback was the low profitability of the operation.

Zimbabwe and Malawi

Zimbabwe and Malawi have a thriving agricultural and forestry industry. The Department of Energy (DoE) through the National Biomass Strategy recommended three potential co-generation projects producing 16.5MW_e in Manicaland to utilize at least 70,000 tpa of saw dust and other unquantified sawmill wastes (Charis et al., 2019b). Currently, Hippo Valley Estates alone produces around 79,056 MWh annually, enough to meet its electricity requirements (14.5MW) and feed a little surplus into the main grid. Triangle estates on the other hand, just meets its requirements of 17.0-21MW, although it can generate up to 35.5MW during harvest time (Mtunzi et al., 2012).

Malawi's sustainable fuel forestry biomass supplies were estimated to be 42.4 million cubic metres of solid wood equivalent (Taulo et al., 2017). Though both Malawi and Zimbabwe use their bagasse for electricity generation, there has also been research around potential valorisation of the resource into lignocellulosic ethanol, which could be used both in vehicles and stationery power engines (Zyl et al., 2011). The abundant agricultural residues in Malawi such as rice straw, sugarcane and cassava pulp have been estimated to produce 46.5 million litres per annum of ethanol. Malawi also has a thriving sugarcane processing industry whose history emanates from the 1980's. The by-product sugarcane bagasse complement of 950 000 tpa is estimated to be able to cater for at least 25% of Malawi's electricity per capita requirement if fully exploited (Taulo et al., 2017). The nation also has a strong initiative for off-grid electricity with a current installed capacity of 532MW comprising 70% of hydroelectricity; 11% solar PV; 3.5% bioenergy and 26.6% largely diesel and other fossil fuel based generators (Taulo et al., 2017). Evidently, there is still a large share of non-renewable energy that biomass can occupy.

Figure 7. Sawmill wastes from Manicaland

The Scope and Potential for Off-Grid Power Systems Based on Biomass Waste: Discussion

The Power Generation Gap Between SSA and Global Best Cases and Opportunities for Closure

Compared to other regions, Africa generally still has a lot of untapped potential in bioelectricity from industrial and municipal waste as well as liquid biofuels. These resources are abundant in Africa. Table 3 and 4 show the amount of electricity generated from biomass fractions in various continents in 2018, in both 'electricity only' and CHP plants.

Table 3. Electricity generation (in TWh) from biomass in 'electricity only' and CHP plants in 2018

	Total power Electricity only/ CHP	*MSW: Electricity only/ CHP*	*Industrial waste Electricity only/ CHP*	*Solid bio-fuels Electricity only/ CHP*	*Biogases Electricity only/ CHP*	*Liquid biofuels Electricity only/ CHP*
Africa	2.85/1.21	0.00	0.00	2.82/1.21	0.03/0.00	0.00
Americas	79.9/75.1	21.0/2.32	1.22/1.14	44.0/69.4	13.0/2.04	0.70/0.21
Asia	310/7.08	8.83/4.13	30.2/1.18	262/0.65	6.57/1.12	2.28/0.00
Europe	79.4/139	23.1/41.9	1.68/10.4	34.6/59.0	18.2/26.6	1.88/1.47
Oceania	1.69/2.86	0.00	0.00	0.56/2.52	1.13/0.34	0.00
EU-28	79.0/125	22.8/37.4	1.68/2.13	34.5/57.4	18.1/26.2	1.88/1.47

Source: (World Bioenergy Association, 2020)

Table 3 shows that Africa as a whole had a total production capacity of 4.06TWh of electricity from both 'electricity only' and CHP plants, compared to 4.55TWh in Oceania and at least 155TWh in other regions. The region with the highest total production of electricity was Asia at 317.8TWh. Despite the fact that Africa is the second most populous country after Asia, its power generation capacity from biomass is disproportionately lower. This low exploitation of biomass resources cascades to SSA, where it has already been demonstrated that there is a large inventory of waste, unutilized biomass.

One of the main drivers of low electrification and resultant energy poverty in SSA are the relatively high average electricity costs of US$ 0.2 – $0.5 per kWh compared to US$ 0.1 per kWh global average. Sound pricing, tariff policy and social protection are essential to support biomass off grid power generation. Unfortunately, there are still a number of factors hindering the development of biomass off grid power generation include feed-in tariff policies, weak regulatory capacity and framework coupled with inefficient parastatal monopolies (World Bank, 2013). As already demonstrated, Africa has abundant biomass, sometimes referred to as the green gold of Africa. Presently this biomass is being used to a large extent in an inefficient way. Bioenergy is therefore a huge opportunity in SSA though sustainability must be ensured for both environmental and social issues. Some of the factors to be considered in sustainability include, the availability of land, the productivity of biomass to be harvested from the land, competition for alternative use of the biomass and sustainable residues harvest rates (Belward et al., 2011). Generally it is imperative to ensure cost effective supply chain logistics for WtE projects compared to biogas plants because of their higher rates of biomass consumption. This challenge was reflected in the Finland case study due to the inconsistency of feedstock supply from forest biomass influenced by the seasonal nature of feedstock and its availability in multiple locations.

Biogas production is another great opportunity with a number of substrates having been researched on by various authors with promising results (Dahunsi et al., 2018). There is a need however for support from the government, donor and private sectors. Government support should include tax incentives and clear, supportive policy guideline and regulatory frameworks. The World Bank highlighted the need for proper technical training of farmers, Agritex officers and staff then after sales support. As costs of technology go down, it would be convenient, also, to have lower maintenance costs which can be sustained by support from relevant stakeholders (World Bank, 2019).

Technologies and Approaches That Could Be Adopted In SSA

The adoption of technologies and approaches to be used in SSA will largely depend on their ease of uptake, which hinges on the CAPEX, OPEX and technical expertise requirements, assuming availability of biomass for any scale of operation. The most popular boiler types include: pile burners, suspension fired, stoker fired and fluidized bed (FB) boiler types (Bauen et al., 2004). FB boilers are fast becoming a preferred technology for plants >10 MWe because of their cleaner and more efficient combustion profile; however, they are not economic for lower capacities (IRENA, 2012). This means that feasibility studies need to first be done in choosing the right technology, to ensure that the supply capacity and radius are compatible with the preferred technology. The other technologies are however, more flexible in terms of capacities although a proper techno-economic feasibility study should be done for every use case given varying feedstock supplies and costs of obtaining them. Sensitivity analyses on the supply capacities should be conducted to establish if the plant will still operate well given possible fluctuations in supply.

Fig 8 shows the capital costs for various power generation technology avenues including bubble fluidized bed (BFB) or circulating fluidized bed (CFB); stoker; stoker CHP; land fill gas (LFG); anaerobic digester (AD) and co-firing with coal; only a few additional adjustments have to be made.. The co-firing option has the lowest costs due to the fact that the infrastructure is already available. IRENA, (2012) also calculated the levelized cost of electricity (LCOE) from biomass for these technologies after factoring in key aspects on CAPEX, OPEX, discount rate (10%), economic life (20yrs) and a net electrical efficiency of 31-36% depending on the particular technology. The LCOEs for these technologies are shown in Fig 9. The stoker and LFG technologies are the next most affordable at around \$1,750/kWh and \$2,050/kWh respectively. The LFG technology is followed closely by the various gasifier types for electricity only at around \$2,100/kWh. In terms of capital therefore, the most attractive and economic could be the conventional stoker fired boilers and they also have the least LCOE after co-firing, at around \$0.06/kWh.

Figure 8. Capital costs for various power generation technologies in 2010. Source (IRENA, 2012).

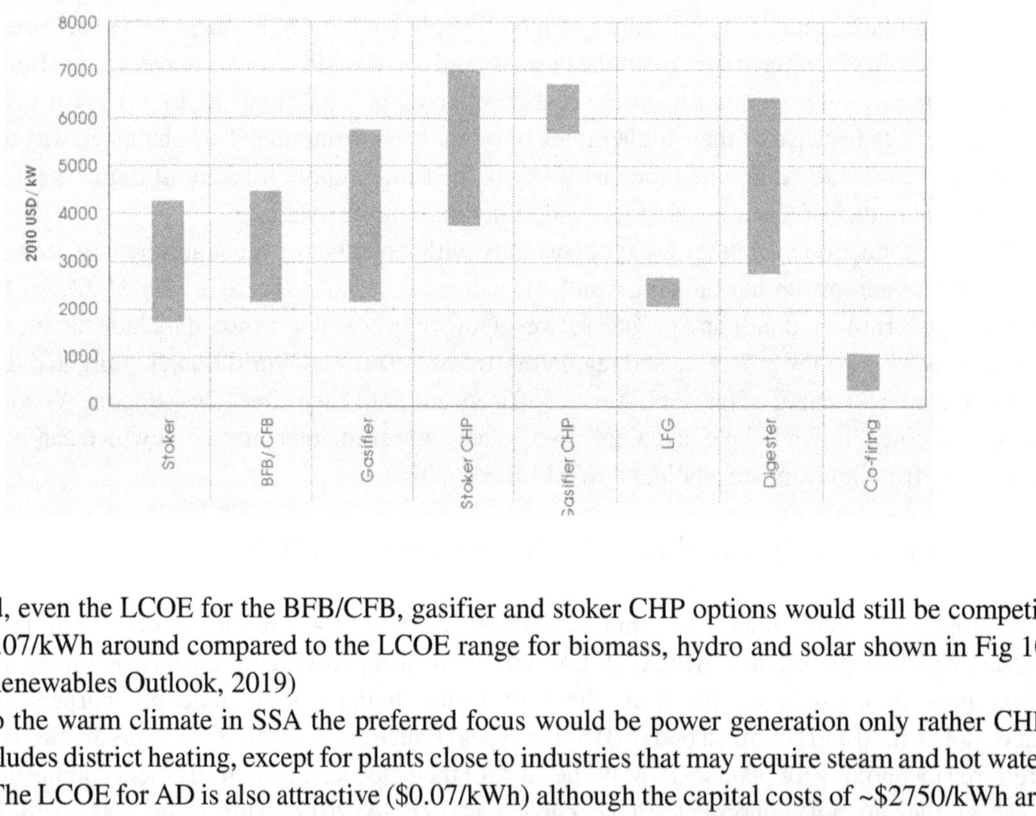

Indeed, even the LCOE for the BFB/CFB, gasifier and stoker CHP options would still be competitive at \$0.07/kWh around compared to the LCOE range for biomass, hydro and solar shown in Fig 10 (Global Renewables Outlook, 2019)

Due to the warm climate in SSA the preferred focus would be power generation only rather CHP which includes district heating, except for plants close to industries that may require steam and hot water utilities. The LCOE for AD is also attractive (\$0.07/kWh) although the capital costs of ~\$2750/kWh are somewhat prohibitive. This is probably due to the complex steps of purifying biogas into biomethane, before attempting to utilize it for power generation with the internal combustion engine.

In terms of technical expertise, SSA should be able to support the operations of these plants, although there will be need for capacity building especially on the maintenance aspects since these plants would have been assembled abroad. After sales support would then be required and it should be such that the SSA purchaser can run the plant independently five to ten years after installation.

Figure 9. LCOE for various biomass fired power generation avenues. Source (IRENA, 2012)

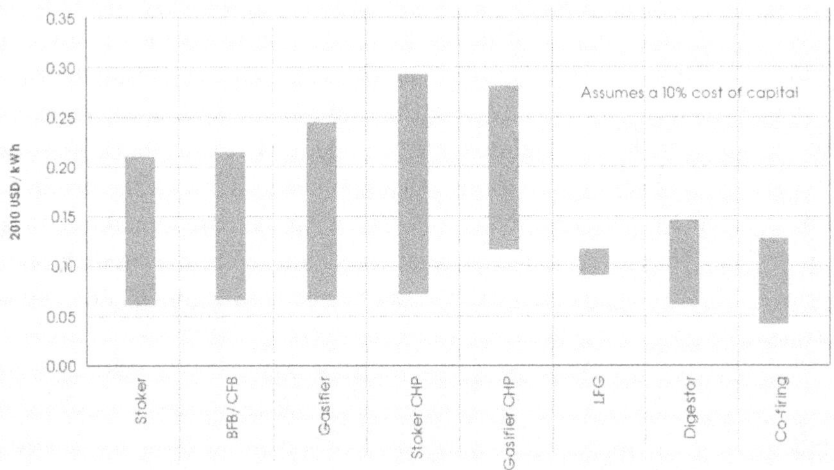

Figure 10. LCOE for various electricity sources in SSA. Source (Global Renewables Outlook, 2019)

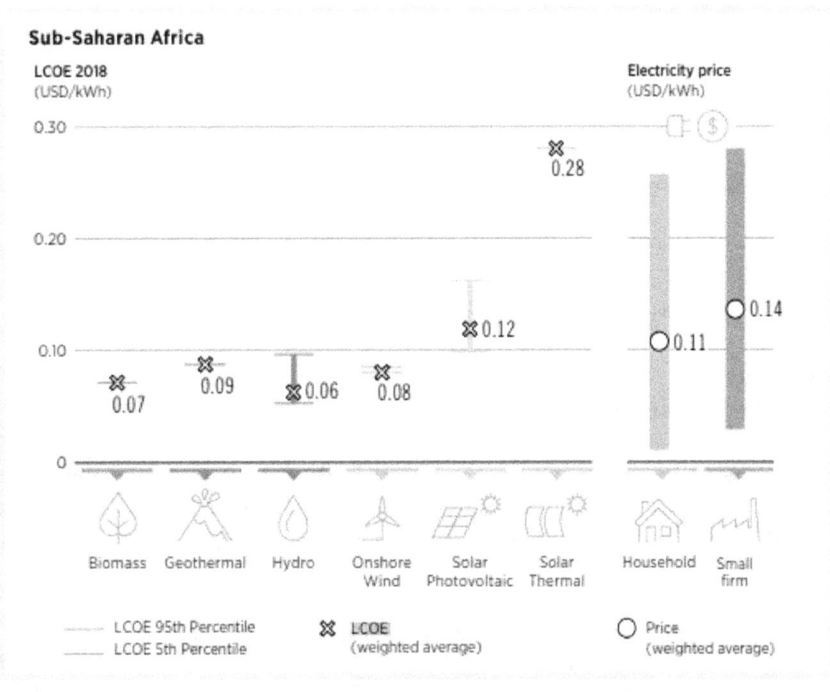

CONCLUSION

Given the abundant supply of unutilized biomass, it can be concluded that there is a tremendous Opportunity for biomass off-grid power generation within SSA with the adoption of relevant, easy to use technologies. A widespread installation of biomass mini-grid systems has the potential to reduce energy poverty in SSA and improve the quality of life for remote areas. Already, farming and forestry practices in SSA are characterized by post-harvest losses and large yields of unutilized residues. There is also many other biomass wastes generated as MSW, empty fruit branches, tree tops and lops, fruit residues, encroacher and invasive species. An integrated approach incorporating energy, environment and agricultural interests could accelerate the production of biomass off-grid power. There are good examples of biomass utilization that have been demonstrated in Germany, Sweden, Finland and the US, where SSA can benchmark in terms of approach and technologies. The stoker fired boiler and LFG technologies would be the most affordable 'biomass only' technologies in terms of capital costs, although the gasification routes and AD would also be competitive with regards to the LCOE. An increased uptake of these technologies for mini-grid power would then increase the share of renewables in power production and also reduce the carbon footprint of SSA. Such potential socio-economic and environmental benefits underscore the importance of sustainable production of energy from biomass.

REFERENCES

Amiandamhen, S. O., Kumar, A., Adamopoulos, S., Jones, D., & Nilsson, B. (2020). Bioenergy production and utilization in different sectors in Sweden: A state of the art review. *BioResources*, *15*(4), 9834–9857. doi:10.15376/biores.15.4.Amiandamhen

Arimoro, F. O., Ikomi, R. B., & Osalor, E. C. (2006). The impact of sawmill wood wastes on the water quality and fish communities of Benin River, Niger Delta area, Nigeria. *World Journal of Zoology*, *1*(2), 94–102.

Bauen, A., Woods, J., & Hailes, R. (2004). *BioPowerswitch*. Academic Press.

Belward, A., Bisselink, B., Bódis, K., Brink, A., Dallemand, J., De Roo, A., Huld, T., Kayitakire, F., Mayaux, P., Ossenbrink, H., Pinedo, I., Sint, H., Thielen, J., Szabó, S., Tromboni, U., Willemen, L., & Monforti, F. (2011). Renewable energies in Africa. JRC Scientific and Technical Reports. doi:10.2788/1881

Berahap, R. (2019). *Thinking outside the Grid: The role of decentralized power systems in electrifying Sub Saharan Africa*. www.policycenter.ma

Blimpo, M. P., & Cosgrove-Davies, M. (2019). Electricity Access in Sub-Saharan Africa: Uptake, Reliability, and Complementary Factors for Economic Impact. In *Electricity Access in Sub-Saharan Africa: Uptake, Reliability, and Complementary Factors for Economic Impact*. The World Bank. doi:10.1596/978-1-4648-1361-0

Bloomberg NEF. (2020). *Sub-Saharan Africa Market Outlook 2020*. Author.

Campbell, R. M., Anderson, N. M., Daugaard, D. E., & Naughton, I. D. (2018). Technoeconomic and policy drivers of project performance for bioenergy alternatives using biomass from beetle-killed trees. *Energies*, *11*(2), 293. Advance online publication. doi:10.3390/en11020293

Charis, G., Danha, G., & Muzenda, E. (2019a). *A review of the application of GIS in biomass and solid waste supply chain optimization: Gaps and opportunities for developing nations*. Academic Press.

Charis, G., Danha, G., & Muzenda, E. (2019b). A review of timber waste utilization: Challenges and opportunities in Zimbabwe. *Procedia Manufacturing*, *35*, 419–429. doi:10.1016/j.promfg.2019.07.005

Charis, G., Danha, G., & Muzenda, E. (2019c). Waste valorisation opportunities for bush encroacher biomass in savannah ecosystems: A comparative case analysis of Botswana and Namibia. *Procedia Manufacturing, 35*, 974–979. doi:10.1016/j.promfg.2019.06.044

Cudjoe, E., Kemausuor, F., Miezah, K., & Mensah, M. (2015). *African perspective on cellulosic ethanol production.* doi:10.1016/j.rser.2015.04.059

Dahunsi, S. O., Shoyombo, A., & Fagbiele, O. (2018). Biofuel Development in Sub-Saharan Africa. *Anaerobic Digestion, 32*, 137–144. https://www.intechopen.com/books/advanced-biometric-technologies/liveness-detection-in-biometrics

DECOSA. (2015). *Support to De-Bushing Project Value Added End-Use Opportunities for Namibian Encroacher Bush* (Vol. 264). https://www.dasnamibia.org/download/value-added-end-use-opportunities-for-namibian-encroacher-bush/

Dimpl, E. (2010). *Small-scale Electricity Generation from Biomass Part II : Biogas.* Academic Press.

Global Renewables Outlook. (2019). *Energy Transformation Sub-saharan Africa.* https://www.irena.org/-/media/Files/IRENA/Agency/Publication/2020/Apr/IRENA_GRO_R10_Sub-Saharan_Africa.pdf?la=en&hash=DB067EF85E0FDB6B8762833E77CC80F3975E46DC

Harper-Simmonds, L., Mendelsohn, J., Roux, J.-P., Pallet, J., Brown, C., Middleton, A., & Kruse, J. (2012). *Development of an inventory of ecosystem services in Namibia.* www.met.gov.na/files/downloads/179_ResMob-Inventory report2-final.pdf

Iakovou, E., Karagiannidis, A., Vlachos, D., Toka, A., & Malamakis, A. (2010). Waste biomass-to-energy supply chain management : A critical synthesis. *Waste Management, 30*(10), 1860–1870. doi:10.1016/j.wasman.2010.02.030 PMID:20231084

IEA. (2010). SuStainable Production of Second Generation Biofuels. *Renewable Energy*, 1–39. 9789461739698

IEA. (2018). *Country Report. Sweden – 2018 update. Bioenergy policies and status of implementation.* IEA.

International Energy Agency. (2015). *Africa Energy Outlook.* https://www.iea.org/PUBLICATIONS/FREEPUBLICATIONS/PUBLICATION/WEO2014_AFRICAENERGYOUTLOOK.PDF

IRENA. (2012). Renewable Energy Technologies: Cost Analysis Series. In *Zeitschrift fur Geomorphologie, Supplementband* (Vol. 1). www.irena.org

IRENA. (2016). *Innovation Outlook Advanced Liquid Biofuels.* http://www.irena.org/DocumentDownloads/Publications/IRENA_Innovation_Outlook_Advanced_Liquid_Biofuels_2016.pdf

IRENA. (2018). *Bioenergy from Finnish forests: Sustainable, efficient and modern use of wood.* www.irena.org

IRENA. (2020). *Scaling up renewable energy deployment in Africa.* www.irena.org/publications

Mtunzi, B., Mampwheli, N., Meyer, E., & Mungwena, W. (2012). Bagasse-based co-generation at Hippo Valley Estates sugar factory in Zimbabwe. *Journal of Energy in Southern Africa, 23*(1), 15–22. doi:10.17159/2413-3051/2012/v23i1a3158

Nyika, J., Adediran, A. A., Olayanju, A., Adesina, O. S., & Edoziuono, F. O. (2020). The Potential of Biomass in Africa and the Debate on its Carbon Neutrality. In *Biomass*. IntechOpen. doi:10.5772/intechopen.93615

Ogunwusi, A. A. (2014). Wood Waste Generation in the Forest Industry in Nigeria and Prospects for Its Industrial Utilization. *Civil and Environmental Research, 6*(9), 62–69. https://www.iiste.org/Journals/index.php/CER/article/view/15450

Pradhan, A., & Mbohwa, C. (2014). Development of biofuels in South Africa: Challenges and opportunities. *Renewable & Sustainable Energy Reviews, 39*, 1089–1100. doi:10.1016/j.rser.2014.07.131

Purkus, A., Gawel, E., Szarka, N., Lauer, M., Lenz, V., Ortwein, A., Tafarte, P., Eichhorn, M., & Thrän, D. (2018). *Contributions of flexible power generation from biomass to a secure and cost-effective electricity supply — A review of potentials, incentives and obstacles in Germany*. Academic Press.

Searle, S., & Bitnere, K. (2017). *Review of the impact of crop residue management on soil organic carbon in Europe*. Academic Press.

Stecher, K., Brosowski, A., & Thrän, D. (2013). Biomass potential in Africa. *Irena*, 44. www.dbfz.de

Taulo, J. L., Gondwe, K. J., & Ben Sebitosi, A. (2017). Energy supply in Malawi: Options and issues. *Journal of Energy in Southern Africa, 26*(2), 19. doi:10.17159/2413-3051/2015/v26i2a2192

UNEP. (2017). *Atlas of Africa Energy Resources*. http://www.unep.org

UNEP. (2018). *Africa Waste Management Summary for Decision-makers*. https://wedocs.unep.org/bitstream/handle/20.500.11822/25515/Africa_WMO_Summary.pdf?sequence=1&isAllowed=y

Van Zyl, W. H., Chimphango, A. F. A., den Haan, R., Görgens, J. F., & Chirwa, P. W. C. (2011). Next-generation cellulosic ethanol technologies and their contribution to a sustainable Africa. *Interface Focus, 1*(2), 196–211. doi:10.1098/rsfs.2010.0017 PMID:22482027

World Bank. (2013). *Toward a Sustainable Energy Future for All : Directions for the World Bank Group's Energy Sector*. Author.

World Bank. (2019). *The power of Dung: Lessons learned from on-farm biodigester programs in Africa*. Author.

World Bioenergy Association. (2020). *Global Bioenergy Statistics 2020*. https://worldbioenergy.org/uploads/201210 WBA GBS 2020.pdf

Chapter 14
Climate Justice in Durban:
State Contradictions and Grass-Roots Action

Innocent Simphiwe Nojiyeza
https://orcid.org/0000-0002-9948-2147
University of Zululand, South Africa

ABSTRACT

This chapter situates itself in the climate justice discourse and unpacks the paradoxes in state and grass-roots action. It argues that due to market-orientated and neo-liberal water and sanitation governance, the failure to take into account water and climate change governance as well as ecological and institutional economics resulted in climate change chaos. Due to this paralysis, communities failed to adapt to climate variability. The illogicalities and ambiguities in what the municipality says and does leaves local communities at the vagaries of nature and to deal with the resulting chaos albeit with some doing so in quite inspiring ways. The empirical data is collected through 120 interviews and four participatory action workshops conducted with civil society actors, government officials, academics, and communities of AmaQadi situated in North Central Durban, Umbumbulu in South Durban, Wood Glen and Zamani B in Durban Outer West, Ntuzuma G and Soweto in North Central Durban. The key finding is that the municipal Climate Change Protection Unit adopted a 'climate-resilient' strategy on the one hand, but on the other, city electricity officials built power stations in flood plains, the Economic Development Unit promoted a shopping mall in a wetland, and eThekwini Water is introducing unpopular UD toilets to deal with water scarcity, without looking at externalities in wetlands and the discharge of faecal waste from broken sewers that compromise biodiversity and ecosystem balances. This chapter is contributing knowledge of climate change resilience and adaptation by communities within the framework of new institutional and ecological economics.

DOI: 10.4018/978-1-7998-8809-3.ch014

INTRODUCTION

This paper is a review of water, health and sanitation related climate change adaptation strategies and institutional mechanisms that are being implemented in Durban. In this paper water and sanitation governance within eThekwini, eThekwini Water and Health Climate Change Municipal Adaptation plan (2009), CSIR Headline Climate Change Adaptation Strategy (2006), eThekwini Water Conservation Guideline (2010) and other policy guidelines that eThekwini Municipality developed as attempts to mitigate and adapt to climate change and the recent Durban Climate Change Partnership established through an initiative by eThekwini Planning and Climate Protection Department in 2010, is reviewed (Roberts, 2008, Hounsome and Iyer, 2006, Constable and Cartwright, 2009). In attempting to understand and analyse water and sanitation institutional mechanisms, impacts of climate change and adaptation in these sectors, an eclectic approach drawing environmental and ecological economics framework on externality, sustainability and property rights perspectives will be used (Quiggin, 2000:2, van den Bergh, 2007). The paper will outline background information about institutions and policies associated with water, sanitation, health and climate change adaptation, climate change hotspots in Durban, impacts of climate change on water availability and future access and externalities associated with floods, sea level rise and other disasters, preparedness of the municipality and its people to adapt to climate change, resources allocated to adaptation in Durban and the experiences and the role of civil society actors in adaptation (Roberts, 2008). This article is divided into contradictions in the water and climate change adaptation within eThekwini Municipality, theoretical framework, methodology, thematic areas and key observations on how communities deal with the negative impacts of climate change that are not mitigated by the Durban state, but on the 'opposite sponsored'.

Contradictions in the Water and Climate Adaptation Governance

In this article water and climate change governance, socio and political theories are utilised to unpack contradictions that characterise the efforts of the state in dealing with the negative impacts of climate change. Within the water and climate change governance, the institutions and structures which play an important role are outlined. One of the neo-liberal institutions established by the pollution industry and co-opted state institutions and NGOs is the Durban Climate Change Partnership. One of the contradictions associated with DCCP is class dimensions. The Durban peasants and workers are alienated, and their space is occupied by the 'green petit bourgeosie', co-opted traditional leaders, state institutions and NGOs that support the neo-liberal agenda of conservationists. It is noted that many people believe that policies of the Durban state are progressive, but one of the contradictions is conventions such as Ramsar Convention on Wetlands that the Republic of South Africa is a signatory of, but pay a lip service in the protection of wetlands. The Durban State organised a Bisasar Road public participation process whose aim was to comment on the suitability of generating gas from the landfill site. The public participation process became a circus because the Durban politicians wanted this carbon trade project to be implemented at all cost, even if the environmental impact assessment become a 'tick a box exercise'. It was established recently by the Manase Commission that the then Mayor Councillor Obed Mlaba solicited bribes and kickbacks in a project that was perceived as a demonstration of a political will to deal with anthropogenic causes of climate change. During the era of Mayor Mlaba a number of climate change initiatives such as the introduction of urine diversion toilets (UDT) to deal with the 'scarcity' of water were introduced. It was established that the introduction of UD toilets led to the pollution of water resources in Vina Road Wetland in Hammarsdale. The spillage of faecal matter from a combination of UDs, and VIPs were also associated with the environmental degradation of another wetland at Mbumbulu. All of these initiatives were introduced to institute debt control among indigent households and not to deal with the negative impacts of climate change as most citizens were initially told.

Figure 1. Map of Rural Areas in Durban
Source: Ethekwini Municipality, 2010

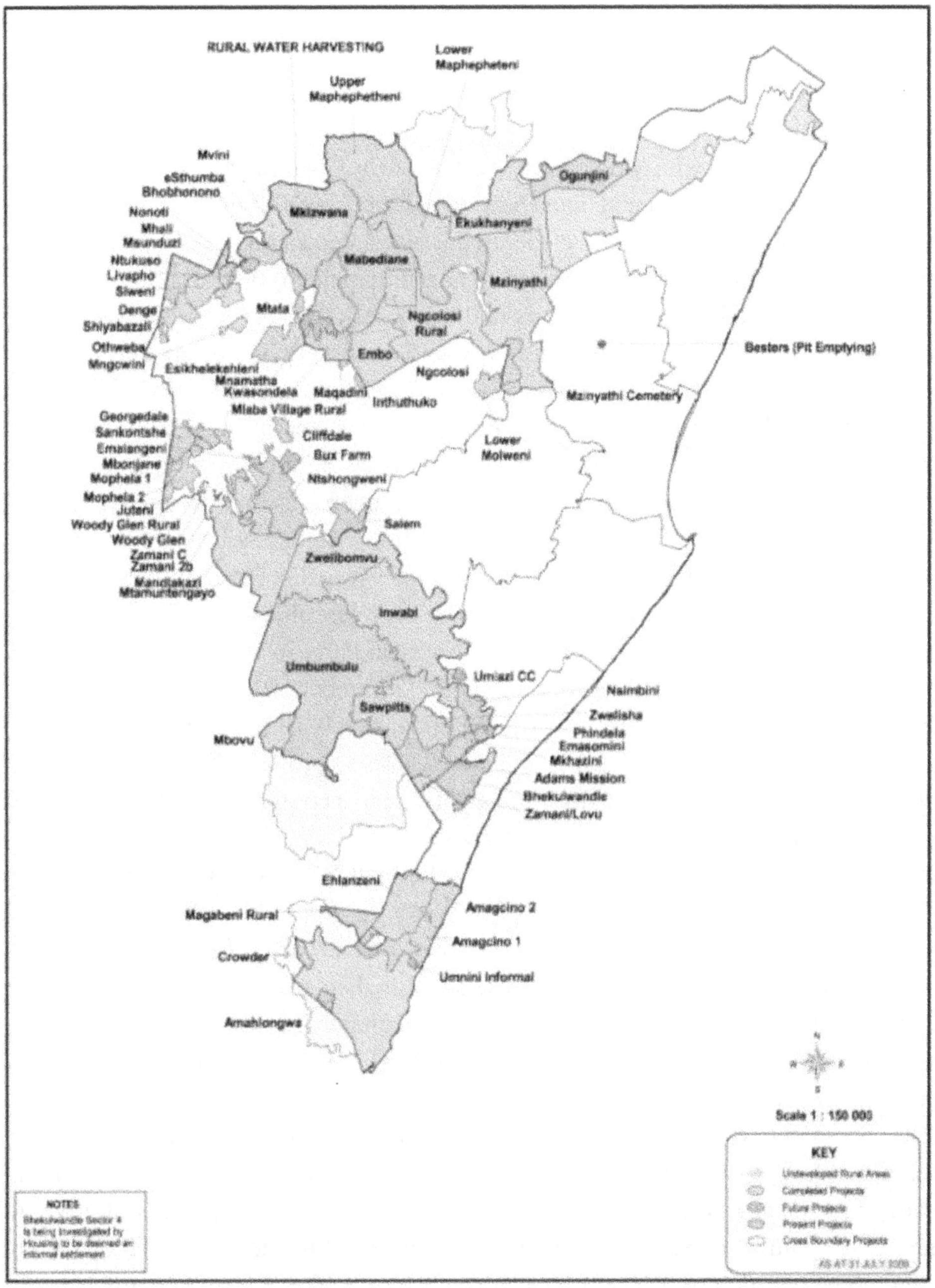

It was established that the Clean Development Mechanism (CDM), Carbon Capturing and Storage (CCS) that dominated President Zuma's New Year's Eve message(December 2012) to South Africans is far from achieving a climate resilient country. The then director general of the KZN government Mr Hamilton Ngidi reported in the KZN Council on Climate Change and Sustainable Development that carbon markets are on the crossroad whilst the Director for Pollution Control, Dr Timothy Fasheun in the KZN Department of Environment was planning a conference to invite the private sector to discuss carbon markets. This is a contradiction to President Zuma's statement that CDM and CCS remains the future of SA's climate change adaptation. The other contradiction is that of the distribution of energy saving electricity globes by the then Minister of Public Enterprises, Malusi Gigaba and Electricity Supply Commission (ESKOM). The challenge is to strengthen spending on renewable energy technologies as a step in mitigating climate change and the focus on the distribution of energy efficient globes seems to be misplaced in a bigger picture.

Studies conducted by eThekwini Municipality, supported by CSIR (eThekwini, 2009) are highlighting sea level rise by 2050 and industries and households are discouraged from building houses and business premises within the vicinity of 100 metres closer to the sea. A contradiction is that there is an increasing construction of additional buildings right in the sea. The municipality is discouraging further development in theory, but in practice the Durban Harbour is in the process of being expanded despite opposition by residents of South Durban mobilised by the South Durban Community Environmental Alliance (SDCEA). It was unexpected of the Economic Development Unit of eThekwini Municipality to be encouraging the construction of Ushaka Marine World and a number of businesses situated in less than 100 metres from the sea. What even complicates matters even further is the South African Government's renewable energy 'menu card' which includes even dangerous nuclear energy as part of renewable energy technology mix. The Treasurer General of the ruling African National Congress (ANC), Dr Zweli Mkhize told members of the KZN Council on Climate Change when he was still a Premier that it is the policy of the South African Government to include nuclear energy in the renewable and non-fossil fuel menu. This is despite the fact that the ANC refused to speak to the National Party in 1990 unless the then Government put a moratorium on all forms of nuclear technology. In December 2012, the former commander of the military wing of the ANC, Umkhonto Wesizwe, President Jacob Zuma, cited the attack of the nuclear power station at Koeberg by MK combatants in the 1980s "as a revolutionary deed of the glorious army of the glorious movement". The former MK combatants are in a counter-revolutionary move embracing nuclear energy which they rejected as guerrilla fighters.

It is also unexpected for the country that was at the helm of World Commission on Dams during the tenure of Professor Kader Asmal as the Minister of Water Affairs (1994-1999) to promote large dams and augmentation of dam walls of smaller dams as part of renewable energy technology mix. This is happening against the backdrop that communities that were displaced when large dams such as Inanda Dam situated in Durban are still in 2016 waiting for their compensation for losses suffered in their livelihoods, land, housing, and other development imperatives. The World Commission on Dam's report highlighted social, environmental, and economic impacts of large dams that remain unresolved in the world and Durban remain with a fair share of social, economic, environmental and human right impacts. The 1008 households that were displaced when Inanda Dam was constructed in 1984 remain with an expectation of getting their land back, being prioritised in economic development initiatives taking place in the vicinity of the dam. The benefit sharing that formed part of former Water Affairs Minister Ronnie Kasrils' resettlement programmes as a response to negative social and economic impacts of large dams are still expected to be realised. Out of the four communities (AmaQadi, AmaNgcolosi, AmaNyuswa and AmaPhephethwa) that were directly affected when Inanda Dam was constructed, only AmaQadi have confirmed land resettlement process in the form of compensation. Even prioritising dam affected communities in housing development initiatives as a step to make displaced households better off than the 'corrugated iron tin shacks' they were allocated in 1987 only benefitted households that relocated to Ntuzuma G as well as KwaDinabakubo Townships. Given this type of background the South African Government should tread carefully when it comes to augmentation of dam walls and construction of new large dams as part of the infrastructure projects championed by the National Planning Commission.

The other often overlooked problem associated with large dams is siltation and sedimentation associated with the size of dam walls (Heath et al, 2012). In the not so distant past (1999) the Department of Water Affairs managed to enforce the National Environmental Management Act principle of 'polluter pays' to Thor Chemicals (the new name is now Guenicar Chemicals) when they were ordered to pay for the 'clean up' operation at Intshanga where externalities associated with their mercury manufacturing caused harm to workers as well as the environment. They managed to escape another 'clean-up' operation when they were allegedly involved with depositing drums of mercury to the tributaries leading to Inanda Dam. Various KZN Members of the Executive Council (MECs) responsible for Environmental Affairs (Narend Singh, Gabriel Ndabandaba, late Mtholephi Mthimkhulu, Lydia Johnson and Meshack Hadebe) attempted to deal with the mercury poisoning of Inanda Dam, but none of them managed to deal with water pollution around Umngeni Catchment area and even demonstrated a political will to uphold the principles of sustainable development that formed an intrinsic part of the World Summit on Sustainable Development Conference of 2002.

In environmental management initiatives, it is imperative to care for the environment for not only the current generations, but even for future generations. eThekwini Municipality have since 1994 involved environmental NGOs in Durban Metropolitan Open Space System (DMOSS), and various communities are involved in adaptation projects, but the Environmental Branch of eThekwini Municipality is coordinating adaptation projects without the involvement of eThekwini Water and Sanitation Department, KZN Agriculture Department and other governmental stakeholders that are playing an important role in ploughing climate resilient crops. The next section of this article will map out the institutional framework within the municipality that is useful in adaptation to negative impacts of climate change.

Water and Climate Change Governance

Within eThekwini, there are both micro (informal) and meso (formal) levels of water and climate change governance. The Water and Sanitation Department together with other departments are participating in Lower uMngeni and Upper UMngeni Catchment Management Forums which are informal structures that are laying a playing field for the establishment of a KZN statutory Catchment Management Agency. eThekwini Environment Department continue to lead the Durban Climate Change Partnership and this department was also instrumental in setting up a Climate Protection Department within the Environment and Town Planning Department (Roberts, 2008). The local departments of Disaster Management, Electricity, Energy, Water and Sanitation are expected to work together in driving the climate change adaptation initiatives and mainstreaming health, water and disaster management climate change initiatives, but they continue to contradict each other. eThekwini Electricity built an electricity power station at the bank of Umzinyathi River which was grossly affected by Hurricane Dumonia in 1984 and other devastating floods in September 1987 that led to the overflowing of Inanda Dam which was in the process of being completed. The water and climate change adaptation sector is dominated by technical experts that have very little social toolkits in their scope of operation. The water and sanitation governance is in full support of private property right dimension. For an example Commercial Farmers have exclusive access to eThelezini Wetland situated across Umzinyathi River and the municipality faced resistance from peasants when plans to build a shopping mall in a wetland at Matikwe were underway at that time. The local farmers union used collective action to change the rules of governance when they rejected the construction of a shopping mall in a wetland. Another collective action by dam affected communities was to stage protests at Duzi Marathon that took place at Inanda Dam in 2008. Inanda dam social and compensation claims advocated for by Amaqadi, Amaphephethwa, AmaNgcolosi and Amanyuswa led to the review of the land claims that were initially rejected by the Land Claims Commission. There is a realization that wetlands and water resources are common and public good and not commercial and private property that can be exploited by commercial farmers without a water license.

In terms of the National Water Act of 1998 wetlands are regarded as public property and the Minister of Water Affairs is the custodian of such assets on behalf of the government. Only households using a wetland for domestic water consumption or irrigation are allowed to use wetlands without applying for licenses. If any person intends to use water resources in the form of a river, a wetland or aquifer for commercial purposes, such person or institution is required to apply for a license and an amount determined by the Department of Water Affairs is payable per cubic metre diverted. Water resources are regarded as a natural capital and are therefore protected even for future generations. When it comes to the practical operationalization of water licensing, there are many mines, farms that divert water resources and operate without water licenses and eThelezini Wetland situated near Umzinyathi River is just one of the many quasi commercialized wetlands where water resources are utilized for commercial purposes without evidence of licenses being granted to those operators. This really affect the socio-economic wellbeing of the local subsistence farmers that depend on water resources for irrigation and in some instances use such water resources for drinking purposes. To them water is both a natural and a social capital. In a hydro-political environment facing Durban where the municipality is hailed as the most progressive water utility in Africa after winning a Stockholm Best Water Utility Award in 2014 for recycling waste water and exploring desalination plants whilst households are protesting against water cut-offs and being excluded from getting water for more than three months due to mechanical breakdowns.

Even the Environmental Impacts Assessments (EIAs) STUDIES that are being conducted by CSIR are expected to be not different from previous ones where the record of decision is likely going to support the continuation of desalination plants. The voices of communities that are not comfortable with drinking water recycled from water treatment plants and sewerage are already told by the then water and sanitation director, Neil Macleod that ‘drinking water from toilets is a practice that was part of the South African culture for more than 50 years even in Johannesburg water is being recycled’. Umngeni Water which supply bulk water to eThekwini Municipality is in the process of constructing desalination plants at Ilovu River in the South and Othongathi in the North of the City. CSIR is conducting EIAs and public participation processes which are likely going to be similar scams experienced at Bisarsar Road, expansion of the Durban Harbour, construction of King Shaka Airport and Dube Trade Port where the concerns of environmentalists were ignored and ‘forced agreements’ resulted even in construction of houses near Dube Trade Port in a wetland. Officials from the KZN Department of Human Settlements told the Institutional and Regulatory Control Sub-Committee of the KZN Council on Climate Change in 2013 that they are dealing with cracks in houses that were constructed in a wetland and environmentally sensitive area. KZN is earning over 4% Gross Geographic Product derived from the Dube Trade Port and tourists arriving through King Shaka International Airport whilst dealing with externalities and diseconomies of building these economically strategic resources in environmentally sensitive area and ignore SA government's commitment to observe Ramsar Convention on Wetlands.

The contradictions that are characteristic of water and climate change governance in Durban are based on institutional mechanisms, rules of the game, formal rules and informal norms, culture, politics and history that shape such governance model. eThekwini State is a product of decentralization where water and climate change adaptation is devolved to the local authority by Departments of Water Affairs and Environmental Affairs. Both these national departments are sector leaders and perform regulatory functions. eThekwini used to be one of the best performing municipalities when it comes to the management of water resources and progressively providing households with water supply and sanitation services. eThekwini used to get a Blue Drop Award and Green Drop Award from the Department of Water Affairs for providing good water supply and sanitation services and having many best managed waste water treatment plants (11 out of 29 were green drop certified in 2011), but since 2012 eThekwini has since lost that status (Nojiyeza, 2013). Like all other municipalities in South Africa, there was a need for both national and provincial spheres of government to intervene to deal with financial irregularities and poor service delivery. When the Department of Cooperative Governance and Traditional Affairs invoke Section 100 of the SA Constitution, Act 108 of 1996, all decentralized functions are deconcentrated. Delegation, devolution and privatization is reversed, but what is unique about the Durban State is that even contractors that build dilapidated houses with wall cracks are given tenders to continue with their poor workmanship and poor service despite Manase Report's recommendation for their total blacklisting. The centralization and recentralization of municipal services that occurred at uMsunduzi Local Municipality in the Capital City of KwaZulu Natal province was never extended to Durban, despite the fact that financial irregularities where councilors provided tenders to their companies and cooperatives were committed and got away with one month salary being docked by the Council.

Within the neoliberal marketing channel mantra, the Department of Water Affairs through its water board, Umngeni Water is a water producer / regulator whilst Umngeni water is a water wholesaler that is supplying bulk water to many water authorities including eThekwini Municipality. EThekwini Municipality operates more as both a water retailer and a water supermarket, reticulating and distributing directly to water consumers. The citizen voice seems to be missing in this 'channel of distribution' as much as water user associations are encouraged, only community service agents that are quasi municipal employees and ward committees chaired by ward councilors remains the only organs of 'civil society' that the municipality is comfortable to work with in water user platforms. If the channel of distribution is located with the theory of consumption, production and distribution, it is appalling that water is drawn from rural areas (Ngcolosi, Maphephetheni, Maqadini), purified in Wiggins Road, behind University of KwaZulu Natal's Howard College and urban dwellers are the one's getting unending water supply unless they are disconnected due to lack of payment, but rural communities living next to Inanda Dam are the ones that face common mechanical cut-offs and rural based water reservoirs situated at Ekukhanyeni, Ngonweni, Umzinyathi and other areas in the North of Durban are always dry whilst water continue to flow in suburbs and townships. In the next section we continue with our theoretical framework focusing on ecological and environmental economics.

The water and climate change governance within eThekwini is analysed on the basis of externalities associated with building houses in the wetlands, the use of water-less toilets known as urinary diversion toilets and the extent to which faeces are discharged in rivers and water sources as indicated in the result section of this paper. Most writers highlight a relationship between externalities, diseconomies, pervasive social costs, Pigovian Taxes, market instruments (Spash, 1999, Bromley, 1978, Pearce, 2002, Stavins,

2004, Pigou, 1924, Quiggin, 2000) as some of the issues, which I believe affect the Durban State. The state of Rivers Report (2012), highlighted high levels of e-coli in various rivers of Durban especially Umngeni River which is one of the many sources of water supply. The fact that Inanda Dam require more than 20 billion rands being invested in remediation due to mercury poisoning and other externalities, it is clear that pollution abatement which is one of the weaknesses of the Durban State will be more difficult to deal with in climate variability experienced in the city. This will result in pervasive social costs being borne by the rate payers whilst the polluters are likely going to get away since they got the voice in KZN Council on Climate Change as well as Durban Climate Change Partnership. Durban also adopted Pigovian Taxes or penalties for transgressions of Air Quality Act, especially in the South Durban Basin.

Oil refineries based in South Durban were consistently being accused of exceeding acceptable standards of Sulphur Dioxide emissions and fires emanating from technological problems by SDCEA. In her last days, the KZN MEC for Environment, Lydia Johnson accompanied by eThekwini Mayor and departmental officials paid a visit to Engen to hold them accountable for exceeding globally acceptable emissions of Sulphar Dioxide. They faced the arrogance of the polluters and she was demoted to an ordinary Member of the Provincial Legislator, by the then Premier of KZN, Dr Zweli Mkhize. The African National Congress Youth League accused her of being the 'dead wood' in the provincial cabinet and called for her expulsion. She was further contradicted by the then KZN MEC for Education, Senzo Mchunu who went to South Durban Basin to praise companies situated in the South Durban Basin for donating a laboratory to Merebank Secondary School. In a primary school nearby called Settlers Primary, researchers from Nelson Mandela School of Medicine accused polluting companies in the South Durban basin for the fact that more than 50% of the children were diagnosed to suffer from respiratory diseases such as asthma and may not reach secondary schools if they do not get adequate treatment. They recommended polluting industries to assist with the payment of medical bills for such children. Former MEC for Education was not so long ago the KZN Premier after the 2014 General election that took place on the 7th of May and he was the chairperson of the KZN Council on Climate Change. The polluters that he praised for being socially responsible when they paid for the construction of a science laboratory enjoyed his state protection and were never accountable.

For many years the highest amount that eThekwini Environmental Health Department can charge for transgressing environmental bylaws remains as low as R20 000 and this is not serving as a deterrent from continuing with their pollution growth path since penalties are very low. They are likely to follow in the footsteps of Thor Chemicals and Cape PLC that were ordered to pay compensation for the loss of lives and were further ordered to pay for clean-up operations in areas where they initially operated. They continued to operate, since penalties were meagre and did not affect their operational profits and in some instances the state also pays for clean-up operations. More than 20 kilometres of Inanda Dam is infected with hyacinths and the remediation process is likely going to be paid for by the state that appear willing and able to sponsor pollution and climate change chaos. The failure to have decisive steps to deal with climate change related externalities creates serious sustainability challenges for the State. In the next section, the methodology adopted in conducting this study is presented.

METHODOLOGY

The primary data is collected using interviews conducted with civil society actors, government, academics and communities living in rural and peri urban areas in Durban. The communities that participated in the study are Amaqadi situated in the North Central, Umbumbulu situated in the South Central, Zamani B and Wood Glen situated in the Outer West, near Hammarsdale, Ntuzuma G and Soweto situated in North central. The communities were chosen based on their various climate change adaptation experiences. Their experiences includes river health campaigning, rainwater harvesting project implementation, flood mitigation and dealing with externalities associated with effluents drawn from using inadequate sanitation facilities. All four areas do not have a very strong civil society and social movement campaigning background and all did not participate in various processes that resulted with the adoption of eThekwini Water and Health climate change adaptation strategy (2009) or the earlier CSIR Headline Climate Change Adaptation Strategy (2006), Durban Climate Change Partnership (2010) or South Durban Climate Change Initiative (2006) and the recent National Climate Change Response Green Paper Process (2010). The secondary data is collected through reviews of documents such as National Climate Change Response Green Paper, Headline Climate Change Adaptation Strategy, eThekwini Municipal Adaptation Plan for health and water sectors, review of climate change adaptation literature and environmental and ecological economics theories (Asafu-Adiaye, 2000, Bromly, 1978, Hahn, 2000, Schaltegger and Synnestvedt, 2002, Saeed, 2004).

In addition to interviews, Participatory Action Research (PAR) Framework was adopted through workshops and site visits to various hotspots in the four research sites. There were 40 participants in each of the 4 workshops conducted at Maqadini Traditional Court (Umzinyathi), Sobonakhona Traditional Court (uMbumbulu), Old Council Chambers at Hammarsdale, and in a tent located at the bank of Piesang River hired for the duration of the research process. In all workshops participants that attended were drawn from youth, traditional leaders and community people of all ages. All 40 participants that attended each workshop were divided into four groups. Each group was given a task to do time Trends, time lines, Venn Diagrams and community mapping reflecting on the climate change scenarios they experienced recently and in the past 10 years or more. In four tasks mapping of Climate Change and water adaptation problems and scenarios and other approaches resulted in securing community driven knowledge production. As groups reported back to the plenary session on each of their tasks, open ended Interviews with households and leaders, farmers, peasants were further conducted. The themes that were covered in each of the workshops included but not limited to lightening, pollution of wetlands (through the use of VIP and UD Toilets), building houses in wetlands, rehabilitation of wetlands, water restrictions by eThekwini Water and Sanitation, sea level rise tool kits and integrated assessment tools and floods attenuation.

RESULTS

Sea Level Rise Scenarios and Climate Adaptation Projects

eThekwini Climate Change Protection Unit's former researcher Meggan Lewis (interview held on 20 July 2011) highlighted that her department worked with CSIR and UKZN School of Hydrology and Biological Sciences in modelling various sea level scenarios. When developing the toolbox they conducted a sea-level rise modelling and established that the current sea-level rise in the east coast of SA is 2.74 mm per year.

The sea-level rise scenarios anticipated in future are:

- *" 300 mm (90% chance of happening by 2100)*
- *600 mm (10% chance of happening by 2100)*
- *1000 mm (1% chance of happening by 2100)"(Lewis, 2011 Interview)*

If the above scenarios are something to consider, businesses situated closer to the sea should be relocating instead of being attracted to establish businesses at Ushaka Marine, eThekwini Beach Front, Blue Lagoon and other areas within 100 metres away from the sea waters. In 2007 many houses situated in Ballito, Brighton Beach and Isiphingo were washed away when unexpected tsunami emerged. In all these areas the sea level rose beyond 1000 mm which is expected to occur by 2100 and according to modelling conducted by UKZN School of Bio Resources these scenarios have 1% probability and chance. In anticipating climate change variability eThekwini Climate Change Protection Unit is already implementing various climate change adaptation projects. There are already three community adaptation projects that are being implemented in Ntuzuma, Buffelsdraai and Ntshongweni. In both Ntuzuma and Ntshongweni community gardens, where water saving crops are being cultivated are implemented. Crops that are climate resilient such as Amadumbe, Casava and others are being cultivated by communities who are sponsored by the municipality. The Department of Agriculture that is a critical stakeholder was still being invited to provide agricultural extension services at the end of 2011. Some of the 160 participants of the study invited officials of the Department of Agriculture in their community gardens. In both Piesang River and Hammarsdale the agricultural extension services were provided by the Department of Agriculture. In Buffeldraai eThekwini Municipality implemented community reforestation project whose main focus area is carbon sequestration. Other initiatives involve providing green roofs in housing projects which is recommended because it reduce temperatures, contributes in storm water runoff harvesting and promotion of inner city biodiversity. The project is still limited and situated at eThekwini Climate Protection Unit.

Other initiatives of the eThekwini Climate Protection Unit include conducting studies and crafting strategies such as eThekwini Water and Health climate change adaptation strategy (2009), CSIR Headline Climate Change Adaptation Strategy (2006), contributed towards the establishment of the Durban Climate Change Partnership (2010) and South Durban Climate Change Initiative (2006) and the recently facilitated governmental responses towards the National Climate Change Response Green Paper Process. Other eThekwini Climate Protection Unit Initiatives include the reduction of risk to developments in flood plains through amendments of bylaws to the extent that the scale of 1:100 metres in flood plains is complied with by all who wish to develop in flood lines. They further developed a master drainage plans for all river catchments within eThekwini Municipal boundaries. They also conducted a multi criteria analysis with a focus on impacts "*on risk, ancillary benefits, reversible or flexibility, impacts on emissions, allows complementary options, ease of implementation, institutional complexity, cost benefit analysis and risk of mal-adaptation.*" eThekwini Climate Protection Unit also use rainfall data projections developed since 30 September 2009 and throughout the period of 5 years in modelling impacts. They elevated flood annexure to Council Policy. The eThekwini Climate Protection Unit is one of the junior departments within the municipality and have no control and regulatory functions that can prevent businesses and electricity department from constructing premises in the flood plains.

The plans of eThekwini Water and Sanitation are not complementing those of eThekwini Climate Protection Unit. For instance they are planning to re-introduce water restrictions similar to the 1983 era where households are not allowed to wash cars using hosepipes and further encouraged to use grey water in watering gardens and lawns. The building of Spring Grove Dam went ahead and it was 80%

full in 2014 according to Neil Macleod and if no drought is experienced, therefore water rationing and restrictions were not be implemented until 2015. uMkhomazi Dam is expected to be completed by 2024 as per recommendations of the Water reconciliation study of the Department of Water Affairs (2010). The process of the augmentation of Hazelmere Dam wall announced by President Zuma in January 2011 started. This was informed by water crisis facing the areas near Ilembe District Municipality. Water from Hazelmere Dam is transported to Stanger, Mandeni and Maphumulo areas of Ilembe District. In line with the theory of production and consumption, water is developed in Durban and is consumed at Ilembe even if there are water shortages in Durban. To compensate Durban, Umngeni Water which was expected to have completed the construction of two desalination plants by 2014 is still awaiting for a record of decision from the designated authority after CSIR conducted an EIA. eThekwini Water and Sanitation donated 200 rainwater harvesting tanks to many parts of Durban. According to the community leader based at uMzinyathi, Mama Dudu Gcabashe the number of Jojo tanks donated at Umzinyathi, Gugwini, Esikebheni and other areas within AmaQadi traditional Authority is a drop in an ocean since most households were without water supply since January 2013 due to mechanical breakdowns and most of the households rely on water harvesting tanks.

eThekwini Water and Sanitation Department is planning to reduce non-revenue water from 40% to 25% over 5 years. They are hoping to drastically reduce the amount of unaccounted for water through fixing water leaks and charging 20 000 households for water theft. As much as some of the households were initially charged for water theft, cases were dropped at Ntuzuma Court due to lack of capacity by the magistrate courts in dealing with such cases. Councillors were generally not willing to support Neil Macleod and his staff when prosecuting households implicated in water theft. The challenge is that big companies operate without water licenses and the municipality decided to lay charges on households that illegally connected water after being disconnected. These households are not using water for commercial purposes, but solely for domestic consumption. In theory water is recognised as a human right in the Constitution of South Africa, but poor households that re-connected their water supply faced the might of the law when their commercial counterparts are not facing similar pressure. eThekwini Water and Sanitation Department was sponsored by Melinda and Bill Gates Foundation to conduct Nutrient Recovery Project. The components of this project include river water quality control, removal of nutrients and pathogens from treated effluents at conventional sewage treatment plants, urine separation and nutrient recovery. Studies are being conducted by the Pollution Research Group of UKZN School of Engineering and School of Built Environment and Development Studies respectively on how to harvest urine from UD toilets and use such nutrients to manufacture fertiliser. eThekwini Water and Sanitation Department is also using water resources for energy efficiency and development purposes. Mini turbines in Western aqueduct, solar panels, solar water heaters, gas to energy, oil from algae projects are contributing towards energy generation.

Voices and Experiences on the Ground: Traditional Leaders Speak OUT

At uMzinyathi there were leaks in municipal pipes, rivers that are sources of drip irrigation are dry. There are 4 dry reservoirs that are situated at Khusane, Mngadi Store, eThelezini and Ntabende. Houses are built in wetlands at Ntabende, KwaMthethwa, and 50 houses are now built on the bank of Umzinyathi River within the flood plain. eThekwini Electricity built a power station in an area that is a flood plain-situated within 100 meters from Umzinyathi River that was flooded in 1984 and 1987 respectively. It is appalling to establish that wetlands are not protected by the relevant authorities. DWA granted water extraction rights and license to commercial farmers and (peasants) farmers at eThelezini are without water for cattle drinking and irrigation. Ethekwini Economic Development and Rural Development are planning to build a shopping mall at Matikwe which is the wetland used by peasants for irrigation purposes. After sending many deputations to the municipality it turns out that it is no longer the site for a new shopping mall. There were 7 more wetlands which were under threat from privatization.

At Nsimbini, there is a wetland that has more mud than water in it. The cause of this situation is effluents from urine diversion toilets that are discharged into the wetlands during rainy seasons. Initially the wetland was a source of drinking water and also used to water community gardens. The state of the wetland began to deteriorate since the introduction of urine diversion toilets in 2007. The wetland is situated at the access route to Isiphingo and in the absence of a bridge that links the two areas, people are forced to walk across this polluted wetland and during rainy seasons they are compelled to use alternative routes and mud with faecal matter threatens the health of the people.

With all of the above initiatives of eThekwini Departments, communities have a different experience. One of the traditional leaders from Umbumbulu stated the following:

"Our experience at Umbumbulu is that we had to cope with drought, seasons are starting differently from what they used to be before, for instance the spring season used to start on the 15th of August, but the start date has shifted to the 15th of September and summer season is now in the middle of December. The delay in seasons affect the timing of ploughing and harvesting, the ploughing time is dependent on rain seasons. There were heavy rains that were experienced in the last five years. The rains we have now are abnormal, ranging from storms, heavy rains to destruction of houses and bridges. The local bridge at Umbumbulu was devastated by heavy rains and it took almost a year for eThekwini to refurbish it. Heat is experienced in winter even in coastal areas, it is now colder in summer than we used to experience before" according to the late induna (headman in IsiZulu) WM.

What the municipality perceived as solutions such as constructing UD toilets is not welcome in communities. The fact that human waste found itself in a wetland situated at Sobonakhona area in Umbumbulu, subsistence agricultural practices in rural areas are affected since wetlands are used as traditional sources of irrigation, remains a cause for great concern. Another community leader Mr G agrees with headman WM regarding negative impacts of climate change that they are experiencing:

"The changes in temperature and climatic conditions affect our water supply. Our people do not have toilets and they defecate openly. During heavy rains the faecal matter found itself in our river system and that caused diseases such as cholera and others. The strong wind blew away houses during November of 2009. Several houses were blown away at KwaMakhutha area and the church building affected is still not repaired as we speak." concludes Mr G in support of the late induna WM. Another traditional leader from uMzinyathi share similar experiences on what they view as key climate change problems:

"Our experience as Amaqadi is that we used to have winter which was known for very cold temperature but now we are experiencing winter throughout the year. As a community we discussed climate change before the current inkosi (traditional leader) came to power. Our great concern is that our people build houses in wetlands, and a place with reeds is generally a wetland and traditionally speaking no one is allowed to build a house there. Our izinduna (headmen) sell land that is zoned for wetlands that need to be protected because they want to make quick cash for themselves" according to Amaqadi Traditional Council member, Induna MM. Other participants from uMzinyathi tend to agree with their headman regarding their experience of climate change related challenges:

"The late inkosi Mzonjani Ngcobo was very strict when it comes to land allocation to new households, but the tendency now is that people build houses everywhere and there is no control anymore. The other practice that is making us more vulnerable is the cutting of trees even those that are sacred like umndoni which is very important to the Zulu culture. The late inkosi Ngcobo was very strict in terms of granting permission regarding cutting of the trees. We believe we can still do something to prevent deaths related to global warming" conclude SP in support of Induna MM.

When listening to people's experiences it is clear that their challenges are not only limited to water related climatic conditions but lightening as well. Lightening destroyed 50 houses each at Khusane and Hammarsdale respectively. The lightening conductors promised by MEC for Cooperative Governance and Traditional Affairs were still awaited in these communities by 2016. Even in Northern KwaZulu Natal 5 people that attended a church tent crusade at Phongola were struck by lightening in December 2011 and a similar incidence took place at Eshowe where four members of the households were also struck by lightening. Both the Zululand District Municipality where Phongolo is situated and Uthungulu District Municipality where Eshowe is situated were unable to assist communities to adapt to climate change. In December 2011 President Zuma was busy promoting carbon trade at the 17th Conference of the Parties (COP17) in Durban when rural households at Msinga had to deal with broken bridges as a result of floods. His view is that "the more South Africans get Clean Development Mechanism (CDM) Projects, the greater are the chances to get green funds to re-build bridges and purchase lightening conductors".

Soweto's Collapsing Houses, Piesang Toilets and Court Cases

The first part of this section is a reflection of community approaches during the days in which they were undergoing training and participating in PAR and the second section is focusing on follow up that took place in 2013. Soweto is a township situated on the bank of Piesang River and closer to Ntuzuma E Section and Inanda Glebe. The township was established in 1984 and experienced flash floods during Hurricane Dumonia of 1984 and was flooded again in September 1987. When there are heavy rains, Piesang River overflows and destroyed houses that are already built on a flood plain. There are a number of homes built on natural springs as this area is a wetland. Generally, all houses in this area were constructed by Umfelandawonye Housing NGO established by local households. The Department of Human Settlements supported the NGO in the construction of social houses. Houses needing urgent attention are A1100 situated next to two natural springs, one in the yard and the other one on the road, House number E45 at Mfino Walk and D593 have natural springs running through the yard since the 1980s, other houses are E29, Mfino Walk, E17 Piesang / Mfino Walk, E92 Mfino Walk, E89, Mfino Walk, E763 Sicelulwazi Road. All these houses were reported to the local Councillor Mabizela and various departments of the municipality were notified and no assistance was provided. The state of the above houses is deteriorating and they are almost collapsing. What is encouraging is how these households adapt to the flooding as well as living in a wetland. Members of the cooperative, Umfelandawonye Housing NGO and the community at large started a community garden situated on the bank of Piesang River to provide food and raise money for coping with floods. They started a project of removing alien vegetation on the bank of Piesang River. This project was supported by the Expanded Public Works Project section of the Department of Public Works through providing them with protection clothes, hats, gloves and other implements. They are paid a stipend by the government and they sell their surplus vegetables in an open market as a cooperative. They constructed a drainage system to channel various springs away from their houses, they planted trees within their yards so that most of the water from springs can be consumed by trees instead of damaging houses.

The community seem to have a variety of sanitation technologies. There are flush toilets in a formal settlement and VIP toilets in an informal settlement. The pollution associated with toilets is discussed in the section below. There are households that have water meters whilst others do not have water meters. The municipal officials and the companies that read meters on their behalf used a rule of thumb because it is not clear when the amount of water used is due to domestic usage or leaks that are all over the place.

Soweto Residents Picking up the Climate Change Pieces After the December 2012 Floods

The state of the Piesang River ecosystem is very bad because of effluent discharges emanating from government activities. eThekwini Water and Sanitation Department is implicated in making pollution worse in the area. Various sewerage pipes are broken and sewerage waste is disposed in the river. The other source of pollution is from Inanda Glebe area which is upstream. At Inanda Glebe there is a taxi rank. Taxi operators use the river to wash their cars. Oil leaks and residues of petroleum products are being disposed in the river. The high levels of pollution even affect the wetland and natural springs. The poor state of Piesang River was also part of the State of the Rivers Report (2012) and a waste water treatment plant nearby is one of those that did not get a Green Drop certification from the Department of Water Affairs. The community gardeners clear un-kept verges throughout the year as there are alien plants that grow quickly. The eThekwini Water's Sewerage Pipes that were broken in 2010 were still not fixed in January 2012. Members of the community engaged eThekwini Water and Sanitation since then all broken sewer pipes were fixed. The officials of the Department of Cooperative Governance and Traditional Affairs visited the area in January 2012 and by March of the same year, roads that were devastated by floods were subsequently fixed. The Piesang community appear to have learned how to get results by engaging various government departments.

The sewerage pipes broke in March 2011 and the owner of House A2142 has been complaining to eThekwini Water and Sanitation Department without any help. All attempts to report the matter in local Sizakala (get help) Centres as well as the local office of Parks and Gardens situated at New Town A was unsuccessful. It was only fixed on the 24th of December 2012. According to Mrs M, it is the preparations for the Africa Cup of Nations Soccer games that took place in Durban and other cities in January 2013 that prompted the repair of the sewerage pipes that were the source of smell for almost 2 years.

In December 2012 there were heavy rains in most parts of Durban. House BB53 situated at Soweto Block BB off Mzinyathi Road is structurally damaged. There are cracks on the wall because the municipal drainage system could not cope with heavy rain and floods were directed towards this house. The drainage and storm water pipes are designed in such a manner that water flow directly towards the house and water enters through the kitchen. On the 16th of December 2012 Masicebe Walk in the D section of Soweto was damaged. House D135 owned by K is almost falling apart due to flooding. The drainage system that the Storm Water Department designed collapsed and water was unofficially directed towards D135. The electric power lines and street lights in this road are damaged and fell off the road. The greater part of K's yard was washed away.

There was uncontrollable flow of water on the direction of Nkalakatha Passage. In both directions of Masicebe Road water was directed towards D135. House D78 situated at Masicebe Walk was built on an aquifer. The boundary that separate D78 from neighbouring houses is also an aquifer. There was still water flowing as late as January 8, 2013 when heavy rains were over. Both D and C sections of Soweto are situated in a wetland with more than four aquifers in each house which result in structural defects in most of the houses. The trend of water flow during floods in F section of Soweto has a pattern whereby water flow from F165 through a passage towards F278. A sewerage pipe in the F165 yard is damaged and most of the water meters in this area are leaking. The boundary between F165 and F278 is built on tyres and not in a position to stop the flow of water during flood seasons. House F165 has cracks in all four walls and the water meter in this house is falling apart and water is leaking day and night. House F168 collapsed in March 2012 and Mfelandawonye NGO was requested to rebuild this house. This house was badly damaged again in December 2012, cracks that led to the total collapse of this house has since resurfaced. Water flows from F172 down the passage and form a pond. The drainage system is poorly built and all water meters in this section are broken and water leaks day and night and put pressure on the houses.

The aquifer situated in House F172 flows towards F168 and cause cracks in this property. There is a water leak in F163's pipe system which results in uncontrollable flow of water. In house F169 water flows from the back yard and created a pond at the back of the house. Even weeks after the December 2012 heavy rains are over this house remains damp and beginning to crack. At the corner of Masisebenze and Senzakahle Walk there is an aquifer which reduced Senzakahle into half and it is dangerous for cars using this road. Senzakahle Walk has since become a stream instead of a road or a passage. An aquifer has since been identified and it is recharging at a very fast rate. Water is no longer flowing according to storm water drains, but created two paths that damages Senzakahle Walk. A situation similar to Masicebe Walk is likely going to happen in this road as well. During heavy rains in December 2012 a road linking Soweto and Inanda Glebe was damaged. The Soweto children that attend Zuzokuhle Primary School situated at Inanda Glebe were supposed to find alternative routes such as Newtown A or Lindelani since the bridge linking these two areas collapsed.

On the bank of Piesang River there are two houses in C section of Soweto that are situated in a distance that is less than 100 metres. This is a flood plain according to eThekwini Climate Protection Unit and occupants of houses situated in a flood plain are encouraged to relocate. In house C77 the drainage system is good, but water flows through the kitchen during flood seasons. The back wall and the kitchen wall have cracks since overflowing Piesang River was directed towards the house. At the time of visit in January 8, 2013 the house was still damp although water was removed from the house. It is not just water that was flowing inside the house but faecal matter from the direction of broken sewerage pipes. A similar situation was also found in house C75 where the dining room and bed room walls have cracks and households spend the Christmas holidays in 2012 removing water from the house. According to LG and PN who are volunteers at uMfelandawonye NGO that constructed these houses the majority of whom are double storey, many attempts were made to encourage households to relocate from the flood plain, but households are refusing. The flood plain has been a problem since 1984 during Hurricane Dumonia, the September 1987 floods and recent flash floods. Many attempts were made to ask eThekwini Municipality to assist, but Storm Water Department, Human Settlement Departments and even Parks and Gardens situated at Inanda Newtown were difficult to access. It was only the workshop conducted by NGOs and university researchers in 2011 that empowered them to take action.

"Through association with researchers we managed to complain about the sewerage pipes that were broken, even if it took long for eThekwini Water and Sanitation to fix, but a project to fix all broken pipes and water meters is under way. In the beginning we did not have information about which department can help us, but now we managed to stop eThekwini Water and Sanitation from taking us to court for non- payment of water bills. We met Neil Macleod at a workshop and forced him to commit to fixing leaks before taking us to court. Our participation in this workshop assisted us in getting information on how to adapt to climate change" states L and P.

A follow up workshop that took place at Piesang ended up with an action plan to clear un-kept alien vegetation on the bank of the river. The Parks and Gardens Department, Department of Water Affairs Working for Water Project, the Department of Agriculture and Climate Protection Unit are already working in this community. A community project was given a Durban Mayoral Award in 2011 for being one of the best water projects that created over 50 climate jobs. The community people were given tools by the Department of Public Works and Agriculture as well as financial support from eThekwini Grant in Aid Department. The community is working with various government stakeholders in creating awareness about flood plains and river health issues.

It is acknowledged that in the past communities were temporally relocated in schools and community halls during floods, but since participating in climate change adaptation projects, they managed to adapt by involving stakeholders that can provide them with help. Plumbers that are members of uMfelandawonye NGO are already working on fixing water leaks and diverting water from the wetland and aquifers to protect houses from further damage. There is an understanding that plumbers associated with uMfelandawonye are better placed to deal with poor drainage in the area since they know the area well. The community is engaging with the KZN Department of Cooperative Governance and Traditional Affairs and sensitising communities about flood plains and protection of wetlands. The community is already working with Disaster Management Team of COGTA in creating awareness. In repairing the infrastructure that is damaged after the December 2012 floods, such as Masicebe Walk, Senzakahle Walk, and houses affected by floods, uMfelandawonye is actively participating.

Sterk Spruit River and Vinah Road Wetland Rehabilitated Through a Climate Change Adaptation Project

The experience of various communities in George-dale, Hammarsdale and surroundings is flash flooding of July 2009. Strong winds blew roofs away which prompted politicians and KZN Disaster Management Team to visit the area and promised relief which is still expected. Lightening destroyed houses and trees in 2010, lightening conductors were promised by Cooperative Governance and Traditional Affairs (COGTA) Department and were never delivered. At Wood glen a combination of UD and VIP toilets at Vinah Road collapsed due to heavy rains and faecal matter contaminated the wetland and nearby primary school. Mosquitos emanating from leaking standpipes are irritating although there are no cases of malaria fever reported in this area. One of the strategies of eThekwini Water and Sanitation Department was to introduce ecologically friendly and water efficient toilets, i.e. UDs which were perceived as being imposed by community people.

According to the local branch of SANCO, the wetland in Vinah Road was contaminated with faecal matter. The wetland has since been polluted when EWSD introduced UD toilets that were resisted by households in Wood Glen, Unit 4 of Mpumalanga Township. In a toilet war that took place in 2008, households demolished their UD toilets because they felt it was an insult to be served with UD toilets when they grew up flushing toilets throughout their lives since most of the residents grew up at Hammarsdale with modern day toilet amenities built during the apartheid era. The toilet revolts as SANCO, Mpumalanga Branch call them resulted in many protests directed at former ward 91 Councillor Njabulo Ntanzi who spent his entire term of office putting down toilet fires together with his colleague Councillor Ngwengwe responsible for Mpumalanga Township, Unit 6 and other areas. Councillor Ntanzi was not re-elected as a councillor in 2009 and toilet war has since been replaced by collaboration amongst many stakeholders.

Various houses were devastated by the hail storm that took place in December 2011. The officials of the Department of Cooperative Governance, visited this area after many houses were destroyed, recommended that the Disaster Management Team and Human Settlement should refurbish all houses devastated by the hail storm. The UD toilets that were constructed by EWSD have since been replaced by flush toilets as part of reconstruction and development programme (RDP). In Mpumalanga Township and surrounding rural areas, there are households served with RDP houses that come with flush toilets and there are those that are still served with UD and VIP toilets.

The removal of alien vegetation commenced within 67 kilometres from the source of the river at Mkhambathini up to Georgedale factories. Many industries are committing funds and technical support towards the rehabilitation of Vinah Road Wetland as well as Sterkspruit River. The process of rehabilitating waste water treatment plants within industries that Bokamoso and stakeholders including the River Health Unit of eThekwini Water and Sanitation Department were pressurising them to do, is bearing fruits. All WWTPs are now operational and industries that were responsible are coming on board to pay for the pollution they caused as in the spirit of the "Polluter Pay Principle" enshrined in the National Environmental Management Act. The project focus on the removal of alien vegetation, rehabilitation of rivers and wetlands, establishment of the buy-back centre for solid waste removed from wetlands and rivers, promote ecological clubs in schools and plough indigenous plants and crops on the river banks.

Bokamoso Cooparative working with SANCO Mpumalanga and Malangeni Community put pressure on Councillor Mngwengwe to voluntarily remove UD toilets in this area. In a petition sent to Ward Councillor Mngwengwe by Malangeni Community in December 2012 they complained about insects that are attracted in this type of toilets of which many people in this area converted into flush toilets. They called on Mngwengwe to empty toilets in the interim, repair all the blocked pipes and redo all the plumbing that was not properly done. The majority of the households have since installed VIP toilets since UDs are not user-friendly and dysfunctional. As much as UD toilets were identified as ecological sanitation and useful during the era of climate disruptions, rehabilitation of Vina Road Wetland undertaken is because of poor implementation of ECOSAN by eThekwini Municipality. The eradication of UD toilets at Wood Glen when they were replaced by flush toilets in RDP houses inspired communities to take further action.

It is a lesson that one can draw from the Sterkspruit River, Vinah Road Wetland rehabilitation and collective action by Malangeni Community in frustrating the roll-out of UD toilets that power remains in the communities. It is members of the community that demolished their UD toilets who have since been served with flush toilets. It is members of the community that mobilised stakeholders to participate in the rehabilitation of wetlands. Wood Glen that was a site of toilet wars has since become a place of cooperation where government, business, CSOs and communities are working together in rehabilitating a wetland. It is collective action by members of the community that resulted in WWTPs that are the source of effluent discharges in Sterkspruit River being repaired. eThekwini Water and Sanitation Department that attempted to hold polluters accountable as a regulator of river health in the past failed to do so and communities took action after a climate change workshop and tangible results are visible. Government stakeholders are relying on Bokamoso, and CSOs in taking action against polluting industries. The role of EWSD as both a pollution controller and a polluter is a lesson to SA government. The focus tends to fall on private sector polluters and very less to state polluters whose WWTPs did not receive a single Green Drop certificate in KZN in 2012. What the Georgedale industries did by repairing their WWTPs should be followed by state institutions in KZN where not a single water and sanitation scheme received a Blue Drop certificate from the Department of Water Affairs in 2012 (Nojiyeza, 2013).

The rehabilitation of wetlands and rivers took place because Hammarsdale, Georgedale and surrounding areas were blessed with peasants who hold polluters accountable without any fear or favour. The private sector and government are aware that members of the community are on their toes if they violate their fundamental rights enshrined in Act 108 of 1996, the South African Constitution. South Africans are "entitled to an environment that is not harmful to their well- being" and they are ready to take action if either the state or the private sector violate this fundamental right. Some of the sentiments were shared with the Human Rights Commission in their KZN hearing that took place in September 2012 at KwaHaza, Nxamalala, near Howick.

CONCLUSION

It can be concluded that what started as state sponsored climate change chaos ended up as a shift from confrontation to cooperation by communities who were equipped with PAR. The success stories at the local level where the local Durban State is cooperating with NGOs and communities in river health, in rehabilitation of wetlands, in refurbishment of UD toilets and protection of wetlands, is encouraging. It is encouraging to see solutions to problems that communities experienced prior to the participation in PAR processes. Even EWSD that initially took communities to court for illegally connecting water after being disconnected, eventually became prepared to negotiate with households. Communities and government are working together to deal with externalities, rehabilitation of rivers and refurbishment of roads. Communities who were unable to adapt to various water related impacts of climate change, are now able to use various strategies to cope with lightening, water scarcity and making UD toilets to work in circumstances that were initially not conducive to do so. There is room for the sustainability of UD toilets, and the roll out of rain water harvesting tanks is contributing in addressing water shortages. It is clear that not all problems were solved at once, but through collective action, the water and climate change governance models established by both communities and government are becoming conducive to benefit them. Communities are being empowered although they initially used chaotic methods such as demolition of UD toilets at Wood Glen and threatening to remove Councillors Mngwengwe, Mabizela and others in a similar manner that they created problems for former Councillor Njabulo Ntanzi, it has become clear that communities have power to change the institutional framework, rules of the game and the governance model. EWSD that initially chose to work with community service agents as well as ward committees were forced through mobilisation and collective action by ordinary peasants to work with them. Through collective action ordinary peasants from uMzinyathi managed to stop the building of a shopping mall at Matikwe. It is clear that PAR result in increasing assertiveness on the part of the communities and this model assisted communities to change the rules of the game in their favour. Under the new institutional mechanism created through PAR, departments that never used to work together are seen working together to rehabilitate a wetland situated at Vina Road, Sterkspruit River, Piesang River and uMzinyathi River. People with disabilities benefitted through their sanitation problems being addressed and development is truly becoming "a people centred development".

REFERENCES

Amir, S. (1997). The Role of Thermodynamics for Ecological Economics. *Ecological Economics, 27*, 213–214.

Asafu-Adjaye, J. (2000). *Environmental Economics for Non-Economists. Asia-Pacific Development Journal, 7*(2), 121–123.

Ayres, R. U. (1998). Eco-thermodynamics: Economics and the second law. *Ecological Economics, 26*(2), 189–209. doi:10.1016/S0921-8009(97)00101-8

Berg, H., Ohman, M. C., Troeng, S., & Linden, O. (1998). Environmental Economics of Coral Reef Destruction in Sri Lanka. Ambio, 27(8), 627-634.

Bromly, D. W. (1978). *Property Rules, Liability Rules, and Environmental Economics. Journal of Economic Issues, 12*(1), 43–60.

Brown, O., Hammill, A., & Mcleman, R. (2007). Climate Change as the 'new' security threat: implications for Africa. *International Affairs Journal, 83*(6), 1141-1154.

Champ, P. A., Bishop, R. C., Brown, T. C., & McCollum, D. W. (1997). *Using Donation Mechanisms to Value Nonuse Benefits from Public Goods. Journal of Environmental Economics and Management, 33*, 151–162.

Chapman, D. (1973). Review. *Environmental Economics Ecology, 54*(6), 1414–1415.

Constable, L., & Cartwright, A. (2009). eThekwini Municipality Municipal Adaptation Plan. Health and Water eThekwini Municipality.

Cropper, M. L., & Oates, W. E. (1992). Environmental Economics: A Survey. *Journal of Economic Literature, 30*(2), 675–740.

Deacon, R. T., Brookshire, D. S., Fisher, A. C., Kneese, A. V., Kolstad, C. D., Scrogin, D., Smith, V. K., Ward, M., & Wilen, J. (1998). *Research Trends and Opportunities in Environmental and Natural Resource Economics. Environmental and Resource Economics, 11*(3-4), 383–397.

Department of Environmental Affairs. (2010). *National Climate Change response Green Paper*. Department of Environmental Affairs.

Dietz, F.J., & van der Straten, J. (1992). Rethinking Environmental Economics: Missing links between Economic Theory and Environmental Policy. *Journal of Economic Issues, 26*(1), 27-51.

Dodman, D., & Satterthwaite, D. (2008). Institutional Capacity, Climate Change Adaptation and the Urban Poor. *Institute of Development Studies Bulletin, 39*(4).

Hahn, R. W. (2000). The Impact of Economics on Environmental Policy. *Journal of Environmental Economics and Management, 39*(3), 375–399. doi:10.1006/jeem.1999.1119

Heath, T., Parker, A., & Weatherhead, E. K. (2010). *How to Climate Proof Water and Sanitation Services for the Urban Poor. Cranfield University*.

Heath, T., Parker, A., & Weatherhead, E. K. (2012). Testing a rapid climate change adaptation assessment for water and sanitation providers in informal settlements in three cities in Sub-Saharan Africa. *Environment and Urbanization*, *24*(2), 619–637. doi:10.1177/0956247812453540

Hounsome, R., & Iyer, K. (2006). Headline Climate Change Adaptation Strategy for Durban Council for Scientific and Industrial Research. Academic Press.

Karin, B., & Eva, L. (2006). Planting Trees to Mitigate Climate Change: Contested Discourses of Ecological Modernization. *Green Governmentality and Civic Environmentalism Global Environmental Politics*, *6*(1), 50–75.

Leggett, C. G., & Bockstael, N. E. (2000). Evidence of the Effects of Water Quality on Residential land Prices. *Journal of Environmental Economics and Management*, *39*(2), 121–144. doi:10.1006/jeem.1999.1096

Massarutto, A. (2008). *Urban Water conflicts: an ecological –Economic approach Paris*. UNESCO.

Mukheibir, P., & Sparks, D. (2003). Water resource management and climate change in South Africa: Visions, driving factors and sustainable development indicators. Energy and Development Research Centre.

Munasinghe, M. (1993). Environmental Economics and Biodiversity Management in Developing Countries. Ambio, 22(2-3), 126-135.

Nojiyeza, S. (2013). Community river restoration in South Africa. In Water and Climate Change in Africa- Challenges and community initiatives in Durban, Maputo and Nairobi. Routledge.

Oates, W. E., & Portney, P. R. (2003). The Political Economy of Environmental Policy. Handbook of Environmental Economics, 1, 450-600.

Patz, J. A., & Kovats, R. S. (2002). Hotspots in Climate change and human health. *BMJ (Clinical Research Ed.)*, *325*(7372), 1094–1098. doi:10.1136/bmj.325.7372.1094 PMID:12424173

Pearce, D. (2002). An Intellectual History of Environmental Economics. *Annual Review of Energy and Environment*, *27*(1), 57–81. doi:10.1146/annurev.energy.27.122001.083429

Pindyck, R. S. (2007). Uncertainty in Environmental Economics. *Review of Environmental Economics and Policy*, *1*(1), 45–65. doi:10.1093/reep/rem002

Price, G. (2009). *eThekwini Municipality Water Conservation Guideline*. Durban: eThekwini Municipality.

Quiggin, J. (2000). *Environmental Economics and the Murray-Darling river system. Australian Research Council.*

Roberts, D. (2008). Thinking globally, acting locally- institutionalizing climate change at the local government level in Durban, South Africa. *Environment and Urbanization*, *20*(2), 521–537. doi:10.1177/0956247808096126

Saed, K. (2004). Designing an Environmental Mitigation Banking institution for Linking the Size of Economic Activity to Environmental Capacity. *Journal of Economic Issues*, *38*(4), 909-937.

Sagoff, M. (2000). Environmental Economics and the Conflation of Value and Benefit. *Environmental Science & Technology*, *34*(8), 1426–1432. doi:10.1021/es990674d

Schaltergger, S., & Synnestvedt, T. (2002). The link between 'green' and economic success: Environmental management as the crucial trigger between environmental and economic performance. *Journal of Environmental Management*, *65*, 339–346. PMID:12369398

Soderbaum, P. (1990). Neoclassical and Institutional Approaches to Environmental Economics. *Journal of Economic Issues*, *24*(2), 481–492. doi:10.1080/00213624.1990.11505047

Spash, C. L. (1993). Economics, Ethics, and Long- Term Environmental Damages. *Environmental Ethics*, *15*(2), 117–132. doi:10.5840/enviroethics199315227

Spash, C. L. (1999). The Development of Environmental Thinking in Economics. *Environmental Values Journal*, *8*(4), 413–435. doi:10.3197/096327199129341897

Stavins, R. (2005). Environmental Economics. Social Science Research Network of Harvard University.

Stavins, R. N. (2002). *Experience with Market-Based Environmental Policy Instruments. The Fondazione Eni Enrico Mattei Nota Di Lavoro*, *52*, 1–88.

Swaney, J. A. (1987). Elements of a Neoinstitutional Environmental Economics. *Journal of Economic Issues*, *21*(4), 1739–1779. doi:10.1080/00213624.1987.11504721

Van den Bergh, J. C. J. M. (2007). Evolutionary thinking in environmental economics. *Journal of Evolutionary Economics*, *17*(5), 521–549. doi:10.100700191-006-0054-0

Van den Bergh, J. C. J. M., & Gowdy, J. M. (1998). Evolutionary Theories in Environmental and Resource Economics: Approaches and Applications. Free University.

van Lieshout, M., Kovats, R. S., Livermore, M. T. J., & Martens, P. (2004). Climate change and malaria: Analysis of the SRES climate and socio-economic scenarios. *Global Environmental Change*, *14*(1), 87–99. doi:10.1016/j.gloenvcha.2003.10.009

Weber, M. L. (2001). Markets for Water Rights under Environmental Constraints. *Journal of Environmental Economics and Management*, *42*(1), 53–64. doi:10.1006/jeem.2000.1149

Wisner, B., Fordham, M., Kelman, I., Johnston, B. R., Simon, D., Lavell, A., Brauch, H. G., Spring, U. O., Wilches-Chaux, G., Moench, M., & Weiner, D. (2007). *Climate Change and Human Security*. Oberlin College.

Chapter 15
Unpacking the Resource Curse and Realism Challenges on Economic Development in the Democratic Republic of Congo (DRC):
Case of Gecamines

Germain Miteu Tshinu
University of Zululand, South Africa

ABSTRACT

The third theme of the book seeks to understand the role of mineral resources in the African continent's socio-economic development. The case study of the DRC is used to understand the role played by the state-owned mining company Gecamines to socio-economic conditions of employees and the country's development at large. Resource curse and realism theories are utilised in this case study to unpack the role of mineral resources to Gecamines employees' access to education and healthcare in particular and to the entire country in general. The chapter employed a case study design with a qualitative approach research in its endeavour of exploring Gecamines' socio-economic contribution to its employees. Semi-structured interviews were conducted with Gecamines' managers, unskilled mine workers, and the Provincial Department of Mines' officers.

INTRODUCTION

Mining communities situated in rural areas are supposed to benefit from the operations especially through quality public education, access to health services and food security. In Katanga region of the DRC, the outcome of mining operations proved to be counterproductive. The reasons for such poor access to quality education, health care and food insecurity is associated with resources curse and realism as use-

DOI: 10.4018/978-1-7998-8809-3.ch015

ful pointers to this conundrum. In this chapter, the main focus will be on two specific theories known as resource curse and realism in order to grasp a good understanding of the relationship between natural resources and socio-economic development in the DRC. Resource curse refers to a situation where a country experiences slow economic growth, incidence of civil war and limited social development in spite of having abundance of mineral resources (Sachs, 1995; Warner, 2001). Usually, the benefit of resource bonanza is enjoyed by government officials and few privileged individuals acting to benefit themselves at the expense of the majority of the population. Resource curse attempts to explain the reasons behind poor economic growth experienced by some resource rich countries when compared to some resource poor countries. Realism as theory attempts to emphasise on personal interests which prevails on other people's socio-economic development interests. This implies that, individuals' interests are pursued to the detriment of other's interests in a given society. Gecamines as a mining company is confronted with the above-mentioned two theories in the sense that its socio-economic development initiatives are not benefiting a huge segment of the employees, the province of Katanga and the DRC population as a whole. The second section will underline the research question and the rationale for the study. The third section will outline the methodology employed to conduct this research. The fourth section will look at the two specific theories mentioned above in order to better understand why Gecamines' employees in the city of Lubumbashi and the population of the DRC at large remain poor despite the abundance of natural resources critical to unleash a durable development in the country. The last section will conclude by summarising the main themes that emerged throughout the chapter.

RESEARCH QUESTION AND RATIONALE FOR THE STUDY

Thus, the paper was guided by the following questions:

- What kind of socio-economic programmes does Gecamines have to contribute to socio-economic development of its employees and Katanga province to become relevant again in the lives of Katanga people?
- What challenges, if any, are being faced by Gecamines in implementing its socio-economic programmes and how can these challenges be addressed in order to improve the programmes' effectiveness?
- What are the opportunities available for Gecamines' socio-economic programmes in order to improve its socio-economic contribution to its employees, the population of Lubumbashi and the entire Province of Katanga?

The study is of paramount importance as it has the potential of improving the effectiveness of Gecamines' socio-economic programme and socio-economic conditions of its employees as well as the people in Katanga Province. This study was also important in the sense that it raised awareness by looking at the living conditions of some remaining 10. 000 Gecamines' employees and former employees who left the company due to the World Bank Voluntary Departure Programme after the company went into partnerships with private firms to exploit copper and cobalt. Furthermore, by scrutinising Gecamines and the impact of its joint ventures on the socio-economic conditions of its employees in particular and Katanga province in general, this study attempted to extend its scope on these joint ventures' contribution to the national treasury as well. This was important to see whether Gecamines and its partners can be relevant again in the Congolese economy as it was the case in the 1960s and 1980s. It was also possible that, lessons could be learnt from the findings of the study that could be utilised by other mining companies in improving their socio-economic contributions in the interest of the employees and the entire population through generated revenues for the state.

METHODOLOGY

This paper employed an exploratory approach to qualitative research design in its endeavour of exploring Gecamines' socio-economic contribution to its employees in the city of Lubumbashi and the Katanga region as a whole. The use of this approach was important as it allowed us to conduct a very close, detailed conversation or interviews to acquire a profound and deep understanding of the issue under investigation (Neuman, 2014:103). Through this approach, it was possible for researchers to deeply engage and interact with major stakeholders to capture their own subjective views on the company's contribution to socio-economic development of its employees and the mining community in Lubumbashi and the province of Katanga as a whole.

This paper's population derived from key stakeholders that include Gecamines' managers in charge of socio-economic development, beneficiaries such as the field workers and the local government officials in Katanga Province. Gecamines' Managers, field and underground workers were chosen because they have the first-hand experience and knowledge of the Gecamines' socio-economic programme that could help the researcher obtain necessary information to answer research questions identified above. The rationale for choosing Provincial Ministry of Mines' social officials was based on the fact that they are not only the gatekeepers, but also are well vested in the socio-economic developmental issues of Katanga Province. The fieldwork was conducted in the city of Lubumbashi in the Katanga province, Democratic Republic of Congo from 12 January 2017 to 22 February 2017. According to the Congolese *Division Provinciale de l'Intérieur et Sécurité* (2015), the city of Lubumbashi covers 747 km^2 of land with a population estimated at 2 million (Division provinciale de l'intérieur et sécurité, 2015).

Purposive sampling, a non-probability sampling technique, was utilised to select 4 Gecamines' social programme managers out of 7 decision-makers, 17 Gecamines' socio-economic programme beneficiaries from the approximately 9000 employees who work on the surface and underground and 4 social officials from the mining division at the Provincial Mining Department in Katanga. This purposive sampling technique was chosen amongst many others because it allowed the researcher to sample participants on the basis of their knowledge and capacity to respond to research questions (Babbie, 1990). In MacNealy's (1999: 157) words, "the individual characteristics, knowledge and ability are selected to answer necessary questions about a certain matter or product". This sampling technique was also important to identify the study population in a consistent, reliable, uniform and objective manner based on their knowledge of the subject under study. The sample of 25 participants was considered as enough since qualitative research make use of small samples in order to collect rich and thick data. Triangulation of data by interviewing different stakeholders helped ensure rigour, trustworthiness and obtain different perspectives on data collected.

The research used semi-structured interviews to enable participants to express their subjective views on the subject of interests (Bless and Smith, 2000). this technique of data collection was essential as it enabled the interviewer to direct the flow of ideas, intervene and ask questions for more clarity when necessary. In addition, This technique of data collection was chosen to gather valuable information as respondents were free to expand on the topic, focus on particular aspects and to some extent relate to their own experiences as decision-makers in their respective fields of expertise (Bless and Smith, 2000:30).

Beside this technique, observation was also utilised as a data collection method during the field work. In order to use observation as a data collection method, the researcher remembered his primary role as a researcher and remained detached enough to collect and analyse data relevant to the problem under investigation (Baker, 2006). The technique was important to observe the living conditions of Gecamines' workers and the community. The conditions of the mine's non-current assets such as buildings, hospital facilities, schools, vehicles, abandoned shafts were also observed to have a better understanding of the socio-economic conditions of the workers. In addition, the technique was crucial to verify nonverbal expression of feelings, grasp how participants responded to questions, and check how much time was spent on responding to questions enquiring on their living conditions (Baker, 2006; Kaulich, 2005; Schmuck, 1997). In brief, this technique was used as a way to increase the validity of the data and enabled the researcher to have a better understanding of the context and phenomenon under study (Kaulich, B. 2005).

The interviews were conducted in French given that the majority of the population speak that language in Katanga province. Interviews were audio recorded, with the consent of participants, to save time from detailed note taking and to make sure that all responses were captured. Pilot study was also conducted in the DRC two weeks before the real interviews in order to get an idea on respondents' ability to respond to research questions and ensure that there were no ambiguous questions.

Given that this was a qualitative research, thematic analysis was utilised to analyse data collected through the interviews. The interview audio records were transcribed and then analysed according to recurring themes, concepts or similar features using the thematic analysis method identified by Newman (2014). For instance, themes related to Gecamines' contribution to socio-economic development in terms of education and training, housing, food distribution, job creation, sport and recreation, farming and health care were identified and analysed separately through rigorous process known as data familiarisation, data coding, theme development and revision (Braun, and Clarke, 2006).

The researchers fully notified the participants about the purpose of the study, ensuring that they understand that their participation was voluntary, and could also freely withdraw from participating in the study at any stage if they wished to do so and a joint decision was going to be taken on what to do with data already collected on withdrawal from the study. The participants were then asked to complete and sign written consent forms to make sure that they participate on their own free will and agree to audio record interviews. Participants' identities were anonymous, and information provided was protected or treated as confidential as possible to protect them from any harm that may arise if their identities were divulged. Pseudonyms such as names of people from the Western part of the DRC were used to hide the identity of the participants. The next section will focus on resource curse and realism as the theoretical framework to understand the effect of mineral resources on socio-economic development of Gecamines employees in the city of Lubumbashi in particular and the DRC as a whole.

RESOURCE CURSE THEORY

Resource curse is a theory that became popular in the 1990s after several studies conducted by Auty (1993, 2001) and later by Sachs and Warner (1995) found or discovered poor economic performance in some resource rich countries. It was discovered throughout several researches that some resource rich countries performed poorly in terms of development when compared to certain resource poor countries.

For instance, the current DRC, former Zaire at the time, experienced slow economic development as a result of production decline in industrial mining activities at the parastatal *La Générale des Carrières et des Mines* (Gécamines). In 1994, the DRC's industrial mining production at Gecamines declined to 29 300 tonnes from 500 000 tonnes of copper and cobalt per annum (Kalenga, 2013). This poor performance led to unprecedented socio-economic crisis in which the living standard of 35 000 employees of Gecamines and their dependents estimated at over 250 000 deteriorated drastically (DSCRP, 2006).

Moreover, resource curse theory suggests that natural resource abundance generates a series of economic and political distortions or policies which ultimately undermine the contributions of extractive industry to development (Bebbington, *et al.* 2007). In other words, the abundance of natural resources creates a type of myopia among public or private actors (Ross, 1999). This is a situation where political and business elite are characterised by short visions of satisfying current interests to the detriment of long-term development of the country. In most cases, resource boom pushes policy makers to come up with economic policies that do not enable forward and backward linkages important for sustainable development and diversification of the economy. For instance, during the resource boom at Gecamines in Katanga province from the 1960s to 1990, the DRC's policy makers failed to optimally invest much of the resources in other productive sectors of the economy such as manufacturing and agriculture as well as education and public infrastructure as important foundations for sustainable development. As a result of this poor planning, the collapse of industrial production at the parastatal Gecamines in the 1990s did not affect only the living conditions of the workers, small firms that were doing business with Gecamines, but it also decreased government revenues as mining contribution was only around 2% in 1994 from 67% in 1970 (BCC, 1995, Marysse, 2009, Kalenga, 2013).

This argument demonstrates that reliance on natural resources without allocating substantial amount of money to the diversification of the economy can impede long term development in a country. The reason why one made it clear that "resource rent or reliance on revenues from natural resources lead to irrational excitement, producing a get rich quick mentality among businessmen and a boom-and-bust psychology among policymakers, marked by bouts of excessive optimism and frantic retrenchments" (Ross, 1999). Wallich and Levin (1960) contend that, "the development path of commodity exporting countries was distorted by a resource mentality that led to poor economic planning, lack of diversification and widespread poverty. Empirical evidence demonstrates that, the DRC's reliance on volatile natural resources without sound macro-economic policies and planning produced harmful effects on socio-economic indicators which are below the Sub-Saharan standards (UNDP, 2010).

For instance, it has been observed in the DRC that since the collapse of industrial mining activities in the 1990s, budgets allocated to health and education sectors have drastically declined below Sub-Saharan standards between 0.8% and 0.1% respectively (World Bank, 2007:36; DRC Ministry of Environmental Affairs, 2012; DRC Ministry of Budget, 2011). According to UNDP (2010), the DRC has been doing poorly in human development index below the minimum standard of 0,389 and 0,393 respectively for Sub-Saharan Africa and countries with low human development index. Furthermore, social indicators on the DRC are in red as the unemployment rate was between 50.3% in 2012 against 51.4% in 2011 (Central Bank of Congo, 2012).

With demographic growth reaching over 3% according to the World Bank (2009), life expectancy at 49 years in 2012, it was unlikely for the DRC to even meet the 2015 Millennium Development Goals deadline. Therefore, it is clear from the above statistics that the living conditions of the majority of the population in the DRC does not reflect the richness of their soil, thus, very difficult to meet the targets of Agenda 2030 - the Sustainable Development Goals (SDGs). This argument is based on the fact that, an estimated 87.7 percent of the population, about 67.5 million people, remained under the international poverty line, which is currently $1.25 per day. (McFerrin, 2017; Omoyi, 2010). Figures are even worse when one makes a comparison between the level of poverty in urban and rural areas. Since most of Congolese live in rural areas where there are no real linkages with the formal economy in urban areas, poverty seems to be more acute there than it is in urban areas. For instance, in 2018, the World Bank estimated that 72% of the population, especially in the North West and Kasaï regions, was living in extreme poverty on less than $1.90 a day (World Bank, 2018).

In the same vein, Pegg (2006) contends that, the abundance of mineral resources discourage investment to increase labour productivity in non-mineral sectors, leading to under-investment in key sectors such as education. However, Stijns (2006) argues with different empirical evidence that mineral resources are associated with greater investment in education. The two divergent arguments have been proven in the DRC as figures demonstrate positive trends in terms of investment in the country's education system during the resource boom, following the creation of the Belgian *Union Minière du Haut Katanga* (UMHK) in 1906 to the 1990s under the auspices of Gecamines created in 1967 as a result of nationalisation by President Mobutu's administration.

For instance, in the 1960s up to the 1990s, 15% of the country's budget was allocated to the education sector, thus improving the literacy rate above most of developing countries in Sub-Saharan Africa (Gann and Duignan, 1973). This argument is corroborated in Henk's study when he indicated that under the auspices of Gecamines that inherited UMHK some sixty-four primary schools, thirty-six secondary schools, some fourteen technical schools at the secondary level and a major technical training institute which were managed by the company (Henk, 1988:276). This particular attention to education system helped maintain the DRC's literacy rate at 67% despite other human development indicators being in red (UNDP, 2011). However, the DRC's robust education system inherited from colonisation could have enabled the country to have one of the best literacy rates in the world if good governance of resources as well as their allocation gave a particular attention to the education sector. Poor governance of the mining sector for example led to a decline in mining production which contributed substantially to a decline in funding of the education sector. This is exemplified by the low level of primary and secondary school enrolment throughout the country, especially in mineral rich provinces such as Katanga, Kasai and others. For example, the collapse of industrial production at Gecamines in Katanga created a situation where Gecamines as a company had difficulties to fund its schools as it was the culture under the paternalistic system (Rubbers, 2010).

In addition to deteriorated living conditions, rising inflation, irregularly paid wages and the abolition of several types of benefit in kind, Gecamines' employees could no longer afford the education of their children (Dibwe dia Mwembu 2001: 157–195). This jeopardised the future of these young people and their social wellbeing as well. That is why, a group of former Gecamines' employees identified malnutrition, divorce, withdrawal of children from school, delinquency among boys, girls forced into prostitution, and in particular increased rates of illness and death as a result of the pauperisation process. This was caused by production decline and the World Bank Voluntary Departure Programme which ended their contract with the company in return of severance compensation (Rubbers, 2010).

As socio-economic crisis intensified in Katanga province, it was found that by 2003, the standard of living of Gecamines' employees was found to have fallen below that of the urban population of Haut-Katanga as a whole (University of Lubumbashi, 2005). These had direct negative effects on different socio-economic indicators as well. For instance, the education system and the future of young generation suffered the most. Statistics show that in the artisanal sector, the number of workers ranged from 500,000 to 2 million people for the entire mining sector (IMF, 2015). In Katanga province alone, it is believed that 50 000 to 70 000 illegal miners are operational in artisanal production within Gecamines' premises (RAID, 2004: 63). It can be argued that, this artisanal exploitation of resources is taking place at the expense of thousands of young people that would have benefited from education for the development of manufacturing and other key important sectors of the economy.

Based on the above argument one can argue that, poor governance and control of young people in terms of allocation of natural resources' revenues can divert the attention of the youth from education to the exploitation of natural resources in order to make quick money. Given the unequal redistribution of natural resources' revenues in resource dependent countries such as the DRC where political institutions are in total ruin, youth which constitute the majority of the country's population without adequate means to afford school fees and meet daily needs to survive will have no alternatives, but to embark on the extraction of mineral resources at their reach. As Congolese, natural resources such as copper, cobalt, diamond, coltan, gold and others do not require technical expertise or heavy machines for extraction, the bulk of young population without viable options to survive easily embark on the extraction of those resources with simple tools such as axes, pickaxes, spades and others shown below in images to satisfy their present needs. Consequently, in the absence of resource boom or depletion of natural resources, this group of population will face dire socio-economic crisis which will have negative ramifications in the country's economic growth and development as a whole (Garrett and Lintzer, 2010).

Figure 1.

Resource curse can be confirmed by the unprecedented crisis that is currently striking the two Kasai provinces after a rapid decline in diamond production at the Minière de Bakwanga (MIBA). A large number of young people without a considerable level of education stagnate in acute poverty which also led thousands of them to migrate to other cities of the country in search for a better life. Consequently, this has increased competition over scarce resources in those cities as well as other socio-economic challenges such as security in areas such as Kinshasa and Lubumbashi. Beside Kasai province, Katanga considered as the economic hub of the DRC given the presence of several mineral resources in large industrial quantities was also struck badly with the 2008 global financial crisis. This crisis led to the closure of several mining companies, which also affected the jobs of thousands of people negatively. Some reports have asserted that, close to 300,000 employees lost their jobs in the copper and cobalt-rich Katanga province in 2008 and 2009. The BBC reported in March 2009 that in the town of Likasi alone, almost 60 mining companies had shut their doors over the previous three months (Mining Journal Special Publication, 2010).

Scholars who are sceptical about the existence of resource curse theory also assert that, the concentration of activity in one sector of the economy brings with it vulnerabilities or negative externalities associated with export dependence (Bebbington, *et al.* 2007). According to Bebbington, et al (2007), this "vulnerability is from mineral price volatility, and over-dependence is reinforced by those upstream economic actors that control processing and marketing of final products derived from the minerals in question. In the case of the DRC this argument has been proven with empirical facts which demonstrate that, 75% of the government's revenue depends on the exportation of natural resources (NiZA, 2006). Despite the relevance of the mining sector to the economy of the DRC, blind reliance without sound macro-economic policies like the case in Botswana will lead to atrocious results during an external shock, change in terms of trade or depletion of natural resources. The 1990s crisis characterised with political instability and the decline of copper production and prices in the international market, diminished the government revenues drastically and led to a very disastrous socio-economic crisis in the country.

Furthermore, the reliance of the government on mineral resources without sound macro-economic policies to efficiently use resources' windfalls to diversify the economy has been demonstrated by the alarming lack of backward and forward linkages. More often mining complexes take the form of enclave economies, developing few links to domestic suppliers (to the point where many modern transnational owned mines bring in food from the capital or overseas through contracts with international catering enterprises (Bebbington, *et al.* 2007). As a result of poor economic policies and management of natural resources, the multiplier effects the mining sector has to play in the local and regional economy remain weak, reducing the possibilities of job creation for thousands of people in desperate need. Based on these observations on the effects of poor management and reliance on these non-renewable commodities one can claim that "mining has a dismal track record to date in poverty reduction" (Pegg, 2006: 376).

However, these 'dismal' effects on poverty are explained in several ways. One interpretation departs from the position that mining without good and transparent political institutions that can hold the elite accountable for their actions are bad for growth and poverty reduction. For instance, the political scientist Michael Ross (2001), asserted while writing in an influential report for Oxfam America that if growth is good for the poor, oil and minerals exports are bad for growth and hence, bad for the poor when poorly managed. The poor management of the DRC's mining sector which has the potential to significantly reduce poverty and absorb over 50% of the population that mires in unemployment, is a challenge to tackle without rational economic and political policies. Lack of economic diversification and solid economic foundations together with heavy reliance on natural resources exacerbated the economic situation of the DRC during the 2008 global financial crisis.

In the same vein, International Financial Institutions (IFI) argued that the DRC has been significantly and negatively impacted by the global financial crisis due to its over-reliance on non-renewable natural resources for foreign exchange and revenues, and its limited capacity to protect against external shocks (Hinde, 2010). A report published by the US Department of State recorded the former managing director of the International Monetary Fund (IMF), Dominique Strauss-Kahn, stating in May 2009 that the DRC is one of the most severely affected countries in Africa due to its over-reliance on volatile and non-renewable resources (Hinde, 2010).

Beside the argument that extractive industries are quite often associated with poor economic performance (Auty, 1993; Freudenburg and Wilson, 2002; Sachs and Warner, 1995; Weber-Fahr, 2002), another important argument related to the resource curse theory is that of a Dutch disease. This argument contends that mineral resources lead to levels of consumption and investment during boom periods that cannot be maintained through subsequent downswings (Bebbington, *et al.* 2007, Daniele, 2010, McFerson, 2010, Hesselbein, 2007, Sachs and Warner, 1995, Auty, 1993). In Bebbington's (2007) words, the Dutch disease named after the discovery of oil in Holland, increases the exchange rate and wage effects that cripple the growth of non-mineral tradable sectors such as agriculture and manufacturing, leading to an economic structure dominated by enclave economies linked to mineral resource extraction.

In other words, during a resource boom local currency appreciate given the large inflow of foreign currencies in the economy, therefore, negatively affecting the competitiveness of the manufacturing and other non-exporting sectors in the international market. As the local currency appreciates, domestic goods are substituted with imported goods as they become cheaper and affordable to all. As a result of this appreciation, the economy will face a rapid deindustrialisation, leaving the country without manufacturing companies of its own to supply the local market (McFerson, 2010). This phenomenon has occurred in the DRC as resource boom from the 1950 to late 1980s paved the way for rapid deindustrialisation of the country's infant industries that could not face competition brought in by foreign companies through cheap goods. It is argued that, some of the manufacturing companies and industries that were flourishing during the 1950s and 1980s closed down their businesses as the resource boom provoked high level of competition with cheap foreign goods.

This deindustrialisation of the Congolese economy is exemplified today with increased levels of imports. In other words, as the country does not have manufacturing companies of its own, it is relying on import for any articles ranging from manufactured to agricultural products. In the long term, this reliance on imports based on volatile and non-renewable resources will come with dire consequences in the country's economy. For instance, statistics show huge trade imbalances in favour of the DRC's several trade partners. While the country exports raw materials, it imports finished goods and other consumable goods that would have been produced locally if not for international competition and lack of sound macro-economic policies to stabilise the currency. Mikesell (1997) makes it clear that, such effects of deindustrialisation and its negative ramifications are commonly observed in mineral-dependent economies though it is likely that the extractive sector is not the only factor limiting diversification. This argument demonstrates that, it is not only the extractive industry or mineral resources that can limit economic diversification, but other factors such as political and economic institutions that can guarantee good governance and management of those resources can also play a certain role. This role may either enable or inhibit economic diversification from taking place. That is why one can argue that, political and economic institutions can either enforce good governance, transparency and good allocation of revenues to other sectors of the economy for economic diversification essential for sustainable growth and development (Bebbington, *et al.* 2007).

This is the reason why; the resource curse theory emphasises that bad governance of natural resources' revenues together with poor planning and allocation of resources leads to the increase of unemployment as domestic companies are driven out of the market by cheap goods and services during a boom. As a result of deindustrialisation and poor institutions, poverty and inequality are likely to take place after the resource boom or depletion of resources (ICMM, 2006). However, it is crucial to keep in mind that, these are not the only factors that take place in poorly managed resource rich countries. The blind concentration of economic activity in one sector of the economy without proper policies in place elicits socio-political and institutional relationships that undermine sustainable and inclusive development (Auty, 2008).

Concentration on natural resources combined with poor economic structures inhibiting equal distribution of revenues lead to revenue stream that are large and easily identifiable, activating struggles and conflict over their control (Auty, 2008; Auty and Gelb, 2001). Both historical and empirical evidence demonstrate that, the DRC has fallen victim of its natural resources since their discovery in the 1870s by the British explorer Sir Stanley (D'Ydewalle, 1960). With a huge potential in terms of untapped natural resources estimated at $24 trillion, the DRC's lack of good economic structures for an equal redistribution of resource's revenues paved the way for intensified competition to access those natural resources (IMF, 2015). Several conflicts that have taken place in the country illustrate the role natural resources have played in the absence of sound macro-economic policies and political institutions. The disastrous 1998-2003 conflict dubbed as the African war in which the country lost close to 6 million lives and billions of US dollars through the plundering of natural resources by government troops and several rebel movements has been considered as a direct effect of poor management of minerals' revenues (Kuditshinyi, 2011; Lujala, 2010). Therefore, one can argue that lack of transparency and misallocation of state revenues are the main elements demonstrating the negative relationship between good governance and mineral wealth (Bebbington, *et al.* 2007).

In addition, it is possible that, there is sufficient literature of political corruption involving the allocation of resources to favoured constituents or group of people who, in turn, favour the politicians currently in power (Bebbington, *et al.* 2007). This argument emphasises on the allocation of natural resources, which in the absence of check and balances can enable the embezzlement of resources' revenues for political gains. Empirical evidence produced during Mobutu Sese Seko's era demonstrates that public resources were confused with private resources as President Mobutu and his family embezzled some $6 billion from the state treasury (NiZA, 2006; Wedeman, 1997). In fact, Mobutu and his inner circle sucked dry mining companies such as Gecamines until it could no longer afford to pay the salaries of its 35 000 employees regularly and also buy equipment and spare parts to keep production afloat. It is that mismanagement alongside the embezzlement of Gecamines' resources, which led to the collapse of industrial production in the 1990s. Another example of mismanagement and allocation of natural resources can be drawn from the 1998-2003 war in which both the DRC's government and rebel groups utilised natural resources in the territory under their control to fund the war and the political supporters at the expense of the country's growth and development (Kuditshinyi, 2011; Lujala, 2010).

Moreover, Karl (2007) asserted that, government revenue from extractive industries could undermine broader based taxation systems that play a vital role in establishing a broad fiscal social contract that ties together citizens and governors by giving citizens certain authority to hold government accountable. Despite the large potential of the country in natural resources, taxation system has been weak in the DRC as the government lacks the financial capacity to collect billions of dollars from the extractive industries. In addition, since the introduction of the new liberal mining code in 2002, several mining companies that entered in joint ventures with Gecamines are exempted from paying taxes and loyalties to the state (NiZA, 2006). As a result, the government is losing billions of dollars from taxation and in the case of external shocks the consequences are often worse. For instance, in 2004 a period when Gecamines was already in joint-venture operations, figures show that $400 000, representing only 1.18% were collected in the national treasury in terms of taxes and royalties. Another example is drawn from the 2008 global financial crisis, which demonstrated how the decline in the prices of diamond, copper and cobalt affected the government revenues and overall economic growth. All growth forecast and development plans had to be reviewed as the government's revenues shrunk due to the 2008 global financial crisis. These bad effects would have remedied by tax money if the government introduced a viable, transparent and feasible taxation system in the country.

The absence of a viable taxation system in the DRC generates negative effects where the government will not be accountable to the people. This implies that, the system of taxation enables citizens to demand answers to their leaders on how their money is used and allocated to different sectors of the economy. Karl

(2007) contends that in the absence of such a taxation system, the possibilities for unchecked corruption and poor use of public resources are far greater in certain resource rich countries. Generally speaking it is stated that, "if state authority is historically constructed through a series of exchange of resources for institutions then in mineral-dependent states these exchanges are not between state and citizenry but rather between the state and those corporations, foreign powers and financial institutions whose activities generate resources for the state" (Karl, 2007). As a result of these privileged relationship between local elites and external partners, there is an emergence of institutional arrangements designed to attend to the needs and demands of these external actors rather than citizens, and thus lead to an extraverted state with more legitimacy *vis-a-vis* international interests than the citizens (Karl, 2007).

The same scenario of privileged relationship between the DRC's politicians and external actors has been ongoing since independence to receive international recognition or legitimacy. The exception was noticeable with Patrice Emery Lumumba who demonstrated high level of nationalism and enjoyed privileged relationship with the majority of the Congolese population. Because of this privileged relationship with the people of the DRC, Lumumba's life was taken so early. This is in sharp contrast with regimes which followed under Mobutu and the two Kabilas who rely heavily on foreign legitimacy for the survival of their respective regimes instead of seeking domestic support. This is exemplified during elections period where candidates to top positions prioritise international support or recognition rather than focussing on internal support. In principle, the lack of trust between the governed and governors combined with poor management of minerals' revenues made it possible for corruption, civil wars and lack of development in the DRC. This is one of the reasons why in the absence of good institutions, natural resource wealth has been identified as a major cause of armed civil conflict, widespread corruption, and limited degree of development (Collier and Hoeffler, 2001&2002; Ross, 2008).

Despite the negative effects of natural resource wealth on economic development in given circumstances as illustrated above, it is crucial to emphasise that the said curse afflicting resource rich countries such as the DRC can be escaped via good governance and transparent institutions. As these debates have unfolded over the resource curse theory there appears to have been convergence among the views of critics and boosters. For instance, Auty seems to see more scope for avoiding the curse (1993, 2001, and 2008); while Pegg (2006: 377) accepts the fact that mining is potentially a great source of wealth which could generate tremendous economic benefits for poor nations.

Moreover, some authors have questioned the existence of a resource curse (Davis, 1995), and others challenge the methodologies and indicators that have been used to demonstrate it (Brunnschweiler and Bulte, 2008), or argue that if mining has been associated with poor economic and political performance this has been due to pre-existing political institutions rather than mining activities per se (Davis and Tilton, 2002; Humphreys et al., 2007). The case of an African country such as Botswana illustrates the role of natural resources in development of a country. Botswana has been enjoying positive economic development since the discovery of diamond by De Beers Company in 1967(Daniele, 2011). Rational decisions, combined with good governance of the country's revenues largely from extractive sector-maintained Botswana's impressive economic performance since the 1970s up to current days where the 2007 income per capita was around $13,405 (Dunning, 2008).

REALISM THEORY

After utilising the resource curse theory to explain why the DRC remains poor despite the presence of untapped natural resources in large quantity, this section will help us with the realism theory to understand the human nature in the management of the country's revenues from extractive industries. Realist scholars blame greed and egoism as strange human nature for the reason of competition that often culminate in conflict. For instance, Nicolo Machiavelli, author of *The Prince (1532)* and Thomas Hobbes, author of *Leviathan* (1651), figured out human nature as a root-cause of the conflict. They "presumed that human beings are basically driven by their self-interest, human beings are bound to satisfy their appetites, intrinsic in their instinct" (Machiavelli, 1532 and Hobbes, 1651). Furthermore, realists believe that, the most dangerous appetite among others is human being's lust or envy for power which can safely enable them to meet their appetites of glory, wealth and prestige (Donelly, 2000).

The above statement from fervent realists is valid and can be proven with empirical evidence in the DRC. The lust of power to access natural resources' revenues for selfish interests characterise Congolese leadership since the discovery of mineral resources in 1871. Starting with the Belgian King Leopold II, the discovery of rubber enabled him to make use of his military power to extract rubber for personal profit and glory. From 1870s to 1908 the DRC was considered as the king's private garden to satisfy his selfish and egoistic needs to the detriment of a large bulk of the population (Kors, 2010). This is one of the reasons why realists bluntly explain this phenomenon by focusing on the position of potential conqueror, who always tries to maximise his power even at the expense of others (Donelly, 2000). In the same vein, Machiavelli (1532:736) characterises human nature as insatiable, arrogant, crafty and shifting and above all else malignant, iniquitous, violent, and savage. This implies that, men are ungrateful, fickle, pretenders and dissemblers, evaders of danger, eager for gain; they never do good things unless necessity drives them to it (Machiavelli, 1561. 1965).

In order for Leopold II to access and extract rubber in large quantity, between 5 and 8 million people perished under atrocious working conditions to meet the needs of an individual and his entourage. For instance, it has been argued that people were whipped to death; hands were chopped off if quotas of rubber were not met to satisfy the selfish interest of the king and his entourage (Colman, 2014 and Hochschild, 2001). In Machiavelli's words, human beings do wrong and to the same extent when there is nothing to prevent them doing wrong, even the good people know how to enter onto evil, when forced by necessity or selfish interest (Machiavelli, 1965). Whereas the premise of this argument is complete and valid, the last part seems incomplete when Machiavelli states that, even good people know how to enter onto evil, when forced by necessity instead of stating that good people become evil when forced by selfish and egoistic necessity. The poor management of the country's mineral revenues that led to the collapse of the formal economy and industrial exploitation of minerals at Gecamines in the 1990s illustrate how President Mobutu Sese Seko performed well between 1966 and late 1980s and could suddenly become predatory or evil for personal interests at the expense of the country's growth and development in the 1990s (Wautelet, 1981).

However, whereas some realists believe that human nature can change to secure selfish interests, biological realists emphasise on a fixed human nature (Donelly, 2000). This argument seems to have serious problems as human beings can change their nature and pose any action to secure selfish interests even at the expense of others. During the 1998-2003 war Laurent Kabila used the diamond producing company known as Minière de Bakwanga (Miba) to finance the war through direct appropriation of the company's revenues after diamond has been sold in the international market. According to the UN expert report published in 2001, the government of the DRC deducted in advance 2/5 of Miba's profit, while Lambert Kandala the representative of Miba in Brussels talked about 50% of Miba's profit levied by the government in 1998 to help fund the war against aggression. In fact, from 1997 to 2002, the total amount deducted by the government of the DRC from Miba reached $100 million, leaving over 7000 employees of the company unpaid and the company itself facing enormous difficulties in terms of production and meeting its social obligation in the city of Mbuji-Mayi, Kasai Oriental province (Renauld, 2005).

Despite the relevance of the war against aggression, egoistic human nature exemplified itself in the management of the company's revenues. The government determined to hold on to power for personal interests turned a blind eye to the survival of Miba and its thousands of employees. Miba that was considered as the mother and the father of 7 thousand employees and all their dependents in the two Kasai provinces found itself in dire socio-economic situations as war erupted in 1998 since much of the government revenues to finance the war and enrich the political elites came from the diamond producing company. This is the reason why, one made it clear that the Congolese army survived to the detriment of Miba as the latter received several instructions to deposit cash into defence industries across the world to purchase military equipment for the army (Renauld, 2005).

Furthermore, Morgenthau (1948: 4, 1962a:7, 312) argues that "social forces are the product of human nature in action, the social world is but a projection of human nature onto the collective plane, political problems are projections of human nature into society". This implies that, human nature is the key element in shaping the societies in which we live. This can however result in positive or negative outcomes for certain class of people in the society. Niebuhr (1932) contends that, "the ultimate sources of social conflicts and injustices are to be found in the ignorance and selfishness of men". This ignorance of the consequences of selfish actions posed by certain people leads directly to conflict, competition over natural resources for glory and pride (Waltz, 1991: 35). Selfish attitudes of human beings characterised by lust, pride and the quest for glory would cause the war of all against all to continue indefinitely (Waltz, 1991). It is, however, important to bear in mind that all of these negative sides of human nature manifest themselves in an environment of anarchy; lacking good institutions to regulate the behaviour of people.

In the similar vein, Butterfield (1960:25, 1949:33) argued that "the essential nature of man may not be altered, but human behaviour in general is sometimes improved, by the establishment of an order of things (rule of law and institutions) which has the effect of reducing temptation and in some instances a healthy disposition of forces can be attained for long periods which, so to speak, makes human nature better than it really is". For instance, the introduction of good and transparent institutions in the DRC will restrain political leaders from embezzling the country's revenues to fund their supporters to remain in power for personal interests. In other words, these good institutions will limit evil behaviours such as the utilisation of public money to take care of political supporters in order to remain in power at the expense of the majority.

These good institutions will then be responsible for rational utilisation and allocation of the state's revenues to more productive and growth conducive sectors of the economy. According to realist, in the world of anarchy (lack of good institutions) and scarcity of resources to acquire anything of use is to tempt others to come prepared with forces united, to dispossess and deprive him not only of the fruit of his labour but also of his life or liberty (Hobbes, 1651). In such an anarchic environment, individuals, especially the ruling elites motivated by selfish interests and driven by glory can dispossess anyone of their rights for personal gains. This prompts one to stress that for personal financial safety, reputation, or gain men will endeavour to destroy or subdue one another (Donelly, 2000).

The harsh treatment of civilians against their will by rebel groups and the government troops in the exploitation of mineral resources in the eastern part of the DRC demonstrates how human beings are prepared to destroy or subdue one another. The panel of United Nations experts has found several cases of civilian exploitation to generate wealth either for rebel groups or government troops operating in eastern Congo (UN Security Council, 2001; 2002; 2003). Moreover, the noticeable huge gap in terms of living standards between the governors and the governed in the DRC demonstrates the evil side of human nature in the absence of equitable, good and transparent institutions. Whereas the minority ruling class lives in good conditions, 70% of the population lives with $1 dollar a day (IMF, 2015). Figures stand at 80% for rural areas against 60% for urban areas, painting a dark picture of the Congolese society despite the presence of valuable mineral resources in considerable quantity (DRC Ministry of Environmental Affairs, 2012). Without political will to enforce good institutions and rules, the DRC remains vulnerable to the evil side of human nature and victim of the resource curse thesis.

Furthermore, one made it clear that, without institutions that restrain the power of the elites, human beings cannot control their desires to take from others, to react with excessive power, or to request greater respect than others are willing to give freely (Hobbes, 1651). This demonstrates again that the absence of good institutions will lead to misallocation of resources that would have benefited the majority of the population if invested in infrastructure, health and education. For instance, figures demonstrate that the budget allocated to the health and education sectors remains below Sub Saharan African standards while they are very important sectors for economic growth and development of the DRC (Omoyi, 2014). In fact, good governance of the country's resources will inhibit corruption, selfish behaviour and pave the way for more efficient investment of the rents arising from primary commodity exports in profit generating infrastructure, industries and education to produce well educated and productive work force for the economy.

CONCLUSION

In this paper I utilised two theories known as the resource curse and realism to help me and the readers understand clearly how and why mineral resources have not helped the DRC lift its 70% of the population out of poverty. Since the discovery of these potential mineral resources by the British explorer Stanley in 1971, these resources have not contributed much to the upliftment of the Congolese population out of poverty and misery. Social indicators in the DRC are in red for years as 70% of the population lives on $1 a day despite the abundance of untapped resources estimated at $24 trillion equivalent to the combined US and Euro Zone countries' GDP.

This is what the proponents of the resource curse theory have emphasised on the negative relationship between natural resources and development. The theory stresses that, certain countries rich in natural resources perform poorly when compared with their resource poor countries. The reason for the poor economic performance is due to several reasons ranging from political and economic factors. Political explanations indicate that, natural resources discourage the political elites to take rational decisions regarding the management and allocation of commodities' revenues. In most cases, these revenues are embezzled to finance occult activities, pay supporters for political votes during elections time and maintain the clientelist network in place. In other words, mineral rents feed the over-expansion of bureaucracy, and induce patronage, clientelism and graft that corrode the quality of government in resource rich countries. I have demonstrated how rebel groups and several governments that succeeded in power since Mobutu to Kabila's era utilised the countries resources largely from mineral sector to fund supporters and the over-expanded bureaucratic system for personal interests and regime survival.

Moreover, mineral resources' wealth has been considered by the resource curse theory as the main factor diverting the interests or attention of political actors from other vital sectors of the economy such as agriculture and manufacturing. In other words, political actors from resource rich countries suffer from myopia or short-sightedness during the resource boom as they neglect other sectors of the economy that can contribute to sustainable development in the absence of resources or their depletion. The culture of making quick money characterises most of political actors in some resource's rich countries, therefore, jeopardising their respective countries economic growth and development. President Mobutu and the two Kabila's reliance on these resources to make quick money and maintain their respective regimes in power has been used and demonstrated as a good example of myopia and short-term planning without long term projections of their actions' consequences on economic development. Under these regimes, mining companies such as Gecamines and Miba were used as the government financier at the expense of vital investment in both physical and human capitals.

The lack of management precipitated the collapse of those companies, leaving space for artisanal activity, which is complicated to control given the government incapacity to effectively control the country's porous borders. This implies that, the government's incapacity to control the country's borders along with intensified artisanal mining enables smugglers to illicitly export mineral resources without paying taxes and other duties to the government. The DRC has lost billions of dollars as the result of illegal smuggling of mineral resources to neighbouring countries. Given all these political problems together with mismanagement of mineral wealth, corruption and rent seeking behaviour became a way of life to make quick money. Rebel groups became active since the central government was weak to enforce its authority all over the country.

The economic explanation of the resource curse contends that, during resource boom mineral resources enables the inflow of foreign currencies which then lead to the appreciation of the local currency. This appreciation makes the non-tradable goods expensive when compared to imported goods, thus, occasioning the deindustrialisation of the country's manufacturing and agricultural sector. As it is well known, deindustrialisation leads to increased unemployment and other negative ramifications in the economy when the extraction of natural resources reaches the end. Beside this phenomenon known as the Dutch disease, mineral resources lead to low level of revenues collected through taxes. The fiscal system is poorly developed and neglected as the mineral sector generates quick and easy wealth for the regime in place. In this regard, it has been demonstrated that, different governments since Mobutu Sese Seko to current days have neglected to enforce taxation system or regime as they relied on revenues from mining sector. This reliance on the mining sector by neglecting to levy taxes from the citizens reduced the government's responsibility and accountability towards their citizens.

The last theory is realism which contends that, the human nature is egoistic and after personal interests at the expense of the country's growth and development. Mineral resources and the anarchy that characterises a country such as the DRC can easily pave the way for more pillaging, corruption, and misallocation of the country's resources for personal gains. To remedy the resource curse and realism theories, one can argue that greater transparency can provide opportunities for the public and civil society to hold the government accountable for sound managing of revenues and their better utilisation to benefit the population, the majority of whom live in dire poverty.

Given the fact that, natural resource's revenues are volatile and finite, greater transparency to ensure that windfalls are deployed for development purposes is required. Because resource extraction can lead to exhaustion of the revenue source, governments in resource-dependent countries must be prepared to replace revenues from resource production and export with other sources of income. During periods of booming prices, lack of transparency and sound management removes the incentive for governments to prioritise expenditures, diversify the revenue base of the economy, and design plans to mitigate the effects of a sudden change in resource prices. When prices drop, budget opacity hinders public discussion of available policy responses to the fiscal squeeze, often resulting in cuts in social programme and public investment that can undermine poverty reduction efforts.

REFERENCES

Auty, M. R. (1993). *Sustainable development in the mineral economies: The resource curse thesis*. London: Routledge Press.

Babbie, E. (2005). *The basics of social research*. Thomson Wadsworth.

Babbie, E. (1990). *Survey Research Methods*. Wadsworth Publishing Company.

Baker, L. (2006, Summer). Observation: A Complex Research Method. *Library Trends*, *55*(1), 171–189. doi:10.1353/lib.2006.0045

Bebbington, Hinojosa, L., Bebbington, D. H., Burneo, M. L., & Warnaars, X. (2008). Contention and Ambiguity: Mining and the Possibilities of Development. *Development and Change*, *39*(6), 887–914. doi:10.1111/j.1467-7660.2008.00517.x

Bhattacherjee, A. (2012). *Social science research: principles, methods and practices*. University of South Florida.

Bless, C., & Smith, C. H. (2000). *Fundamentals of social research methods: an African perspective*. Juta Education.

Braun, V., & Clarke, V. (2006). Using thematic analysis in psychology. *Qualitative Research in Psychology*, *3*(2), 77–101. doi:10.1191/1478088706qp063oa

Brunnschweiler, C., & Bulte, E. (2008). The resource curse revisited and revised: A tale of paradoxes and red herrings. *Journal of Environmental Economics and Management*, *55*(3), 248–264. doi:10.1016/j.jeem.2007.08.004

Bulte, E., Damania, R., & Deacon, R. T. (2005). Resource Intensity, Institutions, and Development. *World Development*, *33*(7), 1029–1044. doi:10.1016/j.worlddev.2005.04.004

Carpenter, L. (2012). *Conflict minerals in the Congo: blood minerals and Africa's under reported First World War*. Working paper. Suffolk University.

Coleman, D. Y. (Ed.). (2004). *Congo (DRC)*. Country Watch Review.

Collier, P., & Hoeffler, A. (2002). On the Incidence of Civil War in Africa. *The Journal of Conflict Resolution*, *46*(1), 13–28. doi:10.1177/0022002702046001002

Collier, P. & Hoeffler, A. (2001). *Greed and Grievance in Civil War*. World Bank.

Congo Central Bank. (2012). Condense hebdomadaire d' informations statistiques No 3. Congo Central Bank.

Daniele, V. (2011). Natural resources and the "quality" of economic development. *The Journal of Development Studies, 47*(4), 545–573. doi:10.1080/00220388.2010.506915

Davis, G. A., & Tilton, J. E. (2005). The resource curse. *Natural Resources Forum, 29*(3), 233–242. doi:10.1111/j.1477-8947.2005.00133.x

Democratic Republic of Congo. (2006). Document de Stratégie de la Croissance et de la Réduction de la Pauvreté (DSCRP) [Strategic document on growth and reduction of poverty]. Kinshasa: DRC.

Dibwe dia Mwembu, D. (2001). *Bana Shaba abandonnés par leur père: structures de l'autorité et histoire sociale de la famille ouvrière au Katanga 1910–1997.* Paris: L'Harmattan.

Division provinciale de l'intérieur et sécurité. (2015). *Cellule d'Analyses des Indicateurs de Développement.* Available at https://www.caid.cd/index.php/donnees-par-villes/ville-de-lubumbashidomaine=fiche

Donelly, J. (2000). *Realism and international relations*. Cambridge University Press. doi:10.1017/CBO9780511612510

Dunning, T. (2008). *Crude Democracy. Natural Resource Wealth and Political Regimes*. Cambridge University Press. doi:10.1017/CBO9780511510052

D'Ydewalle, C. (1960). L'Union Minière du Haut Katanga de l'âge colonial a l'indépendance. Plon, 65-80.

Gann, L. H., & Duignan, P. (n.d.). *Burden of Empire.* An appraisal of Western Colonialism in Africa South of the Sahara. *Hoover Institution Press.*

Garrett, N. (2010). 'Mining as a Source of Growth. In J. Herderschee & J. E. Mans (Eds.), *Obstacles to Growth in the Democratic Republic of Congo*. World Bank.

Garrett, N., & Lintzer, M. (2010). Can Katanga's mining sector drive growth and development in the DRC? *Journal of Eastern African Studies: the Journal of the British Institute in Eastern Africa, 4*(3), 400–424. doi:10.1080/17531055.2010.517408

Hesselbein, G. (2007). Essor et déclin de l'État Congolais. No 21. London: Crisis States Research Centre.

International Monetary Fund (IMF). (2005). *Republic of the Congo: Selected Issues and Statistical Appendix*. IMF Country Report No. 05/373. http://www.lepotentiel.com/afficher_article.php?id_edition=&id_article=12707

McFerrin, H. (2017). *News from Africa: The DRC poverty rate continues to stagnate*. Available at https://borgenproject.org/the-drc-poverty-rate/

Kalenga, J. N. (2013). Economic and toxicological aspects of copper industry in Katanga, DR Congo. *The Japanese Journal of Veterinary Research, 61*(Supplement), S23–S32. PMID:23631149

MacNealy and Mary Sue. (1999). *Strategies for Empirical Research in Writing*. Longman.

Marysse, S., & Geenen, S. (2009). *Win-win or unequal exchange? The case of the Sino-Congolese cooperation agreements*. Cambridge University Press. doi:10.1017/S0022278X09003978

Marysse, S., & Andre, C. (2001). Guerre et pillage economique en RDC [War and looting in the DRC]. Paris: L'harmattan.

Mehlum, H., Moene, K., & Torvik, R. (2006). Cursed by resources or institutions? *World Economy, 29*(8), 1117–1131. doi:10.1111/j.1467-9701.2006.00808.x

McFerson, H. M. (2010). Extractive industries and African democracy: Can the "resource curse" be exorcised? *International Studies Perspectives, 11*(4), 335–353. doi:10.1111/j.1528-3585.2010.00410.x

Ministry of Environment, Conservation of Nature and Tourism of the Democratic Republic of Congo. (2012). Rapport national synthèse sur le développement durable en RDC [National report on the sustainable development in the DRC]. Author.

Neuman, W. L. (2014). *Social research methods: qualitative and quantitative approaches*. Pearson Education Limited.

Omoyi, K. (2014). *Ressources minieres, croissance economique et guerres civiles en RD.Congo* [Mineral resources, economic growth and civil wars in the DRC] (Masters Dissertation). University of Yaounde II.

Pegg, S. (2006). Mining and Poverty Reduction: Transforming Rhetoric into Reality. *Journal of Cleaner Production, 14*(3–4), 376–387. doi:10.1016/j.jclepro.2004.06.006

Union Minière du Haut Katanga. (1957). Production from 1906-1956. *Cuyprius*, 39-40.

University of Lubumbashi. (2005). *The restructuring of Gecamines, impacts on poverty and social issues.* The future of social services and reintegration strategies, final report of the University of Lubumbashi.

RAID. (2004). Unanswered questions. Companies, conflict and the Democratic Republic of Congo. RAID.

Renauld, A. (2005). *Ressources naturelles et transferts d'armes* [Natural resources and transfer of weapons] (Report No 1). Brussels: GRIP.

Ross, M. L. (1999). The political economy of the resource curse. *World Politics, 51*(2), 297–322. doi:10.1017/S0043887100008200

Ross, M. L. (2014). What have we learned about the resource curse? *The American Political Science Review, 102*(1), 107–123. doi:10.1017/S0003055408080040

Robinson, A., Torvik, R., & Verdier, T. (2006). Political foundations of the resource curse. *Journal of Development Economics, 79*(2), 447–468. doi:10.1016/j.jdeveco.2006.01.008

Rubbers, B. (2009). The story of a tragedy: How people in haut-Katanga interpret the post-colonial history of Congo. *The Journal of Modern African Studies, 47*(2), 267–289. doi:10.1017/S0022278X09003838

Rubbers, B. (2010). Claiming workers' rights in the Democratic Republic of Congo: The case of the collectif des ex-agents de la Gécamines. *Review of African Political Economy, 37*(125), 329–344. doi:10.1080/03056244.2010.510629

Sachs, J., & Warner, A. (1995). *Natural resource abundance and economic growth.* National Bureau of Economic Research. doi:10.3386/w5398

Stiling, J. A. (2001). *Thematic networks: An analytic tool for qualitative research.* Sage Publications.

Strauss, A. L., & Corbin, J. M. (1998). *Basics of qualitative research: Techniques and procedures for developing grounded theory*. Sage Publications.

van der Ploeg, F. (2010). *Natural resources: Curse or blessing?* CESifo working paper Resources and Environment, No. 3125.

Wautelet, J. M. (1981A). Pouvoir d'Etat et formation du capital. In Accumulation et sous-développement au Zaïre 1960-1980. Presses Universitaires de Louvain.

Weber, M. (1973). *The protestant ethic and the spirit of capitalism*. Routledge.

Weber-Fahr, M. (2002). *Treasure or trouble? Mining in developing countries*. The World Bank and International Finance Corporation.

World Bank. (2018). *Democratic Republic of the Congo Overview*. Available at https://www.worldbank.org/en/country/drc/overview

World Bank. (2012). *World Development Report*. World Bank.

Chapter 16
Contribution of Mining Operations Towards Education, Healthcare, Food Security, Housing, Sports, and Recreation in Katanga Province of the DRC

Germain Miteu Tshinu
University of Zululand, South Africa

ABSTRACT

Some resource-rich countries face challenges when it comes to the socio-economic development of its employees and the community at large. This study seeks to examine Gecamines' contribution to the socio-economic conditions of its employees based in the Lubumbashi area and to some extent the province of Katanga under the rubric of resource curse theory. Through this study, lessons can be learnt from the findings on how state-owned mining companies could work on improving their socio-economic contributions in the interest of the employees and the entire population through generated revenues for the state. The study employed a case study design with a qualitative approach research in its endeavour of exploring Gecamines' socio-economic contribution to its employees in the city of Lubumbashi and the development of Katanga Province as a whole. Semi-structured interviews were conducted with Gecamines' managers, unskilled mine workers, and the Provincial Department of Mines' officers in the city of Lubumbashi. The key findings from these interactions are that the company's low level of production caused by poor management, political interference, corruption, and poor institutions had negative ramifications on the socio-economic conditions of the workers. A remedy to these challenges lies in strengthening the country's institutions to fight corruption and poor management of parastatals. Autonomy of parastatals from political influence and control is also important to enable the company to play a crucial role in the socio-economic development of employees.

DOI: 10.4018/978-1-7998-8809-3.ch016

INTRODUCTION

Empirical findings have demonstrated that some resource poor countries are performing better than certain resource rich countries. On one hand, Mehlum, Moene and Torvik, (2005) argued that, countries' endowment in resources is a positive asset in socio-economic development of citizens and overall development of the country, while on the other hand, the same resources are considered as the source of economic stagnation, war, corruption and institutional failure (Auty, 2001; Sachs and Warner 1995, 1997a, b). For instance, in certain countries such as Canada, Norway, Australia, the United States of America and Botswana, natural resources have contributed to positive economic growth, which prompted some researchers to consider these countries as growth winners (Mehlum *et al,* 2005). Whereas on the other hand, a considerable number of countries on the African continent and Latin America such as Nigeria, Zambia, Sierra Leone, Angola, the Democratic Republic of the Congo (DRC), Venezuela, and so many more have performed poorly despite the presence of strategic natural resources (Auty, 2001; Boschini; Pettersson; and Roine, 2007; Sachs and Warner 1995, 1997a, b). Mehlum *et al* (2005) coined a term of growth losers for these countries given their failure to economically perform well despite the abundance of natural resources.

The DRC is one of the countries endowed in natural resources, however it has been doing badly in terms of the socio-economic development as a considerable number of its population lives in acute poverty. As of 2011, an estimated 87.7 percent of the population, about 67.5 million people, remained under the international poverty line, which is currently $1.25 per day. (McFerrin, 2017; Omoyi, 2010). In 2018, the World Bank estimated that 72% of the population, especially in the North West and Kasaï regions, was living in extreme poverty on less than $1.90 a day (World Bank, 2018). Despite high level of poverty and precarious conditions of the population, the DRC possesses mining giants such as *la Générale des Carrières et des Mines* (Gecamines) that played a key role in socio-economic conditions of its employees and the entire nation. At its creation in 1906 by the colonial masters, the company was known as *Union Minière du Haut Katanga* (UMHK). This company first changed to *La Générale Congolaise des Minerais,* (Gecomin) and then Gecamines in 1967 following its nationalisation by the regime of President Mobutu Sese Seko.

The company produces strategic minerals known as copper, cobalt, zinc, uranium and manganese (Autesserre, 2012; NiZA, 2006; Rubbers, 2009). With its concessions of 30 000 kilometre square scattered in the towns of Kolwezi, Likasi and Lubumbashi, Gecamines' global reserves in copper and cobalt is evaluated at 70 million and 10 million tonnes respectively (Global Witness, 2004) and 3.4 million tonnes of germanium and 6.4 million tonnes of zinc as well (Kilondo, 2010:188; World Bank quoted by Omoyi, 2014:27;Autesserre, 2012; Kalenga, 2013; NiZA, 2006; Rubbers, 2009). Gecamine's industrial production improved the welfare of close to 35 000 employees and 250 000 dependants (Kilondo, 2010; Omoyi, 2014:29). Gecamines reached its highest production in 1986 when 476 000 tonnes of copper, 14 500 tonnes of cobalt and 64 000 tonnes of zinc were produced (Gecamines, 2015:1; World Bank, 2012). This production did not only improve the living conditions of the employees and their dependants but made it possible for the mining sector and Gecamines in particular to become the cornerstone of the Congolese economy. In other words, the improved production of copper and cobalt enabled Gecamines to put in place an unprecedented socio-economic programme which benefitted thousands of its employees and their dependants. For instance, through the established socio-economic programme, Gecamines employees' children could access quality education at Gecamines' schools for free, and the entire family had access to Gecamines hospitals for free health care (Kilondo, 2010) and these will be covered in this article. Besides these services, Gecamines was providing its employees with houses and food intake or ration at the end of each month based on the size of the family.

However, since the beginning of the 1990s, Gecamines' output and prices of metals declined drastically, leading to intensified misery of the population and deterioration of economic activities (Garrett and Lintzer 2010). It is for instance argued that, in 1994 there was a sharp decline of production from 500 000 tonnes to only 29,300 tonnes of copper metals annually (Gecamines, 1994: 8). This decline had a negative impact in the socio-economic conditions of Gecamines' workers, their dependants and to some extent the entire country (Gecamines, 2015:1; Marysse and Geenen, 2009). It is against this deteriorated socio-economic conditions that the intended paper sought to explore Gecamines' current socio-economic contribution to its employees and the population of Katanga to come up with recommendations a.nnccto address any challenges the company might be facing as well as to identify any opportunities to improve its contribution. The second section will outline the research question and rationale for the study. The third section will emphasise on the methodology used in this paper to address the problem statement the fourth section will underline the theoretical framework and debate around natural resources and their effect on socio-economic development. The fifth section will discuss the findings of the field mission. The sixth section will conclude by summarising discussed issues while the last section will come up with recommendations derived from the findings.

RESEARCH QUESTION AND RATIONALE FOR THE STUDY

Thus, the paper was guided by the following questions:

- What kind of socio-economic programmes does Gecamines have to contribute to socio-economic development of its employees and Katanga province to become relevant again in the lives of Katanga people?
- What challenges, if any, are being faced by Gecamines in implementing its socio-economic programmes and how can these challenges be addressed in order to improve the programmes' effectiveness?
- What are the opportunities available for Gecamines' socio-economic programmes in order to improve its socio-economic contribution to its employees, the population of Lubumbashi and the entire Province of Katanga?

The study is of paramount importance as it has the potential of improving the effectiveness of Gecamines' socio-economic programme and socio-economic conditions of its employees as well as the people in Katanga Province. This study was also important in the sense that it raised awareness by looking at the living conditions of some remaining 10. 000 Gecamines' employees and former employees who left the company due to the World Bank Voluntary Departure Programme after the company has gone into partnerships with private firms to exploit copper and cobalt. Furthermore, by scrutinising Gecamines and the impact of its joint ventures on the socio-economic conditions of its employees in particular and Katanga province in general, this study attempted to extend its scope on these joint ventures' contribution to the national treasury as well. This was important to see whether Gecamines and its partners can be relevant again in the Congolese economy as it was the case in the 1960s and 1980s. It was also possible that, lessons could be learnt from the findings of the study that could be utilised by other mining companies in improving their socio-economic contributions in the interest of the employees and the entire population through generated revenues for the state.

METHODOLOGY

This paper employed an exploratory approach to qualitative research design in its endeavour of exploring Gecamines' socio-economic contribution to its employees in the city of Lubumbashi and the Katanga region as a whole. The use of this approach was important as it allowed us to conduct a very close, detailed conversation or interviews to acquire a profound and deep understanding of the issue under investigation (Neuman, 2014:103). Through this approach, it was possible for researchers to deeply engage and interact with major stakeholders to capture their own subjective views on the company's contribution to socio-economic development of its employees and the mining community in Lubumbashi and the province of Katanga as a whole.

This paper's population derived from key stakeholders that include Gecamines' managers in charge of socio-economic development, beneficiaries such as the field workers and the local government officials in Katanga Province. Gecamines' Managers, field and underground workers were chosen because they have the first-hand experience and knowledge of the Gecamines' socio-economic programme that could help the researcher obtain necessary information to answer research questions identified above. The rationale for choosing Provincial Ministry of Mines' social officials was based on the fact that they are not only the gatekeepers, but also are well vested in the socio-economic developmental issues of Katanga Province. The fieldwork was conducted in the city of Lubumbashi in the Katanga province, Democratic Republic of Congo from 12 January 2017 to 22 February 2017. According to the Congolese *Division Provinciale de l'Intérieur et Sécurité (*2015), the city of Lubumbashi covers 747 km^2 of land with a population estimated at 2 million (Division provinciale de l'intérieur et sécurité, 2015).

Purposive sampling, a non-probability sampling technique, was utilised to select 4 Gecamines' social programme managers out of 7 decision-makers, 17 Gecamines' socio-economic programme beneficiaries from the approximately 9000 employees who work on the surface and underground and 4 social officials from the mining division at the Provincial Mining Department in Katanga. This purposive sampling technique was chosen amongst many others because it allowed the researcher to sample participants on the basis of their knowledge and capacity to respond to research questions (Babbie, 1990). In MacNealy's (1999: 157) words, "the individual characteristics, knowledge and ability are selected to answer necessary questions about a certain matter or product". This sampling technique was also important to identify the study population in a consistent, reliable, uniform and objective manner based on their knowledge of the subject under study. The sample of 25 participants was considered as enough since qualitative research make use of small samples in order to collect rich and thick data. Triangulation of data by interviewing different stakeholders helped ensure rigour, trustworthiness and obtain different perspectives on data collected.

The research used semi-structured interviews to enable participants to express their subjective views on the subject of interests (Bless and Smith, 2000). this technique of data collection was essential as it enabled the interviewer to direct the flow of ideas, intervene and ask questions for more clarity when necessary. In addition, This technique of data collection was chosen to gather valuable information as respondents were free to expand on the topic, focus on particular aspects and to some extent relate to their own experiences as decision-makers in their respective fields of expertise (Bless and Smith, 2000:30).

Beside this technique, observation was also utilised as a data collection method during the field work. In order to use observation as a data collection method, the researcher remembered his primary role as a researcher and remained detached enough to collect and analyse data relevant to the problem under investigation (Baker, 2006). The technique was important to observe the living conditions of Gecamines' workers and the community. The conditions of the mine's non-current assets such as buildings, hospital facilities, schools, vehicles, abandoned shafts were also observed to have a better understanding of the socio-economic conditions of the workers. In addition, the technique was crucial to verify nonverbal expression of feelings, grasp how participants responded to questions, and check how much time was spent on responding to questions enquiring on their living conditions (Baker, 2006; Kaulich, 2005; Schmuck, 1997). In brief, this technique was used as a way to increase the validity of the data and enabled the researcher to have a better understanding of the context and phenomenon under study (Kaulich, B. 2005).

The interviews were conducted in French given that the majority of the population speak that language in Katanga province. Interviews were audio recorded, with the consent of participants, to save time from detailed note taking and to make sure that all responses were captured. Pilot study was also conducted in the DRC two weeks before the real interviews in order to get an idea on respondents' ability to respond to research questions and ensure that there were no ambiguous questions.

Given that this was a qualitative research, thematic analysis was utilised to analyse data collected through the interviews. The interview audio records were transcribed and then analysed according to recurring themes, concepts or similar features using the thematic analysis method identified by Newman (2014). For instance, themes related to Gecamines' contribution to socio-economic development in terms of education and training, housing, food distribution, job creation, sport and recreation, farming and health care were identified and analysed separately through rigorous process known as data familiarisation, data coding, theme development and revision (Braun, and Clarke, 2006).

The researchers fully notified the participants about the purpose of the study, ensuring that they understand that their participation was voluntary, and could also freely withdraw from participating in the study at any stage if they wished to do so and a joint decision was going to be taken on what to do with data already collected on withdrawal from the study. The participants were then asked to complete and sign written consent forms to make sure that they participate on their own free will and agree to audio record interviews. Participants' identities were anonymous, and information provided was protected or treated as confidential as possible to protect them from any harm that may arise if their identities were divulged. Pseudonyms such as names of people from the Western part of the DRC were used to hide the identity of the participants. The next section will focus on the theoretical framework and debate around natural resources and their effect on socio-economic development.

THEORETICAL FRAMEWORK AND DEBATE ON NATURAL RESOURCES REGARDING SOCIO-ECONOMIC DEVELOPMENT

It has been discovered throughout several researches that a considerable number of resource rich countries tend to perform poorly in terms of development when compared to resource poor countries. This is known as resource curse which refers to a situation where a country experiences slow economic growth, incidence of civil war and limited social development in spite of having abundance of mineral resources (Sachs, 1995; Warner, 2001). For instance, the DRC, has been experiencing slow economic development as a result of production decline in industrial mining activities at the parastatal *La Générale des Carrières et des Mines* (Gécamines). In 1994, the DRC's industrial mining production at Gecamines declined to 29 300 tonnes from 500 000 tonnes of copper and cobalt per annum (Kalenga, 2013). This poor performance led to unprecedented socio-economic crisis in which the living standard of 35 000 employees of Gecamines and their dependants estimated at over 250 000 deteriorated drastically (DSCRP, 2006).

Moreover, resource curse theory suggests that natural resource abundance generates a series of economic and political distortions or policies which ultimately undermine the contributions of extractive industry to development (Bebbington, *et al.* 2007). In other words, the abundance of natural resources creates a type of myopia among public or private actors (Ross, 1999). This is a situation where political and business elite are characterised by short visions of satisfying current interests to the detriment of long-term development of the country. In most cases, resource boom pushes policy makers to come up with economic policies that do not enable forward and backward linkages important for sustainable development and diversification of the economy (Okubo, 2007). For instance, during the resource boom at Gecamines in Katanga province from the 1960s to 1990, the DRC's policy makers failed to optimally invest much of the resources in other productive sectors of the economy such as manufacturing and agriculture as well as education and public infrastructure as important foundations for sustainable development. As a result of this poor planning, the collapse of industrial production at the parastatal Gecamines in the 1990s had a negative impact on the living conditions of the workers, small firms that were doing business with Gecamines. In addition, the government revenues decreased as mining contribution was only around 2% in 1994 from 67% in 1970 (BCC, 1995, Marysse, 2009, Kalenga, 2013).

This argument demonstrates that reliance on natural resources without allocating substantial amount of money to the diversification of the economy can impede long term development in a country. This the reason why one made it clear that "resource rent or reliance on revenues from natural resources lead to irrational excitement, producing a get rich quick mentality among businessmen and a boom-and-bust psychology among policymakers, marked by bouts of excessive optimism and frantic retrenchments" (Ross, 1999). Wallich and Levin (1960) contend that, "the development path of commodity exporting countries was distorted by a resource mentality that led to poor economic planning, lack of diversification and widespread poverty. Evidence in the DRC demonstrates that, the country's reliance on volatile natural resources without sound macro-economic policies and planning produced harmful effects on socio-economic indicators which are below the Sub-Saharan standards (UNDP, 2010).

In addition, it has been observed in the DRC that since the collapse of industrial mining activities in the 1990s, budgets allocated to health and education sectors have drastically declined below Sub-Saharan standards between 0.8% and 0.1% respectively (World Bank, 2007:36; DRC Ministry of Environmental Affairs, 2012; DRC Ministry of Budget, 2011). According to UNDP (2010), the DRC has been doing poorly in human development index below the minimum standard of 0,389 and 0,393 respectively for Sub-Saharan Africa and countries with low human development index. Furthermore, social indicators on the DRC are in the red as the unemployment rate was between 50.3% in 2012 against 51.4% in 2011 (Central Bank of Congo, 2012).

FINDINGS AND DISCUSSION ON SOCIAL GIVING INITIATIVES AT GECAMINES

Themes related to Gecamines' contribution to socio-economic development in terms of access to education and professional training and health care, were identified and the data collected were analysed separately through rigorous process which linked the main findings to the literature review.

EDUCATION

It is crucial to recall that Gecamines' managers or skilled workers came from different departments that deal with socio-economic conditions of employees based on their competences. The departments of Medical and Social Programme, General and Technical Education as well as Management of Development and Qualification of Human Resources were consulted to acquire information for this article. Whereas beneficiaries within the category of unskilled workers were composed of current or active employees and former employees. Most of the respondents from the aforementioned categories recognised that Gecamines used to play a critical role in the living conditions of its employees and their respective families. Access to education and health care, were guaranteed by the company. The current situation on socio-economic conditions of workers based on the above indicators have worsened as the company is unable to meet the need and aspiration of its employees due to several financial constraints.

With regards to education it was confirmed by Mr. Mavuzi from Gecamines' department of General and Technical Education that "Gecamines' schools were created for the children of workers to help reduce the cost of studies which was high in private and other schools within the province". In order to achieve this, Mr. Mavuzi continued stating that "Gecamines started with the creation of primary schools for workers' children". Given the fact that Gecamines main objective was to keep the company afloat and increase production, Mr. Mavuzi stated that "the company created technical and professional schools to equip its employees and their family members with certain skills in order to maintain high production at the best interests of the company". Another respondent named Lumpungu confirmed that: the initiative of creating technical schools made the training of technicians who worked in Gecamines' metal plants possible while young ladies were trained in sewing, tourism and hospitality and life skills." It is clear that this was a good programme put in place by the company to empower people through certain technical training which had positive ramifications on the company and living conditions of the beneficiaries as well. This argument is corroborated by the literature which demonstrated that technical schools of high quality in major mining towns of Lubumbashi, Likasi and Kolwezi were created to enable children of UMHK's workers to have access to quality education with an opportunity to be hired by the company (Kilondo, 2010).

Furthermore, the respondent by the name of Lumpungu continued by stating that once the company noticed that not every child was going to technical and vocational schools, secondary or high schools were created to enable these children to further their education. Up to now Gecamines has not received any authorisation to create institutions of higher education. One may argue that this is not a function of a company to establish schools or universities, but the case of Gecamines is a special one since the company is considered as a state within a state with its own schools, hospitals, etc. According to Kilondo (2010) the rationale behind the setting up of these facilities was to ensure the provision of the workforce with necessary skills and to ensure that they are in good health, in the interests of the company. It is equally important to argue that, Gecamines came up with these learning institutions to cover a void since public schools were considered unable to provide necessary technical skills to the company. In other words, the creation of Gecamines' schools is important given the fact that, the DRC has been unstable from independence to current days and this has negative effects on the government provision of basic quality education to the citizens.

In fact, to emphasise on the importance of schools, a skilled worker by the name of Nkufi stated that "the company has its own nursery schools, primary schools, secondary schools and professional or technical schools which are still providing quality general and technical education to the workers' children free of charge despite some deplorable conditions of infrastructure and lack of laboratories and certain equipment". For instance, Nkufi "mentioned the case of Madini as one of the high schools providing quality education in the province and the country as a whole while Kitendo is one of the technical schools teaching fitting, general mechanic, electricity and so on in the city of Lubumbashi, and Amali is a female school training young ladies and wives of employees on sewing, tourism and hospitality, introduction to social life and other related areas.

Moreover, a skilled worker by the name of Kazadi also indicated that "Gecamines' schools are not benefiting only the company's personnel and their respective family members, but they also benefit the community within different mining towns such as Likasi, Kambove, Kolwezi, Luena Kipushi and Lubumbashi where Gecamines is operational". The same respondent continued listing the number of schools under Gecamines management in order to demonstrate the positive role the company plays in the entire province of Katanga in the interests of the employees and the Congolese population in general. For instance, it was indicated that "Gecamines has 111 primary, secondary, technical and professional schools throughout its different operational groups within the province of Katanga known as Group South made up of the cities of Lubumbashi and Kipushi, Centre made up of the cities of Likasi and Kambove and lastly the West group made up of the cities of Kolwezi and Luena". In addition, Mr. Amuli stated that Gecamines' schools absorb a considerable number of learners in all 3 operational groups. Mr. Amuli made it clear that these learners are "40 288 of which 19 488 are children of Gecamines' active employees, 1 690 are children of former Gecamines' employees who left due to voluntary departure programme, 19 110 are children of third parties or non-employees of Gecamines".

As Gecamines has not recruited new employees for more than 10 years, the number of children at primary level in Gecamines' schools decreased drastically and the void is covered by children of third parties or non-employees of the company who are paying for the education of their children. Mr. Amuli from the Department of General and Technical Education made it clear that "the high number of third parties' children in our schools is a testimony of quality education provided by the company and the trust parents have in Gecamines' schools". The same person continued stating that, "the quality of their education is above the national standard as Gecamines' schools get high scores at the national level in the state examinations", equivalent of South African matric examination. Moreover, it was also indicated that "this good and quality education is made possible by well qualified, motivated and paid teachers when compared to other schools or public schools within the country". These arguments on the quality of education in Gecamines' schools are true since there is a considerable number of learners that are attracted in those schools and many people within the city of Lubumbashi, we had informal discussions which confirmed the same information. Some of the schools we visited such as Madini, we could see that learners were following teachings with sustained interests and discipline.

In addition, Mr. Amisi from Gecamines' Department of General and Technical Education stated that "the quality of education is very good in Gecamines' schools because there is a follow up to ensure that the national programme is respected and finished in the prescribed time which is not the case in other schools within the country". This information seems to be true since Gecamines' schools are inspected quite often by inspectors from Gecamines' Department of General and Technical Education. However, given the financial difficulties the company is going through and irregular payment of salaries it seems like Gecamines' inspectors can be demotivated to do their job accordingly. Therefore, this may have a negative effect on the quality of education and strict observation of the calendar. For instance, one can argue that, at the peak of its production, the company could easily allocate huge amount of money to schools for their function but with low production and financial crises, the allocation of a certain budget to schools seems to be difficult for the company. Nyembo Shabani (1975:141) made it clear in the literature that:

Forty four (44%) percent of Union Minière du Haut Katanga's (UMHK) revenues were spent on equipment maintenance, social investment such as housing, education, health care, local entrepreneurs and contractors' payment as well as the remuneration of workers to ensure good living conditions as well as quality education to the benefit of workers' children and the community.

Furthermore, another reason that contributed to quality education at Gecamines is the administration of admission test to recruit competent and qualified students in Gecamines' schools. Mr. Tembo from the Gecamines' Department of General and Technical Education argued that "to ensure good quality of education, Gecamines' schools have put in place a system of admission tests. This entails that any child coming from non-Gecamines' schools have to write an admission test before being admitted". This is really a good approach since it will enable the company to have competent candidates that will easily adjust and adapt to Gecamines' teaching methodology and philosophy for better results. Besides admission test, Mr. Tembo argued that "the company put in place a certain type of examination similar to state examination to prepare grade 12 learners for state examination. These examinations are structured in the same structure and format of state examination". From this information one can argue that, all these elements in place are crucial and important to ensure that learners depart from Gecamines' schools with adequate skills for their success and improved socio-economic conditions in the society.

In addition, another factor contributing to quality education and making a huge difference with other schools in the country is lack of corruption in Gecamines' schools. Mr. Tembo strongly attested that, "the word corruption is foreign in our schools and all of these elements enable us to ensure that education of high quality is provided to the society and the DRC as a country". This strong statement can be accepted to some extent since the children of non-employees which constitute a big chunk of the population in Gecamines' schools pay for quality education. However, given the general crisis the country is experiencing with poverty affecting a huge segment of the population, some of the third-party parents may have difficulties in paying school fees of their children in Gecamines' schools and these may lead to undesirable behaviour where teachers may be tempted to accept bribes.

In other words, since corruption is common in all the sectors of the society in the country due to poverty and lack of transparency as well as descent and regular salaries, Gecamines' teachers may find themselves in a weak position where they have to take bribes in order to supplement their salaries. For instance it has been reported that, government corruption, interference and the overall lack of transparency in the mining contracts resulted in the DRC government losing more than \$5 billion in revenue that could have been allotted to fund infrastructure, public health, education or foreign commerce (McFerrin, 2017). The literature demonstrated that poverty is endemic all over the country with GDP per capita estimated at \$100 in 2006, and \$350 for 2007 which is 15 per cent below that of sub-Saharan Africa (Daniele, 2011; Omoyi, 2014). Therefore, relying heavily on non-employees' fees will not help Gecamines fight corruption in its schools.

Furthermore, besides general education schools, Gecamines has vocational schools where young ladies and old women including wives of employees who want to learn any manual job go to. These schools are also open to people with disabilities in order to enable them to learn certain manual jobs and become important members of the society as well. These kinds of initiatives are crucial as they aim to empower women and the vulnerable despite financial challenges Gecamines is facing. This financial challenge does affect the provision of certain essential items such as laboratories and new technology in Gecamines' schools. This is the reason why more emphasis should be put on getting the management of the company right in order to boost its production and make funds available to equip Gecamines' schools with appropriate technology for a better service to the community and workers' children.

While the company is contributing to the education of the community it should be noted that, children of former employees who ended their contract with the company due to World Bank Voluntary Departure Programme are denied free education in Gecamines' schools. All of the former employees confirmed that since 2003, their children do not have access to free education in Gecamines' schools as if their parents never raised the flag of Gecamines high by contributing to its production. This is the case of Mr. Tango who made it clear that, "his children are staying at home since he cannot afford to pay for their studies and was concerned about their future which already looks blink due to lack of education". This finding corroborates the literature, where Kilondo (2010) made it clear that, the provision of free education to Gecamines' employees ended since the year 2000. However, this literature finding was not confirmed since several interviews with active Gecamines' employees elucidated things that, Gecamines is still providing quality free education to the children of its employees in all its three operational groups. But, the provision of this free education is specifically reserved for the children of Gecamines' employees.

HEALTH CARE

Regarding access to health care Madam Kanda from Gecamines' Medical and Social Programme Department indicated that since the creation of the company, it became crucial to have medical facilities in order to meet medical needs of the workers. Madam Kanda made it clear that "Gecamines has different health facilities in place but major ones in the city of Lubumbashi are South Clinic for skilled workers and South Hospital for unskilled workers". She also indicated that "these medical facilities provide or administer free medical services to all of its employees and their children, wives and both the husband and wife". Madam Lukovi gave a historical overview of Gecamines' hospitals by stating that, they "have been in place since 1989 with 6 big hospitals and clinics, 14 community clinics, 12 medical centres, 17 permanent medical centres and 2 medical technical institutes within the three main groups (South, Centre and West) where Gecamines is operational".

Furthermore, Madam Manga clarified that services rendered depends on the ranking of workers. For instance, "skilled workers are treated at South Clinic while unskilled workers are treated at the South Hospital". However, it was clarified by another participant by the name of Malev that "for certain specialised examination, skilled workers can get that kind of service from South Hospital where most of the equipment is installed and thereafter can continue receiving treatment at the South Clinic". The separation of medical services into two different categories does not mean that Gecamines is discriminating unskilled workers against skilled ones. But it was just a way to follow the structure or hierarchy of employees within the company to maintain discipline and respect between the workers. At the company's headquarters in Lubumbashi, it is easy to observe how hierarchical order was important for Gecamines' employees as superiors were always addressed as chief with all due respect by the subaltern.

Despite the establishment of these two hospitals, the quality of services provided to the employees was excellent in both South Clinic and South Hospital. This is exemplified by the installation of advanced equipment and machines at South Hospital instead of South Clinic where skilled employees are treated. In addition, Madam Manga continued arguing that, "the services provided in both hospitals are of high quality in the country since Gecamines has good infrastructure in place and got a good number of specialists". These comparative advantages "attract many people amongst them dignitaries and other officials who seek medical services from Gecamines' medical facilities". This information indicates that Gecamines' medical facilities play a key role in the employees of the company and the entire community within the province of Katanga as community members are not denied medical treatment even though they have to pay.

Notwithstanding the quality of medical services at Gecamines' hospitals, Mr. Banza raised a concern over lack of medicine or drugs and certain medical equipment in Gecamines' medical facilities and this is largely attributed to production and financial crises the company is going through. For instance, Gecamines' crisis is affecting the provision of certain pharmaceutical products that the company used to provide to its medical facilities throughout the three groups. Mr. Banza stated that, "Gecamines used to provide financial assistance to purchase pharmaceutical products and this is no longer coming due to poor production and difficult circumstances the company is facing". Mr. Banza continued to state that, "current contribution of the company towards pharmaceutical products and medical bills hardly reach 50% due to the crisis the company is facing and the company is unable with its current financial capacity to ensure the replacement and maintenance of certain medical equipment".

In addition, Mr. Mutombo emphasised that "Gecamines' hospitals are lacking operational budget as they are no longer provided for by the company. Some medical personnel left the company due to the crisis as their salaries were not coming or not paid on time". From these challenges one can simply argue that Gecamines' financial constraints are big obstacles in the implementation of the company's medical programme. In other words, the company still has good medical facilities in place to ensure the wellbeing of its employees and the community at large, but lack of funding caused by lower production impede a successful implementation of medico-social programme. This is the reason why Mr. Mukadi from Gecamines' Medical and Social Programme stated that, several opportunities to the workers, the company and the society at large are available if Gecamines' socio-economic and medical programme are executed accordingly. This means that, the company will benefit by having healthy personnel that will enable it to increase production as well as become competitive in the international arena.

Furthermore, the fact that family members of the personnel or employees will be taken care of by the company is a big relief for many workers. They will not have much concern regarding their family members' access to medical care and this will have a positive effect on the production side. This argument was corroborated by Mr. Mukadi who argued that "despite financial crisis, Gecamines funds its medical services to ensure that its employees are in good health and are able to do the job. Shutting down this service will be a suicide for thousands of Gecamines' employees who are not well paid". The reason why the provision of more funding is crucial for Gecamines' Medical and Social Department to flourish and do a good job by providing quality service to the company's employees. Mr. Mukadi continued raising the importance of Gecamines' medical services to the community by giving an example of thousands of Congolese protesters who were treated in Gecamines' facilities for several injuries on 19 December 2016 in the city of Lubumbashi. Given that their facilities are also providing medical services to the community like in cases of protests, Mr. Mukadi said that "Gecamines' Medical and Social department is ready to receive any donation in kind to continue helping the needy population".

Whereas the company's hospitals can administer medical treatment to needy people during protests, Mr. Mavuzi a former employee affected by the World Bank Voluntary Departure Programme stated that, "since 2003 his family members including himself do not have access to any social service from the company and it is badly affecting him". It is true that by terminating their contract with the company as part of the voluntary departure programme, these workers were no longer entitled to receiving any social service such as medical care from Gecamines' hospitals. But since they were put in a situation where they could not say no to the proposed amount due to many months of arrears, access to primary care could have been granted. Despite financial challenges that Gecamines is going through, one can strongly believe that primary medical care could have been granted to these former employees who spent many years working for the interests of the company, the province and the country.

On the question to find out the effect of the new mining code and joint ventures on Gecamines' medical facilities, Mr. Munganga from Gecamines' Medical and Social Programme stated that:

This never affected Gecamines' hospitals as the latter has its own medical facilities in place for a long period of time. However, joint ventures enabled some private companies to sign contracts with Gecamines so that their employees can be medically taken in charge through their contribution to the company. In cases where these employees need advanced medical attention abroad, the company facilitates that transfer and if the employee works for Gecamines, the company takes all medical treatment and travel expenses into consideration.

However, this option cannot be available to all workers as the company is facing a financial crisis. That is why Mr. Mutombo argued that "to avoid sending its employees abroad where treatment may cost a lot of money, the company brings in specialists from abroad to help". In other words, the company is working in collaboration with certain specialists to seek their services if need be. Despite the company's willingness to satisfy the needs of its employees in terms of access to quality education and good primary healthcare, the opinion of one of the Governors Advisers at the Provincial Department of Mines was that the company deals with the issue of lack of transparency in the management of resources to ensure the provision of socio-economic services to the workers, the community and the entire country which used to depend on Gecamines for its revenues".

CONCLUSION

This article attempted to answer main research questions regarding socio-economic conditions of Gecamines' workers in the city of Lubumbashi and the province of Katanga in the Democratic Republic of Congo (DRC). In other words, the study was of paramount importance as it has the potential of making a contribution in improving the effectiveness of Gecamines' socio-economic programme and socio-economic conditions of its employees as well as the people in Katanga Province. This study was also important in the sense that it raised awareness by looking at the living conditions of some remaining 10 000 Gecamines' employees and former employees affected by the World Bank Voluntary Departure Programme after the company went into partnerships with private firms to mine copper and cobalt. By scrutinising Gecamines and the impact of its joint ventures on the socio-economic conditions of its employees in particular and the province in general, this study attempted to extend its scope on these joint ventures' contribution to the national treasury as well. This was important to see whether Gecamines and its partners can be relevant again in the Congolese economy as it was the case in the 1960s and 1980s. It was also possible that, lessons could be learnt from the findings of the study that could be utilised by other mining companies in improving their socio-economic contributions in the interest of the employees and the entire population through generated revenues for the state.

However, it was discovered with empirical data that, . This group of former employees do not have access to Gecamines' medical and education facilities on the ground that they are no longer part of the company.

The example of joint ventures was used to demonstrate that, Gecamines generates revenues with its partners but the effect of these benefits is not felt at the lower level since individual interests predominate those of the company. All of these challenges and deplorable conditions are happening due to Political interference in contract negotiation as they were raised as the main concern leading to poor governance of the company's resources for personal interests.

RECOMMENDATIONS

From these findings, it is clear to understand that, socio-economic conditions of workers at Gecamines are deplorable and require good governance of resources from partners for the company to contribute positively to the living conditions of its employees. This good governance of resources will enable good practice over personal interests for the interests of a large group of the workers, the province and the country. For instance, proper management of the company's resources will pave a way for Gecamines to provide necessary resources to its schools and hospitals for the best interests of the employees, their dependants and to some extent the entire province. Besides the company's putting in place proper management, the government of the DRC has also a critical role to play to ensure that the country's mining sector is run accordingly to the best interest of the sector. For instance, the government should reduce or stop its interference in the management of the company by remaining as the regulator of the sector in the interest of all. In other words, public companies should be autonomous and independent from state intervention since government intervention may affect the efficiency of the company. The Government of the DRC should play a regulatory role in the mining sector to avoid cases of corruption which may have negative ramifications on socio-economic development of Gecamines employees. The case of former employees should be relooked at to rectify where errors and mistakes or allegations of corruption happened. This is crucial for these workers as they spent years working for the company and thus contributing to its glory and growth.

REFERENCES

Autesserre, S. (2012). Dangerous tales: Dominant narratives on the Congo and their unintended consequences. *Journal of African Affairs*, *111*(443), 202–222. doi:10.1093/afraf/adr080

Auty, M. R. (1993). *Sustainable development in the mineral economies: the resource curse thesis*. London: Routledge Press.

Babbie, E. (1990). *Survey Research Methods*. Wadsworth Publishing Company.

Babbie, E. (2005). *The basics of social research*. Thomson Wadsworth.

Baker, L. (2006, Summer). Observation: A Complex Research Method. *Library Trends*, *55*(1), 171–189. doi:10.1353/lib.2006.0045

Bebbington, Hinojosa, L., Bebbington, D. H., Burneo, M. L., & Warnaars, X. (2008). Contention and Ambiguity: Mining and the Possibilities of Development. *Development and Change*, *39*(6), 887–914. doi:10.1111/j.1467-7660.2008.00517.x

Bhattacherjee, A. (2012). *Social science research: Principles, methods and practices*. University of South Florida.

Bless, C., & Smith, C. H. (2000). *Fundamentals of social research methods: An African perspective*. Juta Education.

Boschini, A. D., Pettersson, J., & Roine, J. (2007). Resource Curse or Not: A Question of Appropriability. *The Scandinavian Journal of Economics*, *109*(3), 593–617. doi:10.1111/j.1467-9442.2007.00509.x

Braun, V., & Clarke, V. (2006). Using thematic analysis in psychology. *Qualitative Research in Psychology*, *3*(2), 77–101. doi:10.1191/1478088706qp063oa

Brunnschweiler, C., & Bulte, E. (2008). The resource curse revisited and revised: A tale of paradoxes and red herrings. *Journal of Environmental Economics and Management*, *55*(3), 248–264. doi:10.1016/j.jeem.2007.08.004

Bulte, E., Damania, R., & Deacon, R. T. (2005). Resource Intensity, Institutions, and Development. *World Development*, *33*(7), 1029–1044. doi:10.1016/j.worlddev.2005.04.004

Burns, N., & Grove, S. K. (2001). *The practice of nursing research: Conduct, critique & utilisation* (4th ed.). WB Saunders.

Carpenter, L. (2012). *Conflict minerals in the Congo: Blood minerals and Africa's under reported First World War*. Working paper. Suffolk University.

Coleman, D. Y. (Ed.). (2004). *Congo (DRC)*. Country Watch Review.

Congo Central Bank. (2012). Condense hebdomadaire d' informations statistiques No 03. Congo Central Bank.

D'Ydewalle, C. (1960). L'Union Minière du Haut Katanga de l'âge colonial a l'indépendance. Plon, 65-80.

Daniele, V. (2011). Natural resources and the "quality" of economic development. *The Journal of Development Studies*, *47*(4), 545–573. doi:10.1080/00220388.2010.506915

Democratic Republic of Congo. (2006). Document de Stratégie de la Croissance et de la Réduction de la Pauvreté (DSCRP) [strategic document on growth and reduction of poverty]. DRC.

Dibwe dia Mwembu, D. (2001). *Bana Shaba abandonnés par leur père: structures de l'autorité et histoire sociale de la famille ouvrière au Katanga 1910–1997*. Paris: L'Harmattan.

Division provinciale de l'intérieur et sécurité. (2015). *Cellule d'Analyses des Indicateurs de Développement*. Available at https://www.caid.cd/index.php/donnees-par-villes/ville-de-lubumbashidomaine=fiche

Donelly, J. (2000). *Realism and international relations*. Cambridge University Press. doi:10.1017/CBO9780511612510

Garrett, N. (2010). Mining as a Source of Growth. In J. Herderschee & J. E. Mans (Eds.), *Obstacles to Growth in the Democratic Republic of Congo*. World Bank.

Garrett, N., & Lintzer, M. (2010). Can Katanga's mining sector drive growth and development in the DRC? *Journal of Eastern African Studies: the Journal of the British Institute in Eastern Africa, 4*(3), 400–424. doi:10.1080/17531055.2010.517408

Hamann, R. (2003). Mining companies' role in sustainable development: Why and how of corporate social responsibility from a business perspective. *Development Southern Africa, 20*(2), 1470–3637. doi:10.1080/03768350302957

Hesselbein, G. (2007). Essor et déclin de l'État Congolais. No 21. London: Crisis States Research Centre.

Hoenke, J. (2010). New Political Topographies: Mining Companies and Indirect Discharge in Southern Katanga (DRC). *Politique Africaine (Paris, France), 120*(4), 105–127. doi:10.3917/polaf.120.0105

Kalenga, J. N. (2013). Economic and toxicological aspects of copper industry in Katanga, DR Congo. *The Japanese Journal of Veterinary Research, 61*(Supplement), S23–S32. PMID:23631149

Kilondo, N.D (2010). *Territoires produits par l'industrie minière et rupture de filiation ouvrière: dynamique populaire et territoriale dans le cas des cités ouvrières de la Gécamines (RDC)* [Territories produced by mining industries and failure of labour filiation: popular dynamic and territorial in the case of Gecamines' labour cities]. Universite Catholique de Louvain.

Kuediasala, F. (2009). *Mission au Katanga d'une équipe d'experts de la Banque mondiale avec a l'agenda la relance de la Gecamines*. Le Potentiel.

La Générale des Carrières et des Mines (GECAMINES). (1967). *Annual Report of Gécamines*. Lubumbashi: Gecamines.

La Générale des Carrières et des Mines (GECAMINES). (1974). *Annual Report of Gécamine*. Lubumbashi: Gécamines.

La Générale des Carrières et des Mines (GECAMINES). (2015a). Available from: http://gecamines.cd/html/histoire.html

La Générale des Carrières et des Mines (GECAMINES). (2015b). Available from: http://gecamines.cd/html/social.html

MacNealy, M. S. (1999). *Strategies for Empirical Research in Writing*. Longman.

Marysse, S., & Andre, C. (2001). Guerre et pillage economique en RDC [War and looting in the DRC]. Paris: L'harmattan.

Marysse, S., & Geenen, S. (2009). *Win-win or unequal exchange? The case of the Sino-Congolese cooperation agreements*. Cambridge University Press. doi:10.1017/S0022278X09003978

McFerrin, H. (2017). *News from Africa: The DRC poverty rate continues to stagnate*. Available at https://borgenproject.org/the-drc-poverty-rate/

McFerson, H. M. (2010). Extractive industries and African democracy: Can the "resource curse" be exorcised? *International Studies Perspectives*, *11*(4), 335–353. doi:10.1111/j.1528-3585.2010.00410.x

Mehlum, H., Moene, K., & Torvik, R. (2006). Cursed by resources or institutions? *World Economy*, *29*(8), 1117–1131. doi:10.1111/j.1467-9701.2006.00808.x

Montague, D. (2002). Stolen goods: Coltan and conflict in the Democratic Republic of Congo. *SAIS Review (Paul H. Nitze School of Advanced International Studies)*, *22*(1), 103–118. doi:10.1353ais.2002.0016

Murshed, M. S. (2004). When does natural resource abundance lead to a resource curse? *Environmental Economics Programme*, *4*(1), 1–58.

Neuman, W. L. (2014). *Social research methods: Qualitative and quantitative approaches*. Pearson Education Limited.

Nsakila, M. G. (2008). *The Chinese Presence in Lubumbashi*. Centre for Chinese Studies.

Okubo, T. (2007). *Trade Liberalisation and Agglomeration with Firm Heterogeneity-Forward and Backward Linkages*. University of Manchester.

Olsson, O., & Congdon, F. H. (2004). Congo: The Prize of Predation. *Journal of Peace Research*, *3*(41), 321–336. doi:10.1177/0022343304043772

Omoyi, K. (2014). *Ressources minieres, croissance economique et guerres civiles en RD.Congo* [Mineral resources, economic growth and civil wars in the DRC] (Master's Dissertation). University of Yaounde II.

Robinson, A., Torvik, R., & Verdier, T. (2006). Political foundations of the resource curse. *Journal of Development Economics*, *79*(2), 447–468. doi:10.1016/j.jdeveco.2006.01.008

Ross, M. L. (1999). The political economy of the resource curse. *World Politics*, *51*(2), 297–322. doi:10.1017/S0043887100008200

Ross, M. L. (2014). What have we learned about the resource curse? *The American Political Science Review*, *102*(1), 107–123. doi:10.1017/S0003055408080040

Rubbers, B. (2009). The story of a tragedy: How people in haut-Katanga interpret the post-colonial history of Congo. *The Journal of Modern African Studies*, *47*(2), 267–289. doi:10.1017/S0022278X09003838

Rubbers, B. (2010). *Claiming workers' rights in the Democratic Republic of Congo:* The case of the collectif des ex-agents de la Gécamines. *Review of African Political Economy*, *37*(125), 329–344. doi :10.1080/03056244.2010.510629

Sachs, J., & Warner, A. (1995). *Natural resource abundance and economic growth*. National Bureau of Economic Research. doi:10.3386/w5398

Stiling, J. A. (2001). *Thematic networks: An analytic tool for qualitative research.* Sage Publications.

Strauss, A. L., & Corbin, J. M. (1998). *Basics of qualitative research: Techniques and procedures for developing grounded theory*. Sage Publications.

Union Minière du Haut Katanga. (1957). Production from 1906-1956. *Cuyprius*, 39-40.

University of Lubumbashi. (2005). *The restructuring of Gecamines, impacts on poverty and social issues.* The future of social services and reintegration strategies, final report of the University of Lubumbashi.

van der Ploeg, F. (2010). *Natural resources: Curse or blessing?* CESifo working paper Resources and Environment, No. 3125.

Weber, M. (1973). *The protestant ethic and the spirit of capitalism.* Routledge.

Weber-Fahr, M. (2002). *Treasure or trouble? Mining in developing countries.* The World Bank and International Finance Corporation.

World Bank. (2012). *World Development Report.* World Bank.

World Bank. (2018). *Democratic Republic of the Congo Overview.* Available at https://www.worldbank.org/en/country/drc/overview

Chapter 17

Mining and Water Supply in Marginalised Communities:
The Case of Anglo-Platinum Mogalakwena and Ga-Molekane Village, Limpopo Province, South Africa

Tariro Kamuti
https://orcid.org/0000-0002-4769-9037
Center for Gender and Africa Studies, University of the Free State, South Africa

ABSTRACT

The extractive industry and water security nexus for vulnerable communities that surround a mine has been a subject of debate in the Global South. The study aimed to identify and explain the impact of mining on the ability of vulnerable and marginalized communities to secure adequate and safe water. The study focused on Anglo-Platinum Mogalakwena Platinum on Ga-Molekana Village in Mogalakwena Municipality, Limpopo Province of South Africa. Key informant interviews were administered with community leaders, local community-based organizations, local councilors, and public officials at the national level. Focus group discussions were also conducted for older women and youth to share current experiences. The Mogalakwena Municipality provides water to the Ga-Molekane Community and the Mogalakwena Platinum Mine also intervenes by providing support in this regard. Perceptions about the adequacy of water to meet their household needs were varied, with some being positive considering their situation to be better than others, while some said water had become inadequate.

INTRODUCTION

The nexus between the extractive industry and the welfare of vulnerable communities that surround a mine has been a subject of debate especially in the global South (Conde and Le Billon, 2017; South African Human Rights Commission, 2008; 2016; 2018). There is a growing voice particularly demanding for socio-economic justice and thus questioning the impact that extractive industries have on surrounding

DOI: 10.4018/978-1-7998-8809-3.ch017

communities. Mines develop where a viable mineral resource is found, and this may be in the periphery in terms of infrastructural development which may contribute to their marginality if they affect surrounding communities. This is often what happens with the extractive industry where economic benefits derived from the exploitation of the mineral resources are moved elsewhere leaving the surrounding communities in a less advantaged position (Ye, et al., 2019). Ironically, the same contiguous communities are supposed to benefit from the proceeds of the minerals being exploited from their land. One of those benefits relates to socio-economic development like the provision of basic needs like the supply of adequate and safe water as a basic human right (Mehta, 2014) and this is the context through which this study was formulated. This is palpable, especially in cases where the communities surrounding these mines in South Africa raised issues that show entrenched inequality (South African Human Rights Commission, 2008). Yet, the transformation that is brought by the extractive industry, such as mining, may not be the one that has been envisaged by the affected communities (Leifsen, 2020).

The study aimed to identify and explain the impact of mining (as an extractive industry) on the capacity of vulnerable and marginalized communities to secure adequate and safe water. The case of Anglo-Platinum Mogalakwena Platinum on Ga-Molekana Village in Mogalakwena Municipality, Limpopo Province of South Africa was chosen for this study. The next section covers the background to mining and water supply followed by the theoretical framework that was used to contextualise the impact of mining on the supply of water to the surrounding communities. Next is the methods of data collection. This will be followed by a treatise of the link between mining establishments and surrounding communities and then a look at the case of Anglo-Platinum Mogalakwena Platinum and Ga-Molekane Village concerning the issues of water supply. The paper goes on to discuss the issues raised in the study followed lastly by concluding remarks.

BACKGROUND TO MINING AND WATER SUPPLY

In 2015, it was estimated that about 2.1 billion people around the world did not have good water provision facilities, while 844 million did not have access to basic water facilities at all with most of the challenges found in the rural areas (Koehler et al., 2018) and 337 million people are found in the rural areas of sub-Saharan Africa (Carter, 2021). In the same year, the United Nations said that around 1 million people from Africa were dying per year due to diseases connected to sanitation, hygiene and water (Fagan et al., 2015). Worldwide, about 1500 children below the age of five are said to be dying every day due to diarrhea-related diseases mostly connected to impoverished conditions around water and sanitation (Carter, 2021). The challenges of provision of adequate, just, clean, accessible water have been seen to do with the institutional arrangements governing the supply of water coupled with a diversity of end-user views associated with the risk of the services (South African Human Rights Commission, 2008; Koehler et al., 2018; Carter, 2021).

Water is a basic human right which is enshrined in Section 27 of the Constitution of the Republic of South Africa Act, No. 108 of 1996 (Republic of South Africa, 1996) and therefore its supply to communities must take care of this important requirement. Mining activities do have an impact and they are also influenced by issues related to water. For example, mining operations themselves require huge amounts of water, so the establishment of a mine requires making provisions for a reliable source of water for these purposes. People settling around and/or working at the mine and those related to the mining value chain, for example, those supplying various goods and services are all going to need safe and adequate water for their households needs. The issue of communities surrounding the mine itself needs to be catered for as well when it comes to their need for safe and adequate water. Some of these communities would have been displaced from their land to pave way for the mine or they will simply be adjacent to the mine, and they are in one way or the other affected by the mining operations. Such communities were the focus of this study. Mining operations also discharge effluent that is harmful to the local environment including pollution of water sources around the mine (Conde and Le Billon, 2017).

Section 24 of the Constitution of the Republic of South Africa Act, No. 108 of 1996 guarantees the right to "to an environment that is not harmful to their health or well-being" and provides for safeguarding the environment through the institution of reasonable legislation (Republic of South Africa, 1996). For example, the National Water Act, Act No 36 of 1998 (Republic of South Africa, 1998) paved way for changes in the regulations dealing with water given that it is a scarce resource (Republic of South Africa, 1998). Section 24 also calls for the prevention of pollution and sustainable utilization of natural resources while allowing socio-economic upliftment (Republic of South Africa, 1996). These provisions are relevant to the mining and water sectors as various minerals and water are all natural resources that need to be well managed for the benefit of current and future generations. South Africa is also a signatory to various international treaties, charters, protocols and agreements that cover the mining and water sectors (South African Human Rights Commission, 2018).

THEORETICAL FRAMEWORK

The case was analysed by adapting some criteria from the seminal work on Tanahashi's (1978) concept of coverage of health service and its assessment. The idea of coverage which is defined as the ratio of the recipient population who can take or have taken use of the service (Tanahashi, 1978) is important as it is at the core of the provision of water as a basic service. Tanahashi's (1978) framework includes criteria of availability, accessibility, acceptability and contact coverage of the service in the assessment of the provision of water at the local level. These criteria were also identified to act as phases through which the service can finally reach the end-user. Availability coverage focuses on the resources (which often act as constraints) required as they determine the extent to which a service can be potentially provided to the intended beneficiaries, thus affecting its availability. So, availability coverage is defined as the proportion between the extent to which a service can be potentially provided and the number of the intended beneficiaries (Tanahashi, 1978).

Once the service is made available, it should be within proximity to the intended beneficiaries to use the service. This speaks to the idea of how the service will be delivered to the end-user such that the level of the service is constrained by the number of people who can access and utilize it and this is called accessibility coverage (Tanahashi, 1978). The target population needs to accept the service once they start to access it. The extent to which a service can be provided is constrained by the number of people who intend to utilize the service that is offered, which ultimately defines acceptability coverage (Tanahashi, 1978). Lastly, the provision of service happens when there is an interaction between the service provider and the target population. Therefore, the "number of people who have contacted the service is a measurement of service output; the ratio between this and the size of the target population gives a measurement of coverage that may be called contact coverage" (Tanahashi, 1978: 297). Contact coverage in this case links very much with the idea of stakeholder relations referred to earlier amongst the key players brought together by the quest to provide water to the surrounding communities.

The extent to which people are satisfied with the service when it has been provided is also important as part of the assessment of the coverage of the service against set benchmarks. The number of people who get a service that meets their needs constitutes effective coverage (Tanahashi, 1978). Those people will be happy with the service. In our case, the provision of water should be satisfactory to the consumers, who are the residents of the areas surrounding a mining establishment. Effective coverage becomes the measure of quality and effectiveness of the service (McCollum et al., 2019). This study extended the Tanahashi framework to assess the provision of water to communities because the availability of adequate, safe and clean drinking water together with sanitation contribute towards good health and delivery of health services to any community. The use of the Tanahashi framework is widespread, especially when looking at the provision of health services in marginalized communities. This extension of the framework is relevant given that it has its application on health service must be seen in conjunction with other socio-economic sectors, which in this is water (Christian et al., 1977). The seminal model, having been modified in some cases, has been used in studies of coverage of health services in countries such as India, Kenya, Uganda, Chad, Ghana and Nigeria, which shows its prevalent use (Sharma et al., 2015; Henriksson et al., 2017; Ebhohon, 2018; Lechthaler et al., 2018; McCollum et al., 2019; Sheff et al., 2020).

DATA COLLECTION METHODS

The study sought to answer questions related to the links between the mine and the state in the provision of water to the surrounding community. For instance, there was a need to look at the regulatory links in terms of environmental policies and legislation from the state (at national and local levels) and the mine in the provision of water. This would help to figure out to what extent the mine in conjunction with the state has contributed to the provision of water to the surrounding community that is at the local level. This view is influenced by the idea that there is a disjuncture between where policymaking mainly at global and national levels takes place and the local level, especially at the grassroots where consumption takes place (Koehler et al., 2018). The study also sought to find out the perceptions of the community members on their assessment of the efforts of the mine in providing them with water. This would also be part of the need to establish how the stakeholders are interacting in the interests of providing water to the community.

Initial desktop study on the country's mining and environmental policies and legislation was conducted followed by empirical research of the mining company and the chosen nearby community of Ga-Molekane Village. For example, the South African Human Rights Commission did several studies related to regulations of mining activities, relations between mining operations and communities, including a specific study on the impact of Anglo-American mining company's operations in Limpopo Province in South Africa (South African Human Rights Commission, 2008; 2016; 2018). In this respect, the study adopted an institutional approach by focusing on the governance structures that impinge on the provision of water to communities. Key informant interviews were administered with company officials, community members and leaders from the nearby Ga-Molekane Village, community-based organizations, government officials at local and national levels. Two focus group discussions were also conducted for women and youth to share their experiences. This is because in Africa, there is a prevalent gendered allocation of the role of fetching water to women who mostly bear the burden of securing water for their households, and they often delegate this responsibility to children (Fagan et al., 2015). Apart from the focus group discussions and semi-structured interviews conducted, water test data were collected, and observations were also used to gather data. The study sought to show interactions, and power relations of stakeholders and key actors brought together by the links between the extractive industries and surrounding communities with regards to the provision of safe and adequate water for household use.

This study in South Africa was part of a bigger commissioned research project on mining and supply of water in surrounding communities covering five Southern African Development Community (SADC) countries including Malawi, Mozambique, Zambia and Zimbabwe (Kamuti and Serumaga-Zake, 2017a; 2017b). So, for this South African study, ethical clearance was sought and granted, before fieldwork, by the Ethics Committee at the University of the Free State in 2017. There is a study that impressed that ways of gauging the constraints to access healthcare mainly depend upon probing interviewees in a survey to ascertain if at any moment they required medical attention but were unable to get it or if they let go of such chances and finding out what were the hindrances (Houghton et al., 2020). In the same vein, this study opted to incorporate an empirical element to get first-hand information from the respondents who are the recipients of water provision from the local municipality in conjunction with the mining company. This study had to be evidence-based in seeking a clear picture of the impact of the chosen mine on the supply of water to the affected surrounding communities. Triangulation was then used to increase the validity and reliability of the findings of this study through incorporating data from the interviews with stakeholders, observations made on the ground, focus group discussions and the secondary information emanating from the earlier desktop study.

MINING AND WATER SUPPLY TO SURROUNDING COMMUNITIES

There is a variation in the ability of communities to access water at a worldwide scale (Mehta, 2014). In this regard, a point is made that:

Poor and marginalized people often lack access to safe and adequate water, either because it is too expensive or because they are excluded due to caste, ethnicity, or gender. They also often do not share the gains of large infrastructure projects such as dams, while often negatively affected by them through displacement as well as loss of livelihoods and land. (Mehta, 2014: 59)

Mining developments, need bulk water supply infrastructure to satisfy their water requirements, and this can lead to the displacement of people to create space for such developments. This is because there is a close relationship between mining processes and water (Marston, 2017). The South African Human Rights Commission (2016) notes the existence of what they call "mine-affected communities", that is those communities whose welfare is directly impacted by mining developments. They also point out that since 1994 the South African government has put in place a robust regulatory regime "enable its mining industry to operate in a manner that protects and promotes the well-being and safety of communities affected by its operations" (South African Human Rights Commission, 2016: 1). However, it was noted that:

...many mining-affected communities continue to experience significant levels of poverty and systemic inequality, which reinforces the notion that the benefits of mining operations disproportionately favour mining companies and the State and are often to the detriment of local communities. (South African Human Rights Commission, 2016: 1)

This brings in the question of governance of natural resources especially in a manner that benefits a population that is directly affected by the quest for their sustainable utilisation. In the South African case, these governance issues place primary responsibility on the Department of Water and Sanitation (DWS) in conjunction with local government and the mining companies when it comes to the extractive industry and supply of safe drinking water.

Mining companies do have voluntary programmes about taking care of the environment and communities in areas where they operate but they are not adequate (Perez and Sanchez, 2009). There is a gap between what mining companies should be doing and what they exactly do in areas where they operate. This can be a trigger some resentment of mining corporations in the sense as aptly described that:

The environmental effects of extraction, particularly its impact on water used by surrounding communities for drinking, livestock, and irrigation, sparked widespread antimining mobilization and enabled a close articulation of environmental activism and 'local community,' which was frequently linked to indigenous territorial struggles. (Marston, 2017: 1)

The study, therefore, attempted to contribute to these debates in the context of the provision of water at a local level with regards to Anglo-Platinum Mogalakwena Platinum mine operations.

ANGLO-PLATINUM MOGALAKWENA PLATINUM MINE AND GA-MOLEKANE VILLAGE

Mogalakwena Platinum Mine which started operating in 1993 is the world's largest open-pit platinum mine and it is Anglo American's flagship operation which is also unique as an open-pit mine at its current levels of production when compared to traditional platinum mines which are underground. Now the mine works range at depths of 45 and 245 meters and produced a record 412 000 ounces of platinum in 2016. The mine's use of highly mechanized structures makes it the world's lowest-cost platinum producing mine.

Availability Coverage

For their household needs such as drinking, cooking and washing the community relies mainly on taps that draw water from tanks installed within the mining town. The Mogalakwena Municipality pumps water into the command reservoir which supplies village reservoirs. The village reservoirs are cleaned once a year to ensure the water quality remains at a good level. The community also fetches water from boreholes as an alternative source. To address water shortages the municipality is implementing work to avail bulk water from the Flag Boshielo Dam to cover all the villages in the municipality. Perceptions about the adequacy of water to meet their household needs were varied, with some being positive considering their situation to be better than others, while some said water had become inadequate because of the drought.

Accessibility and Contact Coverage of Service

The village head considered themselves to be lucky because there is adequate infrastructure for water supply. According to the local government official, the infrastructure that delivers water to the villages ensures safety. The water goes from the main reservoir into the village reservoir via reticulation to stand taps. The Mogalakwena Municipality under which Ga-Molokane Village is found has different ways through which they interact with the mine like holding meetings with local mine management. Existing mechanisms that can be used for community members to liaise with your department occur through ward councilors. Meetings or engagements with community members occur twice a year during (Integrated Development Planning) IDP sessions or when there are burning issues for discussion. This is either with community members or a leader from the community. These engagements enable the community members to contact the relevant stakeholders on issues relating to water supply.

Acceptability Coverage of Service

Table 1. Samples of Water Quality for Water Supplied to Ga-Molekane Village

Sample Number	Sample Point	pH pH units @ 25°C	Turbidity NTU units	Conductivity mS/m @ 25°C	*TDS mg/l
		Method: CP001	**Method: CP002**	**Method: CP003**	**Method: CP004**
MLM 1	925Mahwelereng Zone C (Emmanuel Church)	7,13	0,32	57,10	376,86
MLM2	2871Mahwelereng Zone B	7,11	0,19	58,50	386,10
MLM3	Ext 19 RDP(7597 Moss Makgai)	7,36	0,83	20,80	137,28
MLM4	79 Bezuidenhout street	7,09	1,41	71,50	471,90
MLM5	819 Extension 17	7,61	0,74	9,98	65,87
MLM6	701 Zone 1 Moruleng str	7,22	0,35	57,60	380,16
MLM7	Mopani street (Unit D-Public Works)	7,67	0,78	10,39	68,57
MLM8	M21Extension 14 (next to Electrical substation)	7,25	0,48	70,00	462,00
MLM9	Ext2 (1 Second Avenue)	7,86	0,66	10,13	66,86
MLM10	Ext 12(25 Mogol street)	7,74	0,57	72,10	475,86
MLM11	55 Plantina Street (Chroompark)	7,81	0,71	52,90	349,14
MLM12	11 Geyser	7,73	0,42	71,70	473,22

Source: Scientific Services Unit of Lepelle Central Laboratory as of 9 June 2017.

Respondents of the facilitated group discussions indicated that the state of the water is satisfactory. One respondent said that although the water is clean, the roof on the tank fell off after Cyclone Dineo (that struck in February 2017), and this would cause them to believe that water might be dirty. The respondent was not sure about the other water sources. The village and local government officials interviewed also agreed that the water quality is clean and safe as it is chlorinated water that the community has access to. There has not been an issue concerning some illnesses in the community connected to the state of their water. A civil society leader who was interviewed responded that various studies have been conducted around conflict with the mine, land, relocation issues and now water. Their current study focusing on water issues was still underway. The study was looking at the high levels of nitrates in the water and how this has impacted the health of the community members particularly focused on the link between high levels of miscarriage and infant deaths, and the levels of nitrates in the water. There is regular testing of the quality of the community's water monthly by an accredited laboratory. Table 1 shows the results of water testing that was done during the month when this study was conducted, and they showed an acceptable mix of the parameters of pH, turbidity, conductivity and palatability across the sampling points.

Effective Coverage

However, in terms of water quantity, the village head felt that the water they get from the tank is adequate, though in other areas there are water shortages. Despite this positive outlook, however, there is a view that the mine impacted negatively on the water situation around the community. Some respondents from the facilitated group discussions cited several reasons. One respondent said, "it is my understanding that the issues with the borehole water at the local school were because of the mine" while one indicated that the fighting with the water delivery people is bad. Another respondent said that the water where they do fishing is bad; that it has a thin white layer, therefore thought that it is from the mine. One respondent said, "the dust on open water sources, I am scared now since the roof came off the tank that the water may be unsafe" whilst another one said that, "they [the mine] contaminated the borehole water before, I do not trust them." The village head also raised the issue that they were aware that when the Human Rights Commission came in at some time, the water from boreholes was contaminated and it was said to be the mines. So, he was not sure if this is still the case as it was some years back. These concerns have been recently raised by the South African Human Rights Commission (2016: 40) when they reiterated that, water user licenses issued to mining companies should "incorporate more stringent measures to better protect Communities' water rights and the environment." The village head also went on to say that the water body in the area also seems polluted, though he could not say for sure if this is still the case. He commented that the mine is trying to be helpful by delivering water to tanks has caused rifts in the community as only a select few are chosen for these contracts. The local government official said that the mine pays a few locals to deliver water into some of the villages via tanks and pays them individually however because the mine does not have its water source, these individuals steal municipality water and deliver that to the communities.

The civil society leader also agreed that mining activity had an impact on the community's water supply. Mining activity is attributed to having lowered the water table (this is common with open cast mining as they require drier areas to mine). As a result, the community does not have access to enough water for their household needs. Due to a lowered water table, the rivers are drier for longer periods and boreholes and so wells run dry faster. The water sources are hugely stretched especially since it is a generally dry area and with mining, the water table has dropped even lower. Mining activities also affect the quality of other sources such as rivers and boreholes. They also believe mining activity has increased the nitrates in the water and at the time of the research they hinted that they will be testing that assertion in the next few months.

THE NEED FOR EFFECTIVE STAKEHOLDER ENGAGEMENT

Issues of engagement among stakeholders such as the mine, communities surrounding the mine and local government are critical when it comes to the provision of adequate and safe water. This usually hinges on how amenable the mine and the local authority are in ensuring the demand for adequate delivery of water to the communities. For instance, when it comes to engagement between the mine and the Ga-Molekane community, the village head was skeptical. He cited that there is no participation of his community in any key decisions related to mining activities in the area. They are not able to provide direct feedback to the mine as "the mine just does its own thing, we would like to have a good relationship with them though if they would allow." The mine is seen in a different light as they are considered not to be receptive to feedback and that they have not addressed the community about matters concerning clean drinking water, sanitation and hygiene as they only come in to speak on general delivery services. These same views were also echoed by the respondents from the facilitated group discussions with the general perception it appears that the mine would not be receptive anyway, and when the community want to be heard they resort to protest.

On the other hand, the village head indicated that they could provide direct feedback to the local government and local councilors are on hand if there are grievances. However, he commented that the local government is not receptive to their feedback as no action is taken for instance; service delivery is bad in the area. On the ability to provide feedback to local government the facilitated group discussants agreed that yes, they can communicate to local government through the village leader and sometimes the ward councilors. However, local government is not receptive to feedback as no action is taken. There are also meetings or engagements by the local government with community members conducted twice a year during the IDP sessions or when there are burning issues for discussion.

The local government interacts with the mine by holding meetings with local mine management. However, they do not actively engage with the mine on issues of surface water conservation and management. The local government has the idea that policy or legislation relating to water conservation and management is not implemented through the official indicated that as far as he was aware, there is no discharge from the mine, the water is recycled. The municipality is not involved in the monitoring of the recycling process. The civil society leader interviewed also indicated that there is very little engagement between local government and the community. He argued that there is an imbalance of power, whereby communities do not have access to relevant expertise to ask the right questions on their behalf. In terms of community engagement, the company said, "we work with local government, leaders and NGOs to determine how best to contribute to the needs of the local community. Our efforts are fully aligned with the aspirations of the host community to fulfil our social and labor plans."

LOCAL AND HIGH-LEVEL GOVERNANCE ISSUES

The village head said that the mine had impacted negatively on the lives of his community members citing noise, dust, loss of their fields, and unemployment which is still rife. He thought the mine would employ the locals and hinted that some houses crack from the blasts (since the mine is very close to Ga-Molekane). Focus group discussants also shared the same views citing that there is more noise and dust, destruction of the area and taking over of their farming areas, increased suffering because there is no employment, cracks on house walls because of the blasting, fights due to water contracts and general lack of development that is beneficial to the community. Observations by the researcher when she visited the area were that with regards to this community – there are high levels of poverty and unemployment is a big issue. There seems to be anger directed both at the government (for what they feel is a lack of service delivery) and the mining company (for a failure to provide jobs and development). In this regard, there is a call by the respondents for the government and the mine to do more for them, as people have no jobs and service delivery is slacking and poor.

In terms of how effective the government is in ensuring the implementation of legislation the respondent from the civil society organization said that the legislation is clear however, there is wiggle room afforded to the mines by allowing them to have exemptions. For example, as a mining company, they can apply to be exempt from water or air pollution legislation, and this happens to the detriment of communities. Thus, even if punitive measures do exist, they are however difficult to enforce because of the exemptions. The local government seems to be weak as they just rely on whatever they are told by the mine. For example, when asked about the municipality having defined allowable limits for the discharge of mine effluent, they indicated that as far as they are aware, there is no discharge from the mine, since the water is recycled. However, the municipality is not involved in the monitoring of the recycling process.

The different ways of interaction of the national government (Department of Water and Sanitation) with the mining industry include providing authorization for water use and licenses and also hosting forums with the mines in which they discuss any issues. They actively engage with mines on issues of surface water conservation and management. There is a directorate in the Department of Water and Sanitation which handles these issues. They are also in the process of setting up new policies on surface water conservation, and these are still under review. The policies on water management and conservation that apply to the mining industry include the Mine Water Management Policy – which is however currently up for debate. They also have GN704 which regulates the use of water for mining and related activities. They are also trying to set up an environmental levy for operational mines so that in future the state will not have to deal with the damage caused by mines. The respondent believes that national legislation regarding water management and conservation meant for the extractive industry is very clear. Policy or legislation relating to water conservation and management is implemented through inspections and compliance requirements. To punitive measures put in place in case of contravention, the Department is busy with a 'Waste Discharge Charge System' so that the mines will be charged for their waste, and they hope this will be a good mitigation measure. There will be incentives for those who perform well. There are also fines laid out in the license conditions. Monitoring of the implementation of these policies or legislation is done through the compliance enforcement team as well as the Catchment Management Agency (CMA). The compliance teams go in to make sure all is above board on an ad hoc basis. He is of the view that drinking water should be tested daily. The mine ought to test the quality of the water. Their department also does however come in to check on the quality to ensure that standards are met. He is not happy with the situation; it is a water-scarce area and there have previously been issues with the safety of some of the water bodies.

DISCUSSION

Section 24 of the Constitution of South Africa is very specific on the "the right to an environment that is not harmful to health and wellbeing" (Republic of South Africa, 1996; South African Human Rights

Commission, 2016; 2018) and this is in line with international practice (Perez and Sanchez, 2009). The National Environmental Management Act, 1998 (Act No. 107 of 1998) helps in the integration and harmonization of all regulations that relate to the environment and hence it touches on mining operations. South Africa's mining sector uses "about six per cent of the country's allocated water supplies" yet it is a water-scarce country (Tempelhoff et al., 2014). The South African National Water Act, 1998 (No. 36 of 1998) (NWA) suggests effective management of the country's most precious resource in a sustainable way for the benefit of all South Africans. What is particularly encouraging about this legislation is its focus on those who are currently disadvantaged due to former apartheid policies. Positive aspects of the NWA are its four core principles, namely: decentralization, equitable access, efficiency, and sustainability. However, even though the fundamental principles of South Africa's NWA make it an impressive piece of environmental legislation by international standards, many serious challenges to implementation remain, such as lack of human resources, the disconnect between resources management and supply, problems with stakeholder participation, society's interests are not well articulated. It was found that the Mogalakwena Municipality provides water to the Ga-Molekane community and the Mogalakwena platinum mine also intervenes by providing support in this regard. Perceptions about the adequacy of water to meet their household needs were varied, with some being positive considering their situation to be better than others, while some said water had become inadequate because of the drought.

Respondents of the facilitated group discussions indicated that the state of the water is satisfactory. The civil society leader indicated that various studies have been conducted around conflict with the mine, land, relocation issues and now water. Their study focusing on water issues at the time of the interviews in 2017 was still underway. The study was looking at the high levels of nitrates in the water and how this has impacted the health of the community members particularly focused on the link between high levels of miscarriage and infant deaths, and the levels of nitrates in the water. Despite this positive outlook, however, there is a view that the mine impacted negatively on the water situation around the community, for example, respondents cited that there is dust on open water sources, the quality of the water where they fish is bad and that it has a thin white layer. Pollution by mining companies and the need for clean production have driven governments like in the United States of America and Canada to enact stringent environmental laws for mining corporations but it has been noted that in developing countries there are inadequate regulatory arrangements (Nikolaou and Evangelinos, 2010). Issues of engagement among stakeholders such as the mine, communities surrounding the mine and local government are critical when it comes to the provision of adequate and safe water. This usually hinges on how amenable the mine and the local authority are in ensuring the demand for adequate delivery of water to the communities. For instance, when it comes to engagement between the mine and the Ga-Molekane community, the village head was skeptical. In this regard, there is a suggestion to adopt best practices that facilitate enhanced engagement (Lockie et al., 2012) between affected communities, and mining companies, the state and other stakeholders (Kamuti and Serumaga-Zake, 2017a, 2017b). This calls for the mines to neither just use corporate social responsibility (Hamann, 2004; Hamann and Kapelus, 2004; Jenkins and Yakovleva, 2006) nor the social license to operate (Bice, 2014; Nikolaou and Evangelinos, 2010; Conde and Le Billon, 2017), but go beyond and deeply involve themselves in ameliorating the people's living conditions. There should be further integration of traditional structures and institutions, (often regarded as informal and hence peripheral) in the participation and decision-making processes relating to water supply for the communities (Porras, et al., 2020).

The Tanahashi model has been used to study the provision of community health services (that is, at a local level) to attain universal health care in Kenya (McCollum et al., 2019). The provision of community health services was under the auspices of devolution to enhance decision-making at the local level to address issues of social justice and equity among marginalized sections of the country's population. The Tanahashi model enabled the analysis of the changes in local health service provision that came with the implementation of devolution in Kenya as well as identifying the challenges which posed hurdles against quality delivery of health services (McCollum et al., 2019). The Tanahashi model is also flexible and in Uganda, it was tweaked by splitting availability coverage into the availability of medical personnel and availability of medical supplies in the provision of effective healthcare (Henriksson et al., 2017). In that study, the model also enabled the identification of hurdles but its use was noted to be dependent upon the kind of information availed to input the different criteria of assessment (Henriksson et al., 2017). In Chad, the use of the model was done by modifying it again to include affordability and adequacy coverage in the study of the provision of antenatal care in rural areas (Lechthaler et al, 2018). The model, modified by seeing acceptability coverage as perpetual use of the service, was used in the study of the provision of basic healthcare in some districts of the Volta Region of Ghana (Sheff et al., 2020) and other parts of Africa (Mehl and Labrique 2014). The model was altered to what has been termed the Bottleneck Analysis Instrument and used in Nigeria to assess the challenges in the country's basic medical care arrangements (Ebhohon, 2018). These are all examples of the widespread use of the age-old Tanahashi model in contemporary issues of health. This chapter covering the Ga-Molekane community, therefore, shows that despite its widespread use in the health sector, the Tanahashi framework can also be adapted to the assessment of the provision of water, more so in communities that are affected by mining operations within their vicinity. The use of the Tanahashi model is also relevant because the subject of focus, that is water in this case, is an integral part of the environmental health of a community due to its link to sanitation, the occurrence of water-borne diseases, and the possibility of pollution that may all be detrimental to people's health and quality of life when they are not handled well.

CONCLUSION

The South African state has comprehensive regulatory mechanisms which provide for environmentally friendly mining practices which apply from the national to the local level. However, there are some challenges in the implementation of these regulations especially when it comes to the supply of adequate and safe water to local communities that are affected by mining operations. It has been seen that Anglo-Platinum Mogalakwena Platinum in conjunction with Mogalakwena Municipality assists in the provision of water to the nearby Ga-Molekana Village. However, this supply of water has not been to the ultimate satisfaction of the community members as several concerns of irregular provision, pollution and inadequate contact coverage have been raised. Irregular supply cannot be only attributed to the mine as the responsibility of water provision is part of the mandate of the local municipality as part of its service delivery. Nevertheless, the mine is expected to play a critical role in ensuring the provision of water as a basic human right, given the huge profits which they derive from their operations situated in an area where people were displaced to pave way for their operations.

This study feeds into the broad questions about the distribution of mining benefits and voices in this regard are becoming more louder than before as the country is struggling with high levels of unemployment, poverty and inequality. This question of distribution of benefits is even more imperative to a section of the South African population who are often displaced to pave way for mining development and those who find themselves settled next to or surrounding a mining establishment. South Africa's rich mining history provides a good basis for improvements in mining operations to get best practices that are environmentally friendly and socially just. However, one of the major challenges of this legacy of the mining sector is the pollution of water, land and air which have long term effects on the surrounding communities and further afield (Porras et al., 2020). This leaves the surrounding communities even more vulnerable than before. Hence, there has been a deliberate effort on the part of the South African state to develop a proactive policy and legislative framework to protect the environment. These debates can be extended to the need to re-examine the business or operational models to include surrounding communities to be critical stakeholders whose welfare should not be just a matter of corporate social responsibility or a formality of ticking boxes on checklists concerning satisfying regulatory obligations.

ACKNOWLEDGMENT

This major study, from where the material in this chapter was derived, was commissioned by the WaterAid Southern Africa regional office and funded by the Open Society Initiative for Southern Africa (OSISA). Many thanks to the editors and the anonymous reviewers.

REFERENCES

Bice, S. (2014). What gives you a social licence? An exploration of the social licence to operate in the Australian mining industry. *Resources*, *3*(1), 62–80. doi:10.3390/resources3010062

Hamann, R. (2004). Corporate social responsibility, partnerships, and institutional change: The case of mining companies in South Africa. *Natural Resources Forum*, *28*(4), 278–290. doi:10.1111/j.1477-8947.2004.00101.x

Hamann, R., & Kapelus, P. (2004). Corporate Social Responsibility in Mining in Southern Africa: Fair accountability or just greenwash? *Development*, *47*(3), 85–92. doi:10.1057/palgrave.development.1100056

Jenkins, H., & Yakovleva, N. (2006). Corporate social responsibility in the mining industry: Exploring trends in social and environmental disclosure. *Journal of Cleaner Production*, *14*(3-4), 271–284. doi:10.1016/j.jclepro.2004.10.004

Kamuti, T., & Serumaga-Zake, S. (2017a). *Guideline for Good Practices: The Impact of Extractives on Quality and Quantity of Drinking Water*. WaterAid.

Kamuti, T., & Serumaga-Zake, S. (2017b). *The Impact of Extractives on Quality and Quantity of Drinking Water in Malawi, Mozambique, South Africa, Zambia and Zimbabwe*. WaterAid.

Koehler, J., Rayner, S., Katuva, J., Thomson, P., & Hope, B. (2018). A cultural theory of drinking water risks, values and institutional change. *Global Environmental Change*, *50*, 268–277. doi:10.1016/j.gloenvcha.2018.03.006

Lockie, S., Franetovich, M., Sharma, S., & Rolfe, J. (2012). Democratisation versus engagement? Social and economic impact assessment and community participation in the coal mining industry of the Bowen Basin, Australia. *Impact Assessment and Project Appraisal*, *26*(3), 177–187. doi:10.3152/146155108X357257

Marston, A. J. (2017). Alloyed waterscapes: Mining and water at the nexus of corporate social responsibility, resource nationalism, and small-scale mining. *Water (Basel), 4*(1), e1175. doi:10.1002/wat2.1175

Mehta, L. (2014). Water and human development. *World Development, 59*, 59–69. doi:10.1016/j.worlddev.2013.12.018

Nikolaou, I. E., & Evangelinos, K. I. (2010). A SWOT analysis of environmental management practices in Greek Mining and Mineral Industry. *Resources Policy, 35*(3), 226–234. doi:10.1016/j.resourpol.2010.02.002

Perez, F., & Sanchez, L. E. (2009). Assessing the evolution of sustainability reporting in the mining sector. *Environmental Management, 43*(6), 949–961. doi:10.100700267-008-9269-1 PMID:19238481

Republic of South Africa. (1996). Constitution of the Republic of South Africa, 1996 (Act No. 108 of 1996).

South African Human Rights Commission. (2018). *Unregulated artisanal underground and surface mining activities*. https://www.sahrc.org.za/home/21/files/Unregulated%20Artisanal%20Underground%20and%20Surface%20Mining%20Activities%20electronic%20version.pdf

South African Human Rights Commission. (2018). *Investigative hearing report on the underlying socio-economic challenges of mining-affected communities in South Africa*. https://www.sahrc.org.za/home/21/files/SAHRC%20Mining%20communities%20report%20FINAL.pdf

Tanahashi, T. (1978). Health services coverage and its evaluation. *Bulletin of the World Health Organization, 56*(2), 295–303. PMID:96953

Tempelhoff, J. W. N., Ginster, M., Motloung, S., Gouws, C. M., & Strauss, J. S. (2014). The 2012 acid mine drainage (AMD) crisis in Carolina's municipal water supply. *The American Historical Review, 46*(2), 77–107. doi:10.1080/17532523.2014.943978

ENDNOTES

1 See "Take a flight over Mogalakwena platinum mine" [Online] URL: https://www.angloamerican.com/media/our-stories/take-a-trip-to-mogalakwena Accessed: 23/09/2017.

2 Ibid.

3 Ibid.

4 Facilitated group discussions in Ga-Molekane village.

5 Interview with the village head, Ga-Molekane village.

6 See "Take a flight over Mogalakwena platinum mine" [Online] URL: https://www.angloamerican.com/media/our-stories/take-a-trip-to-mogalakwena Accessed: 23/09/2017.

Chapter 18
The Water-Waste Nexus:
Sustainability Implications for Integrated Water Resources and Waste Management in Harare

Trust Nhubu
University of Johannesburg, South Africa

Edison Muzenda
Botswana International University of Science and Technology, Botswana & University of Johannesburg, South Africa

Mohamed Belaid
University of Johannesburg, South Africa

ABSTRACT

The management of water resources and waste is amongst the major challenges facing the majority of urban environments within developing nations due to the rapid population growth and urbanisation as well as improved lifestyles. The Greater Harare Metropolitan Province is not spared by these management challenges. This chapter proposes a number of scenarios that could be implemented in the short to medium terms to address these challenges within the GHMP from a water-waste nexus approach. The water-waste nexus approach is anticipated to significantly reduce the human health impacts and environmental impacts, specifically the water resources pollution which is responsible for the increase in cost of potable water production subsequently resulting in intermittent potable water supply in the GHMP. The study regards the recovery of energy from the waste generated in the GHMP as an integral component of the approach with the energy recovered used for water, wastewater, and waste treatment.

DOI: 10.4018/978-1-7998-8809-3.ch018

INTRODUCTION

The management of waste has been reported and singled out amongst the greatest challenges affecting urban local authorities (ULAs) in developing nations. Such waste includes municipal solid waste (MSW) and wastewater (WW). The challenges are a manifestation of reported astronomical increase in the generation of both MSW and WW within urban environments in developing nations resulting from increased urbanisation due to population increases and rural to urban migration in search of better lifestyles and jobs (Cheng & Urpelainen, 2015; Cohen, 2006; UNDESA, 2018). Increased urbanisation has also brought about the sprouting or emergency of informal settlements outside the urban development master plans which have remained inaccessible to and unserved by the water resources, MSW and WW management services. The majority of MSW generated in these developing nations especially within Africa is unsustainably managed with almost or above 50% being subjected to landfilling with no energy recovery, landfill leachate treatment and landfill gas capture or flaring, whilst the remaining MSW is subjected to dumping at official dumpsites or illegally burnt or buried at source, indiscriminately dumped and strewn in undesignated areas such as vacant spaces, road sides, allays, streets, storm water drains etc. (Hoornweg & Bhada-Tata, 2012). This is so because the increased MSW generation has not been matched with corresponding municipal solid waste management (MSWM) systems developments with regards to infrastructural development (equipment and technology), skills and expertise capacity building and MSWM investments (Yadav, Karmakar, Dikshit, & Vanjari, 2016). The volumes of WW being produced far outstrips the design and commissioned capacities of the wastewater treatment plants (WWTPs) resulting in raw or partially treated WW or sewage being discharged into the environment and water bodies.

Surface run off and on carries with it leachate and other MSW materials from the official dumpsites and undesignated areas depositing them into water bodies. Leachates from the MSW dumpsites and other undesignated areas where MSW is discarded and strewn percolates into the ground polluting the groundwater. The deposition of MSW derived leachate from dumpsites and other MSW materials from the indiscriminately discarded and strewn in runoff during rainfall events together with the discharge of raw or partially treated WW have significantly contributed to the pollution and eutrophication of water bodies. MSW contributes to increased eutrophication of water bodies largely due to the high biodegradable MSW composition in the MSW generated in the developing nations reported to be in excess of 40%, which is anticipated to load nutrients into water bodies when in contact with or laden in run off. The increased urbanisation result in increased demand for portable water which entails increased water pumping power requirements and costs coupled with the need to further expand or extend the potable water supply systems. The MSW derived pollution and eutrophication of water bodies thus is not a welcome phenomenon as it will exacerbates an already strained situation from the resultant increased costs for raw water treatment to potable standards. When surface water supply systems become strained, residents resort to groundwater sources which are unfortunately being laden with leachates from dumpsites polluting the groundwater with heavy metals and other pollutants making the groundwater unfit for potable water uses. A water and waste nexus therefore exists and it is evidently clear that the unsustainable management of MSW and WW impacts heavily on water management thereby bringing in disruptions within the water and waste management nexus. The sustainable management of MSW and WW is thus anticipated to result in the sustainable management of water resources due to the corresponding reduced MSW and WW associated water pollution. The water and waste nexus thus has been negatively impacted with regards to sustainability planning in majority of urban centers within developing nations.

Urban environments within Harare metropolitan province in Zimbabwe like other urban environments within majority of developing nations are not an exception from this waste and water nexus phenomenon. The MSWM practices for Harare unlike having over 50% MSW generated being subjected to landfilling as reported by Hoornweg & Bhada-Tata (2012) for developing nations are not pleasant in that the dorminant practice is dumping at dumpsites responsible for 49% of the MSW generated followed by the burying and burning at source of 38%, recycling of 10% and indiscriminate illegal dumping of 4% of MSW generared in Harare (EMA, 2016; IES, 2013). The WWTPs in Harare have been reported to be receiveing WW volumes far above their design capacities, coupled with their obsolete state have resulted in the discharge of partially treated or raw WW or sewage into water bodies subsequently leading to the super eutrophic status of Lake Chivero, a source of raw water that is treated to supply potable water to Harare (Nhapi, 2014). The urban environments within and surrounding Harare are located within a water catchment which drains into Lake Chivero as shown in Figure 1 entailing that the raw or partially treated WW or sewage and MSW which worsens the situation. These MSWM and WW management challenges have resulted in runoff laden with MSW derived leachate from dumpsites and other MSW materials from the indiscriminately discarded and strewn MSW will find their way into Lake Chivero. Only Norton drains into Lake Manyame another source of potable water for Harare located downstream of Lake Chivero whose reported and observed pollution and eutrophication have been relatively lower than that of Lake Chivero. These have led to the increased water treatment costs and intermittent water supply in Harare reported by Nhubu, Mbohwa, Muzenda, & Patel (2019) with a significant proportion of households resorting to the sinking of boreholes at their households even in high density surbubs where the sinking of such boreholes is prohibited under the Water Act (Chapter 20:24) of Zimbabwe. There are already fears over the possible over-abstraction or

depletion of the groundwater aquifer evidenced by the further deepening of the groundwater table, reduced borehole safe yields and drying up of other boreholes during the dry months.

Figure 1. Location of urban environments in a catchment that drains into Harare's potable water sources

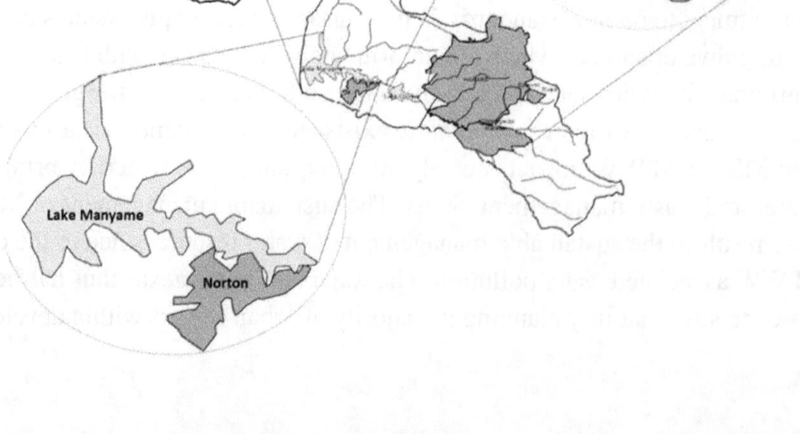

Despite the wastewater (WW) and municipal solid waste (MSW) management challenges characterizing Harare City, Chitungwiza, Ruwa, Epworth and Norton resulting in surface and groundwater pollution evidenced by the perennial outbreaks of water borne diseases (Cholera, Typhoid etc.) and Lake Chivero eutrophication, the WW and MSW could be subjected to any of the following treatment methods namely combined heat and power generation (CHP) thermochemical conversions (incineration, pyrolysis or gasification), landfilling with biogas recovery for CHP, anaerobic digestion (AD) with the biogas and digestate produced used for CHP generation and as biofertiliser respectively and composting or aerobic treatment to produce compost a potential biofertiliser. There is therefore need to use the nexus approach to analyze the water resources and waste management systems in Harare not only from the linkages regarding their associated problems perspective but including the potential use of materials, resources and energy recovered from waste under the IWM in addressing the water resources management challenges under the IWRM. This book chapter, therefore, proposes such a nexus level water and waste systems that will aid and contribute to informing likely pathways towards bringing increased sustainability in and around Harare. The chapter will also attempt to bring the linkages and relatedness between the IWRM and IWM principles together with the principles upon which the Water Act (Chapter 20:24) of Zimbabwe was developed to unpack the water and waste nexus in Harare.

WATER AND WASTE MANAGEMENT IN HARARE

The Greater Harare Metropolitan Province (GHMP) comprises Harare, the Capital City of Zimbabwe and its neighboring urban population centers of Chitungwiza and Epworth. Though Ruwa to the East and Norton to the west of Harare City in Mashonaland East and Mashonaland West Provinces respectively are not part of the GHMP in the strictest of terms, they are generally considered part of the GHMP under this Chapter. The population in the GHMP is increasing with the Worldometer (2020) reporting population figures of 1,542,813 inhabitants giving an increase of 57,582 inhabitants from the reported 2012 census report figure of 1,485,231 people (Zimstat, 2013) signaling an almost 4% increase. However, the actual population figures of 2022 will provide reliable and accurate estimates. The population increases are resultantly putting pressure on the already overwhelmed available water and waste management systems. The urban water, sanitation and hygiene (U-WASH) situation in the GHMP is thus further worsened by the sprouting of informal settlements that are not serviced by the urban local authorities (ULAs) water and waste management services.

By assuming a constant daily per capita MSW generation of 0.6kg estimated by (Makarichi, Kan, Jutidamrongphan, & Techato (2019) and Nhubu and Muzenda (2019), a 4% increase in MSW generation rate to the reported estimate of 467,303 by Nhubu and Muzenda (2019) in the GHMP is also assumed giving an annual MSW generation of 485,995 tons. Worryingly only 10% of the waste is recycled with the rest being poorly managed with 49% subjected to indiscriminate collection and dumping in official dumpsites whilst the remainder is subjected to burying and burning at source as well as illegal indiscriminate dumping in undesignated areas as shown in Figure 2. These MSWM practices contributes to human health and environmental impacts including the pollution of both surface and ground water sources.

Using the same scale of 4% increase, the water demand for the GHMP in 2012 as estimated by the Harare Municipality at 1,200 ML/day despite the noted need for verification (Ndunguru & Hoko, 2016) is assumed to have increased to 1,248 ML/day. Harare Water is responsible for supplying potable water to residents within all the urban centers within the GHMP namely, Harare City, Chitungwiza, Epworth,

Norton and Ruwa. Harare Water owns and operates two water treatment plants namely; Morton Jaffray Water Works which treats raw water from Lakes Chivero and Manyame of 247,180 and 480,240 ML capacities respectively (Nhapi, Hoko, Siebel, & Gijzen, 2002) and Prince Edward Plant which treats raw water from Harava and Seke dams of 9,030 and 3,380 ML capacities respectively (JICA, 1996). Nhapi et al (2002) reported daily design capacities of 614 and 90 ML for Morton Jaffray Water Works and Prince Edward Plant respectively which had been restricted or reduced as a result of the obsolete state of the infrastructure and equipment at the respective water treatment plants to 526 and 74 ML respectively giving a combined capacity of 600 ML against a water demand of 1,200 ML estimated by the Harare Municipality in 2012. This means the water treatment plants can at most supply only half of the population in the GHMP and by factoring in water losses estimated at between 29 and 43% (Ndunguru & Hoko, 2016) which exceeds the World Bank recommended benchmark of 23% and the increased pollution and eutrophic status of the raw water for potable water supply, way less than half of the population is served with potable water from the two water treatment plants. This thus has led to a majority of residents resorting to alternative water sources mainly groundwater through the digging of shallow wells and drilling of boreholes. Boreholes have and are being drilled in high residential areas even if such an act is prohibited.

Figure 2. MSWM practices in GHMP

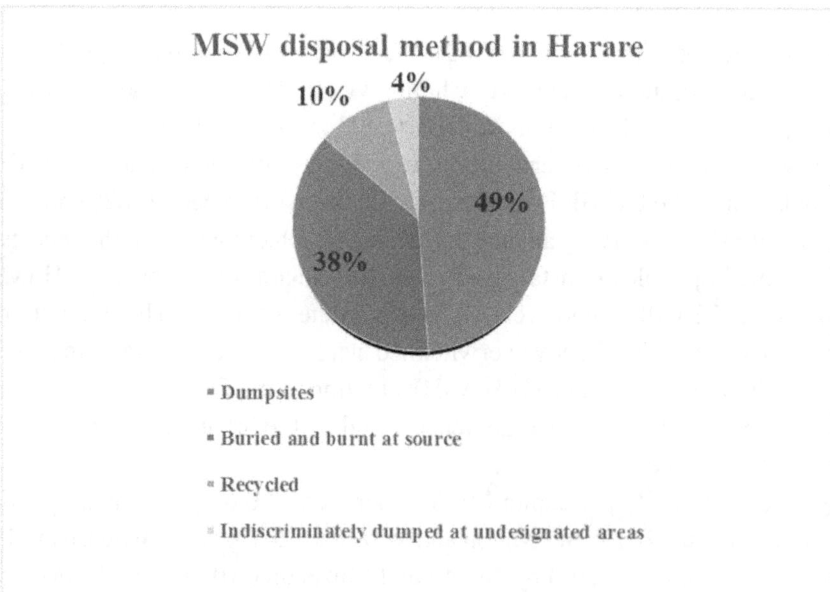

RELIANCE ON GROUNDWATER SOURCES FOR POTABLE WATER IN GHMP

The shortage of potable water supply from surface water sources to meet the demand in the GHMP has thus led to the reliance on improved or unimproved groundwater sources namely boreholes and hand dug wells. Apart from the obsolete infrastructure at the water treatment works and the excessive potable water losses in the distribution networks cited above, the intermittent potable water supply in GHMP is also attributed to the increased costs of potable water production observed by Nhubu et al (2019) emanating from the increased pollution and eutrophication of the main water bodies (Nhapi, 2014) that supply raw water for potable water treatment. Climate change induced low rainfall seasons and increased evaporation from increased temperatures results in increased nutrients and pollutants concentration in the water bodies that supply raw water for the GHMP potable water treatment plants, hence the increased effort and costs for potable water production. The expansion of the residential areas in the GHMP and emergency of informal settlement has not been matched with water supply infrastructure resulting in some sections not receiving water even during the period when the piped water is available. The intermittent potable water supply situation has also brought about quality issues with regards to the water getting contaminated within the distribution networks that would have gone for prolonged periods without water passing through rendering the potable water from the surface water sources unsafe for drinking and other uses.

The dire potable water supply situation in GHMP has thus led to the drilling of community boreholes even in high density suburbs to compliment the intermittent potable water supplies from surface water sources. Though the Harare Municipal by-laws are silent on suburbs that can apply to sink boreholes, the use of the by-laws together with the Regional Town and Country Planning Act [Chapter 29:12] apart from the associated costs of borehole drilling or digging of wells prohibits borehole drilling and well establishment on properties in high density areas. Development partners have partnered government and ULAs in GHMP to drill and equip community boreholes with either hand pumps or solar powered pumps to alleviate the water supply situation. Some communities are however reluctant to have their community boreholes installed with solar powered pumps siting unavailability of water during cloudy days. Individual households who afford have also drilled their household boreholes with those with high yielding boreholes having started or partnered bulk water supply entities resulting in the proliferation of private water suppliers who preys on desperate communities selling them water at exorbitant rates. The selling of water abstracted from these boreholes is a breach of Section 34 subsection (1) of the Water Act (Chapter 20:24) which stipulates that, "No person shall abstract water for any purpose other than primary purposes except in terms of a permit" as these boreholes are specifically meant for primary uses. Boreholes are being drilled without the necessary written authority from the Manyame Catchment Council or Upper Manyame Sub Catchment Council as provided under Section 35 of the Water Act (Chapter 20:24). Therefore, there is no groundwater abstraction monitoring as majority of these boreholes are drilled without the knowledge of the catchment councils and are not equipped with metering devices to measure volumes of water abstracted to avoid the over abstraction or depletion of the aquifer.

The knowledge of boreholes drilled and amount being abstracted from each borehole will provide the basis for Catchment Councils to exercise their powers regarding limiting the quantity of water abstracted for primary purposes as provided under Section 33 of the Water Act (Chapter 20:24). This will go a long way in limiting water abuses such as water selling and aquifer over abstraction and depletion which has seen other boreholes drying up during dry months and the falling of the water table. Only three of the twenty-four boreholes in Chitungwiza shown in Figure 3 and none of the four boreholes in Chitungwiza passed the 48 hour borehole safe yield capacity tests conducted during October of 2020. A minimum safe yield of 3,6 m^3/hr (1l/s) was required as provided under the Zimbabwe protocol on piped water schemes with the best performing borehole having a safe yield of 4.4 m^3/hr. The high borehole failure rate (87.5%) or low pass rate (12.5%) and low safe yields speaks volumes of the anticipated aquifer over abstraction or depletion and falling of groundwater table. Urgent actions are therefore necessary to protect aquifer integrity and its sustainability being threatened with over abstraction and pollution.

Figure 3. Map of boreholes that were tested for their safe yield in Chitungwiza

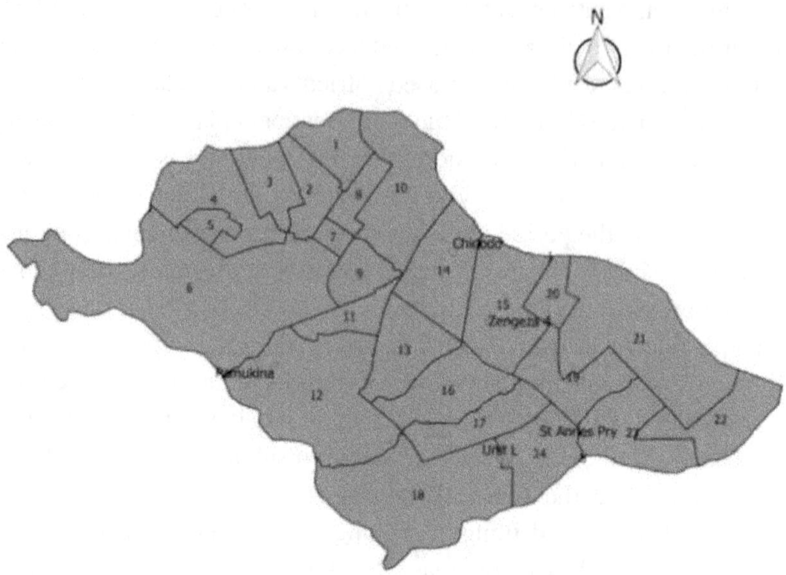

CONTRIBUTION OF MSW AND WW TO WATER POLLUTION IN GHMP

The MSW and WW generated in GHMP have been reported to contribute significantly to the pollution and eutrophication of water bodies that supply potable water to GHMP. Surface runoff carries with it MSW materials and leachate from the official dumpsites and illegal undesignated areas where waste is strewn depositing the leachates and MSW materials in water bodies. Worryingly the MSW generated in the GHMP constitutes in excess of 40% biodegradable matter (EMA, 2016) that decomposes in dumpsites bodies releasing nutrients into dumpsites leachates that contaminates groundwater sources. Love et al (2006) observed higher nutrients, coliforms and metal concentrations than the World Health Organisation quality standards for potable water in the groundwater around Pomona and Golden Quarry (official MSW dumpsites for Harare) which renders the water unsafe for drinking and other potable uses. The nutrients, coliforms and metal concentrations in the groundwater together with their reach around the two dump-sites and other dumpsites is anticipated to have increased since then considering more MSW has been dumped with no remediation efforts have been made regarding the rehabilitation of the dumpsites. It is expected that despite the reliance with groundwater in GHMP, leachate from MSW dumpsites might have compromised a significant portion of the ground water aquifer in GHMP. Muzenda, Masocha, & Misi (2019) recommended the need for residences in Ushewokunze, an informal and semi-formal settlement in GHMP, to treat all the water they use from the different sources before use and be educated on the poor quality status of groundwater in their area after having observed that 90% of groundwater sources in the area do not meet the World Health Organization's Standards for drinking water hence their being unsafe and unsuitable sources for drinking water. Of concern in such informal and semi-formal settlements that have sprouted up within the GHMP is the use of Blair toilets for ablution needs and shallow wells for potable water since these settlements are serviced with municipal MSW and WW management as well as water supply. This provides all the conditions necessary for the likely groundwater contamination with fecal matter and coliforms. Biodegradable MSW decomposition in dumpsites also produces greenhouse gases (especially methane and other volatile organic compounds) that are emitted into the atmosphere contributing to climate change which negatively impacts the water cycle through related climate change induced water disasters such as floods, droughts etc. Biodegradable MSW that is laden in runoff and deposited into water bodies provides a source of nutrients that contributes to eutrophication, the case with Lake Chivero and Manyame. The infrastructure at all the WWTPs within the GHMP have become obsolete just as the water treatment resulting in the discharge of raw or partially treated sewage into the river systems and water bodies having been noted as the major source of Lake Chivero eutrophication (Nhapi, 2014; Nhapi et al., 2002). Considering the combined daily capacities of Morton Jaffray Water Works and Prince Edward Plant of 600 ML and 70% non-consumptive water use, over 150,000 ML of raw or untreated WW is discharged into the environment and water bodies. It is therefore evident that MSW and WW management directly impacts the management of water resources in the GHMP (Nhubu et al., 2019), hence the need for systems analysis that informs the development of a sustainable and integrated waste and water resources management nexus.

WATER AND WASTE NEXUS

The Water-Waste Nexus (WWN) provides the interconnectedness of the water and waste related management systems (processing, treatment, distribution, use and potential recycling) which has received considerable research interests especially considering the role that materials and energy recovered from waste could play in water management systems. This is so because the nexus sustainability has become more critical in light of ever increasing global population, decrease in resource availability and increasing climate change impacts (Garcia, Lovett, & You, 2019). Huge quantities of different waste fractions are being produced daily with the annual MSW generation having been projected by Hoornweg & Bhada (2012) to reach 2.2 x 10^9 tons in 2025 (1.42 kg/ capita/day) from 1.3 x 10^9 tons (1.42kg/capita/day) in 2012. Mateo-Sagasta, Raschid-Sally, & Thebo (2015) approximated an annual global municipal WW generation of 330 km^3 whose significant constituents food and other biodegradable waste fractions. The economic values of the wastes are rarely quantified resulting in the wastes being largely disposed or externalized into the environment without prior appropriate treatment (Garcia et al., 2019), negatively impacting the environment and human health. Systems analysis together with economic approaches can be used to solve the challenges. Thus systems analysis and modelling on the WWN can be used to aid investment promotion, policy formulation and decision making for the development of sustainable systems considering the water-waste-waste derived energy nexus or cross sectoral integration, spatial and temporal resources availability, long term socio-economic and environmental targets and the available technological and skills constraints (Leung Pah Hang, Martinez-Hernandez, Leach, & Yang, 2017). Wang et al (2018) noted in their extensive review that a number of studies have been conducted from a developing nation perspective on energy supply and demand optimization systems analysis with regards to planning and forecasting, renewable energy distribution for remote areas, optimization of bioenergy research and waste to energy (WtE) systems. Though various decision making tools have been developed and effectively applied with those applied in the energy-water-food nexus planning becoming the major focus for the current research agenda on systematic modeling development, Wang et al (2018) noted a vacuum or lag on decision making tools that integrate the water-energy-waste sectors with the supply and demand at the different spatial and temporal scales. A platform to study the urban water-energy-waste systems interactions was developed (Wang et al., 2018). The platform incorporated life cycle sustainability assessment (LCSA), policy support assessment such as Feed in Tariff (FIT), assessment of green financing strategies and decentralised technologies. Significant sustainability benefits in the WWN can be achieved considering the implementation of the IWRM and IWM models incorporating materials and energy recovery from waste together with their subsequent use in water and wastewater treatment.

ANALYSIS OF THE WATER-WASTE MANAGEMENT NEXUS IN GHMP

Figure 4. Schematic of water and waste management institutions, functions and supporting laws and policies

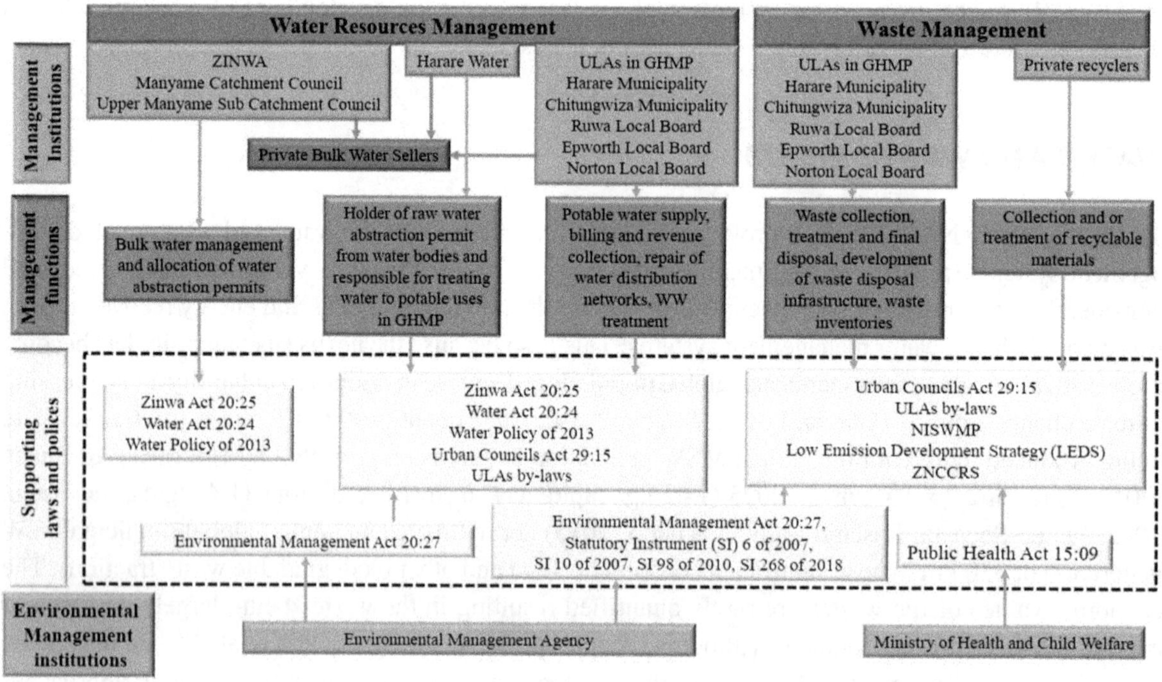

This section covers the water-waste management nexus in GHMP considering the responsible management institutions, their functions and supporting laws and policies so as to highlight and identify any connections and lack thereof with regards to the management of water and waste. The management of water resources, MSW and WW in the GHMP falls under the authority of different institutions with each institutions being responsible for specific functions supported by specific pieces of laws and policies as shown in Figure 4. The institutions responsible for water resources management in GHMP are namely; Zimbabwe National Water Authority (ZINWA), Manyame Catchment Council, the Upper Manyame Sub Catchment Council, Harare Water and all the ULAs in GHMP namely; Harare municipality, Chitungwiza municipality, Ruwa, Epworth and Norton local boards. Illegal private water suppliers have also emerged in GHMP. The management of MSW and WW falls under the authority of ULAs who can engage other private players to manage waste on their behalf. There are also private waste recyclers who also engage in MSW management through the separation and collection of recyclable MSW fraction for reuse, reprocessing and or export. ZINWA and the Environmental Management Agency (EMA) are responsible for environmental monitoring regarding the protection of the environment and water resources from pollution whilst EMA and the Ministry of Health and Child Welfare (MHCW) are responsible for waste management related environmental and human health impacts. The laws and policies that supports the water resources and waste management in GHMP contribute towards the realisation of the right to water and environmental rights as provided under sections 77 and 73 of the Constitution of Zimbabwe Amendment (No.20) Act of 2013.

WATER RESOURCES MANAGEMENT INSTITUTIONS, FUNCTIONS AND SUPPORTING LAWS AND POLICIES IN GHMP

ZINWA was established through the enactment of the Zimbabwe National Water Authority Act [Chapter 20:25] as provided under section 3 of the Act whose functions are provided under section 5 of the same Act. One such a functions that specifically apply to water resources management in GHMP is provided under section 5(e) of the Act, that ZINWA should encourage and assist ULAs in discharging their water resources development and management functions within their jurisdictions with regards to potable water provision and WW disposal as provided in the Urban Councils Act [Chapter 29:15]. ZINWA is also under section 5(f) tasked to ensure that catchment councils (CC) adhere to the provisions of the Water Act [Chapter 20:24] in discharging their duties. Section 5(g) further assigns ZINWA to encourage and assist CC and sub catchment councils (SCC) in planning and coordinating water resources development and management within their jurisdictions or catchment areas. CC and SCC are established through the enactment of the Water Act [Chapter 20:24]. The establishment of CC is as provided under section 20 with their functions and powers provided under sections 21 and 22 respectively. Seven CC namely, Manyame, Sanyati, Gwayi, Runde, Save, Mzingwane and Sanyati were established in Zimbabwe with the GHMP falling under the Manyame SCC. The establishment and functions of SCC is provided under section 24 within the GHMP falling under Upper Manyame SCC. CC are obliged to work with ZINWA under section 21(1) in preparing river outline plans, determining water abstractions permit applications and granting the said permits, regulating and supervising the right to and use of water within the catchment or rive system upon which the CC is established, supervising SCC in performing their functions, ensuring compliance with the Water Act [Chapter 20:24] and performing any other duty conferred or imposed in terms of the Water Act [Chapter 20:24]. Therefore, it is these and other functions as provided in the Water Act [Chapter 20:24] that the Manyame CC perform in conjunction with ZINWA regarding the development and management of water resources in the catchment under which GHMP is located. Harare water thus holds a water abstraction permit from the Manyame and Upper Manyame SCC to abstract raw water from Lake Chivero, Lake Manyame, Harava and Seke dams for the purposes of treating it at Morton Jaffray Water Works and Prince Edward Plant to supply the ULA in GHMP with potable water. The ULAs in GHMP supplies the residents with water in accordance with the Urban Councils Act 29:15 and local by-laws. All these water development and management activities outlined for the various institutions thus are in accordance with the aspirations of the Zimbabwe National Water Policy of 2013, contributing to the realisation of the right to water as enshrined in section 77 of the Constitution of Zimbabwe Amendment (No.20) Act of 2013. EMA through the administration of the provisions of the Environmental Management Act [Chapter 20:27] and supporting Statutory Instruments (SIs) is responsible for ensuring the sustainable use and management of water resources, protection of the environment and administration of the polluter pays principle. EMA is a statutory body that was established through the proclamation of SI 103 of 2003 in accordance with the dictates of the Environmental Management Act [Chapter 20:27] from which its mandate is derived. Amongst its functions, ZINWA together with CC and SCC under section 5(1)(a)(iii) is mandated to advise the responsible member of the executive arm of government on the formulation of national water quality, pollution control and environmental protection policies and standards in accordance with the provisions of the Environmental Management Act [Chapter 20:27] and Water Act [Chapter 20:24].

WASTE MANAGEMENT INSTITUTIONS, FUNCTIONS AND SUPPORTING LAWS AND POLICIES IN GHMP

The Urban Councils Act 29:15 together with accompanying by laws mandate ULAs to be the responsible entities for the management (storage, collection, transportation, processing, treatment, recycling, materials and energy recovery, final disposal) of waste generated within their jurisdictions. Other national policy documents such as the National Integrated Solid Waste Management Plan (NISWMP) National Low Emission Development Strategy (LEDS), Zimbabwe National Climate Change Response Strategy (ZNC-CRS) also promotes the sustainable management of waste by ULAs. EMA through the Environmental Management Act [Chapter 20:27] and a number of SIs namely SI 6 of 2007, SI 10 of 2007, SI 98 of 2010 and SI 268 of 2018 ensures that waste is sustainably managed to avoid the associated human health and environmental impacts. SI 6 of 2007 regulates the polluter pays principle regarding the disposal of solid waste and effluents. SI 10 of 2007 regulates the management of MSW by ULAs as well as the issuance of licenses to waste generators and handlers regarding generation, storage, collection, transportation, use, processing, treatment, recycle, materials and energy recovery and final disposal. Waste generators and handlers are obliged to prepare waste management plans and targets under the same SI 10. SI 98 of 2010 encourages the reuse and reduction of biodegradable and plastic MSW fractions respectively. The Ministry of Health and Child Welfare also regulates the promotion of MSW and WW related public health issues as guided by the provisions of section 83 of the Public Health Act 15:09, mandates ULAs to institute necessary and reasonable practical measures as provided under the law for the provision of clean sanitary environment and prevention of waste accumulation likely to cause harm or threaten the health of the public.

WATER RESOURCES AND WASTE MANAGEMENT NEXUS IN GHMP

The management of water resources and waste in GHMP involves largely the same institutions regarding the need for their sustainable management i.e the cross cutting environmental protection and pollution control. A nexus between the water resources management institutions could easily be established that will lead to the sustainable and integrated management of water resources and waste in GHMP considering the impacts that waste management has on the water resources. The Water Act [Chapter 20:24] was developed based on three main principles namely; sustainability in use, efficiency in use and equity in allocation based on the integrated water resources management model whilst the Zimbabwe NISWMP was developed along the integrated waste management concept to bring about environmental sustainability, economic viability and social acceptability (McDougall, White, Franke, & Hindle, 2001). There is a striking similarity and relation on the IWRM and IWM principles upon which the Water Act [Chapter 20:24] and the Zimbabwe NISWMP were developed. The principles under the Dublin Statement on Water and Sustainable Development generally known as the Dublin principles also have a perfect fit with the IWM and IWRM principles. These interconnectedness and relations are as shown in Table. Differences with regards to the management system boundaries are present since water resources are managed along hydrological boundaries whereas waste is managed along administrative boundaries. The other differences arise on the role of ZINWA, CC and SCC whose mandate is largely centered on raw water supply to the ULAs. It is however important to note that ZINWA, CC and SCC are also mandated to play water quality, pollution control and environmental protection duties which will inform their advice to the responsible member of the executive arm of government on the formulation of respective policies and standards, thus they also play a role in monitoring and controlling waste derived water and other environmental pollution. It is therefore evident that the present institutional arrangements and supporting laws and policies provides a reasonable framework to bring about the water resources and waste management nexus in GHMP and ultimately achieve the needed sustainability.

Table 1. Linkages between the principles of the IWM and IWRM management systems

Management system	IWM (McDougall et al., 2001)	IWRM (Biswas, 2004)	Dublin Principles (IRGCWSS, 1992)
Boundary	Administrative	Hydrological	Hydrological
Principles	Environmental Sustainability	**Ecological Sustainability** -Sustainability in use	Water resources are finite and vulnerable that needs to be sustainably managed
	Economic viability	**Economic Efficiency** - Efficiency in use	Water is an economic good
	Social acceptability	Social Equity - Equity in allocation	Women play an important role in the management of water resources
			Every stakeholder must be involved when making decisions on water resources planning, development and management

SOLUTIONS AND RECOMMENDATIONS

Proposals for Integrated Water Resources and Waste Management Nexus in GHMP are presented. Four likely future WWN scenarios for GHMP that incorporates the utilisation of energy and materials derived from the MSW and WW generated in GHMP are proposed. These proposals were developed considering the findings from an earlier study (Nhubu & Muzenda, 2019) that applied Life Cycle Assessment (LCA) to determine the associated environmental impacts of likely future MSWM options that could be implemented in GHMP. These proposals are anticipated to bring about the needed system-wide environmental and socio-economic sustainability benefits to the WWN in GHMP. The sustainability benefits are anticipated from the recovery of resources and energy from MSW and WW. The proposals thus consider different combinations of waste treatment methods to recover energy and resources. Such waste treatment methods that recover energy and or other resources include landfilling, anaerobic digestion (AD) and thermal treatment (incineration, pyrolysis and gasification). The landfilling of MSW and

WW sludge results in their biodegradation under oxygen limiting environment producing biogas. The AD of MSW organic fraction (MSWOF) and WW sludge to produce biogas has the potential to transform MSW and WW treatment from an energy consumer to a net energy producer resulting from the higher energy recovery capabilities than energy required for AD operations (Amigun, Parawira, & Musango, 2012). The biogas generated from the landfilling and AD of MSW and WW sludge can either be used for combined heat and power (CHP) generation or subjected to further upgrading and compression to a biofuel in the transport industry. Seventy five percent of MSW generated in the GHMP was reported suitable for thermal treatment with a lower heating value of 10.1MJ per kg, capable of producing 3.8 ×106 GJ per annum (112 GWh per annum) (Makarichi et al., 2019). Therefore, since the AD facility is expected to cogenerate more energy than needed for its operations, the excess energy from AD added to the energy from the thermal treatment methods and landfilling can be channeled to the thermal treatment methods with extra energy if any within the WWN systems channeled to WTPs, WWTPs and composting facilities. In the event that the waste derived energy is more than the energy required to sustain all the operations within the WWN system, the extra energy will be sold to the national grid thereby increasing electricity availability in the national grid. However, legislative and policy mechanisms for such tariffs to sell the excess power to the national grid need to be put in place.

WWN Scenario 1

Figure 5. WWN Scenario 1

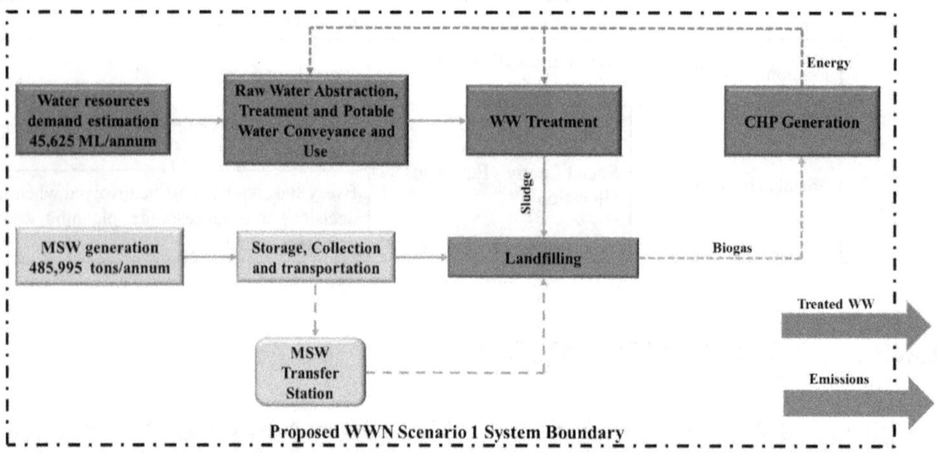

This scenario considers the indiscriminate collection and landfilling of mixed bag MSW and WWT sludge with the recovery of biogas which will be subsequently used for combined heat and power generation (CHP) with the cogenerated heat and electricity used in WTPs and WWTPs. In the event that the cogenerated heat and electricity are more than the WWN system energy needs, the excess heat and electricity can be sent to other uses and sold to the national grid respectively. The landfill leachate (LL) is sent for treatment at the WWTPs. This is anticipated to be the least environmentally performing scenario based on the findings from an environmental life cycle assessment findings by Nhubu and Muzenda (2019) considering the associated impacts associated with the indiscriminate MSW landfilling hence the red color on landfilling. The scenario can also consider the incorporation of MSW transfer stations to bring about the siting and location of large landfills and accompanying MSWM (treatment and processing facilities) distant away from urban residential areas where cheap land is available considering its unavailability due its competing interests in urban areas which makes land expensive. This is anticipated to reduce the MSWM associated human health impacts and pollution of both surface and groundwater resources in GHMP. MSW transfer stations will bring about increased MSW recycling, collection effectiveness and efficiency; reduced MSW collection costs and burden on the overall MSWM budgets resulting in some funds channeled towards other MSWM functions, infrastructural, equipment and human resources needs from MSW collection which is currently consuming the significant portion of the MSWM budget. In addition, MSW transfer stations will result in the reduction of MSW collection derived environmental impacts including greenhouse gases (GHG) emissions and other likely impacts.

WWN Scenario 2

Figure 6. WWN Scenario 2

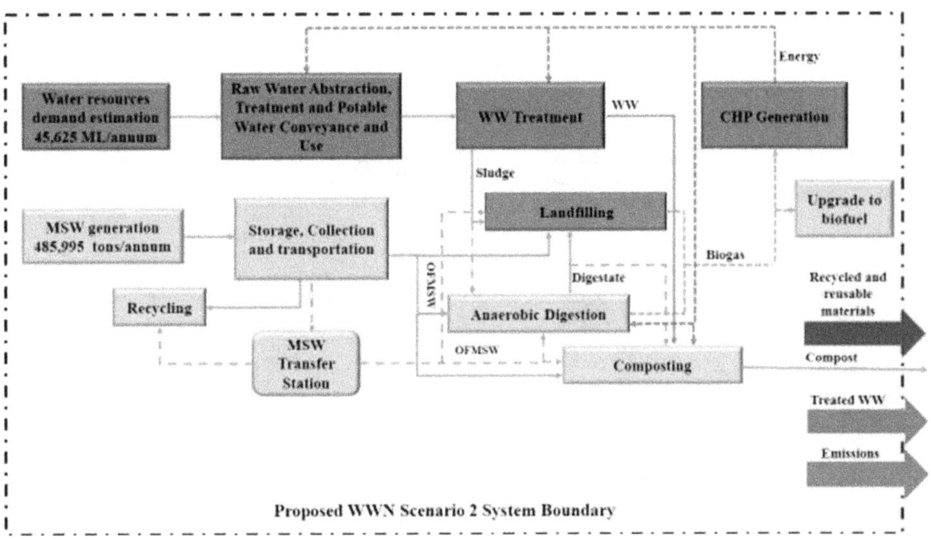

The scenario considers combines the landfilling of mixed bag MSW and WWTP sludge and the anaerobic digestion of the municipal solid waste organic fraction (MSWOF) and WWTP sludge. Alternatively the MSWOF, WWTP sludge and AD digestate can be subjected to composting despite AD showing an overall environmental sustainability edge over composting as observed by Nhubu, Muzenda, Mbohwa, & Agbenyeku (2020). The biogas recovered from landfilling and anaerobic digestion will be used for combined heat and power generation (CHP) with the cogenerated heat and electricity used in WTPs, WWTPs and composting facilities. Alternatively, the biogas can be further upgraded to biofuel. In the event that the cogenerated heat and electricity are more than the WWN system energy needs, the excess heat and electricity can be sent to other uses and sold to the national grid respectively. The LL is sent for treatment at the WWTPs. The composted MSWOF, WWTP sludge and AD digestate together with the AD leachate could be used as potential biofertilisers that will stimulate the Zimbabwean agricultural or crop productivity contributing to the growth of the Zimbabwe's agro based economy. MSW transfer stations and MSW recycling may also be considered under the proposed scenario based on its anticipated benefits.

WWN Scenario 3

Under this scenario, thermal treatment (incineration, pyrolysis and gasification) of MSW and WWTP sludge is considered instead of landfilling under scenario 1. The combined heat and power generation (CHP) derived heat and electricity is used in WTPs and WWTPs. In the event that the cogenerated heat and electricity are more than the WWN system energy needs, the excess heat and electricity can be sent to other uses and sold to the national grid respectively. Inert products from MSW thermal treatment are sent to landfilling. The WWTP sludge may also be sent to landfilling instead of thermal treatment. The scenario is anticipated to environmentally perform better than Scenario 1 based on the better performance of incineration than landfilling (2019). The scenario can also consider the incorporation of MSW transfer stations.

Figure 7. WWN Scenario 3

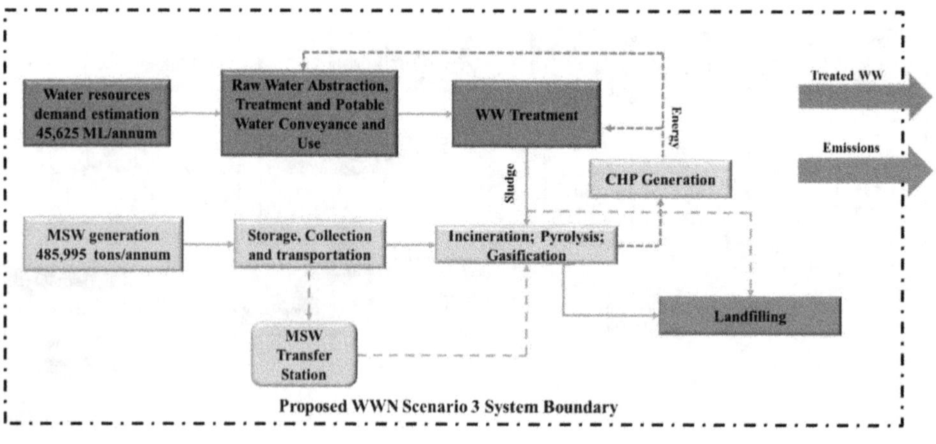

WWN Scenario 4

The scenario combines the thermal treatment of mixed bag MSW and the anaerobic digestion of the municipal solid waste organic fraction (MSWOF) and WWTP sludge. Alternatively, the MSWOF, WWTP sludge and AD digestate can be subjected to composting as under Scenario 2. The biogas recovered from anaerobic digestion will be used for combined heat and power generation (CHP) with the cogenerated heat and electricity used in WTPs, WWTPs and composting facilities. In the event that the cogenerated heat and electricity are more than the WWN system energy needs, the excess heat and electricity can be sent to other uses and sold to the national grid respectively. The LL is sent for treatment at the WWTPs. The composted MSWOF, WWTP sludge and AD digestate together with the AD leachate could be used as potential biofertilisers that will stimulate the Zimbabwean agricultural or crop productivity contributing to the growth of the Zimbabwe's agro based economy. MSW transfer stations and MSW recycling may also be considered under the proposed scenario based on its anticipated benefits.

Figure 8. WWN Scenario 4

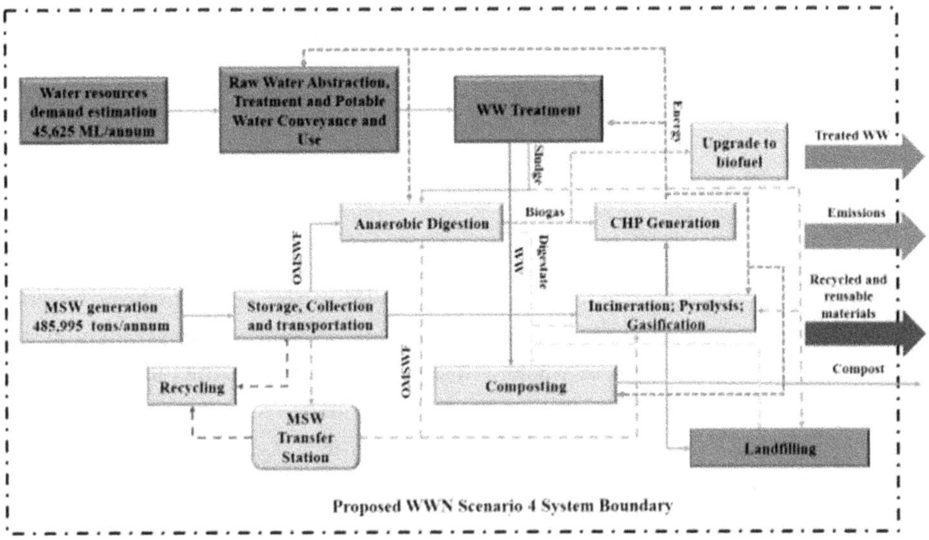

FUTURE RESEARCH DIRECTIONS

Building future waste and water management under the proposed WWN scenarios could bring about significant reductions in human health impacts and environmental impacts mainly the water resources pollution which is responsible for the increase in cost of potable water production subsequently resulting in intermittent potable water supply in the GHMP. There is, however, need to undertake holistic environmental impacts, social acceptance, economic viability and technical feasibility assessments of the proposed WWN scenarios to identify the best environmentally performing, socially acceptable, economically viable and technically feasible and sound scenario or combination of scenario processes. Considering the difficulties associated with optimizing more than one variable as in the case with these WWN scenarios and the need for some trade-offs to be made, sustainability LCA could be employed to assess the environmental, economic and social impacts using the environmental LCA, life cycle costing and social LCA respectively. Other models and tools could also be used such as ASPEN to simulate the performance of the proposed WWN scenarios. There is need for reliable and accurate waste generation and characteristics data which will be input to the WWN simulations hence the need for waste generation estimation and waste characterization studies. Water demand data must be updated to match the rapid population increases that has characterised the GHMP such the water demand projections are simulated against different rainfall forecasts considering the climate change induced low rainfall seasons whose frequencies are on the rise. There is also need to assess the effectiveness of operating the WWN either under small scale decentralised low capacity facilities distributed across the GHMP or large scale centralized facilities. Incorporating the food or agricultural production component within the proposed WWN could also be considered to bring other aspects regarding the water competing interests within the nexus so as to have an understanding on the water balance equation within the catchment. This is so because part of the water in the reservoirs that supply the GHMP with potable water has been allocated to agricultural farmers in the catchment. Proposals for reserving the water in the eutrophic Lakes reserved for agricultural production should also be considered whilst other potable water sources such as the proposed Muda Nyatsime and Kunzvi dams are developed whilst environmentally sound waste management practices under the proposed WWN are pursued to allow for the eutrophic Lakes to self-purify after which they can be reconnected back to the potable water supply system. Above all the WWN legislative and institutional arrangements need to be synchronized such that sustainability and maximum benefits could be derived from the integrated waste and water resources management in the GHMP.

CONCLUSION

The water resources and waste management challenges within the GHMP could be addressed if a WWN approach is taken in developing future waste and water resources management systems. This is so because the sustainable management of waste will significantly reduce the human health impacts and environmental impacts mainly the water resources pollution which is responsible for the increase in cost of potable water production subsequently resulting in intermittent potable water supply in the GHMP. Recovering energy from the waste generated in the GHMP will be an integral component of the WWN as the energy recovered will be used in WTPS and WWTPs. The anaerobic digestion or composting of the organic MSW fraction generated in the GHMP will divert a significant amount of nutrients from water bodies thereby limiting their rate of the eutrophication. The same applies to the anaerobic co-digestion of sewage sludge and the organic MSW fraction generated in the GHMP. Composting and AD results in the production of an organic fertiliser that could be a potential game changer replacing chemical fertilizer in the Zimbabwean agricultural sectors since Zimbabwe is an agro based economy. There seem to be some clarity regarding the legislative and institutional arrangements for the promotion of a WWN in the GHMP though more effort towards synchronization is needed.

ACKNOWLEDGMENT

This research was supported by the University of Johannesburg (Postdoctoral fellowship).

REFERENCES

Amigun, B., Parawira, W., & Musango, J. K. (2012). Anaerobic biogas generation for rural area energy provision in Africa. *Biogas*, 35-62.

Biswas, A. K. (2004). Integrated water resources management: a reassessment: a water forum contribution. *Water International*, *29*(2), 248–256. doi:10.1080/02508060408691775

Cheng, C. Y., & Urpelainen, J. (2015). Who should take the garbage out? Public opinion on waste management in Dar es Salaam, Tanzania. *Habitat International*, *46*, 111–118. doi:10.1016/j.habitatint.2014.11.001

Cohen, B. (2006). Urbanization in developing countries: Current trends, future projections, and key challenges for sustainability. *Technology in Society*, *28*(1-2), 63–80. doi:10.1016/j.techsoc.2005.10.005

EMA. (2016). *Waste generation and management in Harare, Zimbabwe: Residential areas, commercial areas and schools*. Environmental Management Agency.

Garcia, D. J., Lovett, B. M., & You, F. (2019). Considering agricultural wastes and ecosystem services in Food-Energy-Water-Waste Nexus system design. *Journal of Cleaner Production*, *228*, 941–955. doi:10.1016/j.jclepro.2019.04.314

Hoornweg, D., & Bhada-Tata, P. (2012). *What a Waste: A Global Review of Solid Waste Management*. Retrieved from https://openknowledge.worldbank.org/handle/10986/17388

IES. (2013). A situational Analysis of Solid Waste Management in Zimbabwe's Urban Centres. IES.

IRGCWSS. (1992). *The Dublin Statement on Water and Sustainable Development*. Paper presented at the International Conference on Water and the Environment: Development Issues for the 2lst century, Dublin, Ireland.

JICA. (1996). *The Study of Water Pollution Control in Upper Manyame River Basin in the Republic of Zimbabwe*. JICA.

Leung Pah Hang, M. Y., Martinez-Hernandez, E., Leach, M., & Yang, A. (2017). Insight-Based Approach for the Design of Integrated Local Food-Energy-Water Systems. *Environmental Science & Technology*, *51*(15), 8643–8653.

Love, D., Zingoni, E., Ravengai, S., Owen, R., Moyce, W., Mangeya, P., & Hranova, R. (2006). Characterization of diffuse pollution of shallow groundwater in the Harare urban area, Zimbabwe. In *Groundwater pollution in Africa*. CRC Press.

Makarichi, L., Kan, R., Jutidamrongphan, W., & Techato, K. A. (2019). Suitability of municipal solid waste in African cities for thermochemical waste-to-energy conversion: The case of Harare Metropolitan City, Zimbabwe. *Waste Management & Research*, *37*(1), 83–94. doi:10.1177/0734242X18804029 PMID:30355247

Mateo-Sagasta, J., Raschid-Sally, L., & Thebo, A. (2015). Global wastewater and sludgeproduction, treatment and use. In *Wastewater* (pp. 15–38). Springer.

McDougall, F., White, P. R., Franke, M., & Hindle, P. (2001). *Integrated Solid Waste Management: A Lifecycle Inventory*. Blackwell Science. doi:10.1002/9780470999677

Muzenda, F., Masocha, M., & Misi, S. N. (2019). Groundwater quality assessment using a water quality index and GIS: A case of Ushewokunze Settlement, Harare, Zimbabwe. *Physics and Chemistry of the Earth Parts A/B/C*, *112*, 134–140. doi:10.1016/j.pce.2019.02.011

Ndunguru, M. G., & Hoko, Z. (2016). Assessment of water loss in Harare, Zimbabwe. *Journal of Water, Sanitation, and Hygiene for Development: a Journal of the International Water Association*, *6*(4), 519–533. doi:10.2166/washdev.2016.157

Nhapi, I. (2014). *Options for wastewater management in Harare*. CRC Press. doi:10.1201/9781482284041

Nhapi, I., Hoko, Z., Siebel, M., & Gijzen, H. (2002). Assessment of the major water and nutrient flows in the Chivero catchment area, Zimbabwe. *Physics and Chemistry of the Earth, Parts A/B/C, 27*(11-22), 783-792.

Nhubu, T., Mbohwa, C., Muzenda, E., & Patel, B. (2019). *Impacts of waste management practices on water resources in Harare.* Paper presented at the Wastes: Solutions, Treatments and Opportunities III: Selected Papers from the 5th International Conference Wastes 2019, Lisbon, Portugal. 10.1201/9780429289798-84

Nhubu, T., & Muzenda, E. (2019). Determination of the Least Impactful Municipal Solid Waste Management Option in Harare, Zimbabwe. *Processes (Basel, Switzerland)*, *7*(11), 785. doi:10.3390/pr7110785

Nhubu, T., Muzenda, E., Mbohwa, C., & Agbenyeku, E. (2020). Comparative assessment of compositing and anaerobic digestion of municipal biodegradable waste in Harare, Zimbabwe. *Environmental Progress & Sustainable Energy*, *39*(4), e13376. doi:10.1002/ep.13376

UNDESA. (2018). *World Urbanization Prospects: The 2018 Revision*. Retrieved from https://population.un.org/wup/Publications/Files/WUP2018-Report.pdf

Wang, X., Guo, M., Koppelaar, R. H., van Dam, K. H., Triantafyllidis, C. P., & Shah, N. (2018). A nexus approach for sustainable urban energy-water-waste systems planning and operation. *Environmental Science & Technology*, *52*(5), 3257–3266. doi:10.1021/acs.est.7b04659 PMID:29385332

Worldometer. (2020). *Zimbabwe Population*. Retrieved 16 March 2021 https://www.worldometers.info/world-population/zimbabwe-population/

Yadav, V., Karmakar, S., Dikshit, A. K., & Vanjari, S. (2016). A feasibility study for the locations of waste transfer stations in urban centers: A case study on the city of Nashik, India. *Journal of Cleaner Production*, *126*, 191–205. doi:10.1016/j.jclepro.2016.03.017

Zimstat. (2013). *2012 Zimbabwe Census National Report*. Harare, Zimbabwe: Zimbabwe National Statistics Agency. Retrieved from www.zimstat.co.zw/sites/default/files/img/publications/Population/National_Report.pdf

ADDITIONAL READING

Bidoglio, G., Vanham, D., Bouraoui, F., & Barchiesi, S. (2019). *The Water-Energy-Food-Ecosystems (WEFE) Nexus*. European Commission, Joint Research Centre. doi:10.1016/B978-0-12-409548-9.11036-X

Bradshaw, C. J. A., & Brook, B. W. (2014). Human population reduction is not a quick fix for environmental problems. *Proceedings of the National Academy of Sciences of the United States of America*, *111*(46), 16610–16615. doi:10.1073/pnas.1410465111 PMID:25349398

Ikhlayel, M., & Nguyen, L. H. (2017). Integrated approaches to water resource and solid waste management for sustainable development. *Sustainable Development*, *25*(6), 467–481. doi:10.1002d.1683

Karabulut, A., Egoh, B. N., Lanzanova, D., Grizzetti, B., Bidoglio, G., Pagliero, L., Bouraoui, F., Aloe, A., Reynaud, A., Maes, J., Vandecasteele, I., & Mubareka, S. (2016). Mapping water provisioning services to support the ecosystem–water–food–energy nexus in the Danube river basin. *Ecosystem Services, 17,* 278–292. doi:10.1016/j.ecoser.2015.08.002

Kibler, K. M., Reinhart, D., Hawkins, C., Motlagh, A. M., & Wright, J. (2018). Food waste and the food-energy-water nexus: A review of food waste management alternatives. *Waste Management (New York, N.Y.), 74,* 52–62. doi:10.1016/j.wasman.2018.01.014 PMID:29366796

Mancini, G., Luciano, A., Bolzonella, D., Fatone, F., Viotti, P., & Fino, D. (2020). A water-waste-energy nexus approach to bridge the sustainability gap in landfill-based waste management regions. *Renewable & Sustainable Energy Reviews,* 110441.

Russi, D., Ten Brink, P., Farmer, A., Badura, T., Coates, D., Förster, J., Kumar, R., & Davidson, N. (2013). The economics of ecosystems and biodiversity for water and wetlands. In *IEEP. London and Brussels.* Ramsar Secretariat.

Strzelecka, A., Ulanicki, B., Koop, S., Koetsier, L., Van Leeuwen, K., & Elelman, R. (2017). Integrating water, waste, energy, transport and ICT aspects into the smart city concept. *Procedia Engineering, 186,* 609–616. doi:10.1016/j.proeng.2017.03.277

KEY TERMS AND DEFINITIONS

Anaerobic Digestion: The decomposition of organic matter by microorganisms in oxygen starved environments.

Biological Treatment: The use of organisms to decompose organic wastes.

Composting: Natural decomposition and recycling of biodegradable waste fraction to produce compost.

Integrated Waste Management: A comprehensive waste prevention, reduction, reuse and recycle that incorporates the recovery of materials and energy from the unavoidable waste.

Integrated Water Resources Management: A process that promotes coordinated water resources development and management with land and other related resources to bring about sustainability in the use and management of water resources.

Landfilling: Long-term waste storage with the unregulated decomposition of organic waste fractions sometimes associate with landfill gas recovery and flaring as well as the capture and treatment of landfill leachate.

Nexus: The connection, linkages, or interrelations between two aspects in this case integrated water resources management and integrated waste management in GHMP.

Sustainable: The use of the earth's resources to meet the needs of current generations without compromising the needs of generations to come.

Thermal Treatment: The use of heat to treat waste including incineration and thermochemical conversion processes such as pyrolysis and gasification.

Waste to Energy or Energy Recovery: The production of energy from waste.

Compilation of References

Abel, R., & Thieme, M. (2009). Integrated Water Resources Management and Integrated River Basin Management. Encyclopedia of inland waters.

Adams, W. W. (1990). *Green Development: Environment and Sustainability in the Third World*. Routledge.

Addo, L. Y. (2010). Institutional analysis of urban water supply in Ghana: The case of Accra Metropolitan assembly. Academic Press.

African Union Assembly. (2008). *Sharm El-Sheikh Water and Sanitation Commitments*. Accessed from: http://ielrc.org/Content/e0825.pdf

African Union. (2008). *The eThekwini Declaration and AfricaSan Action Plan*. Accessed from; https://www.wsp.org/sites/wsp/files/publications/eThekwiniAfricaSan.pdf

Africon. (2008). *Unit Costs of Infrastructure Projects in Sub-Saharan Africa*. The World Bank. https://www.eu-africa-infrastructure-tf.net/attachments/library/aicd-background-paper-11-unit-costs-summary-en.pdf

Ahwoi, K. (2010). *Local government and decentralisation in Ghana*. Unimax Macmilan.

Akphakli, D. E., Manyeh, A. K., Akpakli, J., Kukula, V., & Gyapong, M. (2018). Determinants of access to improved sanitation facilities in rural districts of southern Ghana: Evidence from Dodowa health and demographic surveillance site. *BMC Research Notes*, *2018*(11), 473. doi:10.118613104-018-3572-6 PMID:30005694

Alba, R., & Bolding, A. (2016). IWRM Avant la Lettre? *Four Key Episodes in the Policy Articulation of IWRM in Downstream Mozambique Water Alternatives*, *9*(3), 549–568.

Alba, R., Bolding, A., & Ducrot, R. (2016). The Politics of Water Payments and Stakeholder Participation in the Limpopo River Basin, Mozambique. *Water Alternatives*, *9*(3), 569–587.

Alberti, M., Marzluff, J. M., Shulenberger, E., Bradley, G., Ryan, C., & Zumbrunnen, C. (2003). Integrating humans into ecology: Opportunities and challenges for studying urban ecosystems. *Bioscience*, *53*(12), 1169–1179. doi:10.1641/0006-3568(2003)053[1169:IHIEOA]2.0.CO;2

Alfred Duma Local Municipality. (2019/2020). *Integrated Development Plan*. Available: www.Ihlm.org.za/annualdocuments

Allouche, J. (2016). *The Birth and Spread of IWRM-A Case Study of Global Policy Diffusion and Translation Water Alternatives*. Academic Press.

Amiandamhen, S. O., Kumar, A., Adamopoulos, S., Jones, D., & Nilsson, B. (2020). Bioenergy production and utilization in different sectors in Sweden: A state of the art review. *BioResources*, *15*(4), 9834–9857. doi:10.15376/biores.15.4.Amiandamhen

Amigun, B., Parawira, W., & Musango, J. K. (2012). Anaerobic biogas generation for rural area energy provision in Africa. *Biogas*, 35-62.

Amir, S. (1997). The Role of Thermodynamics for Ecological Economics. *Ecological Economics*, *27*, 213–214.

Arimoro, F. O., Ikomi, R. B., & Osalor, E. C. (2006). The impact of sawmill wood wastes on the water quality and fish communities of Benin River, Niger Delta area, Nigeria. *World Journal of Zoology*, *1*(2), 94–102.

Arlosoroff, S. (1998). Water demand management. *UNCHS (Habitat) Regional Conference on Sustainable Consumption Patterns in Asian Cities*. http://www.fukuoka.unhabitat.org/docs/occasional_papers/project_a/09/water-e.html

Armstrong, J. W., & Graham, M. C. (1972). Comparative Study of Methods for Long-Range Market Forecasting. *Management Science*, *19*(2), 211–221. doi:10.1287/mnsc.19.2.211

Arnstein, S. R. (1969). A Ladder Of Citizen Participation. *Journal of the American Institute of Planners*, *35*(4), 216–224. doi:10.1080/01944366908977225

Arnstein, S. R. (1969). A ladder of citizen participation. *Journal of the American Planning Association*, *35*(4), 216–224.

Asafu-Adjaye, J. (2000). *Environmental Economics for Non-Economists. Asia-Pacific Development Journal*, *7*(2), 121–123.

Ashton, P. J., Hardwick, D., & Breen, C. M. (2008). Changes in water availability and demand within South Africa's shared river basins as determinants of regional social-ecological resilience. In M. J. Burns & A. B. V. Weaver (Eds.), *Advancing Sustainability Science in South Africa*. African Sun Media.

Asian Development Bank. (2011). *Sanitation and Water Management in Asia and the Pacific: Asia-Pacific*. Asian Development Bank.

Attanasi, E.D. Close E. R., & Lopez, M.A. (1975). *Techniques for Water Demand Analysis and Forecasting: A Case Study of Puerto Rico*. Academic Press.

Austin, L. M. (2006). *Operational safety of urine diversion toilets in Durban, South Africa*. In *32nd WEDC International Conference*, Colombo, Sri Lanka.

Autesserre, S. (2012). Dangerous tales: Dominant narratives on the Congo and their unintended consequences. *Journal of African Affairs*, *111*(443), 202–222. doi:10.1093/afraf/adr080

Auty, M. R. (1993). *Sustainable development in the mineral economies: the resource curse thesis*. London: Routledge Press.

Auty, M. R. (1993). *Sustainable development in the mineral economies: The resource curse thesis*. London: Routledge Press.

Awedoba, A. K. (2015). *The peoples of Northern Ghana*. National Commission on Culture.

Awoke, Z. A. (2012). *Assessment of challenges of sustainable rural water supply: Quarit Woreda, Amhara region: A Project Paper Presented to the Faculty of the Graduate School of Cornell University*. Cornell University.

Ayee, J. R. A. (2008). The balance sheet of decentralisation in Ghana. In F. Saito (Ed.), *Foundations for Local Governance*. Physica-Verlag HD. doi:10.1007/978-3-7908-2006-5_11

Ayres, R. U. (1998). Eco-thermodynamics: Economics and the second law. *Ecological Economics*, *26*(2), 189–209. doi:10.1016/S0921-8009(97)00101-8

Azimoh, C. L. (2016). *Sustainability and development impacts of off-grid electrification in developing countries: An assessment of South Africa's rural electrification program*. Mälardalen University.

Baazanah, P. (2020). Political Ecology of Water Resource Governance in Ghana: Towards Sustainable Pathway for Decentralisation and Participatory Water Supply in Rural Communities of the Savannah Region. *International Journal of Political Science and Development*, *8*(6), 252–279.

Babbie, E. (1990). *Survey Research Methods*. Wadsworth Publishing Company.

Babbie, E. (2005). *The basics of social research*. Thomson Wadsworth.

Babbie, E. (2012). *The practice of social research* (13th ed.). Wadsworth Cengage Learning.

Bailey, D. (2010). *Free Basic Services in Rural Kwazulu-Natal: The Built Environment Support Group*. https://www.ggln.org.za/media/k2/attachments/SoLG.2010-BESG.pdf

Baker, L. (2006, Summer). Observation: A Complex Research Method. *Library Trends*, *55*(1), 171–189. doi:10.1353/lib.2006.0045

Banerjee, S., Foster, V., Ying, Y., Skilling, H., & Wodon, Q. (2010). *Cost Recovery, Equity, and Efficiency in Water Tariffs Evidence from African Utilities*. Policy Research Working Paper 5384, The World Bank.

Banerjee, A., Chaudhury, S., Singh, D. K., Banerjee, I., Mahato, A. K., & Haldar, S. (2007). *Statistics Without Tears*. Inputs for Sample Size Calculations.

Banerjee, S., & Morella, E. (2011). *Africa's Water and Sanitation Infrastructure. Access Affordability and Alternatives*. The World Bank. doi:10.1596/978-0-8213-8457-2

Barber, B. (2003). Strong democracy: Participatory politics for a new age (20th anniversary ed.). University of California Press.

Barnard, S., Routray, P., Majorin, F., Peletz, R., Boisson, S., Sinha, A., & Clasen, T. (2013). Impact of Indian Total Sanitation Campaign on latrine coverage and use: A cross-sectional study in Orissa three years following programme implementation. *PLoS One*, *8*(8), e71438. doi:10.1371/journal.pone.0071438 PMID:23990955

Bartram, J., & Cairncross, S. (2010). Hygiene, sanitation, and water: Forgotten foundations of health. *PLoS Medicine*, *7*(11), e1000367. doi:10.1371/journal.pmed.1000367 PMID:21085694

Bassett, T. J., & Zueli, K. B. (2000). Environmental discourses and the Ivorian Savanna. *Annals of the Association of American Geographers*, *90*(1), 67–95. doi:10.1111/0004-5608.00184

Bastien, S., Hetherington, E., Eggers, M., Wamoyi, J., Hatfield, J., Manyama, M., & Kutz, S. (2017). Participatory science and innovation for improved sanitation and hygiene: Process and outcome evaluation of project SHINE, a school-based intervention in Rural Tanzania. *BMC Public Health*, *17*(1), 172. doi:10.118612889-017-4100-7 PMID:28173789

Batterbury, S. P. J., & Fernando, J. L. (2006). Rescaling Governance and the Impacts of Political and Environmental Decentralisation: An Introduction. *World Development*, *34*(11), 1851–1863. doi:10.1016/j.worlddev.2005.11.019

Bauen, A., Woods, J., & Hailes, R. (2004). *BioPowerswitch*. Academic Press.

Bazaanah, P. (2019). Sustainability of Rural Communities Drinking Water Systems and Local Development Projects in the Bole, West and Central Gonja Districts of the Savannah Region, Ghana. *Review of Social Sciences*, *4*(1), 16–36.

Bebbington, Hinojosa, L., Bebbington, D. H., Burneo, M. L., & Warnaars, X. (2008). Contention and Ambiguity: Mining and the Possibilities of Development. *Development and Change*, *39*(6), 887–914. doi:10.1111/j.1467-7660.2008.00517.x

Belward, A., Bisselink, B., Bódis, K., Brink, A., Dallemand, J., De Roo, A., Huld, T., Kayitakire, F., Mayaux, P., Ossenbrink, H., Pinedo, I., Sint, H., Thielen, J., Szabó, S., Tromboni, U., Willemen, L., & Monforti, F. (2011). Renewable energies in Africa. JRC Scientific and Technical Reports. doi:10.2788/1881

Béné, C., & Arthur, E. N. (2006). *From participation to governance: A critical review of the concept of governance, co-management and participation, and their implementation in Small-Scale Inland Fisheries*. World Fish Center.

Berahap, R. (2019). *Thinking outside the Grid: The role of decentralized power systems in electrifying Sub Saharan Africa*. www.policycenter.ma

Berg, H., Ohman, M. C., Troeng, S., & Linden, O. (1998). Environmental Economics of Coral Reef Destruction in Sri Lanka. Ambio, 27(8), 627-634.

Bernard, H. R. (2011). *Research methods in anthropology: Qualitative and quantitative approaches*. Altamira Press.

Berry, K. A., & Mollard, E. (2010). *Social participation in water governance and management*. Earthscan.

Bertrand-Krajewski, J. L., Barraud, S., & Chocat, B. (2000). Need for improved methodologies and measurements for sustainable management of urban water systems. *Environmental Impact Assessment Review*, *20*(3), 323–331. doi:10.1016/S0195-9255(00)00044-5

Bhattacherjee, A. (2012). *Social science research: principles, methods and practices*. University of South Florida.

Bhattacherjee, A. (2012). *Social science research: Principles, methods and practices*. University of South Florida.

Bice, S. (2014). What gives you a social licence? An exploration of the social licence to operate in the Australian mining industry. *Resources*, *3*(1), 62–80. doi:10.3390/resources3010062

Biju. (2011). Sampling in Social Science Research: Principles, Methods, And Practices used in Social and Health Science Researches. *Nigerian Postgraduate Medical Journal, 22*(4), 195-250.

Biswas, A. K. (2004). Integrated water resources management: A reassessment. *Water International*, *29*(2), 248–256. doi:10.1080/02508060408691775

Biswas, A. K. (2008). Integrated Water Resource Management: Is it Working? *Water: Research and Development*, *24*(1), 5–22.

Biswas, A. K. (2008). Integrated Water Resources Management: Is it Working? *International Journal of Water Resources Development*, *24*(1), 5–22. doi:10.1080/07900620701871718

Biswas, A. K., & Tortajada, C. (2008). Future Water Governance: Problems and Perspectives. *Water: Research and Development*, *2*(2), 129–139.

Biswas, K. (2007). *Water as a focus for regional development. Water Resources Management Series*. University Press.

Bless, C., & Smith, C. H. (2000). *Fundamentals of social research methods: an African perspective*. Juta Education.

Bless, C., & Smith, C. H. (2000). *Fundamentals of social research methods: An African perspective*. Juta Education.

Blimpo, M. P., & Cosgrove-Davies, M. (2019). Electricity Access in Sub-Saharan Africa: Uptake, Reliability, and Complementary Factors for Economic Impact. In *Electricity Access in Sub-Saharan Africa: Uptake, Reliability, and Complementary Factors for Economic Impact*. The World Bank. doi:10.1596/978-1-4648-1361-0

Bloomberg NEF. (2020). *Sub-Saharan Africa Market Outlook 2020*. Author.

Boland, J., & Whittington, D. (1998). *The political economy of increasing block tariffs in developing countries.* Economy and Environment Program for Southeast Asia. Special Paper No. 2. https://idl-bnc-idrc.dspacedirect.org/bitstream/handle/10625/31784/118117.pdf?sequence=8&isAllowed=y

Bolding, A., & Alba, R. (2016). *Viewpoint-IWRM and I: A Reflective Travelogue of the Flows and Practices Research Team Water Alternatives*. Academic Press.

Bole District Assembly-BDA. (2016). Composite budge of the Bole District Assembly for the 2016 fiscal year. Bole District Assembly.

Boschini, A. D., Pettersson, J., & Roine, J. (2007). Resource Curse or Not: A Question of Appropriability. *The Scandinavian Journal of Economics*, *109*(3), 593–617. doi:10.1111/j.1467-9442.2007.00509.x

Bourblanc, M. (2012). Transforming Water Resources Management in South Africa. *'Catchment Management Agencies' and the ideal of Democratic Development Journal of International Development, 24*, 637-648.

Bourblanc, M. (2015). The South African 'Ecological Reserve', A Travelling Concept. *Politikon, 42*(2), 275-292.

Braimah, I., & Fielmua, N. (2011). Community ownership and management of water and sanitation facilities: Issues and prospects in the Nadowli District of the Upper West Region of Ghana. *Journal of Sustainable Development in Africa*, *12*(2), 1520–1550.

Braun, V., & Clarke, V. (2006). Using thematic analysis in psychology. *Qualitative Research in Psychology*, *3*(2), 77–101. doi:10.1191/1478088706qp063oa

Bromley, D. W. (1978, March). Property Rules, Liability Rules, and Environmental Economics. *Journal of Economic Issues*, *12*(1), 43–60. doi:10.1080/00213624.1978.11503504

Bromly, D. W. (1978). *Property Rules, Liability Rules, and Environmental Economics. Journal of Economic Issues*, *12*(1), 43–60.

Brouckaert, C. J., Ridgeway, T., & Buckley, C. A. (2006). *A mathematical analysis of the drying of faeces in an urine diversion toilet- sensitivity analysis*. Water Research Commission.

Brown, O., Hammill, A., & Mcleman, R. (2007). Climate Change as the 'new' security threat: implications for Africa. *International Affairs Journal*, *83*(6), 1141-1154.

Brown, J. C., & Purcell, M. (2005). There's nothing inherent about scale: Political ecology, the local trap, and the politics of development in the Brazilian Amazon. *Geoforum*, *36*(5), 607–626. doi:10.1016/j.geoforum.2004.09.001

Brundtland Commission. (2014). *Defining sustainable development: The World Commission on Environment and Development.* Earthscan/Routledge.

Brunnschweiler, C., & Bulte, E. (2008). The resource curse revisited and revised: A tale of paradoxes and red herrings. *Journal of Environmental Economics and Management*, *55*(3), 248–264. doi:10.1016/j.jeem.2007.08.004

Bryant, R. L., & Bailey, S. (1997). *Third World political ecology*. Routledge.

Bryant, R. L., & Goodman, M. K. (2004). Consuming narratives: The political ecology of alternative consumption. *Transactions of the Institute of British Geographers*, *29*(3), 344–366. doi:10.1111/j.0020-2754.2004.00333.x

Bryman, A. (2003). *Quantity and quality in social research*. Routledge. doi:10.4324/9780203410028

Budds, J., & McGranahan, G. (2003). *Privatization and the provision of urban water and sanitation in Africa, Asia and Latin America*. IIED.

Bukenya, J. O. (2017). Determinants of Food Insecurity in Huntsville, Alabama, Metropolitan Area. *Journal of Food Distribution Research, 48*(1), 73–80.

Bulte, E., Damania, R., & Deacon, R. T. (2005). Resource Intensity, Institutions, and Development. *World Development, 33*(7), 1029–1044. doi:10.1016/j.worlddev.2005.04.004

Burke, E. M. (1968). Citizen participation strategies. *American Institute of Planners, 15*(5), 290–291.

Burns, N., & Grove, S. K. (2001). *The practice of nursing research: Conduct, critique & utilisation* (4th ed.). WB Saunders.

Campbell, R. M., Anderson, N. M., Daugaard, D. E., & Naughton, I. D. (2018). Technoeconomic and policy drivers of project performance for bioenergy alternatives using biomass from beetle-killed trees. *Energies, 11*(2), 293. Advance online publication. doi:10.3390/en11020293

Carpenter, L. (2012). *Conflict minerals in the Congo: blood minerals and Africa's under reported First World War.* Working paper. Suffolk University.

Carpenter, L. (2012). *Conflict minerals in the Congo: Blood minerals and Africa's under reported First World War.* Working paper. Suffolk University.

Carson, E., & Cobelli, E. (2001). *Modelling Methodology for Physiology and Medicine*. Academic Press.

Castro, J. E. (1992). *El conflicto por el agua en México. Los casos de Tuxtla Gutiérrez, Chiapas y Ciudad Juárez, Chihuahua, 1986-1991* (Unpublished thesis). Latin American Faculty of Social Sciences (FLACSO).

Castro, J. C. (2007). Water governance in the twentieth -century. *Ambiente & Sociedade, 10*(2), 1809–4422. doi:10.1590/S1414-753X2007000200007

Central Gonja District Assembly. (2016). *Composite budge of the Central Gonja District Assembly for the 2016 fiscal year*. Central Gonja District Assembly.

Central Gonja District Assembly. (2017). Annual progress report. Central Gonja District Assembly.

Central Gonja District Assembly. (2017). *Draft medium-term development plan (2014-2017) based on national medium-term development framework*. Central Gonja District Assembly.

Centre for Democratic Development & UNICEF. (2015). District League Table on Sanitation. CDD and UNICEF Annual Report on Sanitation: Accra

Centre for Renewable & Sustainable Energy Studies. (2014). *New solar resource maps for South Africa*. Author.

Chambers, R., & Myers, J. (2016). *Sustainable sanitation for all: experiences, challenges and innovations*. Practical Action.

Champ, P. A., Bishop, R. C., Brown, T. C., & McCollum, D. W. (1997). *Using Donation Mechanisms to Value Nonuse Benefits from Public Goods. Journal of Environmental Economics and Management, 33*, 151–162.

Chapman, D. (1973). Review. *Environmental Economics Ecology, 54*(6), 1414–1415.

Chapman, D. (1973, November). Review: Environmental Economics. *Ecological Society of America, 54*(6), 1414–1415.

Charis, G., Danha, G., & Muzenda, E. (2019a). *A review of the application of GIS in biomass and solid waste supply chain optimization: Gaps and opportunities for developing nations*. Academic Press.

Charis, G., Danha, G., & Muzenda, E. (2019b). A review of timber waste utilization: Challenges and opportunities in Zimbabwe. *Procedia Manufacturing, 35*, 419–429. doi:10.1016/j.promfg.2019.07.005

Charis, G., Danha, G., & Muzenda, E. (2019c). Waste valorisation opportunities for bush encroacher biomass in savannah ecosystems: A comparative case analysis of Botswana and Namibia. *Procedia Manufacturing*, *35*, 974–979. doi:10.1016/j.promfg.2019.06.044

Cheng, C. Y., & Urpelainen, J. (2015). Who should take the garbage out? Public opinion on waste management in Dar es Salaam, Tanzania. *Habitat International*, *46*, 111–118. doi:10.1016/j.habitatint.2014.11.001

Chikasema, G. (2009). An assessment of effective metering and billing as a water demand management tool. A case of Area 49 In Lilongwe, Malawi. In *Innovative Practices in the Water Supply and Sanitation Sector*. Sun Press.

Chikozho, C. (2005). *Policy and institutional dimensions of integrated river basin management: Broadening stakeholder participatory processes in the Inkomati River Basin of South Africa and the Pangani River Basin of Tanzania.* Commons Southern Africa Occasional Paper Series No. 12. Pretoria: Centre for Applied Social Sciences/Programme for Land and Agrarian Studies/ International Water Management Institute.

Chilisa, B. (2011). *Indigenous research methodologies*. Sage.

Chipofya, V., Kainja, S., & Bota, S. (2010). Policy harmonisation and collaboration amongst institutions- A strategy towards sustainable development, management and utilisation of water resources: Case of Malawi. *Desalination*, *252*, 261–266.

Chirwa, E., & Junge, N. (2007). *Poverty and Social Impact Analysis: Private Sector Participation in the Distribution and Management of Water Services, A study of Low income Areas in Blantyre and Lilongwe*. Ministry of Economic Planning and Development and Ministry of Irrigation and Water Development.

Christensen, P. P. (1989). Historical Roots for Ecological Economics- Biophysical Versus Allocative Approaches. *Ecological Economics*, *1*(1), 17–36. doi:10.1016/0921-8009(89)90022-0

Christenson, J., & Robinson, W. (1980). Community development in America. Iowa State University Press.

Christiaensen, L. (2017). *Agriculture in Africa - Telling Myths from Facts: A Synthesis.* Policy Research Working Paper No. 7979. Washington, DC: World Bank Publications.

Chuluwe, Q., & Nkhata, B. M. (2014). Analysis of Water Governance in Malawi: Towards a Favourable Enabling Environment? *Journal of Water, Sanitation, and Hygiene for Development: a Journal of the International Water Association*, *1*(2), 313–323. doi:10.2166/washdev.2014.087

Clark, A. (2005). *Resource-based Technology Innovation in South Africa: Innovations in South Africa's Off-grid Concession Programme, Employment Growth & Development Initiative, Human Sciences Resource Council*. HSRC.

Clark, V. L. P., & Cresswell, J. W. (2015). *Understanding Research: A Consumer's Guide*. PEARSON Publications.

Clifford-Holmes, J. K., Slinger, J. H., De Wet, C., & Palmer, C. G. (2017). Modelling in the "Muddled middle": A case study of water service delivery in post-apartheid South Africa. In *Social Systems Engineering: The Design of Complexity*. Wiley.

Cohen, B. (2006). Urbanization in developing countries: Current trends, future projections, and key challenges for sustainability. *Technology in Society*, *28*(1-2), 63–80. doi:10.1016/j.techsoc.2005.10.005

Coleman, D. Y. (Ed.). (2004). *Congo (DRC)*. Country Watch Review.

Coleman, J., & Ferejohn, J. (1986). Democracy and social choice. *Ethics*, *97*(1), 6–25. doi:10.1086/292814

Collier, P. & Hoeffler, A. (2001). *Greed and Grievance in Civil War*. World Bank.

Collier, P., & Hoeffler, A. (2002). On the Incidence of Civil War in Africa. *The Journal of Conflict Resolution*, *46*(1), 13–28. doi:10.1177/0022002702046001002

Collins, D. (2003). Pretesting survey instruments: An overview of cognitive methods. *Quality of Life Research: An International Journal of Quality of Life Aspects of Treatment, Care and Rehabilitation*, *12*(3), 229–238. doi:10.1023/A:1023254226592 PMID:12769135

Community Water and Sanitation Agency-CWSA. (2014). *2014 annual report*. Ministry of Water Resources, Works and Housing.

Community Water and Sanitation Agency-CWSA. (2015). *Fact sheet: Rural and small towns water service*. Ministry of Water Resources, Works and Housing.

Community Water and Sanitation Agency-CWSA. (2017). *Environmental and social management framework (ESMF) for the sustainable rural water and sanitation project- Additional financing addendum report final*. Ministry of Water Resources, Works and Housing.

CONAGUA. (2015). *Governments Contribution to Integrated Water Resource Management: The Scarcity on the Rise*. www.un.org/waterforlifedecade/waterandsustaibledevelopment2015/pdf/SP_Government_IWRM_sancho_FORMAT

Congo Central Bank. (2012). Condense hebdomadaire d' informations statistiques No 03. Congo Central Bank.

Congo Central Bank. (2012). Condense hebdomadaire d' informations statistiques No 3. Congo Central Bank.

Constable, L., & Cartwright, A. (2009). eThekwini Municipality Municipal Adaptation Plan. Health and Water eThekwini Municipality.

Conyers, D. (2003). Decentralization in Zimbabwe: A Local Perspective. *Public Administration and Development*, *23*(1), 115–124. doi:10.1002/pad.265

Corbin, J., & Strauss, A. (2008). *Basics of Qualitative Research*. Sage.

Costanza, R. (1989). What is ecological economics? Coastal and Environmental Policy Program. *Ecological Economics*, *1*(1), 1–7. doi:10.1016/0921-8009(89)90020-7

Costanza, R. (1989). *What is Ecological Economics?* Elsevier Science Publications.

Costanza, R. (1991). *Ecological Economics: The Science and Management of Sustainability*. Columbia University Press.

Creswell, J. W. (2013). *Research design: Qualitative, quantitative, and mixed methods approaches* (3rd ed.). Sage Publications.

Creswell, J. W. (2015). *Research design: Qualitative, quantitative, and mixed method approaches*. Sage publications.

Cropper, M. L., & Oates, W. E. (1992). Environmental Economics: A Survey. *Journal of Economic Literature*, *30*(2), 675–740.

Cropper, M. L., & Oates, W. E. (2002). Environmental Economics. A Survey. *Journal of Economic Literature*, *30*(2), 675–740.

Crow, B., & Sultana, F. (2002). Gender, Class, and Access to Water: Three Cases in a Poor and Crowded Delta. *Society & Natural Resources*, *15*(8), 8. doi:10.1080/08941920290069308

Cudjoe, E., Kemausuor, F., Miezah, K., & Mensah, M. (2015). *African perspective on cellulosic ethanol production*. doi:10.1016/j.rser.2015.04.059

D'Ydewalle, C. (1960). L'Union Minière du Haut Katanga de l'âge colonial a l'indépendance. Plon, 65-80.

Dagdeviren, H., & Robertson, S. A. (2008). *Reforming Without Resourcing: The Case of the Urban Water Supply in Zambia*. International Poverty Centre. https://uhra.herts.ac.uk/bitstream/handle/2299/2697/903025.pdf?sequence=1

Dahunsi, S. O., Shoyombo, A., & Fagbiele, O. (2018). Biofuel Development in Sub-Saharan Africa. *Anaerobic Digestion, 32*, 137–144. https://www.intechopen.com/books/advanced-biometric-technologies/liveness-detection-in-biometrics

Dangah, M. M., (2012). *The role of NGOs in the provision of water and sanitation in Northern Ghana: the case of Pronet in the Nadowli District*. Academic Press.

Daniele, V. (2011). Natural resources and the "quality" of economic development. *The Journal of Development Studies, 47*(4), 545–573. doi:10.1080/00220388.2010.506915

Davis, G. A., & Tilton, J. E. (2005). The resource curse. *Natural Resources Forum, 29*(3), 233–242. doi:10.1111/j.1477-8947.2005.00133.x

De Fraiture, C., & Wichelns, D. (2010). Satisfying Future Water Demands for Agriculture. *Agricultural Water Management, 97*(4), 502–511. doi:10.1016/j.agwat.2009.08.008

De Groot, R. S., Wilson, M. A., & Boumans, R. M. J. (2002). A Typology for the Classification, Description and Valuation of Ecosystem Functions, Goods and Services. *Ecological Economics, 41*(3), 393–408. doi:10.1016/S0921-8009(02)00089-7

Deacon, R. T., Brookshire, D. S., Fisher, A. C., Kneese, A. V., Kolstad, C. D., Scrogin, D., Smith, V. K., Ward, M., & Wilen, J. (1998). *Research Trends and Opportunities in Environmental and Natural Resource Economics. Environmental and Resource Economics, 11*(3-4), 383–397.

DEAT (Department of Environmental Affairs and Tourism). (2004). *A National Climate Change Response Strategy for South Africa*. Author.

DECOSA. (2015). *Support to De-Bushing Project Value Added End-Use Opportunities for Namibian Encroacher Bush* (Vol. 264). https://www.dasnamibia.org/download/value-added-end-use-opportunities-for-namibian-encroacher-bush/

Democratic Republic of Congo. (2006). Document de Stratégie de la Croissance et de la Réduction de la Pauvreté (DSCRP) [strategic document on growth and reduction of poverty]. DRC.

Democratic Republic of Congo. (2006). Document de Stratégie de la Croissance et de la Réduction de la Pauvreté (DSCRP) [Strategic document on growth and reduction of poverty]. Kinshasa: DRC.

Denby, K., Movik, S., Mehta, L., & van Koppen, B. (2016). The 'Trickle Down' of IWRM: A Case Study of Local-Level Realities in the Inkomati Water Management Area, South Africa. *Water Alternatives, 9*(3), 473–492.

Denzin, N. K., & Linoln, Y. S. (Eds.). (2008). *Collecting and Interpreting Qualitative Materials*. Sage.

Department of Energy. (2012). *A survey of energy-related behaviour and perceptions in South Africa: The Residential Sector*. Department of Energy.

Department of Energy. (2012). *Non-grid household electrification policy guideline*. Department of Energy.

Department of Environmental Affairs. (2010). *National Climate Change response Green Paper*. Department of Environmental Affairs.

Department of Minerals and Energy. (2007). *Integrated national electrification programme (INEP) plan to achieve universal access to electricity by 2012 masterplan / business plan*. Department of Minerals and Energy.

Department of Water Affairs and Forestry (DWAF). (2002). *Free Basic Water Implementation Strategy, Version 2*. http://www.dwaf.gov.za/Documents/FBW/FBWImplementationStrategyAug2002.pdf

Department of Water Affairs and Forestry (DWAF). (2004). *National Water Resource Strategy*. Author.

Department of Water Affairs and Forestry. (2002). *Free Basic Water*. Implementation Strategy Document version 2. www.dwaf.gov.za/annualdocuments

Derman, B., Gonese, F., & Ferguson, A. (2001). *Decentralization, Devolution and Development: Reflections on the Water Reform Process in Zimbabwe*. Centre for Applied Social Sciences, University of Zimbabwe.

Derman, B., & Manzungu, E. (2016). The complex politics of water and power in Zimbabwe: IWRM in the Catchment Councils of Manyame, Mazowe and Sanyati (1993-2001). *Water Alternatives*, *9*(3), 513–530.

Derman, B., & Manzungu, E. (2016). The Complex Politics of Water and Power in Zimbabwe: IWRM in the Catchment Councils of Manyame, Mazowe and Sanyati (1993-2001). *Water Alternatives*, *9*(3), 513–530.

Derman, B., & Prabhakaran, P. (2016). *Reflections of the Formulation and Implementation of IWRM in Southern Africa from a Gender Perspective Water Alternatives*. Academic Press.

Dernbach, J. C., & Mintz, J. A. (2011). Environmental Laws and Sustainability: An Introduction. *Sustainability*, *3*(3), 531–540. doi:10.3390u3030531

Dewar College of Education. (2009). *Research rundowns: Uncomplicated Reviews of Educational Research Methods*. https://researchrundowns.wordpress.com/quantitative-methods/instrument-validity-reliability/

Di Baldassarre, G., Sivapalan, M., Rusca, M., Cudenne, C., & Sanderson, M. R. (2019). Challenges in addressing Sustainable Development Goals. *Water Resources Journal*, *55*(23), 6325–6355. PMID:32742038

Dibwe dia Mwembu, D. (2001). *Bana Shaba abandonnés par leur père: structures de l'autorité et histoire sociale de la famille ouvrière au Katanga 1910–1997*. Paris: L'Harmattan.

Dietz, F.J., & van der Straten, J. (1992). Rethinking Environmental Economics: Missing links between Economic Theory and Environmental Policy. *Journal of Economic Issues*, *26*(1), 27-51.

Dietz, F. J., & van der Straaten, J. (1992, March). Rethinking Environmental Economics: Missing Links between Economic Theory and Environmental Policy. *Journal of Economic Issues*, *26*(1), 27–51. doi:10.1080/00213624.1992.11505261

Dill, B., & Crow, B. (2014). The colonial roots of inequality: Access to water. *urban East Africa. Water International*, *39*(2), 187–200. doi:10.1080/02508060.2014.894212

Dimpl, E. (2010). *Small-scale Electricity Generation from Biomass Part II : Biogas*. Academic Press.

Distefano, T., & Kelly, S. (2017). Are we in deep water? Water scarcity and its limits to economic growth. *Journal of Ecological Economics*, *142*, 130–147. doi:10.1016/j.ecolecon.2017.06.019

Division provinciale de l'intérieur et sécurité. (2015). *Cellule d'Analyses des Indicateurs de Développement*. Available at https://www.caid.cd/index.php/donnees-par-villes/ville-de-lubumbashidomaine=fiche

Dodman, D., & Satterthwaite, D. (2008). Institutional Capacity, Climate Change Adaptation and the Urban Poor. *Institute of Development Studies Bulletin*, *39*(4).

Donelly, J. (2000). *Realism and international relations*. Cambridge University Press. doi:10.1017/CBO9780511612510

Dotse, L. N. A. (2016). *Exploring the relationship between water scarcity on food and nutritional security in rural households in the Nqgeleni location, Eastern Cape.* http://wiredspace.wits.ac.za/bitstream/handle/10539/21715/FINAL%20RESEARCH%20REPORT%20LAURA%20CORRECTIONS%20%282%29%20%281%29.pdf?sequence=1&isAllowed=y

Duffy, J. E. (2003). Biodiversity loss, trophic skew and ecosystem functioning. *Ecology Letters*, *6*(8), 680–687. doi:10.1046/j.1461-0248.2003.00494.x

Duncker, L. C. (2008). Prejudices and attitude change to dry toilets in South Africa CSIR. Academic Press.

Duncker, L. C., Matsebe, G. N., & Austin, L. M. (2006). Use and Acceptance of Urine- Diversion Sanitation Systems in South Africa. Water Research Commission.

Dungumaro, E., & Madulu, N. (2003). Public participation in integrated water Resources Management: The case of Tanzania. *Journal of Physics and Chemistry of the Earth*, 1009–1014.

Dunning, T. (2008). *Crude Democracy. Natural Resource Wealth and Political Regimes.* Cambridge University Press. doi:10.1017/CBO9780511510052

EC. (2016). *Supporting decentralisation, local governance and local development through a territorial approach.* https://europa.eu

Eckersley, R. (2004). *The green state: Rethinking democracy and sovereignty.* MIT Press. doi:10.7551/mitpress/3364.001.0001

Eckersley, R. (2019). Ecological democracy and the rise and decline of liberal democracy: Looking back, looking forward. *Environmental Politics*, *29*(2), 214–234. doi:10.1080/09644016.2019.1594536

Eguavoen, I. (2006). Drinking water policy: Water rights and allocation practice in rural Northern Ghana. Centre for Development Research (ZEF).

Elmer, N. (1999). Politics: The Education Effects. Taylor and Francis. *LTD*, *1*(1), 251–273.

Elston, D. M. (2017). Presentation of data. *Journal of the American Academy of Dermatology*. Advance online publication. doi:10.1016/j.jaad.2017.11.009 PMID:29128450

EMA. (2016). *Waste generation and management in Harare, Zimbabwe: Residential areas, commercial areas and schools*. Environmental Management Agency.

Enserink, B., Patel, M., Kranz, N., & Maestu, J. (2007). Cultural factors as co-determinants of participation in river basin management. *Economia e Sociologia*, *12*(2), 24. doi:10.5751/ES-02096-120224

Enzensberger, H.M. (1990). A critique of political ecology. *Kursbuch, 33*, 3-31.

Escobar, A. (1998). Whose knowledge, whose nature? Biodiversity, conservation, and the political ecology of social movements. *Journal of Political Ecology*, *5*, 53–82.

Escobar, A. (2001). Culture sits in places: Reflections on globalism and subaltern strategies of localization. *Political Geography*, *20*(2), 139–124. doi:10.1016/S0962-6298(00)00064-0

Eskom. (2018). *Integrated report 2018.* Retrieved from http://www.eskom.co.za/IR2018/Documents/ Eskom2018IntegratedReport.pdf

Esrey, S. A., Potash, J. B., Roberts, L., & Shiff, C. (1991). Effects of improved water supply and sanitation on ascariasis, diarrhea, dracunculiasis, hookworm infection, schistosomiasis, and trachoma. *Bulletin of the World Health Organization*, *69*(5), 600–609. PMID:1835675

Esrey, S., Gough, J., Rapaport, D., Sawyer, R., Simpson-Hebert, M., Vargus, J., & Winblad, U. (1998). *Ecological Sanitation. Swedish International Development Corporation.* SIDA.

European Commission (EC). (2002). *Water in the European Union.* European Communities Publication.

Fabio, A. D., & Blustein, D. L. (2016). *From Meaning of Working to Meaning Lives: The Challenges of Expanding Decent Work.* Frontiers Publications.

FAO. (2013). *World Food and Agriculture.* Food and Agriculture Organization Statistical Yearbook.

FAO. (2016). *Coping with water scarcity in agriculture: A global framework for action in a changing climate.* https://www.fao.org/3/a-i6459e.pdf

FAO. (2017). *Water at a Glance The relationship between water, agriculture, food security and poverty.* https://www.fao.org/3/ap505e/ap505e.pdf

FAO. (2017). *Water for Sustainable Food and Agriculture: A report produced for the G20 Presidency of Germany.* https://www.fao.org/3/a-i7959e.pdf

Farber, S. C., Costanza, R., & Wilson, M. A. (2002). Economic and Ecological Concepts for Valuing Ecosystem Services. *Ecological Economics, 41*(3), 375–392. doi:10.1016/S0921-8009(02)00088-5

Ferguson, A. E., & Mulwafu, W. O. (2004) *Decentralization, Participation and Access to Water Resources in Malawi.* University of Wincosin-Madison.

Feron, S. (2016). Sustainability of Off-Grid Photovoltaic Systems for Rural Electrification in Developing countries: A Review. *Sustainability, MDPI, 8*(1326), 1–26. doi:10.3390u8121326

Fewtrell, L., Kaufmann, R. B., Kay, D., Enanoria, W., Haller, L., & Colford, J. M. Jr. (2005). Water, sanitation, and hygiene interventions to reduce diarrhea in less developed countries: A systematic review and meta-analysis. *The Lancet. Infectious Diseases, 5*(1), 42–52. doi:10.1016/S1473-3099(04)01253-8 PMID:15620560

Fiala, N. (2008). Measuring Sustainability: Why the Ecological Footprint is Bad Economics and Bad Environmental Science. *Ecological Economics, 67*(4), 1–7. doi:10.1016/j.ecolecon.2008.07.023

Fiani, R. (2004, December). An Evaluation of the Role of the State and Property Rights in Douglas North's Analysis. *Journal of Economic Issues, 38*(4), 1003–1020. doi:10.1080/00213624.2004.11506754

Figueiredo, P., & Perkins, P. E. (2013). Women and water management in times of climate change: Participatory and inclusive processes. *Journal of Water Resources, 60*(3), 188–194. doi:10.1016/j.jclepro.2012.02.025

Finn, J., & Jacobson, M. (2008). Just Practice: A Social Justice Approach to Social Work. Peosta, IL: Eddie Bowers Publishing.

Flores, A., Buckley, C., & Fenner, R. (2008). *Selecting Wastewater Systems for Sustainability in Developing Countries.* In *11th International Conference on Urban drainage*, Edinburgh, UK.

Fonjong, L. N., & Ngekwi, M. A. (2014). Challenges of Water Crisis on Women's Socio-Economic Activities in the Buea Municipality, Cameroon. *Journal of Geography and Geology, 6*(4), 122–131. doi:10.5539/jgg.v6n4p122

Food and Agriculture Organisation- FAO. (2017). *Coping with water scarcity: Challenge of the 21st Century.* https://www.fao.org/3/a-aq444e.pdf

Food and Agriculture Organization (FAO). (2009). *Water at a Glance: The relationship between water, agriculture, food security and poverty.* https://www.fao.org/3/ap505e/ap505e.pdf

Fox, N. J. (2008). Post-positivism. In The SAGE Encyclopedia of Qualitative Methods. Sage.

Fung, A. (2006). Varieties of participation in complex governance. *Public Administration Review*, (66), 66–75.

Galan, D. I., Kim, S. S., & Graham, J. P. (2013). Exploring changes in open defecation prevalence in sub-Saharan Africa based on national level indices. *BMC Public Health*, *13*(1), 500–527. doi:10.1186/1471-2458-13-527 PMID:23721324

Gandhi, V., Crase, L., & Herath, G. (2007). *Determinants of Institutional Success for Water in India: Results from a Study across Three States*. Paper presented at the 51st Annual Conference of the Australian Agricultural and Resource Economics Society (AARES), Queenstown, New Zealand.

Gandhi, V. P., & Namboodiri, N. V. (2009). *Groundwater Irrigation in India: Gains, Costs and Risks*. Indian Institute of Management.

Gann, L. H., & Duignan, P. (n.d.). *Burden of Empire*. An appraisal of Western Colonialism in Africa South of the Sahara. *Hoover Institution Press.*

Garcia, D. J., Lovett, B. M., & You, F. (2019). Considering agricultural wastes and ecosystem services in Food-Energy-Water-Waste Nexus system design. *Journal of Cleaner Production*, *228*, 941–955. doi:10.1016/j.jclepro.2019.04.314

Garrett, N. (2010). 'Mining as a Source of Growth. In J. Herderschee & J. E. Mans (Eds.), *Obstacles to Growth in the Democratic Republic of Congo*. World Bank.

Garrett, N. (2010). Mining as a Source of Growth. In J. Herderschee & J. E. Mans (Eds.), *Obstacles to Growth in the Democratic Republic of Congo*. World Bank.

Garrett, N., & Lintzer, M. (2010). Can Katanga's mining sector drive growth and development in the DRC? *Journal of Eastern African Studies: the Journal of the British Institute in Eastern Africa*, *4*(3), 400–424. doi:10.1080/17531055.2010.517408

Gaventa, J. (2006). Finding the space for change: A power analysis. *Institute of Development Studies Bulletin*, *37*(6), 23–33. doi:10.1111/j.1759-5436.2006.tb00320.x

Gerlach, E., & Franceys, R. (2009). Regulating Water Services for the poor: The case of Amman. *Geoforum*, *40*(3), 431–441. doi:10.1016/j.geoforum.2008.11.002

Gerlach, E., & Franceys, R. (2010). Regulating water services for all in developing economies. *World Development*, *38*(9), 1229–1240. doi:10.1016/j.worlddev.2010.02.006

German Industry and Commerce in Ghana (GICG). (2018). *Access to clean drinking water and water management in Ghana*. Ghana Statistical Service.

Gershman, S. D. (2013). *An evaluation of public participation techniques using Arnstein's ladder: The Portland plan*. University of Florida.

Ghana Meteorological Agency. (2019). *Regional weather, Savannah Region*. Accessed on 05/10/2019 from; https://www.meteo.gov.gh/gmet/regional-weather-savannah-region/

Ghana News Agency. (2017). *Challenges with eradicating illegal mining in Ghana: A perspective from the grassroots*. www.ghanadistricts.com

Ghana News Agency. (2018). *Nadowli District rural water coverage higher than MDGs target*. http://www.ghanabusinessnews.com

Ghana Statistical Service (GSS). (2011). *Ghana-multiple indicator cluster survey-2011*. Government of Ghana.

Ghana Statistical Service (GSS). (2014). *2010 population and housing census: District analytical report, West Gonja District*. Ghana Statistical Service.

Ghana Statistical Service (GSS). (2014a). *2010 population and housing census: District Analytical report of the West Gonja District*. Ghana Statistical Service.

Ghana Statistical Service (GSS). (2018). *Ghana multiple indicator cluster survey; Snapshots on key findings*. Ghana Statistical Service.

Ghana Statistical Service. (2012). *2010 Housing and population census report*. Ghana Statistical Service.

Ghana Statistical Service. (2012b). *2010 Population and Housing Census: Summary of final results*. Ghana Statistical Service.

Ghana Statistical Service. (2014). *2010 Housing and population census report. District analytical report*. Ghana Statistical Service.

Giordano, M., & Shah, T. (2014). From IWRM back to integrated water resources management. *International Journal of Water Resources Development*, *30*(3), 364–376. doi:10.1080/07900627.2013.851521

Gleick, P. H. (1998). Water in Crisis: Paths to Sustainable Water Use. *Ecological Applications*, *8*(3), 571–579. doi:10.1890/1051-0761(1998)008[0571:WICPTS]2.0.CO;2

Gleick, P. H. (2000). The Changing Water Paradigm- A Look at Twenty-first Century Water Resources Development. *Water International*, *25*(1), 127–138. doi:10.1080/02508060008686804

Global Renewables Outlook. (2019). *Energy Transformation Sub-saharan Africa*. https://www.irena.org/-/media/Files/IRENA/Agency/Publication/2020/Apr/IRENA_GRO_R10_Sub-Saharan_Africa.pdf?la=en&hash=DB067EF85E0FDB6B8762833E77CC80F3975E46DC

Global Water Partnership (GWP). (2000). Integrated Water Resources Management. Technical Advisory Committee, Background Paper, no. 4.

Global Water Partnership (GWP). (2004). *Integrated water resources management (IWRM) and water efficiency plans by 2005. Why, What and How? TAC Background Papers No. 10*. Global Water Partnership.

Global Water Partnership. (2000). *Integrated Water Resource Management*. Author.

Goff, M., & Crow, B. (2014). What is water equity? The unfortunate consequences of a global focus on 'drinking water'. *Water International*, *39*(2), 159–171. doi:10.1080/02508060.2014.886355

Goforth, C. (2015). *Using and Interpreting Cronbach's Alpha*. https://data.library.virginia.edu/using-and-interpreting-cronbachs-alpha

Goforth, C. (2015). *Using and Interpreting Cronbach's Alpha*. https://data.library.virginia.edu/using-and-interpreting-cronbachs-alpha/

Goldratt, E. M., & Cox, J. (2016). *The goal*. Routledge. doi:10.4324/9781315270456

Goswami, K. B., & Bisht, D. (2017). The Role of Water Resources in Socio-Economic Development. *International Journal for Research in Applied Science and Engineering Technology*, *5*(12), 1669–1674.

Goswami, K. B., & Bisht, P. S. (2017). The role of water resources in socio economic development. *International Journal for Research in Applied Science and Engineering Technology*, *887*(12), 2321–9653.

Government of Ghana. (1992). 1992 Constitution of the republic of Ghana. Tema: Ghana Publishing Corporation.

Government of Ghana. (1993). Local government Act, 1993 (Act 462). Tema: Ghana Publishing Corporation.

Government of Ghana. (1994). *Structure of Local Government, 1994.* Government of Ghana.

Government of Ghana. (2010). *Community water and sanitation agency sector guidelines-general (rural communities & small towns).* CWSA.

Government of Ghana. (2016). *Local government Act, 2016 (Act 936).* Ghana Publishing Corporation.

Government of Ghana. (2019). *Ministry of Sanitation and Water Resources: Medium-term expenditure framework (2017-2019).* Ministry of Finance.

GoZ (Government of Zimbabwe). (2000). *Water resources management.* The Government Printers.

GoZ. (1994). *Report of the Commission of Inquiry into Appropriate Agricultural Land Tenure Systems. Three volumes.* Government of Zimbabwe.

Gray, D. E. (2009). *Doing Research in the Real World* (2nd ed.). Sage.

Green, K. E. (2016). A political ecology of scaling: Struggles over power, land and authority. *Geoforum, 74*, 88–97. doi:10.1016/j.geoforum.2016.05.007

Grevisse, F., & Brynart, M. (2011). Energy Poverty in Europe: Towards a more global understanding. *ECEEE 2011 Summer Study. Energy Efficiency First: The Foundation of a Low-carbon Society.*

Grey, D., & Sadoff, C. (2008). *Water for Growth and Development: A Theme Document of the 4th World Water Forum.* Commission National del Agua Publications.

Grimm, N. B., Grove, J. M., Pickett, S. T. A., & Redman, C. L. (2000). Integrated approaches to long-term studies of urban ecological systems. *Bioscience, 50*(7), 571–584. doi:10.1641/0006-3568(2000)050[0571:IATLTO]2.0.CO;2

Grimm, N. B., & Redman, C. L. (2004). Approaches to the study of urban ecosystems: The case of Central Arizona – Phoenix. *Urban Ecosystems, 7*(7), 199–213. doi:10.1023/B:UECO.0000044036.59953.a1

Groves, R. M., & Heeringa, S. (2006). *Responsive Design for Household Surveys:* Tools for actively controlling survey errors and costs. *Journal of Royal Statististics Soceiety*, 439–57.

Gumbi, N. (2017). *Challenges of water supply management* (Doctoral dissertation). University of Limpopo.

Gumbo, B., & Van der Zaag, P. (2002). Water losses and the political constraints to demand management: the case of the City of Mutare, Zimbabwe. *Physics and Chemistry of the Earth, Parts A/B/C, 27*(11-22), 805-813.

Guness, M., Jackson, S., Rodda, N., Smith, M., Trotter, D., Buckley, C., & Macleod, N. (2010). *Impact of Buried Urine Diversion Waste on Environmental Quality and Plant Growth.* Water Research Commission.

Guness, M., Pillay, S., Rodda, N., Smith, M., Buckley, C., & Macleod, N. (2009). *Quality of Leacheate from Burried Urine Diversion Toilet Waste.* Water Research Commission.

GWP. (2000). *Integrated water resources management (Background Papers No 4). Technical Advisory Committee (TAC).* Global Water Partnership.

GWP. (2006). *Gender mainstreaming: An essential component of sustainable water management. Policy Brief No. 3.* Global Water Partnership.

Gyapong, M., Sarpong, D., Awini, E., Manyeh, K. A., Tei, D. O. G., & Agyepong, I. A. (2013). Health and demographic surveillance system profile: The Dodowa HDSS. *International Journal of Epidemiology*, *2013*(42), 1686–1696. doi:10.1093/ije/dyt197 PMID:24415607

Habtezion, S. (2012). *Gender, agriculture and food security*. https://gest.unu.edu/static/files/tm4_africa_gender-climatechange-and-food-security.pdf

Hahn, R. W. (2000). The Impact of Economics on Environmental Policy. *Journal of Environmental Economics and Management*, *39*(3), 375–399. doi:10.1006/jeem.1999.1119

Hamann, R. (2003). Mining companies' role in sustainable development: Why and how of corporate social responsibility from a business perspective. *Development Southern Africa*, *20*(2), 1470–3637. doi:10.1080/03768350302957

Hamann, R. (2004). Corporate social responsibility, partnerships, and institutional change: The case of mining companies in South Africa. *Natural Resources Forum*, *28*(4), 278–290. doi:10.1111/j.1477-8947.2004.00101.x

Hamann, R., & Kapelus, P. (2004). Corporate Social Responsibility in Mining in Southern Africa: Fair accountability or just greenwash? *Development*, *47*(3), 85–92. doi:10.1057/palgrave.development.1100056

Hamberto, P. (2011). Social Equity and Integrated Resources Management. *Global Water Partnership*, *3*(15), 5–83.

Hamilton, D. (1989). Thorstein Veblen as The First Professor of Marketing Science. *Journal of Economic Issues*, *23*(4), 1097–1103. doi:10.1080/00213624.1989.11504976

Hammarberg, K., Kirkman, M., & de Lacey, S. (2016). Qualitative research methods: When to use them and how to judge them. *Human Reproduction (Oxford, England)*, *31*(3), 498–501. doi:10.1093/humrep/dev334 PMID:26759142

Hancock, B. (1998). *Trent Focus for Research and Development in Primary Health Care: An Introduction to Qualitative Research*. Trent Focus.

Harper-Simmonds, L., Mendelsohn, J., Roux, J.-P., Pallet, J., Brown, C., Middleton, A., & Kruse, J. (2012). *Development of an inventory of ecosystem services in Namibia*. www.met.gov.na/files/downloads/179_ResMob-Inventory report2-final.pdf

Harris, D., Kleiber, D., Goldin, J., Darkwah, A., & Morinville, C. (2017). Intersections of gender and water: Comparative approaches to everyday gendered negotiations of water access in underserved areas of Accra, Ghana and Cape Town. *South Africa Journal of Gender Studies*, *26*(5), 561–582. doi:10.1080/09589236.2016.1150819

Hart, T. G. B., & Aliber, M. (2009). Should subsistence agriculture be supported as a strategy to address rural food insecurity? *Agrekon*, *48*, 4.

Haruna, M. (2017). *Chieftaincy controversies over the Bole and Mandari skins since 1908*. https://mynewsgh.com/chieftaincy-controversies-over-the-bole-and-mandari-skins-since-1908/

Hawksworth, D., Archer, C., Hanssen, A., Tronnberg, L., Lutchminarayan, R., Knight, S., Smith, M., & Rodda, N. (2008). Occurrence of Helminths and Protozoan Parasites (Giardia and Cryptosporidium) in the Filing Vaults of Urine Diversion Toilets in the eThekwini Municipal Area (Durban, South Africa). *WISA Conference*, Sun City.

Hawksworth, D., Archer, C., Rodda, N., Smith, M., Appleton, C., & Buckley, C. (2009). *An Improved Method for the Recovery of Ascaris Ova from Solid waste from Urine Diversion Toilets*. Water Research Commission.

Heath, T., Parker, A., & Weatherhead, E. K. (2010). *How to Climate Proof Water and Sanitation Services for the Urban Poor. Cranfield University*.

Heath, T., Parker, A., & Weatherhead, E. K. (2012). Testing a rapid climate change adaptation assessment for water and sanitation providers in informal settlements in three cities in Sub-Saharan Africa. *Environment and Urbanization, 24*(2), 619–637. doi:10.1177/0956247812453540

Hecht, S. B., & Cockburn, A. (1990). *Fate of the Forest: Developers, Destroyers, and Defenders of the Amazon.* Harper Collins.

Hellberg, S. (2014). Water, life and politics: Exploring the contested case of eThekwini municipality through a governmentality lens. *Geoforum, 56*, 226–236. doi:10.1016/j.geoforum.2014.02.004

Hellum, A., & Derman, B. (2005). Negotiating water rights in the context of a new political and legal landscape in Zimbabwe. In F. von Benda-Beckmann, K. von Benda-Beckmann, & A. Griffiths (Eds.), *Mobile People, Mobile Law: Expanding Legal Relations in a Contracting World* (pp. 177–198). Ashgate.

Henry, J. G., & Heinke, G. W. (1989). *Water Use in North American Cities: Environmental Science and Engineering.* Prentice-Hall.

Hering, J. G., & Ingold, K. M. (2012). Water Resources Management: What Should Be Integrated? *Journal of Science, 336*(6086), 1234. PMID:22679083

Herman, D. (2006). *Sustainable development: Definition, principles and policies.* Springer.

Herring, R. (2001). Authority and scale in political ecology: some cautions on localism. In L. Buck (Ed.), *Biological Diversity: Balancing Interests Through ACM.* CRC Press. doi:10.1201/9781420042597.ch10

Herselman, J. E. (2013). *Guideline for the Utilisation and Disposal of Water Treatment Residues.* http://www.wrc.org.za/wp-content/uploads/mdocs/TT%20559-13.pdf

Hesselbein, G. (2007). Essor et déclin de l'État Congolais. No 21. London: Crisis States Research Centre.

Heynen, N., Perkins, H. A., & Roy, P. (2006). The political ecology of uneven urban green space: The impact of political economy on race and ethnicity in producing environmental inequality in Milwaukee. *Urban Affairs Review, 42*(1), 3–25. doi:10.1177/1078087406290729

Hoekstra, A. Y. (2006). *The Global Dimension of Water Governance: Nine Reasons for Global Arrangements in order to cope with local water problems.* United Nations Education, Science and Cultural Organization-IHE Institute for Water Education.

Hoenke, J. (2010). New Political Topographies: Mining Companies and Indirect Discharge in Southern Katanga (DRC). *Politique Africaine (Paris, France), 120*(4), 105–127. doi:10.3917/polaf.120.0105

Holstein, J. A., & Gubrium, J. F. (2008). *Handbook of Constructionist Research.* The Guilford Press.

Hoornweg, D., & Bhada-Tata, P. (2012). *What a Waste: A Global Review of Solid Waste Management.* Retrieved from https://openknowledge.worldbank.org/handle/10986/17388

Hoppe, K. P. (2010). Decentralisation and local governance: Theory and the practice in Botswana. *Development Southern Africa, 17*(4), 519–534. doi:10.1080/03768350050173912

Hoppe, K. R. (2009). Capacity development for good governance in developing Countries: Some lessons from the field. *International Journal of Public Administration, 32*(8), 728–740.

Hordijk, M., Scott, L. M., & Sutherland, C. (2015). Resilience, transition or transformation? A comparative analysis of changing water governance systems in four southern cities. *Environment and Urbanization, 26*(1), 130–146. doi:10.1177/0956247813519044

Hornborg, A. (1998). Towards an Ecological Theory of Unequal Exchange: Articulating World System Theory and Ecological Economics. *Ecological Economics*, *25*(1), 127–136. doi:10.1016/S0921-8009(97)00100-6

Hossain, M. Z. (2015). Water: The most precious resource of our life. *Global Journal of Advanced Research*, *2*(9), 1436–1445.

Hounsome, R., & Iyer, K. (2006). Headline Climate Change Adaptation Strategy for Durban Council for Scientific and Industrial Research. Academic Press.

Hove, J., D'Ambruoso, L., Mabetha, D., Peter Byass, P., Kahn, K., Khosa, S., & Witter, S. (2019). 'Water is life': Developing community participation for clean water in rural South Africa. *BMJ Global Health*, *3*(4), 96–105. doi:10.1136/bmjgh-2018-001377 PMID:31263583

Howard, G., & Bartram, J. (2003). *Domestic Water Quantity, Service, Level and Health*. World Health Organization. www.who.int/water_sanitation_health/diseases/WSH03.02.pdf?ua=1

Howard, K. W. F. (1997). Impacts of urban development on ground water. In Environmental Geology of urban areas. Geological Association of Canada.

Human Science Research Council (HSRC). (2013). *Civil Society Organisation' Participation in Food Security Activities in South Africa*. www.hsrc.ac.za/en/research-data/ktree-doc/13815

Hunter, P.R., MacDonald, A.M., & Carter, R.C. (2010). Water supply and Health. *PLoS Med, 7*(11).

Hunt, N., & Tyrrell, S. (2001). *Stratified Sampling*. Webpage at Coventry University.

Hyden, G., & Neef, J. (2004). Comparing governance across Countries and over time: Conceptual challenges. In D. Olowy & S. Sako (Eds.), *Better governance and public policy: Capacity building and democratic renewal in Africa*. Bloomfield Kumarian Press Inc.

Iakovou, E., Karagiannidis, A., Vlachos, D., Toka, A., & Malamakis, A. (2010). Waste biomass-to-energy supply chain management : A critical synthesis. *Waste Management*, *30*(10), 1860–1870. doi:10.1016/j.wasman.2010.02.030 PMID:20231084

IBNET. (2017) *International Benchmarking Network Water and Sanitation Utilities Context Information & Indicators – Definitions*. IBNET. https://www.ib-net.org/toolkit/ibnet-indicators/metering-practices/

IEA. (2010). SuStainable Production of Second Generation Biofuels. *Renewable Energy*, 1–39. 9789461739698

IEA. (2018). *Country Report. Sweden – 2018 update. Bioenergy policies and status of implementation*. IEA.

IES. (2013). A situational Analysis of Solid Waste Management in Zimbabwe's Urban Centres. IES.

IFAD (International Fund for Agricultural Development). (2012). *Agricultural cooperatives: paving the way for food security and rural development*. CapeNSSANews Publications

Igoe, J. (2013). Consume, connect, conserve: Consumer spectacle and the technical mediation of neoliberal conservation's aesthetic of redemption and repair. *Cultural Geographies*, *6*(1), 16–28.

Indranil, D.E. (2009, Jan. 3). Can Decentralisation Improve Rural Water Supply Services. *Economic and Political Weekly*.

International Conference on Water and Environment. (1992). *The Dublin statement and report of the conference*. Author.

International Energy Agency. (2015). *Africa Energy Outlook*. https://www.iea.org/PUBLICATIONS/FREEPUBLICATIONS/PUBLICATION/WEO2014_AFRICAENERGYOUTLOOK.PDF

International Monetary Fund (IMF). (2005). *Republic of the Congo: Selected Issues and Statistical Appendix.* IMF Country Report No. 05/373. http://www.lepotentiel.com/afficher_article.php?id_edition=&id_article=12707

International Water Management Institute (IWMI). (2007). *Water for Food, Water for Life: A Comprehensive Assessment of Water Management in Agriculture.* Earthscan and International Water Management Institute.

IRENA. (2012). Renewable Energy Technologies: Cost Analysis Series. In *Zeitschrift fur Geomorphologie, Supplementband* (Vol. 1). www.irena.org

IRENA. (2016). *Innovation Outlook Advanced Liquid Biofuels.* http://www.irena.org/DocumentDownloads/Publications/IRENA_Innovation_Outlook_Advanced_Liquid_Biofuels_2016.pdf

IRENA. (2018). *Bioenergy from Finnish forests: Sustainable, efficient and modern use of wood.* www.irena.org

IRENA. (2020). *Scaling up renewable energy deployment in Africa.* www.irena.org/publications

IRGCWSS. (1992). *The Dublin Statement on Water and Sustainable Development.* Paper presented at the International Conference on Water and the Environment: Development Issues for the 2lst century, Dublin, Ireland.

Iribarnegaray, M. A., & Seghezzo, L. (2012). Governance, sustainability and decision making in water and sanitation management systems. *Sustainability, 4*(11), 2922–2945. doi:10.3390u4112922

Ismail, Z., & Khembo, P. (2015). Determinants of energy poverty in South Africa. *Journal of Energy in Southern Africa, 26*(3), 66–78. doi:10.17159/2413-3051/2015/v26i3a2130

Jacobs, C., Rivett, U., & Chemisto, M. (2018). Developing capacity through co0design: The case of two municipalities in rural South Africa. *Information Technology for Development, 25*(2), 204–226. doi:10.1080/02681102.2018.1470488

Jakeman, J., & Letcher, R. A. (2003). Integrated assessment and modelling. Features, principles and examples for catchment management. *Environmental Modelling & Software, 181*(6), 491–505. doi:10.1016/S1364-8152(03)00024-0

Jenkins, J. O. (2006). *Water metering: in search of a more critical debate.* University of Hertfordshire. https://uhra.herts.ac.uk/bitstream/handle/2299/3355/903361.pdf?sequence=3

Jenkins, H., & Yakovleva, N. (2006). Corporate social responsibility in the mining industry: Exploring trends in social and environmental disclosure. *Journal of Cleaner Production, 14*(3-4), 271–284. doi:10.1016/j.jclepro.2004.10.004

Jessop, B. (2000). The crisis of the national spatio-temporal fix and the tendential ecological dominance of globalization capitalism. *International Journal of Urban and Regional Research, 24*(2), 323–360. doi:10.1111/1468-2427.00251

JICA. (1996). *The Study of Water Pollution Control in Upper Manyame River Basin in the Republic of Zimbabwe.* JICA.

Jiménez, A., LeDeunff, H., Ricard Giné, R., Sjödin, J., Cron, R., Murad, S., Takane, M., & Bartram, J. (2019). The enabling environment for participation in water and sanitation: A conceptual framework. *Water (Basel), 11*(308), 1–21. doi:10.3390/w11020308

Joint Monitoring Programme for Water Supply, Sanitation, and Hygiene. (2019). *Estimates on the use of water, sanitation and hygiene by region (2000-2017).* UNICEF and WHO. www.washdata.org

Jonch-Clausen, T. (2005). *Integrated Water Resources Management (IWRM) and Water Efficiency Plans by 2005 Why, What and How?* Global Water Partnership.

Jootun, D., McGhee, G., & Marland, G. (2009). Reflexivity: Promoting rigour in qualitative research. *Nursing Standard, 23*(23), 42–46.

Kabir, E., Kim, K., & Szulejko, J. E. (2017). Social Impacts of Solar Home Systems in Rural Areas: A Case Study in Bangladesh. *Energies, 10*(1615), 1–12. doi:10.3390/en10101615

Kalenga, J. N. (2013). Economic and toxicological aspects of copper industry in Katanga, DR Congo. *The Japanese Journal of Veterinary Research, 61*(Supplement), S23–S32. PMID:23631149

Kamuti, T., & Serumaga-Zake, S. (2017a). *Guideline for Good Practices: The Impact of Extractives on Quality and Quantity of Drinking Water.* WaterAid.

Kamuti, T., & Serumaga-Zake, S. (2017b). *The Impact of Extractives on Quality and Quantity of Drinking Water in Malawi, Mozambique, South Africa, Zambia and Zimbabwe.* WaterAid.

Kanyerere, T., & Pietersen, K. (2016). *Assessment of Water Service Delivery in the Municipalities of City of Tshwane, City of Cape Town and Ethekwini.* https://etd.uwc.ac.za/xmlui/bitstream/handle/11394/5332/Makhari_cl_msc_ns_2016.pdf?sequence=1&isAllowed=y

Karin, B., & Eva, L. (2006). Planting Trees to Mitigate Climate Change: Contested Discourses of Ecological Modernization. *Green Governmentality and Civic Environmentalism Global Environmental Politics, 6*(1), 50–75.

Karl, M. (2000). Monitoring and evaluating stakeholder participation in agriculture and rural development projects: A literature review. Food and Agriculture Organization (FAO).

Kasangana, K. K., Makonese, T., & Mbonane, T. P. (2018). Dissemination of Free Basic Electricity in Low-income Settlements. *Conference Paper.* 10.23919/DUE.2018.8384380

Kawulich, B. B. (2004). Qualitative data analysis techniques. *Journal of Research in Education, 14*(1), 96–113.

Khadza, A. K. (2010). The Emergence of Water as a 'Human Right' on the World Stage: Challenges and Opportunities. *International Journal of Water Resources Development, 26*(1), 37–49. doi:10.1080/07900620903391838

Kilondo, N.D (2010). *Territoires produits par l'industrie minière et rupture de filiation ouvrière: dynamique populaire et territoriale dans le cas des cités ouvrières de la Gécamines (RDC)* [Territories produced by mining industries and failure of labour filiation: popular dynamic and territorial in the case of Gecamines' labour cities]. Universite Catholique de Louvain.

Kingdom, B., Liemberger, R., & Marin, P. (2006). *The Challenge of Reducing Non-Revenue Water (NRW) in Developing Countries How the Private Sector Can Help: A Look at Performance-Based Service Contracting.* Water Supply and Sanitation Sector Board Discussion Paper Series, paper n 8

Kjeldsen, T. R., Lundorf, A., & Rosbjerg, D. (1999). Barriers to sustainable water resources management: A Zimbabwean case study. *Hydrological Sciences Journal, 44*(4), 529–539. doi:10.1080/02626669909492250

Klein, P. G. (1999). *New Institutional Economics.* University of Georgia.

Koehler, J., Rayner, S., Katuva, J., Thomson, P., & Hope, B. (2018). A cultural theory of drinking water risks, values and institutional change. *Global Environmental Change, 50*, 268–277. doi:10.1016/j.gloenvcha.2018.03.006

Koolwal, G., & van de Walle, D. (2013). Accesses to water, Women's Work and Child Outcomes. *Economic Development and Cultural Change, 61*(2), 369–405. doi:10.1086/668280

Kramer, A., & Pahl-Wostl, C. (2014). The global policy network behind integrated water resources management: Is it an effective norm diffusor? *Ecology and Society, 19*(4), 11.

Kuediasala, F. (2009). *Mission au Katanga d'une équipe d'experts de la Banque mondiale avec a l'agenda la relance de la Gecamines.* Le Potentiel.

Kulabako, R. N., Nalubega, M., Wozei, E., & Thunvik, R. (2010). Environmental health practices, constraints and possible interventions in peri-urban settlements in developing countries–a review of Kampala, Uganda. *International Journal of Environmental Health Research*, *20*(4), 231–257. doi:10.1080/09603120903545745 PMID:20645199

Kwo, E. (1984). From our correspondent: community education and community development in Cameroon: The British Colonial experience, 1922–1961. *Community Development (Columbus, Ohio)*, *19*(4), 204–213.

Kwo, E. (1986). Community education and community development: Anglophone Cameroon experience since 1961. *Community Development (Columbus, Ohio)*, *21*(3), 61–168.

La Générale des Carrières et des Mines (GECAMINES). (1967). *Annual Report of Gécamines*. Lubumbashi: Gecamines.

La Générale des Carrières et des Mines (GECAMINES). (1974). *Annual Report of Gécamine*. Lubumbashi: Gécamines.

La Générale des Carrières et des Mines (GECAMINES). (2015a). Available from: http://gecamines.cd/html/histoire.html

La Générale des Carrières et des Mines (GECAMINES). (2015b). Available from: http://gecamines.cd/html/social.html

Lago, M., & Möller-Gulland, J. (2012). *WP3 EX-POST Case studies: Comparative Analysis Report*. EPI WATER. http://www.feem-project.net/epiwater/docs/epi-water_dl_3-2.pdf

Lane, J. (2005). *Non-Governmental Organizations: New Players in Water Policy*. http://www.wateraid.org/

Lane, J. (2012). Barriers and opportunities for sanitation and water for all, as envisaged by the New Delhi statement. *IDS Bulletin*, *43*(2), 13–20. doi:10.1111/j.1759-5436.2012.00302.x

Lange, G., & Hassan, R. (2006), Introduction to water issues and water accounting in Southern Africa. *Economics of Water Management in Southern Africa*, 1-11.

Lant, C. (2005). *Water Resources Sustainability: An Ecological Economics Approach*. North-Central Section- 39th Annual Meeting.

Lautze, J., De Silva, S., Giordano, M., & Sanford, L. (2011). Putting the cart before the horse: Water governance and IWRM. *Natural Resources Forum*, *35*(1), 1–8. doi:10.1111/j.1477-8947.2010.01339.x

Le Billion, P. (2000). The political ecology of transition in Cambodia 1989-1999: War, peace and forest exploitation. *Development and Change*, *31*(4), 785–805. doi:10.1111/1467-7660.00177

Leedy, P. D., & Ormrod, J. E. (2005). *Practical Research*. Pearson Custom.

Leff, E. (2017). Power-knowledge relations in the field of political ecology. *Ambiente & Sociedade*, *20*(3), 1809–4422. doi:10.1590/1809-4422asocex0004v2032017

Leggett, C. G., & Bockstael, N. E. (2000). Evidence of the Effects of Water Quality on Residential land Prices. *Journal of Environmental Economics and Management*, *39*(2), 121–144. doi:10.1006/jeem.1999.1096

Lencha, A. (2013). *Rural water supply management and sustainability in Ethiopia with special emphasis on water supply schemes in Adama Area*. Swedish University of Agricultural Sciences.

Leung Pah Hang, M. Y., Martinez-Hernandez, E., Leach, M., & Yang, A. (2017). Insight-Based Approach for the Design of Integrated Local Food-Energy-Water Systems. *Environmental Science & Technology*, *51*(15), 8643–8653.

Lilongwe Water Board (LWB). (2013). *Lilongwe Water Board Annual report 2011-2012*. Author.

Limpopo Department of Economic Development, Environment and Tourism. (2013). *Limpopo Green economy plan including provincial climate change response*. Retrieved from https://www.environment.gov.za/sites/default/files/docs/limpopogreen_economyplan.pdf

Livingston, M. L. (2005). Evaluating Changes in Water Institutions: Methodological Issues at the Micro and Meso levels. *Water Policy*, *7*(1), 21–34. doi:10.2166/wp.2005.0002

Lockie, S., Franetovich, M., Sharma, S., & Rolfe, J. (2012). Democratisation versus engagement? Social and economic impact assessment and community participation in the coal mining industry of the Bowen Basin, Australia. *Impact Assessment and Project Appraisal*, *26*(3), 177–187. doi:10.3152/146155108X357257

Logar, I. (2018). *Environmental Economics (EnvEco)*. https://www.eawag.ch/en/deparment/es/main-focus/environment-economics-enveco/

Logar, I. (2018). *Environmental Economics (EnvEco)*. https://www.eawag.ch/en/department/es/main-focus/environment-economics-enveco

Lohr, S. L. (2010). *Sampling design and Analysis* (2nd ed.). Cengage, Chapman/ Hall Press.

Love, D., Zingoni, E., Ravengai, S., Owen, R., Moyce, W., Mangeya, P., & Hranova, R. (2006). Characterization of diffuse pollution of shallow groundwater in the Harare urban area, Zimbabwe. In *Groundwater pollution in Africa*. CRC Press.

Lssozi, A. (2012). Values and participation: The role of culture in nature preservation and environmental education among the Baganda. *The Journal of Sustainability Education*, *7*(3), 88–99.

Luh, M., Baum, R., & Bartram, J. (2020). Equity in water and sanitation: Developing an index to measure progressive realisation of the human right. *International Journal of Hygiene and Environmental Health*, *216*(6), 662–667. doi:10.1016/j.ijheh.2012.12.007 PMID:23333082

Lutchminarayan, R. D. (2007). Sanitation, Water and Hygiene in eThekwini Municipality, Durban, South Africa: Baseline Cross-Sectional Study. UKZN.

Machethe, E. M. (2011). *The causes and impact of water shortage on the households of Ga-Kgapane Township in the Limpopo Province (Research dissertation)*. University of Limpopo.

MacNealy and Mary Sue. (1999). *Strategies for Empirical Research in Writing*. Longman.

Macy, P. (1999). *Urban water demand management in Southern Africa: the conservation potential. Sida Publications on Water Resources No. 13*. Sida.

Makarichi, L., Kan, R., Jutidamrongphan, W., & Techato, K. A. (2019). Suitability of municipal solid waste in African cities for thermochemical waste-to-energy conversion: The case of Harare Metropolitan City, Zimbabwe. *Waste Management & Research*, *37*(1), 83–94. doi:10.1177/0734242X18804029 PMID:30355247

Malhotra, N. K., & Birks, D. F. (2007). *Marketing research: An applied approach*. Pearson Education. doi:10.1108/S1548-6435(2007)3

Manikandan, S. (2011). Measures of central tendency: The mean. *Journal of Pharmacology & Pharmacotherapeutics*, *2*(2), 140–142. doi:10.4103/0976-500X.81920 PMID:21772786

Manu, P. K. (2015). *Challenges of water supply in urban Ghana: An assessment of the coping strategies in Adentan Municipality*. University of Ghana.

Manzungu, E. (2002). More than a headcount: Towards stakeholder representation in catchment management in South Africa and Zimbabwe. *Physics and Chemistry of the Earth*, *27*(11-22), 927–933. doi:10.1016/S1474-7065(02)00095-5

Manzungu, E., & Derman, B. (2016). Surges and Ebbs. *National Politics and International Influence in the Formulation and Implementation of IWRM in Zimbabwe Water Alternatives, 9*(3), 493–512.

Manzungu, E., & Derman, B. (2016). Surges and ebbs: National politics and international influence in the formulation and implementation of IWRM in Zimbabwe. *Water Alternatives, 9*(3), 493–512.

Mapedza, E., Manzungu, E., Rosen, T., Ncube, P., & Van Koppen, B. (2016). Decentralized water governance in Zimbabwe: disorder within order. Water Resources and Rural Development. doi:10.1016/j.wrr.2016.05.00

Mapedza, E., & Geheb, K. (2010). Power dynamics and water reform in the Zimbabwean context: Implications for the poor. *Water Policy, 12*(4), 517–527. doi:10.2166/wp.2010.141

Mara, D., & Alabaster, G. (2006). A new paradigm for low-cost water supplies and sanitation in developing countries. *Water Policy, 10*(2), 119–129. doi:10.2166/wp.2008.034

Mara, D., Lane, J., Scott, B., & Trouba, D. (2010). Sanitation and health. *PLoS Medicine, 7*(11), e1000363. doi:10.1371/journal.pmed.1000363 PMID:21125018

Maree, K. (2016). First steps in research (2nd ed.). Van Schaik.

Marro, P., & Bertsch, N. (2015). *Energy a Success in Bangladesh: Getting the Business Model Right. Mandaluyong City.* Asian Development Bank.

Marshall, G. (2008). Nesting, subsidiarity, and community-based environmental governance beyond the local level. *International Journal of the Commons, 2*(1), 75-98.

Marston, A. J. (2017). Alloyed waterscapes: Mining and water at the nexus of corporate social responsibility, resource nationalism, and small-scale mining. *Water (Basel), 4*(1), e1175. doi:10.1002/wat2.1175

Martinez-Alier, J., Munda, G., & O'Neill, J. (1998). Weak Comparability of Values as a Foundation for Ecological Economics. *Ecological Economics, 26*(3), 277–286. doi:10.1016/S0921-8009(97)00120-1

Marugwara, W. (2014). *The plight of internally displaced communities: case of Tokwe Mukosi.* http://hdl.handle.net/11408/595

Marysse, S., & Andre, C. (2001). Guerre et pillage economique en RDC [War and looting in the DRC]. Paris: L'harmattan.

Marysse, S., & Geenen, S. (2009). *Win-win or unequal exchange? The case of the Sino-Congolese cooperation agreements.* Cambridge University Press. doi:10.1017/S0022278X09003978

Massarutto, A. (2007). *Urban water conflicts: an ecological – economics approach.* United Nations Educational and Cultural Organisation.

Massarutto, A. (2008). *Urban Water conflicts: an ecological –Economic approach Paris.* UNESCO.

Masuku, M., Selepe, M., & Ngcobo, N. (2017). Status of Household Food Security in Rural Areas at uThungulu District, Kwa-Zulu Natal, South Africa. *African Journal of Hospitality, Tourism and Leisure, 6*(2), 19–38.

Mateo-Sagasta, J., Raschid-Sally, L., & Thebo, A. (2015). Global wastewater and sludgeproduction, treatment and use. In *Wastewater* (pp. 15–38). Springer.

May, T. (Ed.). (2002). Qualitative Research in Action. London: Sage Publications.

Mbua, L. (2013). Water supply in Buea, Cameroon: Analysis and the possibility of rainwater harvesting to stabilize the water demand. University of Germany.

McCarthy, J. (2002). First World political ecology: Lessons from the Wise Use movement. *Environment & Planning, 34*(7), 1281–1302. doi:10.1068/a3526

McCarthy, J. (2005). Scale, sovereignty, and strategy in environmental governance. *Antipode, 37*(4), 731–753. doi:10.1111/j.0066-4812.2005.00523.x

McDougall, F., White, P. R., Franke, M., & Hindle, P. (2001). *Integrated Solid Waste Management: A Lifecycle Inventory*. Blackwell Science. doi:10.1002/9780470999677

McFerrin, H. (2017). *News from Africa: The DRC poverty rate continues to stagnate*. Available at https://borgenproject.org/the-drc-poverty-rate/

McFerson, H. M. (2010). Extractive industries and African democracy: Can the "resource curse" be exorcised? *International Studies Perspectives, 11*(4), 335–353. doi:10.1111/j.1528-3585.2010.00410.x

Medeiros, P., & Sivapalan, M. (2020). A hard-path to soft-path solutions: Slow-fast dynamics of human adaptation to droughts in a water scarce environment. *Hydrological Sciences Journal, 15*(2), 2150–3435. doi:10.1080/02626667.2020.1770258

Mehlum, H., Moene, K., & Torvik, R. (2006). Cursed by resources or institutions? *World Economy, 29*(8), 1117–1131. doi:10.1111/j.1467-9701.2006.00808.x

Mehta, L., Movik, S., Bolding, A., Derman, B., & Manzungu, E. (2016). *Introduction to the Special Issue- Flows and Practices: The Politics of Integrated Water Resources Management (IWRM) in Southern Africa Water Alternatives*. Academic Press.

Mehta, L. (2014). Water and human development. *World Development, 59*, 59–69. doi:10.1016/j.worlddev.2013.12.018

Mensah, J. V., & Antwi, K. B. (2013). *Bridging water and sanitation infrastructure gap in Ghana. Pennsylvania Clarion*. University of Pennsylvania.

Methvin, T., & Philipp, F. (2017). *Energy Provision at the Base of the Pyramid: Are there viable business models to serve South Africa's low-income communities?* Impact Amplifier. Retrieved from http://www.impactamplifier.co.za/wp- content/uploads/2017/02/BOP_Final.pdf

Meyer, W. N. (2007). *The economics of water. Water for life, sanitation for dignity*. Van Schaik Publishers.

Miller, E. S. (2003, March). Evolution and Stasis: The Institutional Economics of David Hamilton. *Journal of Economic Issues, 37*(1), 1045–1059. doi:10.1080/00213624.2003.11506554

Miller, R., & Brewer, J. (2003). *The A-Z of social research*. Sage. doi:10.4135/9780857020024

Ministry of Environment, Conservation of Nature and Tourism of the Democratic Republic of Congo. (2012). Rapport national synthèse sur le développement durable en RDC [National report on the sustainable development in the DRC]. Author.

Ministry of Sanitation and Water Resources. (2019). *Enforce sanitation by-laws on sanitation*. MSWR.

Ministry of Water Resources, Works, and Housing (MWRWH). (2014). Water sector strategic development plan (2012-2025). Ministry of Water Resource, Works, and Housing.

Misra, K., & Malhotra, G. (2011). Water management: The obscurity of demand and supply in Delhi, India. *International Journal of Management of Environmental Quality, 23*(1), 23–35. doi:10.1108/14777831211191575

Mitchell, R., Agle, B., & Sonnenfeld, J. (1999). Who Matters to CEOs? An Investigation of Stakeholder Attributes and Salience, Corporate Performance, and CEO Values. *The Academy of Management Review, 22*(5).

Mitchell, R., Agle, B., & Wood, D. (1997). Toward a Theory of Stakeholder Identification and Salience: Defining the Principle of Who and What Really Counts. *Academy of Management Review, 22*(4), 853–886. doi:10.2307/259247

Mkandawire, T., Mulwafu, W., Chipofya, V., Bota, S., & Kainja, S. (2011). *The Road to a National Integrated Water Resources Management/Water Efficiency (IWRM/WE) Plan: Challenges and Lessons from Malawi*. University of Malawi.

Mnguni, M., Ndlovu, S., Gounden, T., Pfaff, W., Rodda, N., & Buckley, C. (2008). Health Risk Assessment of the Operation and Maintenance of Urine Diversion Toilets in eThekwini Municipality. *WISA Conference*, Sun City.

Mnkeni, P. N. S., & Austin, L. M. (2009). Fertiliser Value of Human Manure from pilot urine-diversion toilets. *Water S.A., 35*(1), 134–138.

Mohammed, A. K. (2015). The politics of municipal fragmentation in Ghana. *Commonwealth Journal of Local Governance, 16*(17), 123–145. doi:10.5130/cjlg.v0i0.4493

Mohammed, I. (2012). Thematic Analysis: A Critical Review of Its Process and Evaluation. *West East Journal of Social Sciences, 1*(1), 39–45.

Mohan, G., & Stokke, K. (2000). Participatory development and empowerment: The dangers of localism. *Third World Quarterly, 21*(2), 247–268. doi:10.1080/01436590050004346

Moilwa, N., & Wilkinson, M. (2006). The Effect of Hygiene Communication on Emptying of Urine Diversion Toilets. *32nd WEDC International Conference*, Colombo, Sri Lanka.

Moloi, D. (2004). The Cholera Outbreak: A 2000-2002 case study of the source outbreak In Madlebe Tribal Authority Areas. UThungulu Region.

Momentum Occupational Care South Africa. (2018). *Water scarcity and associated health risks.* www.ocsa.co.za

Montague, D. (2002). Stolen goods: Coltan and conflict in the Democratic Republic of Congo. *SAIS Review (Paul H. Nitze School of Advanced International Studies), 22*(1), 103–118. doi:10.1353ais.2002.0016

Moriarty, J. (2011). *Qualitative methods overview: improving the evidence base for adult social care practice*. London School of Economics and Political Science Publications.

Moriarty, P., Butterworth, J., & Batchelor, C. (2004). *Integrated water resources management and the domestic water and sanitation sub-sector.* IRC International Water and Sanitation Centre.

Morrison, J. (2004). *Managing leakage by District Metered Areas: a practical approach.* Water 21. IWA. http://www.geocities.ws/kikory2004/39_Water21_5th_article_DMA.pdf

Moser, P. K. (2015). *Epistemology.* https://www.researchgate.net/publication/288829852_Epistemology

Motloung, S. (2010). SIBU and the crisis of water service delivery in Sannieshof, North West Province. *TD: The Journal for Trans disciplinary Research in Southern Africa, 6*(1), 25-56.

Movik, S., Mehta, L., van Koppen, & Denby, K. (2016). *Emergence, Interpretations and Translations of IWRM in South Africa Water Alternatives.* Academic Press.

Movik, S., Mehta, L., & Manzungu, E. (2016). The Flow of IWRM in SADC: The Role of Regional Dynamics. *Advocacy Networks and External Actors Water Alternatives, 9*(3), 434–455.

Mtunzi, B., Mampwheli, N., Meyer, E., & Mungwena, W. (2012). Bagasse-based co-generation at Hippo Valley Estates sugar factory in Zimbabwe. *Journal of Energy in Southern Africa, 23*(1), 15–22. doi:10.17159/2413-3051/2012/v23i1a3158

Mugambe, R. K., Tumwesigye, N. M., & Larkan, F. (2013). Barriers to accessing water, sanitation and hygiene among people living with HIV/AIDS in Gomba and Mpigi districts in Uganda: A qualitative study. *Journal of Public Health, 21*(1), 29–37. doi:10.100710389-012-0515-x PMID:23703127

Mukheibir, P., & Sparks, D. (2003). Water resource management and climate change in South Africa: Visions, driving factors and sustainable development indicators. Energy and Development Research Centre.

Mulder, M. W. (2005). *A political policy analysis of the integrated water resources management approach in South Africa's water policy (1998 - 2001)* (Master's dissertation). University of Pretoria.

Muller, M, (n.d.). *Inter-basin Water Sharing to Achieve Water Security- A South African Perspective.* Department of Water Affairs and Forestry

Muller, M. (2008). Free Basic Water Policy-A sustainable instruments for a sustainable future in South Africa. *Environment and Urbanization, 20*(6), 68–80.

Mulwafu, W., Chipeta, C., Chavula, G., Ferguson, A., Nkhoma, B. G., & Chilima, G. (2003). Water demand management in Malawi: Problems and prospects for its promotion. *Physics and Chemistry of the Earth, 28*(20-27), 787–796. doi:10.1016/j.pce.2003.08.003

Munasinghe, M. (1992). *Environmental economics and sustainable development.* Paper presented at the UN Earth Summit, Rio de Janeiro.

Munasinghe, M. (1993). Environmental Economics and Biodiversity Management in Developing Countries. Ambio, 22(2-3), 126-135.

Munasinghe, M. (1993). Environmental Economics and Biodiversity Management in Developing Countries. *Ambio,* Vol. 22, No 2/3, *Biodiversity: Ecology, Economics. Policy*, (May), 126–135.

Murray, A., & Buckley, C. (2010). *Designing Reuse-Oriented Sanitation Infrastructure: The design for service planning approach.* UKZN.

Murshed, M. S. (2004). When does natural resource abundance lead to a resource curse? *Environmental Economics Programme, 4*(1), 1–58.

Musingafi, M., & Chando, C. (2013). The Legal Framework in the Governance of Potable Water Supply In Zimbabwe: A Global, Regional And National Overview of The IWRM Paradigm. *Journal of Law, Policy, and Globalization.*

Mutikanga, H. E. (2012). *Water Loss Management: Tools and Methods for Developing Countries.* UNESCO-IHE.

Mutikanga, H. E., Vairavamoorthy, K., Sharma, S. K., & Akita, C. S. (2011). Operational Tools for Decision Support in Leakage Control. *Water Practice & Technology, 6*(3), wpt2011057. Advance online publication. doi:10.2166/wpt.2011.057

Muzenda, F., Masocha, M., & Misi, S. N. (2019). Groundwater quality assessment using a water quality index and GIS: A case of Ushewokunze Settlement, Harare, Zimbabwe. *Physics and Chemistry of the Earth Parts A/B/C, 112*, 134–140. doi:10.1016/j.pce.2019.02.011

Myers. (2002). Local communities and the new environmental planning: A case study from Zanzibar. *AREA, 34*(2), 149-159.

Nabli, M. K., & Nugent, J. B. (1989). The New Institutional Economics and its Applicability to Development. *World Development, 17*(9), 1333–1347. doi:10.1016/0305-750X(89)90076-4

Naik, P. K. (2016). Water crisis in Africa: Myths or reality? *International Journal of Water Resources Development, 33*(2), 326–339. doi:10.1080/07900627.2016.1188266

Nastiti, A., Komarulzaman, A., & Sudradjat, A. (2018). The prospect of the domestic water equity indicators in Indonesia: A Review. *IOP Conference Series. Earth and Environmental Science, 106*(1), 1–6. doi:10.1088/1755-1315/106/1/012065

Nathan, S., Rotem, A., & Ritchie, J. (2002). Closing the gap: Building the capacity of non-government organizations as advocates for health equity. *Health Promotion International, 17*(3), 69–78. doi:10.1093/heapro/17.1.69 PMID:11847140

National Consumer Council (NCC). (2002). *Towards a sustainable water charging policy.* A Discussion Paper. NCC.

National Development Planning Commission and United Nations Development Programme. (2010). *Ghana millennium development goals report.* Government of Ghana.

National Planning Commission. (2012). *The South African National Development Plan (Vision 2030).* National Planning Commission.

National Statistical Office of Malawi. (2008). *2008 Population and Housing Census. Zomba.* National Statistical Office.

National Statistical Office of Malawi. (2009). *The Welfare Monitoring Survey. Zomba.* National Statistical Office.

National Statistical Office of Malawi. (2010). *Malawi Demographic and Health Survey 2010: Preliminary Report. Zomba.* National Statistics Office.

Ndunguru, M. G., & Hoko, Z. (2016). Assessment of water loss in Harare, Zimbabwe. *Journal of Water, Sanitation, and Hygiene for Development: a Journal of the International Water Association, 6*(4), 519–533. doi:10.2166/washdev.2016.157

Neil Adger, W. (2001). Advancing a political ecology of global environmental discourses. *Development and Change, 32*(4), 681–715. doi:10.1111/1467-7660.00222

Neil, C. (2018). *The politics of the environment: Ideas, activism, policy* (3rd ed.). Cambridge University Press.

Nelson, N. K., & Shilling, D. (2019). *Traditional ecological knowledge: Learning from indigenous practices for environmental sustainability.* Cambridge University Press.

Nemarwandwe, N., & Kodzanayi, W. (2001). *Institutional arrangements for water resources use: A case study from Southern Zimbabwe.* Institute of Environmental Studies-University of Zimbabwe.

Neuman, W. L. (2010). Social research methods: Qualitative and quantitative approaches (7th ed.). Pearson.

Neuman, W. L. (2014). *Social research methods: qualitative and quantitative approaches.* Pearson Education Limited.

Neuman, W. L. (2014). *Social research methods: Qualitative and quantitative approaches.* Pearson Education Limited.

Ng'ong'ola, D. H. (1999). *Policies Influencing Patterns of use of Water Resources in Malawi.* Ministry of Irrigation and Water Development.

Nhapi, I., Hoko, Z., Siebel, M., & Gijzen, H. (2002). Assessment of the major water and nutrient flows in the Chivero catchment area, Zimbabwe. *Physics and Chemistry of the Earth, Parts A/B/C, 27*(11-22), 783-792.

Nhapi, I. (2014). *Options for wastewater management in Harare.* CRC Press. doi:10.1201/9781482284041

Nhubu, T., Mbohwa, C., Muzenda, E., & Patel, B. (2019). *Impacts of waste management practices on water resources in Harare.* Paper presented at the Wastes: Solutions, Treatments and Opportunities III: Selected Papers from the 5th International Conference Wastes 2019, Lisbon, Portugal. 10.1201/9780429289798-84

Nhubu, T., & Muzenda, E. (2019). Determination of the Least Impactful Municipal Solid Waste Management Option in Harare, Zimbabwe. *Processes (Basel, Switzerland)*, *7*(11), 785. doi:10.3390/pr7110785

Nhubu, T., Muzenda, E., Mbohwa, C., & Agbenyeku, E. (2020). Comparative assessment of compositing and anaerobic digestion of municipal biodegradable waste in Harare, Zimbabwe. *Environmental Progress & Sustainable Energy*, *39*(4), e13376. doi:10.1002/ep.13376

Nicol, A., & Odinga, W. (2016). *IWRM in Uganda- Progress after Decades of Implementation Water Alternatives.* Academic Press.

Nikolaou, I. E., & Evangelinos, K. I. (2010). A SWOT analysis of environmental management practices in Greek Mining and Mineral Industry. *Resources Policy*, *35*(3), 226–234. doi:10.1016/j.resourpol.2010.02.002

Njalam'mano, J. B. J. (2007). *Current Status and Management of Hand Pump Equipped Water Facilities in Blantyre Rural District, Malawi: a case study of Kapeni and Lundu Traditional Authorities.* Linkoping University.

Nkuna, Z. W. (2012). *Water governance challenges for rural water supply: A case study of two local municipalities in South Africa* [Unpublished Masters dissertation]. Faculty of Natural and Agricultural Sciences, University of Pretoria.

Nojiyeza, S. (2013). Community river restoration in South Africa. In Water and Climate Change in Africa- Challenges and community initiatives in Durban, Maputo and Nairobi. Routledge.

North, D. C. (1990). *The New Institutional Economics and Development.* Washington University.

Nsakila, M. G. (2008). *The Chinese Presence in Lubumbashi.* Centre for Chinese Studies.

Nukunya, G. K. (1992). *Tradition and change in Ghana: An introduction to sociology.* Ghana Universities Press.

Nxumalo, Z. F. (2007). *The role of continuous assessment in primary school* (Doctoral dissertation).

Nyarko, K. B., Oduro-Kwarteng, S., & Adama, I. (2006). Cost recovery of community-managed piped water systems in the Ashanti region, Ghana. *Water and Environment Journal: the Journal / the Chartered Institution of Water and Environmental Management*, *21*(2), 92–99. doi:10.1111/j.1747-6593.2006.00051.x

Nyika, J., Adediran, A. A., Olayanju, A., Adesina, O. S., & Edoziuono, F. O. (2020). The Potential of Biomass in Africa and the Debate on its Carbon Neutrality. In *Biomass.* IntechOpen. doi:10.5772/intechopen.93615

Nylen, W., & Dodd, L. (2003). *Participatory democracy versus elitist democracy: Lessons from Brazil.* Palgrave Macmillan. doi:10.1057/9781403980304

Oates, W. E., & Portney, P. R. (2003). The Political Economy of Environmental Policy. Handbook of Environmental Economics, 1, 450-600.

Oates, W. E. (1972). *Decentralisation theorem.* Harcourt Brace Jovanovich.

Obeng, P. A., Keraita, B., Oduro-Kwarteng, S., Abaidoo, R. C., Awuah, E., & Flemming, K. (2015). *Usage and barriers to use of latrines in a ghanaian peri-urban community.* Springer International Publishing. doi:10.100740710-015-0060-z

OFWAT. (2004). *Annex - Future water and sewerage charges 2005 -10 Final determinations.* https://webarchive.nationalarchives.gov.uk/20150604023659/https://www.ofwat.gov.uk/regulating/charges/ltr_rd3003_annex.pdf

Ogula, P. A. (2005). *Research Methods.* CUEA Publications.

Ogunwusi, A. A. (2014). Wood Waste Generation in the Forest Industry in Nigeria and Prospects for Its Industrial Utilization. *Civil and Environmental Research, 6*(9), 62–69. https://www.iiste.org/Journals/index.php/CER/article/view/15450

Oki, T., & Quiocho, E. R. (2020). Economically challenged and water scarce: Identification of global populations most vulnerable to water crises. *International Journal of Water Resources Development*, *36*(3), 416–428. doi:10.1080/07900627.2019.1698413

Okot-Okumu, J., & Oosterveer, P. (2010). Providing sanitation for the urban poor in Uganda. In *Social perspectives on the sanitation challenge*. Springer. doi:10.1007/978-90-481-3721-3_4

Okubo, T. (2007). *Trade Liberalisation and Agglomeration with Firm Heterogeneity-Forward and Backward Linkages*. University of Manchester.

Olsson, O., & Congdon, F. H. (2004). Congo: The Prize of Predation. *Journal of Peace Research*, *3*(41), 321–336. doi:10.1177/0022343304043772

Olum, Y. (2014). Decentralisation in developing countries: Preconditions for successful implementations. *Commonwealth Journal of Local Governance*, *15*(2), 23–37. doi:10.5130/cjlg.v0i0.4061

Omoyi, K. (2014). *Ressources minieres, croissance economique et guerres civiles en RD.Congo* [Mineral resources, economic growth and civil wars in the DRC] (Master's Dissertation). University of Yaounde II.

Omoyi, K. (2014). *Ressources minieres, croissance economique et guerres civiles en RD.Congo* [Mineral resources, economic growth and civil wars in the DRC] (Masters Dissertation). University of Yaounde II.

Ostrom, E. (1990). *Governing the commons: the evolution of institutions for collective action*. Cambridge University Press.

Osunade, M. A. (1994). Community environmental knowledge and land resource surveys in Swaziland. *Singapore Journal of Tropical Geography*, *15*(2), 157–170. doi:10.1111/j.1467-9493.1994.tb00059.x

Overton, D. M., Smith, J. D., Barchiesi, S., Acreman, M. C., Stromberg, J. C., & Kirby, J. M. (2014). Implementing environmental flows in integrated water resources management and the ecosystem approach. *Hydrological Sciences Journal*, *59*(3), 860–877. doi:10.1080/02626667.2014.897408

Paavola, J., & Adger, W. N. (2005). Institutional Ecological Economics. *Ecological Economics*, *53*(3), 353–368. doi:10.1016/j.ecolecon.2004.09.017

Page, B. (2003). Communities as the agents of commodification: The Kumbo Water Authority in Northwest Cameroon. *Geoforum*, *34*(4), 483–498. doi:10.1016/S0016-7185(03)00049-6

Pallant, J. (2003). *SPSS survival manual: A step by step guide to data analysis using SPSS for Windows (Version 10 and 11)*. Open University Press.

Palmer, C. G., & Munnik, V. (2018). *Practising adaptive IWRM. Integrated Water Resources Management (IWRM) in South Africa*. Report No. 2248/1/18. Pretoria: Water Research Commission.

Palmer, R. (1996). *Residue of tradition or adaptation to an arid environment—cosmology, gender and the struggle for water in two rural communities in the Eastern Cape* (Unpublished paper). Anthropology Department, Rhodes University, Grahamstown, Eastern Cape, South Africa.

Panhwar, A. H., Ansari, S., & Shah, A. A. (2017). *Post-positivism: An Effective Paradigm for Social and Educational Research*. https://www.researchgate.net/publication/317605754_Post-positivism_An_Effective_Paradigm_for_Social_and_Educational_Research

Panhwar, A. H., Ansari, S., & Shah, A. A. (2017). Post-positivism: An Effective Paradigm for Social and Educational Research. *Research Gate*, *11*(2), 253–256.

Parker, P., Letcher, R., Jakeman, A., Beck, M. B. G., Harris, R. M., Argent, M., Hare, C., Pahl-Wostl, A., Voinov, M., Janssen, P., Sullivan, M., Scoccimarro, A., Friend, M., Sonnenshein, D., Barker, L., Matejicek, D., Odulaja, P., Deadman, K., Lim, G., ... Bin, S. (2002). Progress in integrated assessment and modelling. *Environmental Modelling & Software, 17*(3), 209–217. doi:10.1016/S1364-8152(01)00059-7

Parr, B., Swilling, M., & Henry, D. (2018). *The Paris Agreement and South Africa's just transition*. Melbourne Sustainable Society Institute. Retrieved from www.sustainable.unimelb.edu.au

Pateman, C. (1970). *Participation and democratic theory*. Cambridge University Press. doi:10.1017/CBO9780511720444

Patz, J. A., & Kovats, R. S. (2002). Hotspots in Climate change and human health. *BMJ (Clinical Research Ed.), 325*(7372), 1094–1098. doi:10.1136/bmj.325.7372.1094 PMID:12424173

Paulson, L. D. (2015). What is water scarcity? *RWL Water, 5*(2), 1–7. PMID:25220873

Paulson, S., Gezon, L., & Watts, M. (2005). *Politics, ecologies, genealogies*. Rutgers University Press.

Pearce, D. (2002). An Intellectual History of Environmental Economics. *Annual Review of Energy and the Environment, 27*(1), 57–81. doi:10.1146/annurev.energy.27.122001.083429

Peet, R., Robbins, P., & Watts, M. (2010), Global political ecology. London: Routledge.

Pegg, S. (2006). Mining and Poverty Reduction: Transforming Rhetoric into Reality. *Journal of Cleaner Production, 14*(3–4), 376–387. doi:10.1016/j.jclepro.2004.06.006

Pelling, M. (1999). The political ecology of flood hazard in urban Guyana. *Geoforum, 30*(3), 249–261. doi:10.1016/S0016-7185(99)00015-9

Penny, S. B. (2001). *Ecological implications of water spirit beliefs in Southern Africa: The need to protect knowledge, nature, and resource rights*. Science and stewardship to protect and sustain wilderness values: Seventh World Wilderness Congress symposium, Port Elizabeth, South Africa. https://www.fs.fed.us/rm/pubs/rmrs_p027/rmrs_p027_148_154.pdf

Perez, F., & Sanchez, L. E. (2009). Assessing the evolution of sustainability reporting in the mining sector. *Environmental Management, 43*(6), 949–961. doi:10.100700267-008-9269-1 PMID:19238481

Peters, S. (2011). *Municipal Consumer Debt in South Africa, in Financial and Fiscal Commission*. 2012/13 Submission for the Division of Revenue Technical Report. South Africa. https://www.ffc.co.za/docs/submissions/technical_report/2011/Financial%20and%20Fiscal%20Commission%202012%20-%202013%20Technical%20Report.pdf

Phansalkar, S. J. (2007). Water, equity and development. *International Journal of Rural Management, 3*(1), 1–25. doi:10.1177/097300520700300101

Pickett, S. T. A., Cadenasso, M. L., & Grove, J. M. (2004). Resilient cities: Meaning, models, and metaphor for integrating the ecological, socio-economic, and planning realms. *Landscape and Urban Planning, 69*(4), 369–384. doi:10.1016/j.landurbplan.2003.10.035

Pillay, S., & Buckley, C. (2005, October). *LCA Studies of Water Delivery Systems*. Water Research Commission.

Pilosoff, R. (2012). *The unbearable whiteness of being: Farmers' voices from Zimbabwe*. Weaver Press.

Pilot, Van Teijlingen, & Hundley. (2001). Research methodology. *Journal of Quantitative Approach, 12*(2), 12-17.

Pindyck, R. S. (2007). *Uncertainty in Environmental Economics*. Oxford University Press.

Pindyck, R. S. (2007). Uncertainty in Environmental Economics. *Review of Environmental Economics and Policy, 1*(1), 45–65. doi:10.1093/reep/rem002

Pink, R. M. (2016). Water Rights in China and India: A Human Security Perspective. *Asian Affairs: An American Review, 43*(2), 19–35. doi:10.1080/00927678.2016.1145987

Pirzdeh, H., Shanian, S., & Hamou-Lhadj, A. (2011). The concepts of stratified sampling of executing traces. Department of Electrical and Computer Engineering, Concordia University.

Poplin, A., Pereira, G. C., & Rocha, M. C. F. (2013). The participatory cube: A framework for analysis of online participation platforms. In S. Geertman, F. Toppen, & J. Stillwell (Eds.), *Planning support systems for sustainable urban development. Lecture Notes in Geoinformation and Cartography* (Vol. 195). Springer. doi:10.1007/978-3-642-37533-0_23

Postel, S. L. (2000). Entering an Era of Water Scarcity: The Challenges Ahead. *Ecological Applications, 10*(4), 941–948. doi:10.1890/1051-0761(2000)010[0941:EAEOWS]2.0.CO;2

Pradhan, A., & Mbohwa, C. (2014). Development of biofuels in South Africa: Challenges and opportunities. *Renewable & Sustainable Energy Reviews, 39*, 1089–1100. doi:10.1016/j.rser.2014.07.131

Price, G. (2009). *eThekwini Municipality Water Conservation Guideline*. Durban: eThekwini Municipality.

Prüss, A., Kay, D., Fewtrell, L., & Bartram, J. (2002). Estimating the burden of disease from water, sanitation, and hygiene at a global level. *Environmental Health Perspectives, 110*(5), 537–542. PMID:12003760

Purcell, M. (2002). The state regulation and global restructuring: Reasserting the political in the political economy. *Review of International Political Economy, 9*(2), 284–318. doi:10.1080/09692290110126128

Purkus, A., Gawel, E., Szarka, N., Lauer, M., Lenz, V., Ortwein, A., Tafarte, P., Eichhorn, M., & Thrän, D. (2018). *Contributions of flexible power generation from biomass to a secure and cost-effective electricity supply — A review of potentials, incentives and obstacles in Germany*. Academic Press.

Pye, S., & Dobbins, A. (2015). *Energy poverty and vulnerable consumers in the energy sector across the EU: Analysis of policies and measures*. Insight Energy. Retrieved from http://www.innoenergy.com/wp-content/uploads/2016/03/INSIGHT_E_PR2_Energy-Poverty-Main-Report.pdf

Quiggin, J. (2000). *Environmental Economics and the Murray-Darling river system. Australian Research Council.*

Quiggin, J. (2000). *Environmental Economics and the Murray-Darling River System*. Australian Research Council.

RAID. (2004). Unanswered questions. Companies, conflict and the Democratic Republic of Congo. RAID.

Ramana, E. K. (2013). *The Role of NGOs in Rural Development with Special Reference to Women*. https://shodhganga.inflibnet.ac.in/bitstream/10603/18591/9/09_chapter%203.pdf

Renauld, A. (2005). *Ressources naturelles et transferts d'armes* [Natural resources and transfer of weapons] (Report No 1). Brussels: GRIP.

Rennings, K. (2000). Redefining Innovation- Eco-Innovation Research and Contribution from Ecological Economics. *Ecological Economics, 32*(2), 319–332. doi:10.1016/S0921-8009(99)00112-3

Republic of South Africa. (1996). Constitution of the Republic of South Africa, 1996 (Act No. 108 of 1996).

Republic of South Africa. (1996). Constitution of the Republic of South Africa, Act 108 of 1996. *Government Gazette*, 378.

Republic of South Africa. (1996). *Constitution of the Republic of South Africa*. Republic of South Africa.

Republic of South Africa. (1996). *The Constitution of the Republic of South Africa Act 1996 (Act 108 of 1996)*. Government Printer.

Richter, R. (2005). The New Institutional Economics – Its Start, Its Meaning, Its Prospects. *The European Business Organisation Law Review*, *6*(2), 161-200.

Rinkesh, K. (2017). *What is Water Scarcity?* https://www.conserve-energy-future.com/causes-effects-solutions-of-water-scarcity.php#abh_about

Robbins, P. (2012). *Political ecology: A critical introduction* (2nd ed.). John Wiley and Sons.

Rober, C. (2015). *Rural Development putting the last first*. Routledge Taylor & Francis Group.

Roberts, N. C. (2015). The age of direct citizen participation. London: Routledge, Taylor and Francis Group.

Roberts, D. (2008). Thinking globally, acting locally- institutionalizing climate change at the local government level in Durban, South Africa. *Environment and Urbanization*, *20*(2), 521–537. doi:10.1177/0956247808096126

Robertson, P. J., & Choi, T. (2010). Ecological Governance: Organizing Principles for an Emerging Era. *Future of Public Administration in 2020, 70*(S1), 89–99.

Robinson, A., Torvik, R., & Verdier, T. (2006). Political foundations of the resource curse. *Journal of Development Economics*, *79*(2), 447–468. doi:10.1016/j.jdeveco.2006.01.008

Roche, R., Bain, R., & Cumming, O. (2017). A long way to go–Estimates of combined water, sanitation and hygiene coverage for 25 sub-Saharan African countries. *PLoS One*, *12*(2), e0171783. doi:10.1371/journal.pone.0171783 PMID:28182796

Rogers, P. (2002). *Water Governance in Latin America and the Caribbean*. Inter-American Development Bank.

Rogers, P., & Hall, A. (2003). *Effective water governance*. Global Water Partnership. http://www.gwp.org/global/toolbox/publications/ background papers effective water governance.pdf

Roma, E., Buckley, C., Jefferson, B., & Jeffrey, P. (2010). Assessing user's experience of shared sanitation facilities: A case study of community ablution blocks in Durban, South Africa. *Water S.A.*, *36*(5).

Roma, E., Philp, K., Buckley, C., Xulu, S., & Scott, D. (2013). User perceptions of urine diversion dehydration toilets: Experiences from a cross-sectional study in eThekwini Municipality. *Water S.A.*, *39*(2), 305–312. doi:10.4314/wsa.v39i2.15

Romano, S. R. (2019). Transforming participation in water governance: The multi sectoral alliances of rural water committees and NGOs in Nicaragua. *International Journal of Water Resources Development*, *35*(3), 430–445. doi:10.1080/07900627.2017.1363722

Romeiro, A. R. (2012). Sustainable development: An ecological economics perspective. *Estudos Avançados*, *26*(74).

Ross, M. L. (1999). The political economy of the resource curse. *World Politics*, *51*(2), 297–322. doi:10.1017/S0043887100008200

Ross, M. L. (2014). What have we learned about the resource curse? *The American Political Science Review*, *102*(1), 107–123. doi:10.1017/S0003055408080040

Rouse, M. (2013). *Pilot program (pilot study)*. https://searchcio.techtarget.com/definitions/pilot-program-pilot-study

Rubbers, B. (2009). The story of a tragedy: How people in haut-Katanga interpret the post-colonial history of Congo. *The Journal of Modern African Studies*, *47*(2), 267–289. doi:10.1017/S0022278X09003838

Rubbers, B. (2010). Claiming workers' rights in the Democratic Republic of Congo: The case of the collectif des ex-agents de la Gécamines. *Review of African Political Economy*, *37*(125), 329–344. doi:10.1080/03056244.2010.510629

Ruiz-Mier, F., & van Ginneken, M. (2006). *Consumer cooperatives: An alternative institutional model for delivery of urban water supply and sanitation services*. The World Bank. https://documents1.worldbank.org/curated/en/131841468137400796/pdf/355470WN50coop1version0January02006.pdf

Sachs, J., & Warner, A. (1995). *Natural resource abundance and economic growth*. National Bureau of Economic Research. doi:10.3386/w5398

Saed, K. (2004). Designing an Environmental Mitigation Banking institution for Linking the Size of Economic Activity to Environmental Capacity. *Journal of Economic Issues*, *38*(4), 909-937.

Sagoff, M. (2000). Environmental Economics and the Conflation of Value and Benefit. *Environmental Science & Technology*, *34*(8), 1426–1437. doi:10.1021/es990674d

Saleth, R. M., & Dinar, A. (2003). *Water Institutions and Sector Performance: A Subjective Theory of Institutional Change*. Paper presented at the 25th Congress of the International Association of Agricultural Economics, Durban, South Africa.

Saleth, R. M. (2004). *Understanding Water Institutions: Structure, Environment, and Change Process*. International Water Management Institute.

Sanctuary, M. (2007). Making water a part of economic development: The economic benefits of improved water management and services. Stockholm International Water Institute (SIWI) Publications.

Saner, R., Yiu, L., Filadoro, M., & Khusainova, V. (2018). Access to water in developing countries: Four options for provision and regulation in the water sector. *Asia Pacific Journal of Public Administration*, *37*(3), 193–206. doi:10.1080/23276665.2015.1075698

Sanoff, H. (2000). *Community participation methods in design and planning*. John Wiley & Sons.

Sarantakos, S. (2005). *Social Research* (3rd ed.). Palgrave MacMillan. doi:10.1007/978-0-230-20901-5

Sarantakos, S. (2013). *Social research* (6th ed.). Macmillan: Red Glob Press. doi:10.1007/978-1-137-29247-6

Sareen, S. (2017). Who Governs Local Access in Jharkhand? Mechanisms of Access to Government Services. *Forum for Development Studies*, *44*(2), 249–274. doi:10.1080/08039410.2016.1233135

Scanlon, T., Uguru, O. P., Jafry, T., Chinsinga, B., Mvula, P., Chunga, J., Zimba, L. M., Mwape, M., Nyundo, L., Mwiinga, B., & Chungu, K. (2016). The Role of Social Actors in Watewr Access in Sub-Saharan Africa: Evidence from Malawi and Zambia. *Water Resources and Rural Development*, *8*, 25–36. doi:10.1016/j.wrr.2016.08.001

Schaltegger, S., & Synnestvedt, T. (2002). The link between "green" and economic success: Environmental management as the crucial trigger between environmental and economic performance. *Journal of Environmental Management*, *65*, 339–346. PMID:12369398

Schaltergger, S., & Synnestvedt, T. (2002). The link between 'green' and economic success: Environmental management as the crucial trigger between environmental and economic performance. *Journal of Environmental Management*, *65*, 339–346. PMID:12369398

Schelly, C. (2015). Framework for Understanding and promoting Solar Energy Technologies. *MDPI Resources*, *4*(1), 55–69. doi:10.3390/resources4010055

Schreier, M. (2012). *Qualitative content analysis in practice*. Sage Publications.

Schroeder, R. A. (1993). Shady practice and the political ecology of resource stabilization in Gambian garden / orchards. *Economic Geography*, *69*(4), 349–365. doi:10.2307/143594

Schroeder, R. A. (1999). *Shady practices: agroforestry and gender politics in The Gambia*. University of California Press.

Schulze, R. E. (2007). Coefficient of Variation of Annual Precipitation: South African Atlas of Climatology and Agro hydrology. Water Research Commission, Pretoria, RSA, WRC Report 1489/1/06, Section 6.3.

Schumacher, E. F. (2011). *Small is beautiful: A study of economics as if people mattered*. Random House.

Scoones, I. (1999). New ecology and the social sciences: What prospects for a fruitful engagement? *Annual Review of Anthropology*, *28*(1), 479–507. doi:10.1146/annurev.anthro.28.1.479

Searle, S., & Bitnere, K. (2017). *Review of the impact of crop residue management on soil organic carbon in Europe*. Academic Press.

Sebova, M. (2015). Impact of Municipality Size on Economic Performance: Evidence from Slovakia. *Journal of Applied Economic Sciences*, *11*(3), 1–14.

Sekaran, U., & Bougie, R. G. (2013). Research methods for business: A Skill Building Approach (6th ed.). Wiley.

Selee, A.D., & Tulchin. (2004). Decentralization and Democratic Governance: Lessons and Challenges. In *Decentralization, Democratic Governance, and Civil Society Perspective- Africa, Asia, and Latin America*. Woodrow Wilson Center Press.

Sen, A. (2001). *Development as freedoms*. Oxford.

Sen, A., & Smith, C. S. (2015). *Economic development* (20th ed.). Pearson.

Shi, F. (2015). *Study on a Stratified Sampling Investigation Method for Resident Travel and the Sampling Rate. Europe*. Hindawi Publications.

Shi, F. (2015). *Study on a Stratified Sampling Investigation Method for Resident Travel and the Sampling Rate*. Hindawi Publications.

Shiklomanov, I. A. (2000). Appraisal and Assessment of World Water Resources. *Water International*, *25*(1), 11–32. doi:10.1080/02508060008686794

Simbota, M. (2013). *LWB losing millions through non-payment bills*. MBC Malawi.

Simukonda, K., Farmani, R., & Butler, D. (2018). Intermittent water supply systems: Causal factors, problems and solution options. *Urban Water Journal*, *15*(5), 488–500. doi:10.1080/1573062X.2018.1483522

Singh, N., Åström, K., Hydén, H., & Wickenberg, P. (2008). Gender And Water From A Human Rights Perspective: The Role Of Context In Translating International Norms Into Local Action. *Rural Society*, *18*(3), 185–193. doi:10.5172/rsj.351.18.3.185

Singh, V. P. (2017). Challenges in meeting water security and resilience. *Water International*, *42*(4), 349–359. doi:10.1080/02508060.2017.1327234

Smith, N. (1993). Homeless/Global: Scaling places. In Mapping the futures: Local cultures global change. Routledge.

Soderbaum, P. (1990, June). Neoclassical and Institutional Approaches to Environmental Economics. *Journal of Economic Issues*, *24*(2), 481–492. doi:10.1080/00213624.1990.11505047

Soderbaum, P. (1999). Values, Ideology and Politics in Ecological Economics. *Ecological Economics*, *28*(2), 161–170. doi:10.1016/S0921-8009(98)00139-6

Solanes, M., & Gonzalez-Villareal, F. (2008). The Dublins Principles for water as reflected in a comparative assessment of institutional and legal arrangements for water resource management. *Global Water Partnership*, *12*(3), 9–37.

Solanes, M., & Gonzalez-Villarreal, F. (1999). *The Dublin Principles for Water as Reflected in a Comparative Assessment of Institutional and Legal Arrangement for Integrated Water Resources Management Sweden.* Global Water Partnership/ Swedish International Development Cooperation Agency.

South Africa (SA). (1997). *Water Services Act 108 of 1997.* Government Printer.

South African Human Rights Commission. (2018). *Investigative hearing report on the underlying socio-economic challenges of mining-affected communities in South Africa.* https://www.sahrc.org.za/home/21/files/SAHRC%20Mining%20 communities%20report%20FINAL.pdf

South African Human Rights Commission. (2018). *Unregulated artisanal underground and surface mining activities.* https://www.sahrc.org.za/home/21/files/Unregulated%20Artisanal%20Underground%20and%20Surface%20Mining%20 Activities%20electronic%20version.pdf

Spash, C. L. (1993). Economics, Ethics, and Long- Term Environmental Damages. *Environmental Ethics, 15*(2), 117–132. doi:10.5840/enviroethics199315227

Spash, C. L. (1999). The Development of Environmental Thinking in Economics. *Environmental Values, 8*(4), 413–435. doi:10.3197/096327199129341897

Staddon, C. (2010). *Do Water Meters Reduce Domestic Consumption? A summary of available literature.* Heednet. http://www.heednet.org/metering-defraHEEDnet.pdf

Stanwell-Smith, R., Bloomfield, S. F., & Rook, G. A. W. (2012). *The hygiene hypothesis and its implications for home hygiene, lifestyle and public health.* International Scientific Forum on Home Hygiene. http://www. ifh-homehygiene. org/reviews-hygiene hypothesis

Statistics South Africa. (2011). *General Household Survey.* Version 2. Author.

Statistics South Africa. (2015). *Electricity redistribution: Which councils are likely to feel the pinch?* Author.

Statistics South Africa. (2016). *Limpopo Community Survey 2016 results.* Retrieved from http://www.statssa.gov.za/?p=7988

Statistics South Africa. (2018). *General Household Survey: 2017.* Statistics South Africa.

Stats, S.A. (2011). *Formal Census.* Statistics South Africa.

Stavins, R. (2005). Environmental Economics. Social Science Research Network of Harvard University.

Stavins, R. (2004). *Environmental Economics.* Harvard University.

Stavins, R. N. (2002). *Experience with Market-Based Environmental Policy Instruments. The Fondazione Eni Enrico Mattei Nota Di Lavoro, 52*, 1–88.

Stecher, K., Brosowski, A., & Thrän, D. (2013). Biomass potential in Africa. *Irena*, 44. www.dbfz.de

Stern, D. I. (1997). Limits to Substitution and Irreversibility in Production and Consumption: A Neoclassical Interpretation of Ecological Economics. *Ecological Economics, 21*(3), 197–215. doi:10.1016/S0921-8009(96)00103-6

Stiling, J. A. (2001). *Thematic networks: An analytic tool for qualitative research.* Sage Publications.

Stockholm International Water Institute (SIWI). (2017). *Governance of water and waste: A key to sustainable development?* https://www.worldwaterweek.org/event/7450-governance-of-water-and-waste-a-key-to-sustainable-development

Stratford Global Intelligence (SGI). (2014). *Water Scarcity: Examining Impacts Around the World.* https://cdn2.hubspot.net/hubfs/515194/B2C_content/B2CWF1/Water_Scarcity-slideshare_-final.pdf

Strauss, A. L., & Corbin, J. M. (1998). *Basics of qualitative research: Techniques and procedures for developing grounded theory*. Sage Publications.

Suchman, M. C. (1995). Managing legitimacy: Strategic and institutional approaches. *Academy of Management Review*, *20*(3), 571–610. doi:10.2307/258788

Sultana, F. (2009). Fluid lives: Subjectivities, Gender and Water in Rural Bangladesh. *A Journal of Feminist Geography*, *16*(4), 427-444.

Sutherland, S., Hordijk, M., Lewis, B., Meyer, C., & Buthelezi, S. (2014). Water and sanitation provision in eThekwini Municipality: A spatially differentiated approach. *Environment and Urbanization*, *26*(2), 469–488. doi:10.1177/0956247814544871

Swaney, J. A. (1987). Elements of a Neoinstitutional Environmental Economics. *Journal of Economic Issues*, *21*(4), 1739–1779. doi:10.1080/00213624.1987.11504721

Swatuk, L. (2002). Water reforms in Zimbabwe: Some observations based on the Save catchment experiences and suggestions for ways forward. In E. Manzungu (Ed.), *The processes and dynamics of catchment management in Zimbabwe* (pp. 45–68). Save Africa Trust.

Swyngedouw, E. (2004). Scaled geographies: nature, place, and the politics of scale. In Scale and geographic enquiry: Nature, society, and method. Blackwell. doi:10.1002/9780470999141.ch7

Swyngedouw, E. (1997). *Neither global nor local, Glocalization and the politics of scale*. Guilford press.

Swyngedouw, E., & Heynen, N. (2003). Urban political ecology, justice and the politics of scale. *Antipode*, *1*(5), 899–918. doi:10.1111/j.1467-8330.2003.00364.x

Tanahashi, T. (1978). Health services coverage and its evaluation. *Bulletin of the World Health Organization*, *56*(2), 295–303. PMID:96953

Tarisayi, K. (2018). Traditional leadership and the Tokwe-Mukosi induced displacements: Finding the missing link. *Journal of Disaster Risk Studies*, *10*(1), 1–22. PMID:30473759

Taulo, J. L., Gondwe, K. J., & Ben Sebitosi, A. (2017). Energy supply in Malawi: Options and issues. *Journal of Energy in Southern Africa*, *26*(2), 19. doi:10.17159/2413-3051/2015/v26i2a2192

Tempelhoff, J. W. N., Ginster, M., Motloung, S., Gouws, C. M., & Strauss, J. S. (2014). The 2012 acid mine drainage (AMD) crisis in Carolina's municipal water supply. *The American Historical Review*, *46*(2), 77–107. doi:10.1080/17532523.2014.943978

Tevera, D., & Crush, J. (2003). *The new brain drain from Zimbabwe. Southern African Migration Project*. SAMP.

Tewksbury, R. (2009). Qualitative versus Quantitative Methods: Understanding Why Qualitative Methods are Superior for Criminology and Criminal Justice. *Journal of Theoretical and Philosophical Criminology*, *1*(1), 38–58.

Thabethe, P. D. (2011). *An evaluation of water and sanitation provision by uThukela water in selected districts, Public Administration and Business Management* [Unpublished doctoral dissertation]. University of KwaZulu-Natal, South Africa.

The Sustainable Livelihoods In Southern Africa. (2008). *The Politics of Decentralisation in Southern Africa*. Future Agriculture Consortium Policy Process Briefing 2008: The Sustainable Livelihood in Southern Africa.

Thomas, D. R. (2003). *A general inductive approach for qualitative data analysis*. Retrieved from https://www.researchgate.net/profile/David_Thomas11/publication/263769109_Thomas_2003_General_Inductive_Analysis_- Original_web_version/links/0a85e53bdc04f64786000000/Thomas-2003-General-Inductive-Analysis-Original-web- version.pdf?origin=publication_detail

Thomas, P. Y. (2010). *Towards developing a web-based blended learning environment at the University of Botswana* (D.Ed. Thesis). University of South Africa.

Thomas, P. Y. (2010). *Research Methodology and Design*. https://uir.unisa.ac.za/bitstream/handle/10500/4245/05Chap%204_Research%20methodology%20and%20design.pdf

Thompson, H. (2006). *Water law: A practical approach to resource management & the provision of services*. Juta & Co Ltd.

Tumwebaze, I. K., Orach, C. G., Niwagaba, C., Luthi, C., & Mosler, H. J. (2013). Sanitation facilities in Kampala slums, Uganda: Users' satisfaction and determinant factors. *International Journal of Environmental Health Research*, *23*(3), 191–204. doi:10.1080/09603123.2012.713095 PMID:22873693

Turner, M. D. (2014). Political ecology I. An alliance with resilience? *Progress in Human Geography*, *38*(4), 616–623. doi:10.1177/0309132513502770

Tyman, M., & Kingdom, B. (2002). *A Water Scorecard: Setting Performance Targets for Water Utilities. Viewpoint*. The World Bank. https://openknowledge.worldbank.org/handle/10986/11351

UCLA Center for Communication Policy. (2005). *The UCLA Internet Report on sanitation Survey in Digital Future*. UCLA Center for Communication Policy.

Umesh, P., & Nagarkatte, N. O. (2017). *Theory of constraints: Creative problem solving*. Taylor and Francis Group.

UN Documents. (2003). *The Dublin Statement on Water and Sustainable Development*. www.un-documents.net/h20-dub.htm

Underwood, D. A. (1999). An Introduction to Ecological Economics. *Journal of Economic Issues*, *33*(1), 196–198. doi:10.1080/00213624.1999.11506145

UNDESA. (2018). *World Urbanization Prospects: The 2018 Revision*. Retrieved from https://population.un.org/wup/Publications/Files/WUP2018-Report.pdf

UNEP. (2017). *Atlas of Africa Energy Resources*. http://www.unep.org

UNEP. (2018). *Africa Waste Management Summary for Decision-makers*. https://wedocs.unep.org/bitstream/handle/20.500.11822/25515/Africa_WMO_Summary.pdf?sequence=1&isAllowed=y

Unger, C. C. (2013). Treating diarrheal disease in children under five: The global picture. Archives of disease in childhood. *Archives of Disease in Childhood*, *99*(3), 273–278. doi:10.1136/archdischild-2013-304765 PMID:24197873

Union Minière du Haut Katanga. (1957). Production from 1906-1956. *Cuyprius*, 39-40.

United Nation Children's Emergency Fund. (2017). *Progress on drinking water, sanitation & hygiene*. https://data.unicef.org/wp-content/.../safely-manageddrinking-water-JMP-2017-1.pdf

United Nation General Assembly. (2016). *Transforming our world: the 2030 Agenda for Sustainable Development*, A/RES/70/1. Available at: https://www.refworld.org/docid/57b6e3e44.html

United Nations Development Programme (UNDP). (2004). *The human right to water and sanitation*. https://www.un.org/waterforlifedecade/human_right_to_water.shtml

United Nations Development Programme and United Nations-Water. (2014). Implementing water sanitation and hygiene (WASH) to improve water quality and supply, sanitation and hygiene. *Cochrane Database of Systematic Reviews*, *8*(8), CD009382.

United Nations Development Programme. (2002). *Zimbabwe Land Reform and Resettlement Assessment and Suggested Framework for the Future*. Interim Mission Report.

United Nations Development Programme. (2007a). *User's guide on assessing water governance.* UNDP.

United Nations Environment Programme. (2011). *Towards a Green Economy: Pathways to Sustainable Development and Poverty Eradication - A Synthesis for Policy Makers.* Retrieved from www.unep.org/greeneconomy

United Nations Human Rights Council (UNHRC). (2014). *Resolution on the Human Rights to Safe Drinking Water and Sanitation.* Geneva: UNHRC. A/HRC.RES/27/7.

United Nations International Children's Emergency Fund and World Health Organization. (2010). *Progress on Drinking Water and Sanitation.* UNICEF.

United Nations International Children's Emergency Fund. (2008b). *Use of improved sanitation facilities in Ghana.* UNICEF.

United Nations International Children's Emergency Fund. (2012). *Eliminating open defecation.* UNICEF.

United Nations International Children's Emergency Fund. (2014). *Water Sanitation and hygiene (WASH).* UNICEF.

United Nations. (2003). *General comment No.15: The right to water (Article 11 and 12 of the covenant).* Adopted at the twenty-ninth session of the Committee on Economic, Social and Cultural Rights. United Nations.

United Nations. (2005). *Valuing water resources. Module 10.* http://www.cap.net.org/node/1551

United Nations. (2007). *Global monitoring of water and sanitation: History, Methods and future challenges.* United Nations.

United Nations. (2014). *United Nations water: Global analysis and assessment of Sanitation and Drinking Water (GLAAS) 2014 Report.* UNDP.

United Nations. (2015). *Transforming our world: The 2030 agenda for sustainable development.* United Nations.

United Nations-Water. (2015). *Eliminating Discrimination and Inequalities in Access to Water and Sanitation.* UN-Water. www.unwater.org/publications/eliminating-discrimination-inequalities-access-water-sanitation

United Nations-Water. (2015). *Water, sanitation and hygiene.* United Nations.

University of Lubumbashi. (2005). *The restructuring of Gecamines, impacts on poverty and social issues.* The future of social services and reintegration strategies, final report of the University of Lubumbashi.

University of Zululand. (2016). *Research proposal guide, R1 P2 DVC Research and Innovation, Senate approved on 17 April 2013, Ref: S625/13.* Author.

UN-Water. (2008). *Integrated Water Resources Management (IWRM).* www.un.org.waterforlifedecade/iwrm.shtml

UREGNI. (2007). *Utility Regulator Strand 1 Submission to the Water Review.* UREGNI. https://www.uregni.gov.uk/sites/uregni.gov.uk/files/media-files/Utility%20Regulator%20Strand%201%20Submission%20to%20Water%20Review%20%20September%202007.pdf

UThukela District Municipality. (2019/2020). *Integrated Development Plan Review.* The Office of the Municipal Manager: IDP Unit.

UThukela District Municipality. (2020). District Growth and Development Plan. Cooperative Government and Traditional Affairs (COGTA).

Van den Bergh, J. C. J. M., & Gowdy, J. M. (1998). Evolutionary Theories in Environmental and Resource Economics: Approaches and Applications. Free University.

Van den Bergh, J. C. J. M. (2007). Evolutionary thinking in environmental economics. *Journal of Evolutionary Economics, 17*(5), 521–549. doi:10.100700191-006-0054-0

Van den Bergh, J. C. J. M., & Verbruggen, H. (1999). Spatial Sustainability, Trade and Indicators: An Evaluation of the 'Ecological Footprint'. *Ecological Economics, 29*(1), 61–72. doi:10.1016/S0921-8009(99)00032-4

Van Der Mescht, J., & Van Jaarsveld, M. (2012). Addressing operations and maintenance challenges in smaller municipalities. Academic Press.

van der Ploeg, F. (2010). *Natural resources: Curse or blessing?* CESifo working paper Resources and Environment, No. 3125.

Van Der Zaag, P. (1999). Water allocation Principles in Catchment Areas. In Water for Agriculture in Zimbabwe: Policy and Management Options for the Smallholder Sector. University of Zimbabwe.

Van Koppen, B., Tarimo, A. K. P. R., van Eeden, A., Manzungu, E., & Sumani, P. M. (2016). Winners and Losers of IWRM in Tanzania. *Water Alternatives, 9*(3), 588–607.

van Koppen, B., Tarimo, A. K. P. R., van Eeden, A., Manzungu, E., & Sumuni, P. M. (2016). Winners and losers of IWRM in Tanzania. *Water Alternatives, 9*(3), 588–607.

Van Koppen, B., Tarino, A. K. P. R., van Eeden, A., Manzungu, E., & Sumuni, P. M. (2016). *Winners and Losers of IWRM in Tanzania Water Alternatives.* Academic Press.

van Lieshout, M., Kovats, R. S., Livermore, M. T. J., & Martens, P. (2004). Climate change and malaria: Analysis of the SRES climate and socio-economic scenarios. *Global Environmental Change, 14*(1), 87–99. doi:10.1016/j.gloenvcha.2003.10.009

Van Minh, H., & Hung, N. V. (2011). Economic aspects of sanitation in developing countries. *Environmental Health Insights, 5*, EHI-S8199.

Van Teijlingen, E.R., & Hundley, V. (2001). *The importance of pilot studies.* Academic Press.

Van Zyl, W. H., Chimphango, A. F. A., den Haan, R., Görgens, J. F., & Chirwa, P. W. C. (2011). Next-generation cellulosic ethanol technologies and their contribution to a sustainable Africa. *Interface Focus, 1*(2), 196–211. doi:10.1098/rsfs.2010.0017 PMID:22482027

VEI. (2013). *Peri-Urban Water and Sanitation Project /Private Sector Service Contract to the Lilongwe Water Board Annual. Progress Report. Contract Year 3: November 2011 - October 2012.* VEI.

Visagie, C. (2010). Medium Term Scenarios for Electricity Supply/Demand Balance. In *The Sustainable Energy Seminar.* Stellenbosch: AMEU.

Vogelstein, R. (2017). *Women around the World.* https://www.cfr.org/blog/women-around-world-week-75

Vogelstein, R. (2017). *Women Around the World.* https://www.cfr.org/blog/women-around-world-week-75

Von Munch, E., & Mayumbelo, K. M. K. (2007, October). Methodology to compare costs of sanitation options for low-income peri-urban areas in Lusaka, Zambia. *Water S.A., 33*(5). doi:10.4314/wsa.v33i5.184017

Wackernagel, M., & Rees, W. E. (1997). Perceptual and Structural Barriers to investing in Natural Capital: Economics from an Ecological Footprint Perspective. *Ecological Economics, 20*(1), 3–24. doi:10.1016/S0921-8009(96)00077-8

Walker, N. J., Still, D. A., Ravenscroft, P., Hazelton, D., King, W., Erskine, S., & Eardley, R. (2010). *Financial Sustainability of Sanitation Services in South Africa.* Water Research Commission.

Walker, P. A. (2003). Reconsidering 'regional' political ecologies: Toward a political ecology of the rural American West. *Progress in Human Geography*, *27*(1), 7–24. doi:10.1191/0309132503ph410oa

Walker, P. A. (2005). Political ecology: Where is the ecology? (2005). *Progress in Human Geography*, *29*(1), 73–82. doi:10.1191/0309132505ph530pr

Walker, P. A. (2006). Political ecology: Where is the policy? *Progress in Human Geography*, *30*(3), 382–395. doi:10.1191/0309132506ph613pr

Walker, P., & Fortmann, L. (2003). Whose landscape? A political ecology of the 'exurban' Sierra. *Cultural Geographies*, *10*(4), 469–491. doi:10.1191/1474474003eu285oa

Wang, X., Guo, M., Koppelaar, R. H., van Dam, K. H., Triantafyllidis, C. P., & Shah, N. (2018). A nexus approach for sustainable urban energy-water-waste systems planning and operation. *Environmental Science & Technology*, *52*(5), 3257–3266. doi:10.1021/acs.est.7b04659 PMID:29385332

Wanjiru, M. M. (2014). *Determinants of sustainability of community water projects in Kieni East District, Nyeri County: Kenyan*. University of Nairobi.

Water and Sanitation Program (WSP). (2008). *Developing Effective Billing and Collection Practices, WSP Field Note 44119 April 2008*. The World Bank. https://www.wsp.org/sites/wsp/files/publications/527200820356_SA_PIP2_May2008.pdf

Water and Sanitation Program (WSP). (2009). *Improving Water Utility Services through Delegated Management, Field Note Serving the Urban Poor*. The World Bank. https://www.wsp.org/sites/wsp/files/publications/Af-imp_through_delegated_mgmt.pdf

WaterAid. (2009). Menstrual hygiene in South Asia: A neglected issue for WASH (water, sanitation and hygiene) programmes. *Gender and Development*, *18*(1), 99–113.

WaterAid. (2011). A condensed version of the WaterAid research report Looking back: Participatory impact assessment of older projects. *African Journal of Environmental Science and Technology*, *5*(13), 1170–1176.

Watts, M., & Peet, R. (2004). Liberating policy ecology. In Liberation ecologies: Environment, development, social movements. New York: Routledge.

Watts, M., Peet, R., & Robbins, P. (2010). Global political ecology. London: Routledge.

Watts, M. (2015). Now and then: the origins of political ecology and the rebirth of adaptation as a form of thought. In T. Perrault, J. McCarthya, & G. Bridge (Eds.), *Handbook of political ecology*. Routledge.

Watts, M. J. (2000). Political ecology. In E. Sheppard & T. Barnes (Eds.), *A companion to economic Geography*. Blackwell.

Wautelet, J. M. (1981A). Pouvoir d'Etat et formation du capital. In Accumulation et sous-développement au Zaïre 1960-1980. Presses Universitaires de Louvain.

Weale, A. (1992). *The new politics of pollution*. Manchester University Press.

Weber-Fahr, M. (2002). *Treasure or trouble? Mining in developing countries*. The World Bank and International Finance Corporation.

Weber, M. (1973). *The protestant ethic and the spirit of capitalism*. Routledge.

Weber, M. L. (2001). Markets for Water Rights under Environmental Constraints. *Journal of Environmental Economics and Management*, *42*(1), 53–64. doi:10.1006/jeem.2000.1149

Welle, K. (2005). *Learning for Advocacy and good practice- Water Aid Water Point Mapping- Report of Findings based on country visits to Malawi and Tanzania*. Overseas Development Institute and Water Aid.

Wentzel, M. (2005). Achieving universal access to electricity in South Africa. *Energize*, 12-15.

Werker, E. D., & Ahmed, F. Z. (2007). *What Do Non-Governmental Organizations Do?* https://www.hbs.edu/faculty/Publication%20Files/08-041.pdf

West Gonja District Assembly. (2016). Composite budge of the West Gonja District Assembly for the 2016 fiscal year. Damango: West Gonja District Assembly.

West Gonja District Assembly. (2017). Composite budge of the West Gonja District Assembly for the 2017 fiscal year. Damango: West Gonja District Assembly.

Wheeler, T., & Kay, M. (2011). Food crop production, water and climate change in the developing. *Journal of Outlook on Agriculture*, *39*(4), 239–244. doi:10.5367/oa.2010.0017

White, C. (2012). *Understanding water scarcity: Definitions and measurements*. Global Water Forum Publications.

William, N. (2010). *UNICEF – "At a glance: Ghana – Improved sanitation catches on in rural Ghana one latrine at a time*. http://www.unicef.org/infobycountry/ghana 55767.html

Williamson, O. E. (1995). *The Institutions and Governance of Economic Development and Reform*. The International Bank for Reconstruction and Development.

Williamson, O. E. (2000). The New Institutional Economics: Taking Stock, Looking Ahead. *Journal of Economic Literature*, *38*(September), 595–613. doi:10.1257/jel.38.3.595

Winblad, U., & Simpson-Hebert, M. (Eds.). (2004). *Ecological Sanitation revised and enlarged edition* (2nd ed.). Stockholm Environment Institute.

Windham, C. (2008). *Description of Target and Sample Populations for Research Methods Sanyati River System Outline Plan (RSOP) 2009*. Academic Press.

Winter, S., Barchi, F., & Dzombo, M. N. (2018). Drivers of women's sanitation practices in informal settlements in sub-Saharan Africa: A qualitative study in Mathare Valley, Kenya. *International Journal of Environmental Health Research*, *28*(6), 609–625. doi:10.1080/09603123.2018.1497778 PMID:30027750

Wisner, B., Fordham, M., Kelman, I., Johnston, B. R., Simon, D., Lavell, A., Brauch, H. G., Spring, U. O., Wilches-Chaux, G., Moench, M., & Weiner, D. (2007). *Climate Change and Human Security*. Oberlin College.

Woolley, O. (2014). Contents. In Ecological Governance: Reappraising Law's Role in Protecting Ecosystem Functionality (pp. v-viii). Cambridge University Press.

Workman, C. L. (2019). Perceptions of drinking water cleanliness and health-seeking behaviours: A qualitative assessment of household water safety in Lesotho, Africa. *Global Public Health: An International Journal for Research, Policy and Practice*, *12*(6), 34–56. doi:10.1080/17441692.2019.1566483 PMID:30657028

World Bank. (2008). *Project Paper on a Proposed Additional Financing Credit to the Republic of Ghana for a Small Towns Water Supply and Sanitation Project in Support of the Second Phase of the CWSP. Report No: 39782 – GH*. World Bank.

World Bank. (2012). *World Development Report*. World Bank.

World Bank. (2013). *Toward a Sustainable Energy Future for All : Directions for the World Bank Group's Energy Sector*. Author.

World Bank. (2015). *Evaluation of Rural Electrification Concessions in sub-Saharan Africa: Detailed Case Study: The South African Solar Home System (SHS)*. Concessions.

World Bank. (2018). *Democratic Republic of the Congo Overview*. Available at https://www.worldbank.org/en/country/drc/overview

World Bank. (2019). *The power of Dung: Lessons learned from on-farm biodigester programs in Africa*. Author.

World Bioenergy Association. (2020). *Global Bioenergy Statistics 2020*. https://worldbioenergy.org/uploads/201210 WBA GBS 2020.pdf

World Economic Forum (WEF). (2015). *Global Risks 2015: 10th Edition*. Insight Report. https://www3.weforum.org/docs/WEF_Global_Risks_2015_Report15.pdf

World Food Programme (WFP). (2012). *Water for food security: Lessons learned from a review of water-related interventions*. https://www.odi.org/sites/odi.org.uk/files/resource-documents/11610.pdf

World Health Organisation (WHO). (2010). *Water for Health: WHO Guidelines for Drinking-water Quality*. https://www.who.int/water_sanitation_health/WHS_WWD2010_guidelines_2010_6_en.pdf?ua=1

World Health Organisation. (2019). A guide to equitable water safety planning: ensuring no one is left behind. Geneva: World Health Organisation, Licence CC BY-NC-SA 3.0 IGO.

World Health Organization (WHO), United Nations Children's Fund (UNICEF), & Joint Monitoring Programme (JMP). (2010). Progress on drinking water, sanitation and hygiene: World Health Organization.

World Health Organization (WHO), United Nations Children's Fund (UNICEF), & Joint Monitoring Programme (JMP). (2017). Progress on drinking water, sanitation and hygiene: 2017 Update and SDG Baselines. Geneva: WHO and UNICEF.

World Health Organization and United Nations International Children's Emergency Fund. (2006a). *Water and sanitation coverage: Drinking water, sanitation and hygiene database*. WHO.

World Health Organization and United Nations International Children's Emergency Fund. (2010). *Progress on sanitation and drinking water: 2010 Update*. WHO.

World Health Organization and United Nations International Children's Emergency Fund. (2010). *Progress on sanitation and drinking water: Update*. WHO.

World Health Organization and United Nations International Children's Emergency Fund. (2012). *Trends in maternal mortality: 1990 to 2010*. WHO.

World Health Organization. (2009). *Towards universal access: Scaling up priority HIV/AIDS interventions in the health sector*. WHO.

World Health Organization. (2014). *Preventing diarrhoea through better water, sanitation and hygiene: Exposures and impacts in low- and middle-income countries*. WHO Press.

World Health Organization. (2021). *The measurement and monitoring of water supply, sanitation and hygiene (WASH) affordability: a missing element of monitoring of sustainable development goal (SDG) targets 6.1 and 6.2*. UNICEF and WHO. https://washdata.org/sites/default/files/2021-05/unicef-who-2021-affordability-of-wash-services-full.pdf

Worldometer. (2020). *Zimbabwe Population*. Retrieved 16 March 2021 https://www.worldometers.info/world-population/zimbabwe-population/

Wunsch, J. S. (2001). Decentralisation, Local Governance and "Recentralization" in Africa, Public Administration and Development. *Public Admin. Dev*, *21*, 277–288.

Yacoob, M., & Walker, J. (1991). Community management in water supply and sanitation projects-costs and implications. *Aqua AQUAAA*, *40*(1).

Yacoob, M. (1990). Community self-financing of water supply and sanitation: What are the promises and pitfalls? *Health Policy and Planning*, *5*(4), 358–366. doi:10.1093/heapol/5.4.358

Yadav, V., Karmakar, S., Dikshit, A. K., & Vanjari, S. (2016). A feasibility study for the locations of waste transfer stations in urban centers: A case study on the city of Nashik, India. *Journal of Cleaner Production*, *126*, 191–205. doi:10.1016/j.jclepro.2016.03.017

Yanmaz, A. M. (2013). Applied water resources engineering (3rd ed.). Odtu: Metu Press.

Yazdi, H. S., Shakouri, B., Salehi, H., & Fashandi, A. (2017). Sustainable Development and Ecological Economics. *Energy Sources. Economic Policy*, *12*(8), 740–748.

Zanu, S. Y. Y. M. (1996). *The institutionalization of local self-government in Ghana's present-day constitutional processes: A paper presented at conference on traditional and contemporary forms of local participation and self-government in Africa.* Nairobi: Konrad Adenauer Foundation.

Zetland, D. (2016). The struggle for residential water metering in England and Wales. *Water Alternatives*, *9*(1), 120–138.

Zimmerer, K. (2000). Rescaling irrigation in Latin America: The cultural images and political ecology of water resources. *Ecumene*, *7*(2), 150–175. doi:10.1177/096746080000700202

Zimstat. (2013). *2012 Zimbabwe Census National Report*. Harare, Zimbabwe: Zimbabwe National Statistics Agency. Retrieved from www.zimstat.co.zw/sites/default/files/img/publications/Population/National_Report.pdf

ZLHR. (2005). *Report on the Impact of Operation Murambatsvina. The Case of Hatcliffe Extension*. Researched and compiled by Department of Sociology, University of Zimbabwe Zimbabwe Lawyers for Human Rights Sponsored by Zimbabwe Defence and Aid Fund. http://archive.kubatana.net/docs/urbdev/zlhr_uz_impact_murambatsvina_hatcliffe_0506.pdf

Zululand District Municipality (ZDM). (2017). *Indigent Policy Status: Final (2017/18).* http://www.zululand.org.za/media/6137/Indigent%20Policy%20-%20Final.pdf

About the Contributors

Innocent Simphiwe Nojiyeza is a senior lecturer and the Head of Department for Anthropology and Development Studies, University of Zululand. His research interest includes water resource management, rural livelihoods, political ecology, environmental/ecological economics.

Prosper Bazaanah is a Lecturer at the Department of Anthropology and Development Studies, University of Johannesburg. He holds a PhD in Development Studies, MPhil in Development Studies and BA in Political Science with Philosophy. He has experience working in fields like environment, water and sanitation, lecturing and other corporate experiences. His doctoral thesis examined "community water governance for sustainable local development in northern Ghana", a project funded by the National Institute for the Humanities and Social Sciences (NIHSS), South Africa, and the Council for the Development of Social Science Research in Africa (CODESRIA), Senegal. He is also a recipient of bursary awards from the University of Johannesburg. His research interests include water and rural livelihoods, environment and sustainable development, decentralisation and local governance, community development, local development and political ecology.

* * *

Mohamed Belaid serves as Associate Professor at the university of Johannesburg, South Africa. He holds a PhD (University of Witwatersrand), and an MSc in Chemical Engineering (University of Kwa-Zulu Natal, Former University of Durban Westville South Africa); and a BSc in Industrial Chemical Engineering (University of Blida, Algeria). A rated researcher by the National Research Foundation (NRF) of South Africa. His research interests and expertise are centred on three areas: separation processes; fluidised bed technology; and renewable/ clean energy. He published over 80 academic papers in these areas and supervised more than a hundred students in chemical engineering at all levels, including postgraduate level. Mohamed Belaid has more than 22 years of academic experience at three higher education institutions in South Africa, including 14 years at University of Johannesburg (UJ). He served as Head of Department (HOD. Member of the Faculty High Degree Committee, Faculty Ethics Committee and represent the faculty in the university forums of First Year Experience and Senior Students Experience. He served at Mangosuthu University of Technology (MUT) for nine months as project leader at Technology Station for Chemicals, which inspired him to be part of the Creation of Process Energy And Environment Technology Station (PEETS) at UJ. He worked for eight years at the University of KwaZulu-Natal (UKZN), serving as first year Mentor, Mentorship coordinator, tutoring coordinator lab manager, contract lecturer, associate lecturer, and lecturer in various courses in chemical and mechanical engineering. Prior to that he worked in industry for 5 years. He led the Department of Chemical Engineering at UJ for more than 6 years during which student throughput was increased by 25% (for the diploma programme from 10% to 35%; and for BTech from 25% to 50%); research output increased by 100% (0 units in 2007 to 36 units in 2014) and postgraduate intake increased by 95% (from 1 student in 2007 to 40 students in 2016). He also drove the transformation agenda in the department where he played an instrumental role in supporting the development, appointment, mentoring and promotion of five former students to full academic members and four staff promoted to associate professor or full professor respectively.

Gwiranai Danha is an Associate Professor in the Department of Chemical, Materials and Metallurgical Engineering at the Botswana International University of Science & Technology.

Thomas Debwe is a PhD Student at the Department of Geography and Environmental Studies at the Midlands State University, Zimbabwe.

Tariro Kamuti is a Research Fellow at the Centre for Gender and Africa Studies of the University of the Free State, Bloemfontein, South Africa. He holds a joint PhD from Vrije University Amsterdam, The Netherlands and the University of the Free State, South Africa. His research interests include land and environmental policy, governance issues surrounding sustainable utilisation of natural resources, the social dimensions of biodiversity conservation, sustainable rural and urban development.

Gordon Nii-Nai Marley is a Research Fellow at the Institute for Interdisciplinary Research and Consultancy Services (IIRaCS, University for Development Studies (UDS), Tamale, Ghana. He holds MPhil in Development Studies and BA in Integrated Development Studies and currently pursuing a PhD in Development Studies. He has in-depth experience in development research. He is also conversant with issues related rural and community development, social issues of alcohol and substance abuse, sanitation, climate and livelihoods, education and literacy issues. He also lectures at the IIRaCS, Institute for Distance and Continuing Education (IDCE), Department of Development Education of the Faculty of Education (FoE) of UDS and College for Community and Organisational Development, (CCOD), Ghana.

Marta Marson is a research fellow and consultant in development issues. She holds a PhD in Economics of Production and Development and is currently Research Fellow at the Turin Centre on Emerging Economies (OEET) and at the University of Turin. She has a long track record of projects' management and consulting in Africa with NGOs, research institutions, international organizations, and firms and worked in seven different countries of Sub-Saharan Africa and in China.

Edison Muzenda is a Full Professor of Chemical and Energy Engineering, previous head of Chemical, Materials and Metallurgical Engineering Department in the Faculty of Engineering and Technology at the Botswana International University of Science and Technology where he frequently serves as Acting Dean of Postgraduate Studies. He is also a Visiting Full Professor of Chemical Engineering at the University of Johannesburg where he previously served as a Full Professor of Chemical Engineering, Research and Postgraduate Coordinator as well as Head of the Environmental and Process Systems Engineering and Bioenergy Research Groups at the University of Johannesburg. Professor Muzenda was also Chair of the Process Energy Environment Technology Station Management Committee at the

University of Johannesburg. Professor Muzenda holds a BSc Hons (Zim) and PhD (Birmingham, UK). He has more than 20 years' experience in academia which he gained at various institutions including the National University of Science and Technology, Zimbabwe, University of Birmingham, University of Witwatersrand, University of South Africa, the University of Johannesburg and the Botswana International University of Science and Technology. Through his academic preparation and career, he has held several management and leadership positions such as Special Adviser to the Deputy Vice Chancellor, Research, Postgraduate Studies and the Library and Staff Qualification Coordinator at the University of Johannesburg. Edison's teaching interests and expertise are in unit operations, separation processes, waste to energy technologies, environmental engineering, chemical engineering thermodynamics, professional engineering skills, research methodology as well as process economics, management and optimization. He is a recipient of several awards and scholarships for academic excellence such as the Nomination as an Outstanding Researcher for an African Researcher Booklet in 2017 by the Department of Science and Technology, South Africa. His research interests are in bioenergy engineering, waste and coal volatilization, sustainable and social engineering, integrated waste management, air pollution, and separation processes and as well as phase equilibrium measurement and computation. His current research activities are mainly focused waste and coal volarization particularly biowaste to biogas for vehicular application in collaboration with SANEDI and City of Johannesburg as well as Morupule coal dust to tar through pyrolysis. He has contributed to more than 400 international peer reviewed and refereed scientific articles in the form of books, book chapters, journals and conference papers. He has supervised more than 30 postgraduate students and over 280 Honours and BTech research students. He serves as reviewer for a number of reputable international conferences and journals. Edison is a member of several academic and scientific organizations including the Institute of Chemical Engineers, UK, the American Institute of Chemical Engineers, South African Institute of Chemical Engineers, International Waste Working Group, International Solid Waste Association and International Society for Development and Sustainability. He is an Editor for a number of Scientific Journals and Conferences. He has organized and chaired several international conferences. He currently serves as an Editor of the South African Journal of Chemical Engineering. Edison was previously member of the South African Government Ministerial Advisory Council on Energy and currently serves as a member of the Steering Committee of City of Johannesburg – University of Johannesburg Biogas Digester Project. He was recently awarded the collaborative grant on Coal Beneficiation for the Metallurgical Industry under the bilateral research collaboration from the Botswana Innovation Hub and Research Council of Zimbabwe Science Granting Councils Initiative.

Nokuphila Ndimande is a Master's student at the Department of Development Studies at University of Zululand. Her research interest focuses on water access and sanitation, water resources management and ecological economics

Trust Nhubu is a sustainable development oriented researcher and practitioner with over ten years' experience covering design and implementation of community water engineering and management projects as well as researching on sustainable water resources management, green energy technologies and integrated solid waste management incorporating integrated strategies on waste reduction, recycle and reuse. A holder of a PhD in Chemical Engineering from the University of Johannesburg, South Africa. He applied the life cycle assessment methodolody during his PhD study to determine the least environmental impactful municipal solid waste management (MSWM) option amongst a number of probable

MSWM options that could be considered for possible future implementation in Harare, the Capital City of Harare. He also holds a master of science in Integrated Water Resources Management and a Bachelor of Engineering Honors in Chemical Engineering. He is a recipient of the Life Cycle Award from the Life Cycle Initiative in partnership with UN Environment Programme, Federation of Indian Chambers of Commerce and Industry and PRé Sustainability. Also a recipient of the Life Cycle Excellence Award from the World Resources Forum in partnership with UN Environment Programme. He has published several journal articles in high impact peer reviewed journals, presented and published at international scientific conferences, a book chapter on biogas production potential from anaerobic digestion of biodegradable MSW fraction generated in Harare and a policy brief on youth employment and empowerment opportunities from green innovations for the Government of Zimbabwe in partnership with UNICEF, SNV and BOOST Fellowship . He is a member of the International Waste Working Group (IWWG). A recipient and reviewer of the early career scientist grant for the United States based National Geographic Society. He is a recipient of the SADC Waternet fellowship that aided his research on sustainable and integrated water resources management options for the Eastern Zambezi region of Namibia. He reviews for a number of journals. His research interests are on integrated waste management, water resources management, waste valorisation, waste to energy and life cycle assessment of waste management systems, pyrolysis and gasification of coal, biomass, waste tyres and plastics.

Patricia Oppong works with the Ghana Education Service as an Assistant Headmistress of the Kanvilli Tunayilli MA Primary School. Her interest is in Gender Advocacy, especially in the welfare of the Girl-Child. She has supported a number of girls at various levels of their education and continues to advocate for the rights of the vulnerable.

Germain Miteu Tshinu is a PhD candidate in Development Studies at the University of Zululand. He is also a researcher at the Africa Peer Review Mechanism. He conducts research on extractive industries and resources curse.

Index

A

access to water 4, 12, 31, 33, 35, 48-49, 56, 65, 78-85, 88, 90-91, 93, 98-99, 103-104, 106, 109, 113, 118, 120, 128-129, 134-135, 149-155, 157, 160, 162-166, 168-174, 176-177, 179-186, 188, 191, 196, 198, 200, 214
affordability 84, 122, 124, 253-256, 261, 263, 265-266, 271, 365
Alfred Duma Local Municipality 149-151, 155-157, 159, 161-165, 168-174
anaerobic digestion 277, 286, 289, 295, 368, 371, 379, 381-382, 384, 386-387

B

bioelectricity 277, 279, 287-288, 290
bioenergy 277-279, 281-284, 287-288, 290-291, 294-296, 375
Biological Treatment 387
biomass 277-284, 286-296

C

catchment management institutions 76
Climate Justice 297
Commercial and industrial water demand 125
Community Water 4-5, 19, 30, 35, 41-42, 45, 47, 49-51, 56, 82-83, 112
composting 368, 371, 380-382, 384, 387
corruption 152, 327-328, 331-332, 337-338, 345, 349

D

decentralization 31, 93, 95, 104, 106, 151, 154, 168, 171, 183, 225-228, 230-232, 234-236, 303, 364
District Municipality 136, 141, 148, 150-151, 156, 164, 166-167, 173, 177-178, 182, 201, 307, 309
Durban 107, 181, 186, 202-205, 207-209, 211-212, 215, 217, 221-224, 297-307, 309-310, 312, 314, 316

E

Ecological Economics 96-97, 105-108, 129, 145-147, 149, 153-154, 170, 177-179, 185-186, 199-200, 253, 255-258, 270-271, 273-274, 297-298, 305, 315
ecological governance 1-6, 8-11, 13, 17-22, 24-25, 27-35, 37-39, 46, 149, 151, 155, 172, 253-259, 261, 270-271, 274-276
energy poverty 253-257, 259, 261-263, 270-271, 275, 279-281, 291, 293
energy recovery 281, 369, 376, 378, 380, 387
equitable access to water 31, 33, 149-155, 157, 162-164, 169-173, 183
equity 6, 10, 29-30, 33, 56, 73, 84, 90, 104, 110, 122, 129-130, 147, 149-150, 152-154, 171, 173-176, 184-186, 202, 204, 210, 212, 214, 222, 227, 231-232, 235, 257, 261, 266, 365, 378

F

food security 127-130, 133-134, 137-138, 140, 142-146, 148, 190, 200, 318, 337

G

Ga-Molekane Village 354-355, 357, 359-360, 367
Gecamines 318-324, 327, 329, 332, 335, 337-349, 351, 353
Grass-root Action 297

H

Harare 368, 370-374, 376-377, 385-386
healthcare 53, 318, 337, 348, 358, 365
Housing 5, 12, 41-43, 51, 55, 72, 106, 116, 134, 300-301, 306, 309, 321, 337, 341, 345

human rights 2, 28, 33, 50, 83, 150, 152-153, 171, 173, 177, 179, 183, 193, 196-198, 201, 219, 238, 314, 354-358, 361, 363, 367

I

Improved and Sustainable Water and Sanitation Facility 75
institutional shift 202
integrated waste management 368, 378, 387
Integrated Water Resources Management (IWRM) 96, 145, 149, 174, 183, 236
Inter-Basin Transfer 239, 247

L

landfilling 369-371, 379-382, 387
legitimacy 8, 21, 27, 31-32, 39, 45-46, 328
Limpopo 92, 200, 253, 255, 259-261, 263-264, 267-269, 273, 275-276, 354-355, 357
Local Municipality 127, 132, 141, 144, 149-151, 155-157, 159, 161-165, 168-174, 179-183, 186-189, 196-199, 259, 262, 266, 303, 358, 365

M

marginalized communities 153, 354-355, 357
Mending 76
mini-grid 277, 279, 281, 293-294
mining 72, 228, 231, 233, 239, 245-247, 249, 318-320, 322-325, 327-328, 332-345, 348-351, 353-359, 361-367
Mogalakwena Municipality 354-355, 359-360, 364-365

N

nexus 3, 52-53, 354, 367-371, 375-376, 378-379, 384-387
NGO Programmes 75
NGOs 4-5, 14, 21, 30, 32, 37, 47, 49-52, 56-57, 59-61, 64-71, 75, 78, 80-81, 83, 85-90, 99, 101, 142, 147, 176, 205, 298, 301, 311, 314, 362
Non-Governmental Organizations 49, 51, 72, 75, 147
Non-Revenue Water (NRW) 123, 125
Northern Region 13, 47, 50-51, 55, 69, 75

O

off-grid 255, 257, 261, 270, 272, 274-275, 277-279, 281-282, 286, 288-290, 293-294

P

political ecology 9-10, 33, 45, 70, 72-74, 174, 222-227, 234, 236, 238
political interference 68, 218-219, 337, 349
pollution control 24, 96-97, 100, 104, 288, 300, 378-379, 385
poor management 134, 325, 327-329, 337
public water points 82, 112-113, 115, 125

R

realism 318-319, 321, 329, 331-332, 334, 350
resource curse 318-319, 321-322, 324-326, 328-329, 331-335, 337, 341-342, 349-350, 352
rural community 20, 24, 46, 50, 75
rural households 1, 9, 13, 17, 24, 27, 34, 37, 55, 102, 131, 138, 144-145, 196, 202, 204, 254, 263, 271, 309

S

sanitation 2-5, 8, 10, 14, 18-21, 26-33, 38, 41-43, 45-57, 59, 61-88, 90-92, 96-100, 102-105, 113-114, 122-125, 128, 135, 139, 147-148, 150, 152-155, 168-169, 171-173, 175, 177-178, 180-182, 184-186, 190, 192-193, 196-198, 202-205, 207-215, 218-219, 221-224, 251, 297-298, 301-303, 305-307, 310-316, 355, 357, 359, 362-363, 365, 371, 385
Sanyati catchment 225-234, 239-244, 246-248, 250
socio-economic development 13, 49, 127-128, 143, 153, 184, 194-196, 200, 243, 278, 281, 318-321, 337-342, 349, 355
solar home systems 253-254, 271, 274-275
sports and recreation 337
stakeholder 21-22, 27, 29, 31, 35, 38, 43-44, 46, 92, 101, 141, 227, 230-232, 237, 251, 306, 356, 362, 364
State Contradictions 297

T

Tamale Metropolis 47, 58
Tariff structure 116, 125
thermal treatment 379-380, 382, 387
Tunayilli 47, 49-51, 54-55, 58-59, 61-64, 69-70

U

Universal water metering 111, 125
Urban Water Services 109
urinary diversion toilets 202, 303

V

village action committees 76-78

W

waste to energy 278, 375, 387
water access 10, 33, 35, 50-52, 68, 85, 112-113, 121, 125, 127, 129, 133, 136, 140-144, 149-150, 152-156, 160, 163, 167, 169-171, 173, 175, 180, 186, 194, 196, 198-199
water and sanitation services 4, 48, 61, 64, 75-78, 81, 83-86, 91, 100, 155, 181, 186, 192, 202, 315
water conservation 1-3, 5-9, 11, 15-19, 21, 23-24, 26-30, 32-35, 37-39, 46, 110, 136, 233, 249-250, 298, 316, 362-363
water demand 2, 8, 101, 109-111, 114, 116-118, 121-123, 125, 130, 149, 151, 200, 239-243, 245-246, 248-250, 371-372, 383
Water Demand Management (WDM) 101, 109-111, 116-117, 121-123, 125
water governance 5, 7, 9-10, 14, 27, 31, 33-34, 40, 43, 79, 91, 93-94, 96, 105-106, 146, 152-154, 172, 174, 176, 179-180, 184-188, 191, 193-194, 198-201, 226, 237
Water Institutions 76, 90-92, 94, 107
water management 41, 43, 51, 64, 70, 93, 107, 112, 142, 147, 150-152, 154-155, 167, 171-175, 180, 183-184, 194, 200, 225-228, 230-231, 234-236, 251, 363, 369, 375, 383
water scarcity 2-5, 7, 33-34, 41, 44, 107, 125, 127-131, 133-134, 137-140, 144-148, 179, 184, 190-191, 194, 196, 198, 240, 297, 314
water use 9, 19, 81, 97, 106, 137, 140, 232, 239-240, 245-247, 250-251, 363, 375
water utilities 77, 84, 109-110, 112, 120, 124, 292

www.ingramcontent.com/pod-product-compliance
Lightning Source LLC
LaVergne TN
LVHW062108300926
849161LV00010B/2402